Multiphoton and Light Driven M
Processes in Organics:
New Phenomena, Materials and Applications

# NATO Science Series

*A Series presenting the results of activities sponsored by the NATO Science Committee. The Series is published by IOS Press and Kluwer Academic Publishers, in conjunction with the NATO Scientific Affairs Division.*

| | |
|---|---|
| A. Life Sciences | IOS Press |
| B. Physics | Kluwer Academic Publishers |
| C. Mathematical and Physical Sciences | Kluwer Academic Publishers |
| D. Behavioural and Social Sciences | Kluwer Academic Publishers |
| E. Applied Sciences | Kluwer Academic Publishers |
| F. Computer and Systems Sciences | IOS Press |

| | |
|---|---|
| 1. Disarmament Technologies | Kluwer Academic Publishers |
| 2. Environmental Security | Kluwer Academic Publishers |
| 3. High Technology | Kluwer Academic Publishers |
| 4. Science and Technology Policy | IOS Press |
| 5. Computer Networking | IOS Press |

**NATO-PCO-DATA BASE**

The NATO Science Series continues the series of books published formerly in the NATO ASI Series. An electronic index to the NATO ASI Series provides full bibliographical references (with keywords and/or abstracts) to more than 50000 contributions from international scientists published in all sections of the NATO ASI Series.
Access to the NATO-PCO-DATA BASE is possible via CD-ROM "NATO-PCO-DATA BASE" with user-friendly retrieval software in English, French and German (WTV GmbH and DATAWARE Technologies Inc. 1989).

The CD-ROM of the NATO ASI Series can be ordered from: PCO, Overijse, Belgium

Series 3. High Technology – Vol. 79

# Multiphoton and Light Driven Multielectron Processes in Organics:
# New Phenomena, Materials and Applications

edited by

## François Kajzar
Atomic Energy Commission,
Direction of Advanced Technology,
LETI, Saclay, France

and

## M. Vladimir Agranovich
Institute of Spectroscopy,
Russian Academy of Sciences,
Troitsk, Russia

**Kluwer Academic Publishers**

Dordrecht / Boston / London

Published in cooperation with NATO Scientific Affairs Division

Proceedings of the NATO Advanced Research Workshop on
Multiphoton and Light Driven Multielectron Processes in Organics: New Phenomena,
Materials and Applications
Menton, France
26-31 August 1999

A C.I.P. Catalogue record for this book is available from the Library of Congress.

ISBN 0-7923-6271-3 (HB)
ISBN 0-7923-6272-1 (PB)

Published by Kluwer Academic Publishers,
P.O. Box 17, 3300 AA Dordrecht, The Netherlands.

Sold and distributed in North, Central and South America
by Kluwer Academic Publishers,
101 Philip Drive, Norwell, MA 02061, U.S.A.

In all other countries, sold and distributed
by Kluwer Academic Publishers,
P.O. Box 322, 3300 AH Dordrecht, The Netherlands.

*Printed on acid-free paper*

# TABLE OF CONTENTS

# PREFACE

The book contains the proceedings of the NATO Advanced Research Workshop "Multiphoton and Light Driven Multielectron Processes. New Phenomena, Materials and Applications", held in the Royal Westminster hotel in Menton (France), August 26 - 31, 1999. The workshop consisted of plenary lectures, given by leading specialist in this field, shorter oral contributions , poster sessions and working groups meetings.

The contributions assembled in this volume give the present state of art and deal with the latest developments and discoveries with photoactive organic materials in view of their applications in photonic devices. Up to now the optical or electronic devices operate with atoms and molecules almost exclusively at fundamental states. With increasing high technology demand, new phenomena and new applications are envisaged involving the use of atoms and molecules at excited states. Written by leading specialists in the corresponding research fields the different, original, contributions address several actual and pertinent problems such as molecular design and synthesis of highly light sensitive molecules and phenomena connected with electron - photon interaction in organic molecules. In particular the topics treated are: nonlinear beam propagation, photorefractivity, multiphoton excitations and absorption, charge photogeneration and mobility, photo - and electroluminescence, photochromism and electrochromism, organic synthesis, material engineering and processing. Such device applications as optical power limiters, optical data storage, light emitting diodes, optical signal processing are described. The present proceedings are addressed not only to people active in this research field, but also to newcomers, graduate and postgraduate students in photonics science and technology. Three working groups were also organized on the subjects which were center of presentations and debates:
(i) multiphoton absorption : science and applications
(ii) organic electroluminescence
(iii) photochromism
The results of discussions are enclosed in this proceedings too as working group reports.

The organizers of this workshop are highly indebted to its main sponsor the NATO Scientific Affairs Division for a generous financial support. Thanks are also due to the other workshop sponsors: Service de Physique Electronique of Commissariat à l'Energie Atomique - LETI, France, Ministère de l'Education Nationale, de la Recherche et de la Technologie, France, Direction Général de l'Armement (DGA) and City of Menton. They would like also to thank the management of the Hotel Royal Westminster for the collaboration and very good working conditions offered to the participants of this workshop.

Thanks are also due to the members of the scientific committee: Prof. Paras N. Prasad and Prof. Carlo Taliani for useful suggestions and help in putting this workshop altogether. Special thanks are due to Dr. Malgorzata Helwig for the help in secretarial tasks and Dr. Pierre-Alain Chollet for his assistance in the organization of this meeting.

The workshop is dedicated to the memory of Dr. Bruce Reinhardt who played a key role in the organization of this meeting and who sadly passed away this spring. He was known not only as an excellent polymer chemist and scientist but also as a valuable friend and colleague. Dr. Reinhardt will be missed very much.

Saclay, December 1999                           Vladimir M. Agranovich
                                                François Kajzar
                                                (workshop co-directors)

# NONLINEAR MATERIALS AND PROCESSES FOR ELECTRONIC DEVICES AND 3D OPTICAL STORAGE MEMORY APPLICATIONS

Y. C. LIANG[#], D. A. OULIANOV[#], A. S. DVORNIKOV[#*],

I. V. TOMOV[#*] AND P. M. RENTZEPIS[#*‡]

[#]*Department of Chemistry, University of California, Irvine, CA 92697*

*Call/Recall Inc. 18017 Suite R, Irvine, CA 92614*

‡ Corresponding author

## 1. Introduction

Many commercial and military applications generate enormous amounts of data which must be stored and be available for rapid parallel access and very fast processing. The major component which is expected to modulate the practical limits of high speed computing, will most probably be the memory. In addition because of the huge data storage requirements, the need for the parallel execution of tasks and necessity of a compact, very high capacity low cost memory is becoming a practical necessity. However, at the present time, CD-ROMs and DVDs and even some of the most advanced optical storage systems, such as the 4.5 GB DVD, are not sufficient to fulfill some of the stringent demands imposed today by multimedia, medical and other applications. This method utilizes the inhomogeniously broadened zero phonon band as the storing medium. Storage media, including magnetic disk, electronic RAM and optical disks are fundamentally limited by their two-dimensional nature. The data capacity is proportional to the storage area divided by the minimum bit size. Three dimensional optical storage surmounts this limitation by extending the storage into the third dimension. They offer, therefore, an attractive possibility for highly parallel access, large storage capacity and high bandwidth memory. To circumvent this deficiency we have developed a high density fast read-out 3D optical storage memory system[1-4]. Our method relies on the non-linear absorption of two photons which causes changes in the structure and spectra of organic molecules.

Three-dimensional memories (3DM), because they extend the storage into the third dimension, make possible the achievement of higher capacities and shorter access times. Storing and retrieving, all of the data of a complete page composed of several megabits, simultaneously in contrast to the single bit access, dramatically increases the usable data rate. In addition, parallel-access optical memories are to a large extent compatible with the next generation of ultra-fast parallel hybrid, opto-electronic, computers which rely on optical interconnections and electronic processing. However, it is doubtful that a single 3D memory system may posses the optimum requirements for high speed computing, therefore tradeoffs between high storage density, access time and data transfer rates need be made. Holographic methods constitute another means for storing data in 3D.[5-7]

*F. Kajzar and M.V. Agranovich (eds.), Multiphoton and Light Driven Multielectron Processes in Organics: New Phenomena, Materials and Applications, 1–19.*

For optical memories, the density of stored information is dependent upon the reciprocal of the wavelength raised to the power of the dimension used to store information. For example, the information density which can be stored in a one-dimensional space, i.e.. tape, is proportionally $1/\_$ This relationship also suggests that the information storage density is much higher for UV rather than IR light. For a two dimensional memory, the maximum theoretical storage density for a 2D storage device which uses light at 200 nm is $2.5 \times 10^9$ bits/cm$^2$ . In the case of a 3D storage memory which utilizes the same wavelength of light, the maximum density which can be stored per cm$^3$ is $1.2 \times 10^{14}$ bits/cm$^3$. This represents an increase of $10^5$ bits/cm$^3$.

The two states of the binary code, 0 and 1, are formed by the photochemical changes which lead to two distinct structures of the molecular species used as the storage medium. The stored information is read by illuminating the written bits with either one photon of the energy necessary to induce fluorescence, or by intersecting two optical beams at that point. The energy of each photon is less than the energy gap between the ground and first allowed excited state, the sum of the energies of the two excited states is, however, larger than the energy needed to excite the molecule. In the case of the photochromic organic materials to be described[8], the information is stored in the form of binary code. The non-linear, i.e. two photon virtual absorption, makes possible the storage of data inside the volume of the 3D device and in fact enables the selection of any arbitrary location within the device volume to where data is to be written. By this means, two-photon 3-D memory devices with over 100 2D planes, within an 8 mm cube, have been stored, at a distance of a few microns between planes.

These optical storage memory devices have the capability of the parallel access needed to accommodate the demands of today's technologies. We expect that by this means it will be possible to construct an optical memory system which may provide a data capacity of 1 Gbit/cm$^3$ with a 10 ns-1 ms access.

## 2. Two-photon Processes

The theoretical bases for two-photon processes were established in the early 1930's[9]. It was shown that the probability for a two-photon transition may be expressed as a function of three parameters: line profile, transition probability fit for all possible two-photon processes (see Fig. 1 for a schematic representation of four such processes), and light intensity. These factors are related to $P_{if}$ by:

$$P_{if} \cong \frac{\gamma_{if}}{\left[w_{if} - w_1 - w_2 - \vec{v} \cdot (\vec{K}_1 + \vec{K}_2)\right]^2 + (\gamma_{if/2})^2} \cdot \left| \sum_k \frac{\vec{R}_{ik} \cdot \vec{\varepsilon}_1 \cdot \vec{R}_{kf} \cdot \vec{\varepsilon}_2}{(w_{ki} - w_1 - \vec{K}_1 \cdot \vec{v})} + \frac{\vec{R}_{ik} \cdot \vec{\varepsilon}_2 \cdot \vec{R}_{kf} \cdot \vec{\varepsilon}_1}{(w_{ki} - w_2 - \vec{K}_2 \cdot \vec{v})} \right|^2 \cdot I_1 I_2$$

(1)

According to this postulate, two photon transitions may also allow for the population of molecular levels which are forbidden for one-photon processes such as g→g and u→u in contrast to the g→u and u→g transitions which are allowed for one-photon processes. In practice, however, when one is concerned with large molecules in condensed media, the density of the states is very large, the levels broadened by collisions and the laser line bandwidth large enough to accommodate many levels. Therefore, there is little, if any, difference in the energy between the levels observed experimentally by one or two photon transitions in the large molecular entities used under the experimental conditions of the experiments presented here. The above expression is composed of three factors: the first factor describes the spectral profile of a two-photon transition. It corresponds to a single-photon transition at a center frequency $\omega_{if} = \omega_1 + \omega_2 + \underline{v}(\underline{k}_1 + \underline{k}_2)$ with a homogeneous width $\gamma_{if}$. If both light waves are parallel, the Doppler width which is proportional to $|\underline{k}_1 + \underline{k}_2|$ becomes maximum. For $\underline{k}_1 = \underline{k}_2$ the Doppler broadening vanishes and we obtain a pure Lorentzian line. The second factor describes the transition probability for the two-photon transition. This is the sum of products of matrix elements $R_{ik}R_{kf}$ for transitions between the initial state i and intermediate molecular levels k and between these levels k and the final state f. The sum extends over all molecular levels. Often a "virtual level" is introduced to describe the two-photon transition by a symbolic two-step transition $E_i \rightarrow E_v \rightarrow E_f$. Since the two possibilities

a)      $E_i + \hbar\, \omega_1 \rightarrow E_v$ ,  $E_v + \hbar\, \omega_2 \rightarrow E_f$      (first term)      (2)

b)      $E_i + \hbar\, \omega_2 \rightarrow E_v$ ,  $E_v + \hbar\, \omega_1 \rightarrow E_f$      (second term)      (3)

lead to the same observable result, the excitation of the real level $E_f$, the total transition probability for $E_i \rightarrow E_f$ is equal to the square of the sum of both probability amplitudes.

The frequencies of $\omega_1$ and $\omega_2$ can be selected in such a way that the virtual level is close to a real molecular state. This greatly enhances the transition probability and it is therefore generally advantageous to populate the final level $E_f$ by means of two different energy photons with $\omega_1 + \omega_2 = (E_f - E_i)/\hbar$ rather than by two equal photons. The third factor shows that the transition probability depends upon the product of the intensities $I_1$, $I_2$. In the case where the photons are of the same wavelength, then the transition probability depends upon $I^2$. It will therefore be advantageous to utilize lasers emitting high intensity light such as picosecond and subpicosecond pulses. For practical use there is a trade off between very short pulses with very high intensity, i.e. femtosecond pulses, but small photon flux and wider pulses, i.e. picosecond pulses, and large number of photons per pulse. Also the damage cause by high intensities is a factor when selecting a laser pulse for optical devises. Organic materials are, in general, damages at lower power densities than inorganic materials.

4

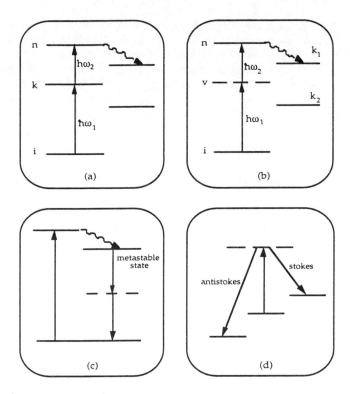

*Figure 1.* Two photon processes

Four possible processes which may operate by means of two photon excitation are shown in Fig. 1. The first corresponds to a stepwise or sequential two photon absorption process (Fig. 1a) in which each photon absorption is allowed (i.e., two allowed one- photon sequential processes). Even though the final state reached by the two photons may be the same as by the simultaneous absorption of two photons via a virtual level (Fig. 4b) the effect towards a volume memory is overwhelmingly different. In the case of the sequential two photon process, the first photon absorption takes place on a real level - by definition - by a molecule, or atom, which has an allowed state at that energy. It will therefore be absorbed by the first such molecule or atom on its path which is usually located on the surface. Subsequently, if sufficient photon intensity is available, several possibilities exist: a) the photons from the same pulse may be absorbed by the same molecule inducing a transition to higher electronic states; b) populate an additional molecule further, inside the volume; c) after energy decay, populate a metastable level such as a triplet. The second photon will encounter the same fate, namely it will be absorbed preferentially by the molecules first encountered in its path which are the molecules at or near the surface, then with decreasing intensity, this beam will propagate and be absorbed by molecules further inside the volume. This second sequential beam may be delayed by a time interval equal to or slightly longer than the time required by the first excited state to decay to a low lying metastable level. If the wavelength of this second beam is adjusted to be longer (smaller in energy) than the energy gap between the ground state and the first allowed state, (wavelength of the first beam) then the second light beam will populate only upper electronic states of the excited metastable level. This is an interesting and important scientific aspect of nonlinear spectroscopy and photophysics, however, it does not result in a true 3D volume memory. This is because, as we mentioned above, there is no means possible by which light can interact preferentially with molecules located inside a volume without interacting first with at least equal efficiency with molecules residing on the surface.

The second scheme for two photon absorption (Fig.1b) makes it possible to excite preferentially, molecules inside a volume in preference to the surface. This may be achieved because the wavelength of each beam is longer, has less energy than the energy gap between the ground state and first allowed electronic level. However, if two beams are used the energy sum of the two laser photons must be equal to or larger than the energy gap of the transition. It is also important to note that there is no real level at the wavelength of either beam, therefore neither beam can be absorbed alone by a one photon mechanism. When two such photons collide within the volume, absorption occurs only within the volume and size defined by the width of the pulses. This is in sharp contrast to the sequential two photon process, fig.1, where the first step involves the absorption of a single photon by a real spectroscopic level and hence is not capable of preferential volume storage. The principal difference in the two cases as far as their suitability for 3D volume memory is concerned, is that the virtual case avails itself to writing and reading in any place within the 3D volume while the sequential excitation is restricted in writing and reading first at the surface. The other two processes shown in Fig. 4 show the possibility of two photon emission, (Fig. 1c) after single photon excitation and the Raman effect (Fig. 1d). Neither of these last two cases are relevant to the topic under discussion and will not be discussed further. We must note at this point that the same physics holds for all two photon transitions regardless of the sequence by which they take place let it be either via sequential steps between real levels or via virtual state interaction. In the case presented here we utilized two-photon process, the photon energy of each beam was smaller than the energy gap between the ground state $S_0$ and the first allowed electronic level $S_1$, therefore such a beam of light propagates though the medium without observable absorption. When two such beams are made to intersect at a point within the memory volume their effective energy is equal to the sum of the two photon energies $E_1 + E_2$ therefore absorption will occur if the $E_{S_1} - E_{S_0}$ energy gap is equal to or smaller than $E_1 + E_2$. At the point where the two beams interact, the absorption induces a physical and/or a chemical change which will distinguish this microvolume area from any other part of the memory volume which has not been excited. These two molecular structures, i.e. the original and the one created by the two photon absorption, are subsequently utilized as the write and read forms of a 3D optical storage memory. For a successful completion of this type of writing and reading, the light beams which perform either function must also be capable of propagating through the medium and be absorbed only at preselected points within the memory volume where the two beams intersect, in time and space without any noticeable effect on other areas of the memory volume in which information maybe written or not.

6

## 3. Two Photon 3D Storage Method

Three dimensional storage memory, using two photon excitation, was first proposed by Parthenopoulos and Rentzepis in 1989[11] and Hunter, Esener, Dvornikov and Rentzepis(12) in 1991. This data storage approach has been under development at UCI ,UCSD and Call/Recall Inc. since 1989. Other research groups which have made significant contribution to this field include P.Prassad[12] and R. Birge[12]. The goal of these efforts is the development of a monolithic terabit capacity 3D storage memory devise. Call/Recall Inc. developed and built a prototype consisting of over 100 planes with 100Mgb/2D plane. The physical process at the heart of the 3DM is a molecular change caused by a two photon optical absorption, as shown in Fig. 2. A molecule in the ground (unwritten) state is excited to a higher energy state by simultaneous absorption of two distinct photons, one red (1064 nm) and one green (532 nm), or two green photons The energy required to reach the excited state is greater than either photon alone can provide, but when two photons interact simultaneously, at a preselected point within the volume, they are absorbed resulting in a bond dissociation. The molecular geometry (structure) is changed into a new, written, molecule with an entirely different absorption spectrum., as shown in Fig. 2. The written molecule, in this example, absorbs two 1064 nm photons, or one green photon and fluoresces in the red region, ~ 700 nm. Accessing of the stored information is achieved, either by the detection of the written' bit fluorescence, absorption, index of refraction or any other property which is preeminent, and can be detected, in either the written or the original form of the molecule

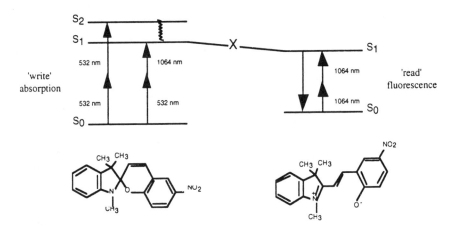

*Figure 2.* Two photon write (left) and readout (right) processes.

The basic importance of the two photon writing is its unique ability to select and write or read a single bit anywhere within a three-dimensional volume by simply intersecting two optical beams at that point. The capacity of such a bit-oriented volume memory is fundamentally limited only by the memory volume divided by the optical spatial resolution.

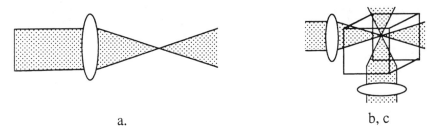

*Figure 3*. Paths for two photon absorption: : a) confocal: b, c) counter propagating and orthogonal

## 4. Materials

Organic photochromic materials have been dispersed in transparent polymer matrices and glasses to form a homogenious, monolythic, block or disk, which upon excitation with two photons allow for storing and accessing information in 3D format. In the case where the organic materials are used for storing information, the two states of the binary code, 0 and 1, are formed by the photochemical changes, which lead to two distinct structures of the molecular species used as the storage medium. Such an example, which is the first ever used in 3D memories, is provided by the changes in molecular structure occurring in photochromic materials such as spirobenzopyrans after the simultaneous absorption of two photons. The two molecular structures, i.e. the original and the one created by the two photon absorption, become in practice the "write" and "read" forms of a 3D optical storage memory.

The procedure used to access the information written within the volume of the memory is similar to the "store" process except that the "reading" form absorbs at longer wavelengths than the "write" form. Under these conditions reading is achieved by the use of longer wavelengths than writing, therefore no writing takes place while reading. After the written molecule is excited it emits fluorescence. The fluorescence spectrum is located at longer wavelengths than the absorption of both the write and read forms. The emission was detected by a photodiode or Charge Coupled Device (CCD) and processed as 1 in the binary code. The proper selection of materials which provide widely separated spectra is extremely important because it assures that only one of the two forms, usually designated as the "written" form, emit light and only from the area of the written memory that is being excited. The spirobenzospirans, although demonstrated very convincintly the two photon 3D method for storage, display and microscopy they were found to be unsuitable for use in practical storage devices because they do not meet some of the strigent requirements imposed. The major one is thermal stability. Similar difficulties have been found for the vast majority of organic compounds synthesized and proposed by several investigators. High two photon absorption crosssection is, undoubdetly, a very desired property, yet several of the molecules proposed as excellent choices for 3D two photon application, because of their very high two photon absorption, we found them to be inadequate because of their poor laser power stability or low read/write fatigue and poor fluorescence quantum yield. In addition many of the very high two photon cross section have weak absorption in the one photon wavelength. Even a very weak one photon absorption is sufficient to overshadow a very high two photon cross section.

8

There is a large number of materials, that change their molecular structure and spectroscopic properties after illumination with light and have relatively high two photon crosssection. However, to be suitable for use in the 3D memory devices the materials should possess certain characteristics, which strongly limit the number of possible candidates:

1) High two-photon absorption cross-section to perform efficient writing of information.
2) The photochemical reaction should have a high conversion yield.
3) One of the forms should emit fluorescence with high quantum yield.
4) Both write and read forms of the material must be stable between- 55 C and + 55 C.
5) High fatigue resistance to perform more than $10^6$ write-read-erase cycles.
6) Wide absorption and emission spectra separation to minimize the crosstalk between thewritten Bits.
7) Capability for non-destructible readout process.
8) Withstand high laser power.

We have found a number of materials which may be used with some success for optical storage and access of information and which satisfy a large fraction if not all of the mendatory characteristics. One type of molecular photochromism, which may be used in 3D memory devices, is light induced dimer - monomer transformation[14]. The process of reversible photodimerization and photodissociation of polycyclic aromatic hydrocarbons, such as anthracene and its derivatives, may be used for developing photochromic materials for optical memory,and other electronic devices:

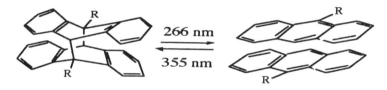

The dissociation of the dimer results in the regeneration of a conjugated double bond system and consequently a red shift of the absorption band of the dimer. The monomer has its long wavelength absorption band in the 300-400 nm region, while the dimer is blue shifted and has practically no absorption at wavelengths longer than 300 nm. The monomer was found to emit with a fluorescence quantum efficiency of approximately 30%, while the dimers are practically void of any fluorescence. The fluorescence spectrum of anthracene dispersed in PMMA is between 380-450 nm. One advantage of the dimer-monomer based 3D storage systems is that both dimer and monomer forms are highly stable at room temperature, unlike the spiropyrans, where the written state reverts spontaneously, within a few hours at room temperature, into the unwritten state and consequently the information previously stored is erased. In addition, the high absorption cross-section and high quantum efficiency of fluorescence suggest that this molecular system is very suitable for utilization in 3D memory optical devices.

A new memory material was designed by us lately for 3D-memory ROM devices, where the information is written once, stored indefinitely, but may be retrieved for very large number of times. It is composed of an organic dye whose structure is drastically different in acidic media than exhibits in basic host media. In addition while the basic form does not fluoresce , the acidic form emits with a quantum efficiency of practically one. For example organic laser dyes, such as Rhodamine B, can exist in two forms, depending on the acidity and polarity of the matrix or solvent. One of these forms, Rhodamine B base, is colorless and has no detectable fluorescence. However, in the presence of acid, this colorless form undergoes a transformation into a colored, strongly fluorescing dye, Rhodamine B, which is well known as a stable and efficient laser dye:

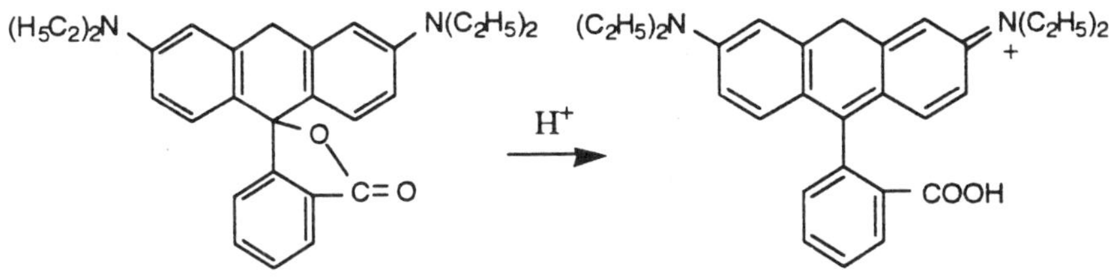

Rhodamine B base                    Rhodamine B

We have used 1-nitro-2-naphthaldehyde (NNA) as the acid generator component, which upon excitation with UV light undergoes phototransformation into the corresponding nitroso acid with quantum efficiency of 50%:

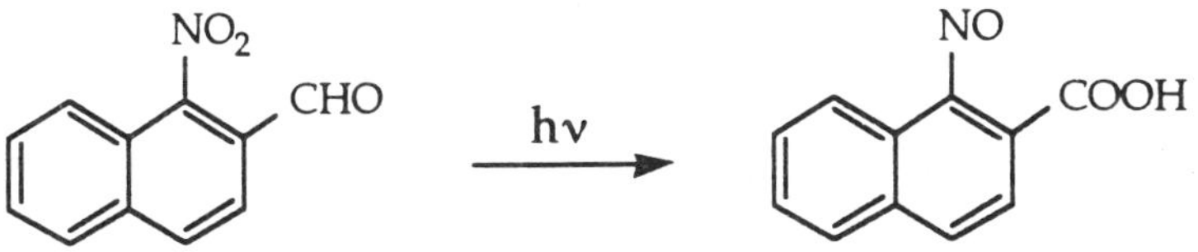

We have successfully utilized this polymer based light sensitive molecular system as a ROM memory material for our 3D optical memory system to store many 2D planes inside the 3D volume of the memory. Other photochromic materials have been studied in our laboratories, which show promise to be much better than the materials presented in this paper. for use as optical memory devices.

## 5. Systems

Two-photon 3-D memories are characterized by high density, data content stored in the form of numerous 2D planes, simple fabrication, parallel access and high transfer rates. Although single bits may be stored and read, parallel access of columns or planes of data is naturally achieved in two-photon 3-D memories, resulting in very high data transfer rates. Diffraction of the addressing beam as it propagates through the data image, imposes limits in the volumetric density of the memory and the parallelism, or data transfer rate. Writing, reading, and erasure of 3-D memory data have been demonstrated, and now we are evaluating the practical potentials associated with this technology.

To demonstrate the writing process we stored over 100, 2D pages, megabit size disks, inside the volume of a 1 cm$^3$ by intersecting the 1064 nm, information beam, and the 532 nm, addressing beam, inside the memory cube, which is composed of the photochromic material dispersed in polymethylmethacrylate (PMMA). The information storing method is shown schematically in Fig. 3. Accessing the stored information is achieved by means of the same optical system except that the wavelength of one or both beams is longer. In the case where the information of an entire page is to be accessed simultaneously, rather than bit by bit mode, a single, low intensity 532 nm plane beam is used to illuminate the written 2D page and induce fluorescence by one photon process. The reading 532 nm beam has the dimensions of the 2D plane to be read and as it propagates through the bulk of the memory device intersects only plane to be read, which emits fluorescence, see Fig. 4, without accessing any of the neighboring stored information.

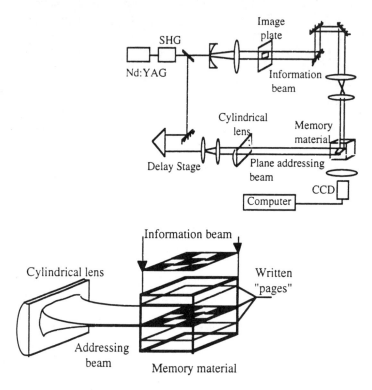

*Figure 4.* a) Experimental system for writing and reading information in 3D.; b) a pattern stored on one 2D plane of a 1 cm$^3$ memory cube

The fluorescence is detected and recorded by a CCD which transmits the signal in digital form directly to the processor. Fig. 4a shows a schematic representation of of the optical path for storing information. Fig. 4b, shows a simple LCD mask image being stored in a 2D plane within the 3D volume of the disk. It is also shown that the complete image, on the mask, was written simultaneously by intersecting, inside a memory cube containing spiropyran in PMMA, the 1064 nm information carrying pulse and the 532 nm addressing beam, shaped to the form of a thin, ~ 20μm, plane, see Fig. 4b. Accessing any 2D pattern within the memory volume is simply achieved by illuminating this area with a thin plane of light of the appropriate wavelength to induce fluorescence.

One advantage of the photochromic materials discussed is that unlike the spiropyrans, where the written form decays spontaneously within a few hours at room temperature into the original write form and effectively erasing the information stored, both dimer and monomer forms are stable at room temperature. In addition, the high absorption cross-section and high quantum efficiency for converting the material to both, writing, and reading forms, suggest that these photochromic systems are potentially attractive for utilization in 3D memory and other optical devices. Several other photochromic materials have been studied in our laboratories which show promise for use in electronic and optical memory devices.

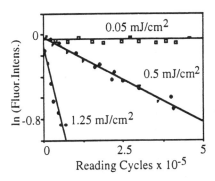

*Figure 5.* Effect of reading laser power on fatigue of the written form.

The 3D devices presented have been tested for fatigue under various changing parameters. The cycle times as a function of laser energy is shown in Fig. 5. Similar studies as a function of temperature and molecular fatigue as a function of cycle times have been performed. At low energies more than $10^6$ cycles of read and write have been performed without noticeable degradation.

## 6. 3D System Prototype

An automated recording and reading system, Fig. 6, has been constructed, at Call/Recall Inc, which is used to store data in $10 \times 10 \times 10$ mm$^3$ cubes or disks, 1"D 0.25" thick. These cubes are fabricated by dispersing the photochromic molecule in a polymethyl methacrylate, PMMA matrix, which is subsequently molded and polished, into 1 cm cube.

Chrome or photographic film masks were illuminated by the 1064 nm beam of a mode-locked Nd:YAG laser and imaged into the media to form 4x4 mm data planes where they were intersected with the SHC, 532 nm, of the same laser. The memory, mask, and imaging lens closest to the 3D storage cube were mounted on motor driven linear stages. Retrieval and analysis of the data is performed by illumination of the plane to be read with the 543 nm beam of a He/Ne laser. The induced fluorescence is imaged onto a simple CCD camera and subsequently processed. Access the images from the memory has also been done using the portable ROM systems shown in Fig. 2. This and more sophisticated systems use a simple stepper motor driven stage, a He/Ne laser or diode laser and a low cost video camera. Fig. 7 shows the data stored in the ROM cube media in the form of 100 data planes recorded on an 80 _m pitch, having $10^4$ random bits/plane.

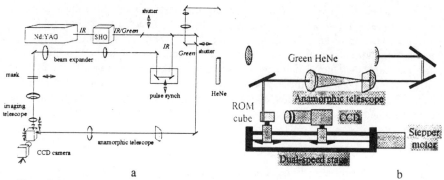

*Figure 6.* Automated two-photon recording/readout setup, a, and portable 3D reading system, b.

*Figure 7.* Results of 100 data layer recording: (a) side-view; (b) image of one data plane.

The system is capable of one micron resolution and with coming improvements to the range of a one micron and a BER of $10^{-13}$. The resolving power is shown in Fig. 8, where the US Airforce resolution plate is stored inside the volume of a disk and displaying micron resolution.

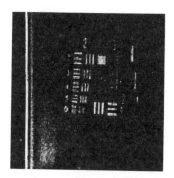

*Figure 8.* 2D image stored inside a 3D memory cube made of 9-methylanthracene dimer dispersed in PMMA.

Even though there are several limitations to the materials and system presently employed, there is strong evidence to suggest that the commercial applications of two-photon, 3D memory devices are feasible.

## 7. Nonlinear Properties of Materials and Use in Optical Limiting Devices

One of the most active areas of nonlinear research is centered on the study of the nonlinear optical, NLO, properties of materials. In addition to the basic science importance of this knowledge their use in technology is becoming very large. Besides their application to electronic switching and 3D storage there are several types of NLO materials that could be utilized for power limiting applications. The most promising type includes compounds with large reversed saturable absorption (RSA) [16-20]. It is possible, however, to have materials with these properties and exhibit a $10^4$ nonlinear attenuation. When a RSA material is exposed to intense pulsed laser radiation the population of the excited state increases very fast; consequently, if the excited state absorption cross section of the material is much higher than that of the ground state, then the effective absorption between excited states will increase significantly and the material will "limit" the transmission of laser light. For materials to be suitable as media for optical limiting devices, several requirements should be met [21]: (1) in order to attenuate the power of the incoming light to a safe level, the excited state absorption cross section $\sigma_{ex}$ should be much larger than $\sigma_{gr}$, the absorption cross section of the ground state. (2) The materials should have broadband spectral response that covers the entire or a very large segment of the visible region. (3) They should have relatively high transmission at low intensity light levels. (4) The response time should be faster than the exciting laser pulse rise time, but the material should become immediately transparent after the pulse. In order to design materials, which satisfy the above requirements, their nonlinear optical properties, as well as the spectra and kinetics of their intermediate states must be determined. Information concerning the nonlinear optical properties, ultrafast spectra and kinetics can be collected with the Z-scan technique [22-23], a simple method that allows one to measure both the nonlinear refraction and nonlinear absorption coefficients, and by means of ultrafast transient spectroscopy. The results of these measurements make possible the selection of the materials which are suitable for power limiting applications and in addition provide guidance for the design and synthesis of new molecules possessing better nonlinear properties.

As an example of this technique we present the experiments performed on four newly synthesized azulenic donor-acceptor compounds[24], which are known to have highly polarizable $\pi$ electron groups. The chemical structures of these compounds are shown on Fig.9. Their ground state electronic spectra show a very intense charge-transfer (CT) band. For most of the compounds the CT band maximum which is located in the red region of the visible spectrum shifts towards the infrared with increase in conjugation. The location of the CT band, as expected, depends highly on solvent polarity. To optimize the optical limiting effect of these materials we must increase the $\sigma_{ex}/\sigma_{gr}$ cross section ratio. Almost all of the azulenic compounds studied, have a high transmission in a broad segment of the visible region as required for optical limiters and in addition strong RSA has been observed for several azulenes which contain CN groups. To determine the suitability of the azulenic compounds for power limiting applications and their optical limiting mechanism, we measured the transient spectra and kinetics of the intermediate states of these materials, after excitation of the CT band with a picosecond pulse. We observed that all of the azulenes, studied, developed a new transient absorption band with $\lambda_{max}$ between 480-560 nm, shortly after excitation of the CT band. The decay lifetime of this band varies from 7 to 30 ps depending on the molecule. For some of the molecules we observed a second transient absorption band with $\lambda_{max}$ between 700-730 nm, whose rise time was the same as a decay of the first transient absorption band. The second band decays, subsequently, with a lifetime of 15-30 ps. Several azulenic molecules were found to have quite strong reversed saturable absorption in a wide region of the visible spectrum. These molecules can be used as the active elements of optical limiting devices. The two-dimensional Z-scan method [25-27] was used for the determination of the nonlinear properties of these molecules. The major component of this technique is a two-dimensional CCD camera. For the interpretation of the data we have used the split step beam propagation method (BPM). The two-dimensional Z-scan method has several advantages over the conventional Z-scan. Among them is its ability to use any arbitrary beam shape and sample thickness, which makes it a perfect diagnostic tool for optical limiters. In addition, the two-dimensional Z-scan technique covers most of the extensions made to the conventional Z-scan, such as the eclipsing Z-scan[28], top-hat beam Z-scan[29-30] and thick sample Z-scan[31-33].

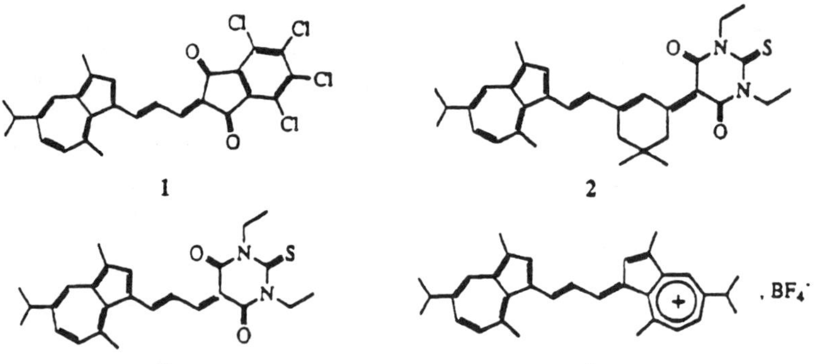

*Figure 9.* Chemical Structures of the azulenic compounds studied

## 7.1 PICOSECOND SPECTROSCOPY AND KINETICS

A schematic representation of the experimental system used for transient absorption spectra and ultrafast kinetics measurements is shown on Fig. 10. For our measurements we employed a subpicosecond laser system, which consists of a mode-locked Ti:sapphire laser (Tsunami from Spectra-Physics) coupled to a regenerative amplifier (Spitfire, Spectra-Physics). The output of the regenerative amplifier consisted of 1.5 ps pulses at 1 kHz repetition rate and a wavelength centered at 840 nm. The fundamental beam was frequency doubled to give the 420 nm pulse used for excitation. The probe beam was focused to a 0.2 mm diameter spot inside a 1 mm thick sample cell by a 250 mm focal length lens and was intersected by the pump beam. The reference beam passed through a 1 mm reference cell, which contained the same solution as the sample cell. The probe and reference beams were focused onto the slit of a monocromator. A 16 bit 256×1024 pixels CCD detector (pixel size 26 μm, Princeton Instrument LN/CCD 1024EUV) was mounted at the output end of the monochromator and was used to monitor the spectra of the probe and reference beams. To measure the kinetics the arrival of the probe pulse in the cell was delayed by a certain period of time with respect to the pump pulse. For each delay time, two consecutive measurements were made, one with excitation and the other without excitation. Together with the reference, four spectra were used to calculate the transient absorption spectrum. This procedure minimized the possible energy fluctuations in the continuum. For the measurement of each spectrum, 10,000-40,000 pulses were averaged. With this experimental system absorbance changes as small as $5 \times 10^{-3}$ were observed and resolved.

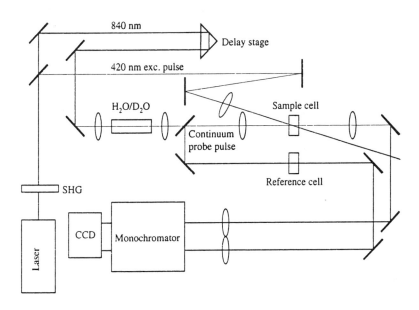

*Figure 10.* Schematic diagram of the experimental system for subpicosecond transient absorption spectra and kinetics measurements

In the experiments described here, the CT absorption band was excited with a 420 nm subpicosecond laser pulse and subsequently the spectra and kinetics of the excited transient states were recorded. All of the compounds formed a new transient absorption band, within 3-5 ps after excitation, with $\lambda_{max}$ at 500 nm and 555 nm. This band subsequently decays within 7-30 ps. For compounds **1-3** we observed a second absorption band maximum located at 700-730 nm. The rise time of this band was found to be the same as a decay time of the first transient absorption band. The second absorption band subsequently decays with a lifetime of 15-30 ps. Fig.11 and Fig.12 show the difference absorption spectra of compound **1** and **4** respectively at several time intervals before and after excitation. Using the chloranil data for calibration we measured the transient

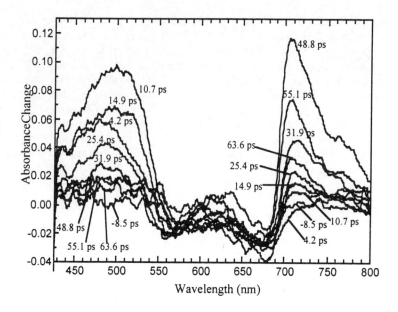

*Figure 11.* Transient spectra of compound 1 in dichloromethane at various delays

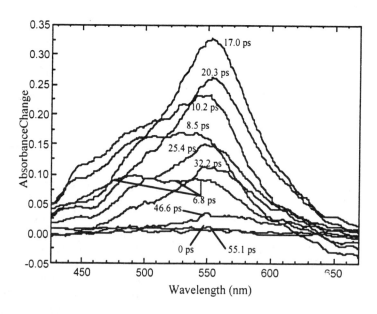

*Figure 12.* Transient spectra of compound 4 in dichloromethane at various delays

ᵢspectra and then calculated the transient absorption cross sections for several azulenic compounds. Compound **4** exhibits the highest cross section, at its absorption maximum 555 nm $\sigma = 1.14 \times 10^{-16}$ cm$^2$. The data of the transient spectra and transient kinetics indicate that compounds **1-4** have strong reversed saturable absorption in the broad region of the visible spectrum, and fast response, time, which makes them suitable for power limiting applications.

## 7.2 TWO-DIMENSIONAL Z-SCAN:

The nonlinear optical properties of the azulenic compounds 1-4 (Fig.9) have been measured by means of our two-dimensional Z-scan technique[25-27]. A schemetic representation of the experimental system is shown on Fig. 13. For the measurements discussed here, we used a single 35 ps (FWHM), 527 nm laser pulse emitted by a cw mode locked Nd:YLF laser coupled to a regenerative amplifier. The spatial shape of the pulse was very close to Gaussian. This pulse was focused by a 220 mm focal length lens, to a spot with a radius of 25 μm. Immediately after the lens the beam was split into two beams. One beam passed through the sample, which was placed into a 1 mm quartz cell and was used as the signal beam. The other beam passed through the air and was used as the reference. Both signal and reference beams were measured simultaneously by a 16 bit, 256x1024 pixels CCD detector located at the far field. For all measurements, a Fig. 12. Transient spectra of compound 4 in dichloromethane at various delays 2x2 binning was used, therefore the effective pixel size was 52 μm. During this experiment the cell was moved along the optical path of the signal beam (Z-direction) in the vicinity of the focus.

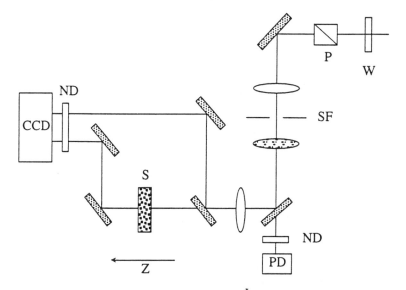

*Figure 13.* Schematic diagram of the two-dimensional Z-scan experimental system: W – half wave plate; P – polarizer; SF – spatial filter; ND – neutral density filter; PD – photodiode; S – sample.

The plot of the normalized energy ratio (normalized transmittance) vs. the Z-position of the cell is referred to as the open aperture result, from which the index of nonlinear absorption can be determined. The index of nonlinear refraction was measured by integrating the image of the signal beam over a circular aperture that contains 40% of the overall energy and then we divide the result by the energy of the reference beam. After averaging over the frames corresponding to the same Z-positions of the cell and normalization we obtain the so-called closed aperture result (with aperture parameter $S=0.4$), from which one can calculate the index of nonlinear refraction. In order to calculate indexes of nonlinear refraction and nonlinear absorption, the closed and open aperture graphs were fitted to the theoretical curves that were calculated using Gaussian decomposition methods[22]. We first calibrated the system with a nonlinear liquid, carbon disulfide, whose nonlinear index of refraction is well known. The indexes of nonlinear refraction and nonlinear absorption for carbon disulfide and azulenic compounds 1-4 were measured using the two-dimensional Z-scan technique. These data are listed in Table 1. It is evident that, at the wavelength of the experiment, $\lambda=527$ nm, the azulenic compounds show positive nonlinear refraction for compound 3 and negative for compounds 1, 2 and 4. The values of the nonlinear index were found to vary from $2.1\times10^{-14}$ cm$^2$/W for compound 3 to $-7.5\times10^{-14}$ cm$^2$/W for compound 2, which is more than twice the magnitude of the nonlinear index of CS$_2$. For compounds 1-3 we did not observe nonlinear absorption, while compound 4 has the nonlinear absorption index of $7.1\times10^{-9}$ cm/W. Further research on azulenic and other highly nonlinear molecules is continuing in order to understand better the structure and properties of molecules with high nonlinear cross sections.

TABLE.1. Nonlinear coefficients measured by 2-D Z-scan method

| Compound | $\gamma$ (nonlinear refraction index) | $\beta$ (nonlinear absorption index) |
|---|---|---|
| CS$_2$ | $3.1\cdot10^{-14}$ cm$^2$/W | ~ |
| 1 in dichloroethane | $-5.2\cdot10^{-14}$ cm$^2$/W | ~ |
| 2 in dichloromethane | $-7.5\cdot10^{-14}$ cm$^2$/W | ~ |
| 3 in dichloromethane | $2.1\cdot10^{-14}$ cm$^2$/W | ~ |
| 4 in methanol | $-2.6\cdot10^{-14}$ cm$^2$/W | $7.1\cdot10^{-9}$ cm/W |

## 8. Acknowledgements

This work was supported in part by ARO Grant DAAH04-96-1-0162 and supported the United States Air Force, Rome Laboratory, under contract F-30602-97-C-0029. We thank Prof. A.E. Asato, University of Hawaii, for the azulenic samples.

## 9. References

1. D.A. Parthenopoulos and P.M. Rentzepis, *J. Appl. Phys.* **68**, 814, 1990.
2. A.S. Dvornikov, I. Cokgor, F. McCormick, R. Piyaket, S. Esener, P.M. Rentzepis, *Opt. Comm.* **128**, 205, 1996.
3. A.S. Dvornikov, S.C. Esener and P.M. Rentzepis, in Optical Computing Hardware, J. Jahns and S.H. Lee eds., Academic, 1993, 287-325.
4. A. S. Dvornikov, I. Cokgor, F. B. McCormick, E. E. Esener, P. M. Rentzepis. IEEE *Memory Tech.* **141**, 68, 1998
5. K. Buse; A Adibi,; D Psaltis,. *J. Opts.* **A**, Pure and Appl. Opts. 1999, **1**, 237, 1999

6. G.A. Rakuljic; V. Layva; A. Yariv. *Opttics Letts*. **17**, 1471, 1992

7. F.H. Mok. *Optics Letts*, **18**, 915, 1993

8. A. S. Dvornikov; P. M. Rentzepis. *Opt.. Comm*. **136**, 1, 1997

9. M. Goeppert-Mayer, *Ann. Phys*. **9**, 273, 1931

10. D. A. Parthenopoulos, P. M. Rentzepis, *Science* **245**, 843, 1989

11. S. Hunter, F. Kramclev, S. Esener, D.M. Parthenopoulos and P. M. Rentzepis, *Appl. Optiics* **29**, 2058, 1990

12. Polymers and other advanced materials : emerging technologies and business opportunities / edited by Paras N. Prasad, James E. Mark, and Ting Joo Fai. New York: Plenum Press, 1995.

13. R. R. Birge, *Ann. Rev. Phys. Chem*. **9**, 683, 1990

14. A. S. Dvornikov, H. Bouas-Laurent, J-P. Desvergne,P. M. Rentzepis. *J. Mater. Chem*. **9**, 1081, 1999

15. A. S. Dvornikov, C. M. Taylor, Y. C. Liang, P. M. Rentzepis, *Photochem. Photobio*. **112**, 39, 1997

16. J.W. Perry, K. Mansour, I.-Y S. Lee, X.-L. Wu, P.V. Bedworth, C.-T. Chen, D.Ng, S.R. Marder, P. Miles, T. Wada, M. Tian, and H. Sasabe, *Science* **273**, 1533-1536, 1996.

17. W. Blau, H. Byrne, W.M. Dennis, J.M. Kelly, *Opt. Comm*. **56**, 25-29, 1985

18. S. Guha, K. Kang, P. Porter, J.E. Roach, D.E. Remy, F.J. Aranda, D.V.G.L.N. Rao, *Opt. Lett*. **17**, 264-266, 1992

19. G.S. He, J.D. Bhawalkar, C.F. Zhao, and P.N. Prasad, *Appl. Phys. Lett*. **67**, 2433-2435, 1995

20. J.S. Shirk, R.G.S Pong, F.J. Bartoli, A.W. Snow, *Appl. Phy. Lett*. **63**, 1880-1882, 1993

21. B.L. Justus, A.L. Huston, A.J. Campillo, *Appl. Phys. Lett*. **63**, 1483-1485, 1993

22. M. Sheik-Bahae, A.A. Said, T.H. Wei, D.J. Hagan, and E.W. Van Stryland, *IEEE J. Quantum Electron*. **26**, 760-769, 1990

23. A.A. Said, M. Sheik-Bahae, D.J. Hagan, T.H. Wei, J. Wong, J. Young and E.W. Van Stryland, *J. Opt. Soc. Am*. **B9**, 405-414, 1992

24. A.E. Asato, R.S.H. Liu, V.P. Rao, Y.M. Cai, *Tetrahedron Letters* **37**, 419-422, 1996

25. P. Chen, I.V. Tomov, P.M. Rentzepis, M. Nakashima, and J.F. Roach, SPIE **3146**, 160-169, 1997

26. P. Chen, D.A. Oulianov, I.V. Tomov, and P.M. Rentzepis, *J. App. Phys*. **85**, 7043-7050, 1999

27. D.A. Oulianov, P. Chen, I.V. Tomov, P.M. Rentzepis, SPIE **3472**, 98-105, 1998

28. T. Xia, D. J. Hagan, M. Sheik-Bahae, and E. W. Van Stryland, *Opt. Lett*. **19**, 317-319, 1994

29. W. Zhao, and P. Palffy-Muhoray, *Appl. Phys. Lett*. **63**, 1613-1615, 1993

30. W. Zhao, and P. Palffy-Muhoray, *Appl. Phys. Lett*. **65**, 673-675, 1994

31. M. Sheik-Bahae, A.A. Said, D.J. Hagan, M.J. Soileau, and E.W. Van Stryland, *Opt. Eng*. **30**, 1228-1235, 1991

32. J.A. Hermann and R.G. McDuff, *J. Opt. Soc. Am*. **B10**, 2056-2064, 1993

33. P.B. Chapple, J. Staromlynska, and R.G. McDuff, *J. Opt. Soc. Am*. **B11**, 975-982, 1994

# TWO-PHOTON PROCESSES: DYNAMICS AND APPLICATIONS

G. S. MACIEL, S.–J. CHUNG, J. SWIATKIEWICZ, X. WANG,
L. J. KREBS, A. BISWAS and P. N. PRASAD
*Institute for Laser, Photonics and Biophotonics*
*State University of New York at Buffalo*
*Buffalo NY 14260-3000*

Abstract - Recent advances in producing highly efficient two-photon up-converters have opened up a great number of potential applications. This paper describes our effort in the development of highly efficient organics with direct two-photon absorption and inorganics with sequential as well as cooperative two-photon processes. An important part of our program is characterization of the two-photon excitation dynamics using femtosecond time-resolved studies for a direct two-photon absorption process in organics. A major effort of our program is developing two-photon technology. In this paper two specific applications are reviewed: (i) Optical power limiting and (ii) Optical tracking of chemotherapy.

## 1. Introduction

There are a multitude of nonlinear optical phenomena, absorptive (resonant) and refractive (nonresonant), classified according to the strength of the response of the material's optical nonlinear polarizability with the external electrical field [1]. Here we focus attention on one specific class of nonlinear phenomenon: energy upconversion (utilization of a nonlinear medium for generation of photons that are more energetic than the absorbed photons via multi-photon absorption). In this case, the overall energy balance is maintained because more than one incident photon is needed to obtain an upconverted photon. These upconverted photons are emitted by the material as isotropic radiation (luminescence). The broad band of applications for a multiphoton process includes optical data storage [2], displays [3], lasers [4], sensors [5], power limiters [6] and imaging [7].

From a microscopic point of view, photons are converted due to electronic transitions occurring in ions, molecules, etc., and due to the interaction of these species with the surroundings (crystal lattice or solvent molecules) and neighbors through multipolar [8] or exchange [9] interactions. A large variety of nonlinear excitation/absorption pathways can produce upconverted luminescence. Accordingly, we highlight and separate these upconversion pathways basically in three categories:
I – Sequential excited state absorption;
II – Simultaneous multi-photon absorption;

*F. Kajzar and M.V. Agranovich (eds.), Multiphoton and Light Driven Multielectron Processes in Organics: New Phenomena, Materials and Applications, 21–30.*

22

III – Cooperative absorption and cooperative luminescence.

These categories are shown in Figs. 1(a) – (d), for the simplest case, i.e., a two-photon absorption process. Figure 1(a) shows sequential two-photon absorption. In this case, a photon with energy matching the gap between the ground state and the excited state i, can be absorbed. If the lifetime of the excited state i is relatively long compared to the excitation exposure time, a second photon can be absorbed from level i' to level ii, after a nonradiative decay takes place from i to i'. The system eventually returns to the configuration of minimum energy (ground state) by emitting phonons (nonradiative decay) or photons (luminescence). These photons present energy that is larger than the energy of the excitation photons. In Figure 1(b), a process of simultaneous two-photon absorption is shown. In this case, the transition takes place between the ground state and the excited state ii without any intermediary step. Because there is no real energy level matching that of the excitation photons, the transition probability is low. Therefore, it is necessary to pump the specie (ion, molecule, etc) with high optical intensities to achieve a considerable population at level ii. Again, photons are emitted from level ii to the ground state with photons of energy higher than that absorbed. Figures 1(c) and 1(d) describe the mechanisms of cooperative absorption and cooperative luminescence respectively. These processes rely on the interaction between specie A and specie B, generally requiring a high density of species because the interaction depends strongly on the relative distance between A and B [8,9]. In contrast to the examples shown in Fig. 1(a) and Fig. 1(b), upconverted photons are observed *only* when a cooperative absorption of the excitation photons takes place, populating level i of both species, as shown in Fig. 1(c). A second photon can be subsequently absorbed by A and B, and eventually more energetic photons are emitted. Figure 1(d) shows another example of a cooperative process. In this case the absorption of photons occurs independently and upconverted emission can take place *only* cooperatively in pairs. This last example demonstrates that upconverted photons can be emitted even when no real excited state matching the transition to the ground state is present [10].

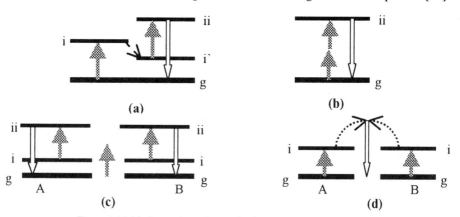

*Figure 1.* Multi-photon absorption mechanisms: (a) sequential two-photon absorption; (b) simultaneous two-photon absorption; (c) cooperative absorption; (d) cooperative luminescence. The gray upward arrows indicate the excitation absorption and the white downward arrow represents the upconversion emission. The electronic energy states of species A and B are represented by a ground state (g) and excited states i, i' and ii.

## 2. Materials

There is a need to develop highly efficient next generation photonics materials which go much beyond simply providing incremental improvements. To achieve this goal, one has to use new design criteria guided by an understanding of structure-properties relations which can come from a careful study of systematically derivatized structures as well as from theoretical modeling. Also, design concepts based upon new physical principles can lead to development of a new class of photonics devices with specific enhanced functionality. New design concepts based on cooperative interaction among several chemical coupled structural units can provide another approach for the development of photonics materials. Another new approach is to use inorganic/organic hybrid materials which combine the merits of these two widely different classes of materials with a nanometer size control of their structure and functionality.

One special class of material that has been extensively studied is based on sol-gel technology [11]. Its main advantages including high purity and homogeneity obtained with it at low temperatures compared to traditional melting processes, simplicity in preparation and more uniform phase distribution for preparing new crystalline and non-crystalline materials (multicomponent systems). As a consequence, sol-gels have been successfully used as a host for inorganic and organic compounds as well [12].

One of the most studied inorganic compounds doped in sol-gels is a rare-earth ion [13]. Rare earths present a large transparency window in the near-infrared and visible range with sharp absorption bands corresponding to intraconfigurational electronic transitions inside the $4f^n$ shell. These electronic states are characterized by long lifetimes ($\mu s$ – ms range) and narrow luminescence emission lines motivating the use of rare earths as suitable elements for photonics applications, for example, solid state lasers [14].

Two-photon organic chromophores belongs to another class of materials that finds applications in photonics, and the major parameter defining a two-photon chromophore as a good candidate for these applications is its two-photon absorption cross-section [15]. To identify structural units with enhanced two-photon activity, as well as to establish symmetry criteria which together can lead to a better understanding of the relationship between molecular structure and two-photon properties, we designed a series of symmetrically and asymmetrically substituted two-photon dyes bearing $\pi$-electron donor groups or electron acceptor groups, as shown in Fig. 2. Type I and II chromophores are symmetrical, consisting of a highly fluorescent $\pi$-electron rich aromatic bridge flanked on both sides with an electron donor group or electron acceptor group. Type III chromophores are asymmetrical molecules consisting of an aromatic bridge flanked on one side by a $\pi$-electron donor and on the other by $\pi$-electron acceptor. As each phenylene group in triphenylamine carries the same two-photon dye, Type IV chromophores have more two-photon dye *per mole* than that of Type I, II and III. This *multibranched* structure is not planar but has a pyramidal geometry because of the lone pair electrons of nitrogen atoms. It is thus reasonable that the chromophore will have *both* dipolar and octupolar contributions to nonlinearity.

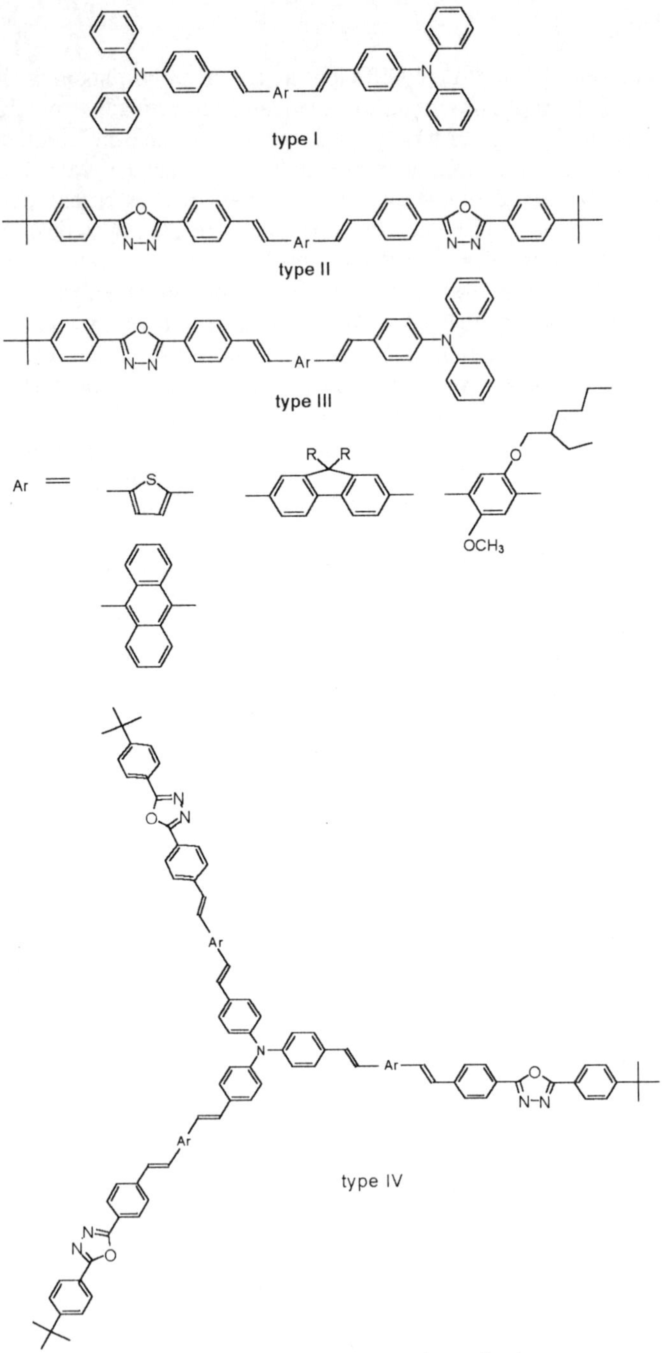

24

*Figure 2.* Two-photon chromophores for photonics applications.

# 3. Characterization

The Z-scan method [16] is a very efficient technique to determine both the cross-section for direct two-photon absorption and that for the excited state absorption. The procedure is the following. The sample is translated around the focal point of the laser beam and a large size detector without an aperture collects the transmitted signal. This arrangement reduces any complication due to self-focusing and defocusing. A typical two-photon absorption curve obtained is shown in Figure 3, with minimum transmission at Z=0 (focal point). A theoretical fit of this curve allows one to obtain the two-photon absorption coefficient and, therefore, the molecular two-photon absorption cross-section by using the number density of the two-photon molecules. If only direct two-photon absorption takes place, the fitted two-photon absorption cross-section should be independent of the irradiance (power density) used. In the presence of any excited state absorption, the effective two-photon absorption cross-section becomes dependent on the power density, Figure 4 shows this power dependence for two-photon chromophores AF-50 and AF-250, provided by the Air Force Research Laboratory at Dayton. The linear dependence reveals a linear absorption by the two-photon excited states. From the slope of this curve, one can get the excited state absorption cross-section and from extrapolation to zero irradiance, one can obtain the true two-photon absorption cross-section.

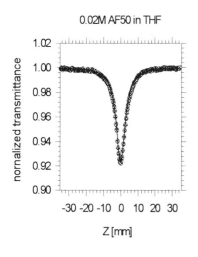

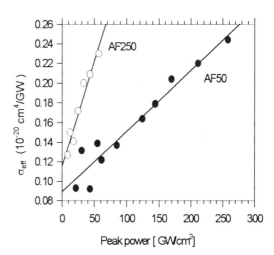

*Figure 3.* Typical Z-scan trace for nonlinear absorption.

*Figure 4.* TPA cross sectio as a function of peak intensity (Z=0).

Another interesting result is observed when the two-photon absorption cross-section is measured in compounds Type III and Type IV of Fig. 2 dissolved in 1,1,2,2-tetrachloroethane. Using femtosecond pulses (~176 fs) at 796 nm the ratio between the two-photon absorption cross section of the trimer (Type IV) and the monomer (Type III) is larger than six and not only three as expected from the number density increase. The cooperative effects responsible for this enhancement are probably due to the delocalization of charges extending along the various arms and from higher-order multipolar contributions.

## 4. Applications

Photonics is expected to play a dominant role in information technology from laser protection, to optical communications, to high capacity 3D data storage, to photodynamic therapy. Here, we will describe a few applications of multi-photon absorption processes.

### 4.1 OPTICAL LIMITING

Optical power limiting is an area of growing interest due to applications such as eye and sensor protection against intense light. Optical power limiters (OPL) are devices currently used for protection of optical systems such as direct viewing devices, focal plane arrays, etc. OPL are used to prevent damage due to high intensity optical exposure. The principle of operation requires transmission of the optical signal at minimum losses up to a certain intensity threshold when a transmission reduction is expected. This nonlinear process generally relies on some material's response to the excitation, e.g., nonlinear absorption, nonlinear refraction and induced scattering [17]. There is a need for new generation materials that can provide broad wavelength coverage as well as protection against both CW and pulsed conditions. Current research is actively pursuing two classes of materials:
(a) Organic structures with enhanced two-photon absorption. This type of nonlinear absorption has the advantage that the transmission at low optical power is very high ($\sim$ 100 %) but effective power limiting behavior is observed only under short pulse conditions (nanosecond or shorter).
(b) Inorganic or organic structures with reverse saturable absorption. This performance relies on sequential two-photon absorption in which the excited state absorption cross section is much larger than the ground state absorption cross section. This mechanism can provide protection against CW lasers but the transmission at low power is not so high because of the significant ground state absorption.

Highly efficient optical power limiting using new organics with strong two-photon absorption has been reported from our group in collaboration with researchers from Air Force Laboratory at Dayton [18].

Here we report a new approach for optical power limiting: CW near infrared optical power limiting at 840 nm using a two-step two-photon absorption. The optical limiting is achieved by reverse saturable absorption (RSA) using a quasi-resonant ground state absorption approach. The sample used was an $Er^{3+}$-doped sol-gel multi-component silica glass and we employed a nonlinear absorption set-up for optical limiting measurements [19]. These measurements are made with the sample at the position of maximum nonlinear absorption (Z=0). Our results showed a decrease in the sample's transmittance by $\sim$ 9 % through a 1mm thick sample with 0.75 % molar concentration of $Er^{3+}$. Nonlinear absorption was observed as the pump intensity increased above a threshold value ($\sim 1.9 \times 10^4$ W/cm$^2$) giving rise to the attenuation. Figure 5 shows the transmittance (defined as the ratio of the transmitted intensity to the incident intensity) as a function of the incident intensity. The value of the ground state absorption cross-section, $\sigma_g$, ($0.3 \times 10^{-20}$ cm$^2$) is obtained from the absorbance spectra of the sample and the best-fit curve shown in Fig. 5 gives $\sigma_e = 2 \times 10^{-20}$ cm$^2$ for the excited state absorption cross-section. It is possible to lower the threshold value and to increase the nonlinear transmittance attenuation (> 9 %) by using a higher Er$^{3+}$ doping concentration favoring cross-relaxation pair interaction ($^4I_{15/2} + \,^4I_{9/2} + $ phonon $\rightarrow \,^4I_{13/2} + \,^4I_{13/2}$). Under these conditions the absorption involving the transition $^4I_{13/2} \rightarrow \,^4S_{3/2}$ will increase. The transmittance at low excitation intensity in such a case is still expected to remain practically at the same level (89 %) because the quasi-resonant ground state absorption process is independent of the Er$^{3+}$ concentration.

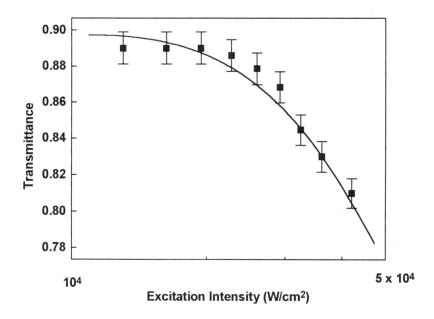

*Figure 5.* Transmittance as a function of intensity in an Er$^{3+}$-doped sol-gel silica glass. Sample thickness: 1 mm.

## 4.2 OPTICAL TRACKING OF CHEMOTHERAPY

Multi-photon microscopy is being successfully used to analyze cellular microanatomy and microphysiology. The laser fluence must be sufficiently high to provide multi-photon excitation at the focused region (Rayleigh range) in a fluorophore probe (dye). The confocal microscopy technique is currently the most frequently used tool for this purpose, allowing monitoring of the fluorescence emitted by the targeted sample with Z-discrimination. Two-photon laser scanning microscopy (TPLSM) was introduced by

Watt Webb's group at Cornell University [7] and it has improved the image signal-to-noise ratio, basically by eliminating auto-fluorescence. Our recent efforts on design and synthesis of two-photon chromophores have produced highly efficient two-photon upconverters, some of them even exhibit up-conversion lasing [20]. We have reported a new two-photon fluorophore (Fig. 6), which was coupled to a chemotherapeutic agent, and used in optical tracking of its interaction and entrance into the target cells by TPLSM [21].

Many cancer patients receive a course of treatment that includes chemotherapy. In many cases, the cellular mechanism of action of chemotherapeutic drugs is unclear. Where in the cell these agents may act is often unknown. It is of great interest for the development of better therapies to know whether an agent acts in the cell nucleus through interruption of DNA replication, or outside the cell through membrane interactions, or through some other means. The ability to understand the cellular and molecular mechanisms of drug action will allow research scientists and physicians to optimize the use of drugs already available and more importantly, to develop better therapies for cancer and other diseases.

In our laboratories, we have utilized our two-photon capabilities to study a targeted chemotherapeutic agent, AN-152 [21]. AN-152 is a conjugate of the widely used cytotoxic agent, doxorubicin, and an analog of Luteinizing Hormone-Releasing Hormone (LH-RH) [22]. LH-RH serves as a carrier of the drug to the selected cancers because receptors for the peptide hormone are expressed in many cancers, but a few in normal tissues [23]. The mechanism of AN-152 action was uncertain. Whether the drug acted by binding to the cell membrane and releasing cytotoxic radaicals, by internalization into the cell, or by interactions in the cell nucleus was a source of speculation [24]. These questions were addressed by linking our two-photon fluorophore, C625, to AN-152 [21]. We succesfully traced its cellular pathway in a human breast cancer cell line that expresses LH-RH receptors (MCF-7 cells) [21]. Using this technology, we were able to visualize the entire process of drug association with the membrane, internalization and localization of drug in real time, with no loss of cell viability [21].

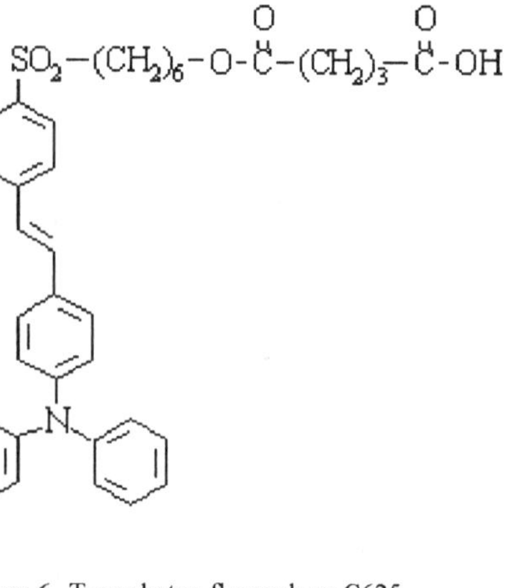

*Figure 6.* Two-photon fluorophore C625.

## 5. Summary

We have demonstrated examples of applications where multi-photon absorption processes are used in photonics. For optical power limiting, two-photon dyes and rare-earth doped sol-gels have been discussed. A new two-photon chromophore has also been introduced to create a better understanding of chemotherapy.

## 6. Acknowledgements

This work was supported by the Air Force Office of Scientific Research, Directorate of Chemistry and Life Science through the contract number F4962093C0017. The work on Optical Tracking of Chemotherapy was done in collaboration with Prof. C. Liebow from the State University of New York at Buffalo and Prof. A. V. Schally from Tulane University and Veterans Affairs Medical Center in New Orleans.

## 7. References

1.  Bloembergen, N. (1996) *Nonlinear Optics*, World Scientific Pub. Co., New York; Butcher, P. N. and Cotter, D. (1991) *The Elements of Nonlinear Optics*, Cambridge Univ. Press, Cambridge; Prasad, P. N. and Williams, D. J. (1991) *Introduction to Nonlinear Optical Effects in Molecules and Polymers*, John Wiley & Sons, New York; Zyss, J. (1994) *Molecular Nonlinear Optics: Materials, Physics and Devices (Quantum Electronics Principles and Applications)*, Academic Press, New York.
2.  Parthenopoulos, D. A. and Rentzepis, P. M. (1989) Three-dimensional optical data storage, *Science* **245**, 843 - 845; Pudavar, H. E., Joshi, M. P., Prasad, P. N. (1999) High-density three-dimensional optical data storage in a stacked compact disk format with two-photon writing and single photon readout, *Applied Physics Letters* **74**, 1338 - 1340.
3.  Downing, E, Hesselink, L., Ralston, J. and MacFarlane, R. (1996) A three-color, solid-state three-dimensional display, *Science* **273**, 1185 - 1189.
4.  He, G. S., Yuan, L., Prasad, P. N., Abbotto, A., Facchetti, A. and Pagani, G. A. (1997) Two-photon pumped frequency-upconversion lasing of a new blue-green dye material, *Optics Communications* **140**, 49 - 52.
5.  Berthou, H. and Jörgensen, C. K. (1990) Optical-fiber temperature sensor based on upconversion-excited fluorescence, *Optics Letters* **15**, 1100 - 1102.
6.  Joshi, M. K., Swiatkiewicz, J., Xu, F., Prasad, P. N., Reinhardt, B. A. and Kannan, R. (1998) Energy transfer coupling of two-photon absorption and reverse saturable absorption for enhanced optical power limiting, *Optics Letters* **23**, 1742 - 1747.
7.  Denk, W., Strickler, J. H., Webb, W. W. (1990) Two-photon laser scanning fluorescence microscopy, *Science* **248**, 73 - 76.
8.  Föster, T. (1965), in O. Sinanoglu (Ed.), *Modern Quantum Chemistry*, Academic Press, New York.
9.  Dexter, D. L. (1953) A theory of sensitized luminescence in solids, *Journal of Chemical Physics* **21**, 836 - 850.
10. Nakazawa, E. and Shionoya, S. (1970) Cooperative luminescence in $YbPO_4$, *Physical Review Letters* **25**, 1710 - 1712.
11. Brinker, C. J. and Scherer, G. W. (1989) *Sol-gel Science*, Academic Press, New York.
12. Messaddeq, Y. (1999), ed., Proceedings of the 3rd Brazilian symposium on glasses and related materials, *Journal of Non-Crystalline Solids* **247**, North-Holland, Amsterdam; Gvishi, R., Narang, U., Ruland, G., Kumar, D. N. and Prasad, P. N. (1997) Novel, organically doped, sol-gel-derived materials for photonics: Multiphasic nanostructured composite monoliths and optical fibers, *Applied organometallic chemistry* **11**, 107 - 127.
13. Wybourne, B. G. (1965) *Spectroscopic Properties of Rare-earths*, Interscience, New york; Reisfeld, R. and Jörgensen, C. K. (1977) *Lasers and Excited States of Rare-earths*, Springer-Verlag, Berlin.

30

14. Digonnet, M. J. (1993), ed., *Rare-Earth Doped Fiber Lasers and Amplifiers*, Dekker, New York.
15. Albota, M., Beljonne, D., Brédas, J-L., Ehrlich, J. E., Fu, J-Y., Heikal, A. A., Hess, S. E., Kogej, T., Levin, M. D., Marder, S. R., McCord-Maughon, D., Perry, J. W., Röckel, H., Rumi, M., Subramaniam, G., Webb, W. W., Wu, X-L. and Xu, C. (1998) Design of organic molecules with large two-photon absorption cross sections, *Science* **281**, 1653 – 1656.
16. Sheik-Bahae, M., Said, A. A., Wei, T-H, Hagan, D. J. and Van Stryland, E. (1990) Sensitive measurements of optical nonlinearities using one beam, *IEEE Journal of Quantum Electronics* **26**, 760 - 769.
17. Tutt, L. W. and Boggess, T. F. (1993) A review of optical limiting mechanisms and devices using organics, fullerenes, semiconductors and other materials, *Progress in Quantum Electronics* **17**, 299 - 338.
18. Bhawalkar, J. D., He, G. S. and Prasad, P. N. (1996) Nonlinear multiphoton processes in organic and polymeric materials, *Reports on Progress in Physics* **59**, 1041 – 1070.
19. Crane, R., Lewis, K., Van Stryland, E., and Khoshnevisan, M. (1995), eds., *Materials for Optical Limiting*, Materials Research Society Symposium Proceedings **374**, Pittsburgh.
20. He, G., Xu, G. C., Prasad, P. N., Reinhardt, B. A., Bhatt, J. C. and Dillard, A. G. (1995) Two-photon absorption and optical-limitng properties of novel organic compounds, *Optics Letters* **20**, 435 - 437; Bhawalkar, J. D., He, G. S., Park, C., Zhao, C. F., Ruland, G. and Prasad, P. N. (1996) Efficient, two-photon pumped green upconversion cavity lasing in a new dye, *Optics Communications* **124**, 33 - 37.
21. Wang, X., Krebs, L. J., Al-Nuri, M., Pudavar, H. E., Ghosal, S., Liebow, C., Nagy, A. A., Schally, A. V. and Prasad, P. N. (1999) A chemically labeled cytotoxic agent: Two-photon fluorophore for optical tracking of cellular pathway in chemotherapy, *Proceedings of the National Academy of Sciences of the United States of America* **96**, 11081 - 11084.
22. Nagy, A., Schally, A. V., Armatis, P., Szepeshazi, K., Halmos, G., Kovacs, M., Zarandi, M., Groot, K., Miyazaki, M., Jungwirth, A. and Horvath, J. (1996) Cytotoxic analogs of luteinizing hormone-releasing hormone containing doxorubicin or 2-pyrrolinodoxorubicin, a derivative 500-1000 times more potent, *Proceedings of the National Academy of Sciences of the United States of America* **93**, 7269 - 7273.
23. Schally, A. V. and Comaru-Schally, A. M. (1997) in J. R. Holland, E. Frei III, R. R. Bast-Jr, D. E. Kufe, D. L. Morton, R. R. Weichselbaum (Eds.), *Cancer Therapy*, Williams & Wilkins, Baltimore, pp. 1067 - 1086.
24. Schally, A. V. and Nagy, A. (in press) Chemotherapy targeted to hormone receptors on tumors, *European Journal of Endocrinology*.

# LINEAR AND NONLINEAR OPTICAL PROPERTIES OF IMPROVED SINGLE CRYSTAL PTS

MINGGUO LIU, SERGEY POLYAKOV, FUMIYO YOSHINO, LARS
FRIEDRICH AND GEORGE STEGEMAN
*School of Optics/Center for Research and Education in Optics and Lasers*
*University of Central Florida*
*4000 Central Florida Boulevard, Orlando, FL 32816-2700, USA*

## Abstract

Improved methods for the growth and polymerization of single crystals of
polydiacetylene PTS, poly bis(p-toluene sulfonate) of 2,4-hexadiyne-1,6-diol, have
allowed new optical measurements to be made. The absorption spectrum, and typical
Z-scans for measuring the optical nonlinearity up to fourth order are described.

## 1. Introduction

PTS, poly bis(p-toluene sulfonate) of 2,4-hexadiyne-1,6-diol, is known for its
outstanding nonlinear optical properties in single crystal form as first reported by
Sauteret et al. [1] in 1976. Based on Third Harmonic Generation (THG) they obtained
the non-resonant, intensity dependent refractive index coefficient $n_2 = 2 \times 10^{-12} cm^2/W$
where the optically induced refractive index change $\Delta n$ is given by $n_2 I$ and $I$ is the local
intensity. This launched many investigations into the optical properties of
polydiacetylenes including the absorption spectrum [2], the excitation of triplet states
[3], the photoconductivity [4], the dispersion of the THG spectrum [5], the lifetime of
the exciton state [6], two photon absorption [7], etc. This also led to many additional
experiments on $n_2$ in single crystal PTS that resulted in a large range of values [8-14],
probably implying a variation in sample quality.

Crystal quality problems have hindered applications of PTS. Typically both the
interior and the surface of the crystals have had defects and scattering centers. We have
improved the growth and polymerization techniques of single crystal PTS as evidenced
by its linear optical properties, and the quality of the Z-scan experiments that show
strong four photon absorption.

*F. Kajzar and M.V. Agranovich (eds.), Multiphoton and Light Driven Multielectron Processes in Organics: New*
*Phenomena, Materials and Applications, 31–38.*
© 2000 *Kluwer Academic Publishers. Printed in the Netherlands.*

## 2. CrystalGrowth and Polymerization

The best known and most frequently investigated compound of the polydiacetylene family is the symmetrical diacetylene referred to as pTS or PTS (thereafter pTS as the monomer, and PTS as the polymer), shown in Figure 1. Centimeter size monomer pTS crystals have been grown in the past by slow solvent evaporation from saturated acetone solution [15-18]. The polymerization of monomer pTS molecules has typically been initiated by thermal annealing or by exposure to radiation (X-ray, γ-ray), even at room temperature and in ambient light. This solid state reaction results in a long conjugated chain along the crystallographic b axis, essentially leading to polymer PTS crystals within the initial monomer pTS crystal structure "template". The resulting polymer crystals were typically full of flaws, cracks and some other defects [19, 20]. This resulted in a large amount of stray scattering when illuminated with laser light for measurements or for device applications. With the exception of the shear growth technique developed by Thakur and co-workers for waveguide applications, and in spite of PTS'es applications potential to optical devices, very little progress has been made towards the improvement of the optical quality of polymer PTS crystals.[21]

*Figure 1.* The structure of single crystal PTS showing the direction of the b-axis.

pTS crystallizes in the monoclinic space group $P2_1/c$ with a = 1.460nm, b = 0.515nm, c=1.502nm and β=118.4°. After the polymerization, the space group remains the same while the cell parameters change to a = 1.448nm, b = 0.493nm, c = 1.491nm and β=118.0° [16]. The greatest structural change during the polymerization occurs in the lattice parameter b, a shrinkage of 5% along the b-axis which induces enormous stresses in the crystals.

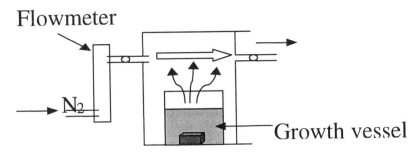

Flowmeter

$N_2$

Growth vessel

*Figure 2.* Experimental apparatus for pTS crystal growth.

Modifications were made to the methods for preparing optical quality PTS polymer crystals and the details will be described in detail elsewhere. The pTS material was synthesized by the usual procedures [15]. The freshly synthesized material has a large amount of impurities and was carefully purified by successive re-crystallization from acetone until the solution became nearly colorless. Crystals were grown in the growth setup illustrated in Figure 2. A controlled nitrogen flow was used to control the evaporation speed of the acetone in the growth vessel housed in a sealed glass jar. In a typical growth batch, one to three nucleation sites were usually created. Each of the nucleated seeds was then grown into a monomer crystal with well-developed crystallographic facets and dimensions greater than 1 cm. The typical as-grown crystals are shown in Figure 3 (after polymerization). The as-grown monomer crystals were very light pink upon completion of the growth. The color quickly changed to red at room temperature indicating the onset of polymerization.

*Figure 3.* PTS single crystal.

There was a strong crystal thickness dependence of the defects produced during polymerization. Thin samples suffered much less cracking than the thick ones. This can be easily interpreted as a size effect where there were more chances to have inhomogeneous regions in the thicker crystals than the thinner ones. Typically, nearly perfect polymer crystals less than 1 mm thick could be obtained after polymerization, where-as samples several mm thick did show evidence of cracks.

*Figure 4.* SEM of the (100) facet of a PTS polymer crystal. The dust particle in the upper third of the surface was used for focusing the electron beam.

A SEM picture of the (100) facet of a typical polymer PTS crystal is shown in Figure 4. The bright junk on the upper part of the picture was intentionally left there to be able to focus the SEM beam without damaging the crystal facet. It can be seen clearly from the figure that the crystal is a single domain crystal, and the surface is optically flat. The optical characterization on the as-prepared PTS confirmed the optical quality unambiguously, which will be discussed in detail in the following characterization sections.

## 3. Optical Characterization

### 3.1. ABSORPTION SPECTRUM

The absorption spectrum was measured on a sample 100s of microns thick using standard ellipsometry. The data was processed in the usual way to yield the absorption coefficient from the measured variables, namely the magnitude of the reflection coefficient and the phase change on reflection.

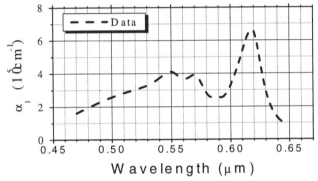

*Figure 5.* Absorption spectrum of single crystal PTS.

The wavelength dispersion of the absorption coefficient is shown in Figure 5. The maximum value of the absorption, of order $7 \times 10^5$ cm$^{-1}$, is indicative of essentially complete polymerization of the crystal. The secondary maxima at higher energies than the principal peak have been identified previously as due to transitions between vibronic sub-levels in the ground ($1A_g$) and exciton excited ($1B_u$) states, as indicated schematically in Figure 6. In summary, the absorption spectrum is indicative of very high quality PTS single crystals.

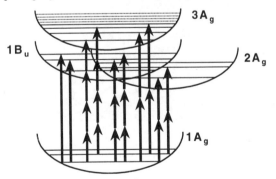

*Figure 6.* Schematic representation of the dominant electronic transitions occurring for one photon, two photon, three photon and four photon absorption.

## 3.2. Z-SCAN CHARACTERIZATION OF PTS CRYSTALS

Amongst the nonlinearity measurement techniques practiced in nonlinear optics, the Z-scan method is a proven high-resolution technique for measuring the absolute value of the intensity dependent refractive index and multi-photon absorption coefficients [22]. However, the Z-scan signal can be easily distorted with only a slight internal inhomogeneity or by rough surfaces. The distorted signal can even mislead the data analysis. High quality samples with good internal homogeneity and optically flat surfaces are therefore essential to achieving the best accuracy in the Z-scan measurements. The quality of the Z-scan, especially the closed aperture one, is a nice characterization approach to assessing the crystal quality.

We performed a series of Z-scan measurements on our PTS crystals with a wide range of input intensities at 1600 nm. [23]. Typical open and closed aperture Z-scan results on our PTS crystals are shown in Figures 7 and 8, for incident light polarization along the polymer chain direction. Both of the figures show perfect Z-scan curves as predicted by theory with very low noise. In Figure 7 the signal-to-noise is sufficient to resolve the difference between three and four photon absorption. Clearly this particular Z-scan result is dominated by four photon absorption. In the closed aperture result of Figure 8, the curve again shows excellent signal-to-noise, smooth evolution of the Z scan from the valley to crest indicating a positive nonlinearity at this wavelength and the corresponding low intensity open aperture case shows a complete absence of stray scattering. The multi-photon absorption minimum in the open aperture can be as low as 1% to be easily distinguished and analyzed with our samples. As a result, the very low noise level and the perfect shapes of the Z-scan strongly suggest the excellent optical quality of the PTS crystals.

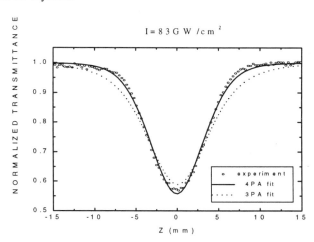

*Figure 7.* Open aperture Z-scan spectrum of PTS at 1600 nm.

Although Figure 7 shows that at 83 GW/cm$^2$ the nonlinear absorption is clearly due to four photon absorption, the range of intensities for which the difference is so clear-cut is rather limited. Instead it is necessary to analyze the Z-scans as being produced by either three or four photon absorption independently, and then plot the resulting coefficients $\alpha_3$ and $\alpha_4$ ($\Delta\alpha = \alpha_3 I^2$ or $\Delta\alpha = \alpha_4 I^3$) versus the peak intensity. The results on the new samples are shown in Figure 9. Clearly four photon absorption is the dominant process.

The deduced $\alpha_4 = 0.3$ GW$^3$/cm$^5$ is very large. The implication is that at only 10 GW/cm$^2$, the induced absorption coefficient is 300 cm$^{-1}$, meaning that even a 1 mm

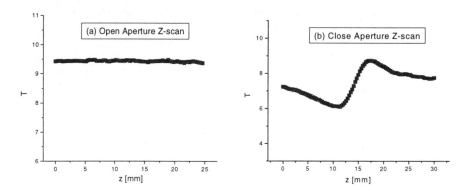

*Figure 8.* Low intensity (~GW/cm$^2$) open (a) and closed (b) aperture Z-scans of a 380 μm thick PTS crystal.

thick crystal can go from essentially nonlinearly transparent ($\Delta\alpha = 0.03$ mm$^{-1}$ at 1 GW/cm$^2$) to opaque. However, this result begs the question as to why the coefficient is so large. There are two possible explanations, both involving the existing accidental degeneracy between three and four photon absorption which occurs in PTS.

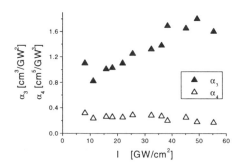

*Figure 9.* Values of $\alpha_3$ and $\alpha_4$ deduced by fitting the Z-scan data to just three photon, or just four photon absorption versus the peak intensity of the input beam at the center of the sample.

A detailed calculation to be published elsewhere shows that

$$\alpha_3 \propto \frac{|\mu_{gu}|^4}{(\omega_{gu} - 3\omega - i\Gamma_{gu})} f(\omega, \omega_{gu})$$

where $\Delta\alpha = \alpha_3 I^2$ and $f(\omega, \omega_{gu})$ is a slowly varying function of its variables. Here $\mu_{gu}$ is the transition dipole moment between the $1A_g$ and $1B_u$ states and $\hbar\omega_{gu}$ is the energy between the two states. Note that the three photon absorption spectrum peaks at $\omega_{gu}/3$. In terms of wavelength, the absorption spectrum in Figure 5 suggests that three photon should occur over the wavelength range 3x500 nm = 1500 nm to 3x650 nm = 1950 nm, peaking around 1860 nm. The same calculation extended to the next highest order for $\alpha_4$ where the increase in the absorption coefficient is $\Delta\alpha = \alpha_4 I^3$ gives

$$\alpha_4 \propto \frac{|\mu_{gu}|^6 |\mu_{ug'}|^2}{(\omega_{gu} - 3\omega - i\Gamma_{gu})^2 (\omega_{gg'} - 4\omega - i\Gamma_{gg'})} F(D, \omega, \omega_{ug}, \omega_{gg'})$$

where $\hbar\omega_{gg'}$ is the energy difference between the ground and excited two photon state, $D = |\mu_{ug'}/\mu_{gu}|$, $\mu_{ug'}$ is the dipole transition moment between the $1B_u$ and even symmetry states and the function $F$ varies only slowly with its variables. The denominators imply that $\alpha_4$ is resonantly enhanced at both $\omega \cong \omega_{gu}/3$ and at $\omega \cong \omega_{gg'}/4$. As noted before, the first corresponds to 1500 nm to 1950 nm, and the second to 2x700 nm = 1400 nm to 2x1000 nm = 2000 nm.[14,17] This double enhancement for $\alpha_4$ is called three photon enhanced, four photon absorption. Note that three photon absorption is not enhanced by the coincidence with four photon absorption. Thus it is possible that four photon absorption could actually dominate three photon absorption, despite being a higher order effect.

Note that an effective four-photon coefficient can also be obtained from three photon absorption into the $1B_u$ state followed by one photon absorption from the $1B_u$ to the $3A_g$ dominant two photon state. Further experiments are underway to resolve this issue.

A series of experiments is currently being conducted in our laboratory towards obtaining a full multi-photon absorption spectrum.

## 4. Conclusions

The growth and polymerization of PTS has led to crystal optical properties which exhibited very large one and four photon absorption coefficients at 620 and 1600 nm respectively. The large $\alpha_4$ was interpreted as due to three photon resonantly enhanced four photon absorption, although an alternative explanation of three photon absorption followed by one photon absorption from the exciton state to the two photon state cannot be excluded at this time.

**Acknowledgements:** The early stages of this research were supported by AFOSR, and later stages by NSF and Lockheed-Martin in conjunction with the State of Florida I4 program.

38

References:
1.  Sauteret, C., Hermann, J.P., Frey, R., Pradere, F., Ducuing, J., Baughman, R.H., Chance, R.R. (1976) *Phys. Rev. Lett.*, **36**, 956.
2.  For example, Blanchard, G.J., Heritage, J.P., Baker, G.L and Etemad, S. (1989) *Chem. Phys. Lett.*, **158**, 329.
3.  For example, Austin, R. H., Baker, G. L., Etemad, S. and. Thompson, R. (1989) *J. Chem. Phys.*, **90**, 6642.
4.  For example, Siddiqui, A. S. (1980) *J. Phys. C: Solid St. Phys.*, **13**, 2147.
5.  Kajzar, F., and Messier, (1987) *J. Polymer J.*, **19**, 275; in (1987) *Nonlinear Optical Properties of Organic Molecules and Crystals*, Vol **2**, edited by D. Chemla and J. Zyss, Academic Press, New York, chapter 3.
6.  Reviewed in Kobayashi, T., (1992) *Synth. Metals*, **49-50**, 565; (1992) *Nonlinear Opt.*, **1**, 101.
7.  Lequime, M., and Hermann, J.P. (1977) *Chem. Phys.*, **26**, 431
8.  Krol, D. M., and Thakur, M., (1990) *Appl. Phys. Lett.*, **56**, 1406.
9.  Gabriel, M. C., Whitaker, N. A., Dirk, C. W., Kuzyk, M. G. and Thakur, M. (1991) *Opt. Lett.*, 16, 1334.
10. Ho, S.-T., Thakur, M. and LaPorta, A. (Optical Society of America, Washington DC 1990) paper *QTUB5*, vol. **8**, , 40-42.
11. Lawrence, B., Torruellas, W. E., Cha, M., Sundheimer, M. L. and Stegeman, G. I., Meth, J., Etemad, S., Baker, G. (1994) *Phys. Rev. Lett.*, **73**(4), 597.
12. Lawrence, B., Cha, M., Kang, J. U., Torruellas, W., Stegeman, G. I., Baker, G. Meth, J., and Etemad, S. (1994) *Electron. Lett.*, **30**(5), 447.
13. Lawrence, B. L., Cha, M., Torruellas, W. E., Stegeman, G. I., Etemad, S. and Baker, G., (1995) *Nonlinear Optics*, **10**, 193.
14. Torruellas, W.E., Lawrence, B.L., Stegeman, G.I. and Baker, G. (1996) *Opt. Lett.*, **21**, 1777.
15. Wegner, G. (1971) *Makromol. Chem.* **145,** 85-94.
16. Bloor, D., Koski, L., Stevens, G. C., Preston, F. H., Ando, D. J. (1975) *J. Mat. Sci.* **10**, 1678-1688.
17. Dudley, M., Sherwood, J. N., (1983) *Mol. Cryst. Liq. Cryst.* **93**, 223-237.
18. Krug, W., Miao, E., Derstine, M. (1989) *J. Opt. Soc. Am.* B**6**(4), 726-732.
19. Schott, M., and Wegner, G. (1987) Nolinear Optical Properties of Organic Molecules and Crystals, Vol.**2**, edited by D. Chemla and J. Zyss, Academic Press, New York, Chapter 3.
20. Bloor, D., Bowen, D. K., Davies, S. T., Roberts, K. J., Sherwood, J. N., (1982) *J. Mat. Sci. Lett.* **1**, 150-152.
21. Thakur, M., Meyler, S.(1985) *Macromolecules* **18**, 2341-2344
22. Sheik-Bahae, M., Said, A. A., Wei, T. H., Hagan, D. J., Van Stryland, E. W. (1990) *IEEE J. Quantum Electron.* **26,** 760.
23. Shim, H., Liu, M., Chang, H. B., Stegeman, G. I. (1998) *Opt Lett.* **23**(6), 430-432.

# OPTICAL LIMITING: CHARACTERIZATION & NUMERICAL MODELING

E. W. VAN STRYLAND, D. I. KOVSH, R. NEGRES, D. J. HAGAN, V. DUBIKOVSKY and K. BELFIELD*
*School of Optics/CREOL, University of Central Florida (*Deprtment of Chemistry), Orlando, FL, USA*

Abstract Progress in making effective optical limiting devices requires careful characterization of the material nonlinearities as well as modeling of the propagation of optical beams through the material. We present a method to study the spectral properties of the nonlinear response as well as the results of modeling nanosecond pulse propagation in optically absorbing media. We specifically look at two-photon absorbing and reverse saturable absorbing materials in liquid hosts. The characterization technique is an excitation- femtosecond continuum probe technique. The modeling includes beam propagation through thick media (i.e. thickness much greater than the diffraction length or depth of focus) and includes the effects of index changes associated with acoustic waves generated by any absorption process. This requires a simultaneous solution to the acoustic and electromagnetic wave equations. A graphical user interface to a C++ numerical code has been developed for modeling such devices including the possibility of multiple nonlinear elements. We have extended this code for a tight focusing geometry beyond the paraxial ray approximation, but assuming cylindrical symmetry.

## INTRODUCTION

Considerable effort has been expended in characterizing nonlinear optical materials for applications such as optical limiting where the transmittance is high for low inputs and becomes low for high inputs. Much of the data, however, has been taken at one or only a few wavelengths. On the other hand, the spectral dependence of the nonlinear response is important for determining the effectiveness of a device and this spectral dependence is often critical in determining the nature of the nonlinear response. An extremely valuable method for determining the spectrum of the nonlinear response is to use a pulsed pump probe technique where the probe is a broadband pulse. A very broadband pulse is produced when an intense femtosecond pulse is focused into a transparent optical material, a so-called femtosecond continuum. We use this technique to measure nondegenerate two-photon absorption (2PA) spectra as well as excited –state absorption spectra.[1] For 2PA, the nonlinearity measured is nondegenerate since the pump (excitation) is at a fixed frequency, $\omega_e$, while the probe has a variable frequency, $\omega_p$. This leads to a measurement of the nondegenerate 2PA coefficient $\beta(\omega_p;\omega_e)$, i.e. the 2PA at $\omega_p$ due to the presence of an intense beam at frequency $\omega_e$. For materials exhibiting excited-state absorption (ESA) where the ESA

*F. Kajzar and M.V. Agranovich (eds.), Multiphoton and Light Driven Multielectron Processes in Organics: New Phenomena, Materials and Applications, 39–52.*
© 2000 *Kluwer Academic Publishers. Printed in the Netherlands.*

cross section is larger than that of the ground state, i.e. reverse saturable absorption (RSA), the absorption measured in this pump-probe experiment is simply the ESA spectrum. Reference [1] explains this method in more detail and shows some examples of the spectra obtained. Figure 1 shows the schematic of the method. In the configuration we use, the excitation energy can be varied from 500 nm to 1.6 μm while the continuum probe can be measured from 300nm to 1.7 μm using a combination of a Si-based dual diode array for the visible and an AlGaAs dual diode array for the infrared. Figure 2 shows results of measurements on an aminobenzothiazolefluorenyl polystyrene copolymer [2] taken with two different excitation wavelengths. Excitation at 388 nm lies within the region of strong linear absorption and directly excites molecules. The probe then measures the spectrum of the excited-state absorption as calculated from the change in transmittance $\Delta T = T_{NL} - T_L$, $T_{NL}$ is the nonlinear transmittance and $T_L$ is the linear transmitance.

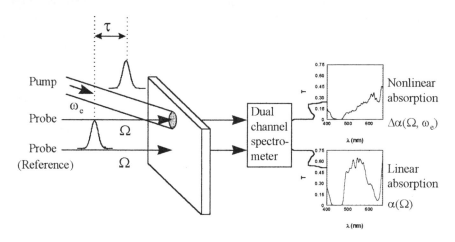

*Figure 1. Schematic of the pump-continuum probe spectroscopic method.*

From data such as these, application of causality allows the dispersion of the nondegenerate nonlinear refraction to be determined via the Kramers-Kronig relation:

$$\Delta n(\omega, \omega_e) = \frac{c}{\pi} \int_0^\infty \frac{\Delta \alpha(\Omega, \omega_e)}{\Omega^2 - \omega^2} d\Omega \ . \tag{1}$$

For 2PA materials this relation becomes:

$$n_2(\omega, \omega_e) = \frac{c}{\pi} \int_0^\infty \frac{\beta(\Omega, \omega_e)}{\Omega^2 - \omega^2} d\Omega \ , \tag{2}$$

where β is the generalized nondegenerate 2PA coefficient,[3] and $n_2$ is the nondegenerate nonlinear refractive index.

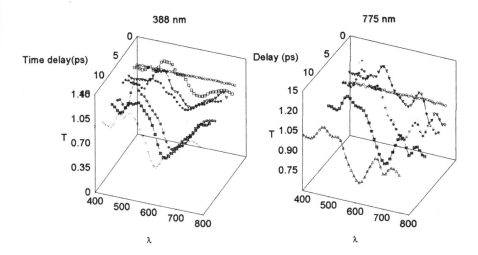

*Figure 2. Left: The normalized change in transmittance (denoted by T in the figure) as a function of wavelength (horizontal axis) and temporal delay, for an excitation wavelength of 388 nm. Right: same as the top except the excitation wavelength is 775 nm. T>1 indicates saturation of the linear absorption while T<0 indicates excited state absorption.*

In the particular examples shown in Fig. 2, the excited-state spectrum is independent of the excitation method and excitation wavelength, i.e. once excited it has no memory of the method of excitation. Thus, the Kramers-Kronig relation reduces to:

$$\sigma_R(\omega) = \frac{c}{\pi} k \int_0^\infty \frac{\sigma_{ex}(\Omega)}{\Omega^2 - \omega^2} d\Omega \ , \tag{3}$$

where $k = \omega/c$ and $\sigma_R$ is the refractive cross section derived from the excited state absorption cross section, $\sigma_{ex}$.[4]

Analysis and modeling of the nonlinear mechanisms allow a material to be characterized as to magnitude and type of nonlinearity and its spectral dependence. Given a full characterization of the nonlinear response to determine nonlinear parameters, e.g. cross sections and/or 2PA coefficients, these parameters can be used in nonlinear propagation codes to determine their potential application to optical limiting devices.

One of the propagation codes that we have developed assumes cylindrically symmetrical beams and works well on PC's. We have also developed a graphical user interface (GUI) for this propagation code that makes it user friendly. This code propagates Gaussian temporal pulses of various spatial profiles (e.g. Gaussian, Bessel-Gauss, flat-top) through thick optical materials. Here «thick» means a thickness large compared to the depth of focus or Rayleigh range, $Z_0$. We have successfully propagated up to 500 $Z_0$. The nonlinearities included are multiphoton absorption, bound electronic nonlinear refraction or optical Kerr effect, electrostriction, thermal acoustic waves and excited state absorption. Excited state absorption models include up to a 5-level model which allows triplet states and multiphoton-induced excited state absorption as shown at the left of Fig. 2, as well as the accompanying refraction from level population changes (characterized by $\sigma_R$). This covers the range of nonlinearities responsible for optical limiting with the one notable exception of nonlinear scattering. Of these nonlinearities, acoustic waves generated by the absorption of light (or by electrostriction), by either linear or nonlinear absorption, is by far the most difficult and computationally intensive problem. Its solution requires simultaneous solution of the acoustic and electromagnetic wave equations.

Thermally induced refractive index changes caused by absorption of light in a material have been intensively investigated both experimentally [5,6,7] and theoretically [8-14] for various time scales of the input laser pulses. Most of these studies addressed the problem of thermal lensing produced inside the material on microsecond and longer (up to CW) time scales [8-10]. The shape of the thermal lens and its impact on the near and far field spatial distributions of the laser beam were analytically estimated for these time scales in references [13-16]. Several authors have considered shorter (nanosecond) pulses to be a source for the refractive index change. In this case the effect of thermally induced index changes can be highly transient for focused beams. The theoretical analysis of such a situation was offered in Refs. [11,12] where the authors derived the coupled hydrodynamic equations defining the local changes in density, pressure and temperature due to the presence of laser radiation. Under certain approximations on the laser beam temporal and spatial profiles (namely Gaussian) the solutions for those equations were obtained. The strong beam self-action due to the thermal lens formed on a nanosecond scale was also considered by Brochard et. al. [17]. The simplified model describing such self-action and comparison to the steady state case was offered and tested against the results of Z-scan and pump-probe experiments. The main difficulty one experiences when trying to analyze thermal refraction in the transient regime is the fact that both wave equations for electromagnetic laser field and acoustic equation must be solved simultaneously. Unless certain simplifications are made it is an extremely computationally intense task. However, with the recent increase of computing power of modern PCs and workstations it becomes possible to model the dynamics of the nonlinear media response to pulsed laser light including spatial beam size changes within the medium from external focusing as well as nonlinear self action.

The state-of-the-art in optical limiter design exhibiting linear absorption is using multiple nonlinear elements in tandem to extend the dynamic range while keeping the limiting threshold low. While 2PA can in principle simply use a thick nonlinear element [18] this tandem design is particularly advantageous for limiters using RSA [4,19,20] where some linear absorption is needed to initiate the nonlinear response. Figure 3 shows a typical design.

The beam diameter may vary by more than an order of magnitude when propagating through such a device so that the irradiance can vary by several orders of magnitude. This requires sophisticated numerical beam propagation algorithms to model the limiter response. In addition, the elements (cells in Fig. 3) are often liquid elements where the thermal nonlinearity can be large. Solid state limiters (e.g. doped polymers) can have significantly reduced thermal nonlinearities, however, their reduced damage threshold significantly lowers the upper limit of fluence that can be used.

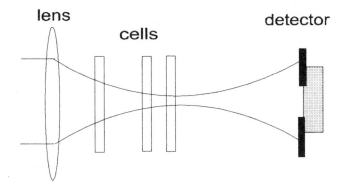

*Figure 3. Typical design for a 3 element optical limiter where the detector is set up to measure the fluence by using an aperture.*

## 1. BEAM PROPAGATION ALGORITHM

The propagation of light through an optical medium can be described by the solution to the vector wave equation [21]:

$$\nabla \times \nabla \times \vec{E}(\vec{r},t) + \frac{1}{c^2}\frac{\partial^2 \vec{E}(\vec{r},t)}{\partial t^2} = -\mu_0 \frac{\partial^2 \vec{P}(\vec{r},t)}{\partial t^2} , \qquad (4)$$

where $\vec{E}(r_\perp,z,t)$ and $\vec{P}(r_\perp,z,t)$ are the electric field and the medium polarization. Making the Slowly Varying Envelope Approximation and assuming that group velocity dispersion can be neglected, this equation can be greatly simplified; and for slow (i.e. large F-number) systems rewritten in a scalar paraxial form:

$$2jk\frac{\partial \Psi(r_\perp,z,t)}{\partial z} = \nabla_\perp^2 \Psi(r_\perp,z,t) + \left(k_0^2 \chi_{NL}(r_\perp,z,t) - jk\alpha_L\right)\Psi(r_\perp,z,t), \qquad (5)$$

where we used the expression for the electric field $E(r_\perp,z,t) = \Psi(r_\perp,z,t)e^{j\omega t - jkz}$ . Here $\nabla_\perp^2$ and $r_\perp$ denote the transverse Laplace operator and radial spatial coordinate, while $k = n_0 k_0 = n_0 \omega/c$ is the wave vector in the media with linear index of refraction $n_0 = \sqrt{1 + \mathrm{Re}\{\chi_L\}}$ and linear absorption coefficient $\alpha_L = -(k_0/n_0)\cdot\mathrm{Im}\{\chi_L\}$. $\chi_{NL}(r_\perp,z,t)$ is the nonlinear susceptibility of the material, which may consist of instantaneous and cumulative parts:

$$\chi_{NL}(r_\perp,z,t) = \chi_{NL}^{ins}(r_\perp,z) + \chi_{NL}^{cum}(r_\perp,z,t). \qquad (6)$$

The nonlinear susceptibility is related to the nonlinear refractive index change, $\Delta n$, and the nonlinear absorption, $\alpha_{NL}$, of the material as:

$$\mathrm{Re}\{\chi_{NL}\} = 2n_0 \Delta n \qquad (7)$$

$$\text{Im}\{\chi_{NL}\} = -\frac{n_0}{k_0}\alpha = -\frac{n_0}{k_0}(\alpha_L + \alpha_{NL}).$$  (8)

Due to the fact that there is no explicit time dependence in Equation (5) (although the field amplitude as well as the nonlinear susceptibility are, in general, functions of time), the modeling of the laser pulse propagation in the nonlinear medium can be split into two separate numerical tasks. The first one is dividing the pulse into a number of time slices and propagating each slice as a CW beam. The second one is computing and storing the cumulative part of the nonlinear susceptibility induced by each slice. Therefore, the solution to the original time-dependent wave equation (2.2) becomes a CW propagation problem. To reduce the computation time and storage memory we assume cylindrical symmetry, thus reducing the spatial 3D problem to 2D. We then implement the so-called Beam Propagation Method (BPM), which requires recomputing the transverse field distribution along the direction of propagation, $z$, using the formal solution to Eq. (5). We use both a spectral method developed by Feit and Fleck [22] and a unitary, finite difference Crank-Nicholson scheme [23] when dealing with propagation through distances of more than a few Rayleigh ranges [24,25]. For tight focusing geometries where the paraxial ray approximation becomes invalid, we use a Pade approximation scheme that has proven surprisingly fast and versatile (Crank-Nicholson is a form of first-order Pade approximation).[26]

## 2. THERMALLY-INDUCED INDEX CHANGES

Upon heating of the medium by absorption of the laser pulse, the medium begins to expand. This expansion propagates outward as an acoustic disturbance. The method of analyzing this problem depends very much on the relative time scales for the acoustic expansion and the optical pulsewidth. If the laser pulse is much longer than the time for the acoustic wave to propagate across the laser beam (the acoustic transit time), the thermal lens is can be assumed to be essentially instantaneous. The refractive index change in this case is linearly proportional to the temperature change and hence the pulse fluence (if the medium exhibits only linear absorption), which greatly simplifies the analysis. However, if the laser beam size is larger than or comparable to the acoustic transit time, then only later portions of the laser pulse can experience index changes induced by earlier portions. Such a situation has been referred to as the transient regime [27].

Sound travels $\cong 1$ $\mu$m per nanosecond so that sound will travel across a typical focused beam of a few microns spot size in a few nanoseconds, i.e. of the order of the pulsewidth of Q-switched Nd:Yag lasers. For large beams the sound wave cannot travel across the beam within the pulse. We have modeled these thermal effects by both solving the acoustic wave equation as driven by the absorption of light (either by linear or nonlinear absorption), as well as by making the thermal lensing approximation. In this approximation, the absorbed heat is assumed to instantaneously change the index according to the thermo-optic coefficient dn/dT. We find that this approximation is only good near the focus of tightly focused beams as expected.

Combining the three main equations of hydrodynamics: continuity, Navier-Stokes and energy transport equations [28], and assuming small changes of temperature, $\Delta T$, density, $\Delta \rho$, and pressure, $\Delta p$, and following the derivation of Ref. [29] we obtain the acoustic equation defining the density perturbation of the material [17, 30,31]:

$$\frac{\partial^2 (\Delta n)}{\partial t^2} - C_S^2 \nabla^2 (\Delta n) = \frac{\gamma^e \beta C_S^2}{2n} \nabla^2 (\Delta T). \tag{9}$$

with thermal expansion coefficient $\beta = -(1/V)(\partial V/\partial T)_p$ and index of refraction $n$. $\gamma^e = \rho(\partial n^2/\partial \rho)_T$ is the electrostrictive coupling constant, which can be estimated using the Lorentz-Lorenz law as $\gamma^e = (n^2-1)(n^2+2)/3$. We have assumed that the light energy absorbed is converted into kinetic energy of the molecules on a time scale shorter than the pulse itself. This is usually true for nanosecond pulses and vibrational relaxation.[31] Electrostriction has been ignored in deriving Eq. 9, and diffusion of heat is also neglected on the nanosecond time scale.

If the pulsewidth, $\tau_p$, is longer than the acoustic transit time, we simplify the numerical modeling of the photo-acoustic effect by parameterizing the index change close to the propagation axis [11] as:

$$\Delta n \cong \left( \frac{dn}{dT} \right) \Delta T, \tag{10}$$

where $(dn/dT) = \gamma^e \beta/(2n)$ is the thermo-optic coefficient. Equation (10) is a commonly used approximation called the thermal lensing effect [6], which is usually applied for much longer time scales (microseconds up to CW) where the index change is governed by thermal diffusion. As seen below the actual thermal lens introduced by acoustic density perturbations in the material is a very aberrated replica of the one in Eq. (10), since this approximation only can be made for near axis index changes because it ignores the acoustic wave propagation. However, such an approximation is attractive because we can significantly reduce the computational time required to numerically solve the acoustic wave equation for each time slice of the pulse. In fact, there are experimental results in the literature where $(dn/dT)$ was calculated in this approximation for different liquids [32-34]. The last approximation requires great care to apply and can lead to erroneous results.

We demonstrate the model using 532 nm, $\tau_p = 7$ ns (FWHM of irradiance) with a Gaussian spatial and temporal profile propagating through an aqueous solution of nigrosine dye. Nigrosine is chosen since it shows very little nonlinear response other than thermal refraction from linear absorption for nanosecond input pulses. The thermo-optic coefficient of water is $-5.7 \times 10^{-4}$ K$^{-1}$.[17] We choose cases where the linear-optics beam waist in the sample $w_0$ is (i) 6 $\mu$m and (ii) 30 $\mu$m (HW1/e$^2$M of irradiance). The acoustic transit times in cases (i) and (ii) are $\tau_{ac} \cong 4$ ns and 20 ns respectively. The sample is 1 mm thick in each case. Figure 4 shows the thermally induced refractive index change in case (i) after the laser pulse passes through the sample. The full modeling using the acoustic equation coupled with propagation is shown on the left, while the thermal lensing approximation is applied on the right. Although this approximation ignores the small index disturbances on the wings of the

46

pulse, which are due to the acoustic wave propagation, it predicts the changes of the refractive index close to the axis reasonably well, and this is where most of the beam energy is concentrated. Moreover, the thickness of the sample is several Rayleigh ranges, which complicates the modeling since beam diffraction is included along with the effect of nonlinear self-action. In fact, the beam profiles with and without the approximation in the near and far field are also very similar.

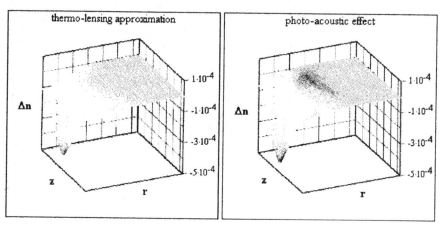

*Figure 4. Spatial distribution of the thermally induced refractive index change ($\tau_p = 7$ ns, $w_0 = 6$ $\mu$m, $T_L = 80\%$, $L = 1$ mm, $E_{IN} = 5$ $\mu J$).*

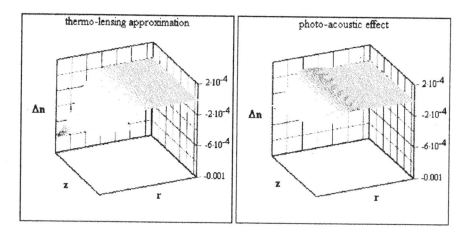

*Figure 5. Spatial distribution of the thermally induced refractive index change ($\tau_p = 7$ ns, $w_0 = 30$ $\mu$m, $T_L = 80\%$, $L = 1$ mm, $E_{IN} = 125$ $\mu J$).*

The situation changes with an increase in beam size in case (ii) to 30 $\mu$m. Here the energy was also increased by a factor of 25 to keep the same value of maximum irradiance and fluence at the focal plane. Now the acoustic transit time is longer than the pulsewidth, and as shown in Fig. 5, the acoustic wave does not have enough time to propagate and Eq. (10) overestimates the effective index change. This can be also seen from the differences in the near and far field radial fluence profiles. It is important to notice that the index change developed as a result of the photo-acoustic effect reaches nearly the same value as the one predicted by Eq. (10), but only for later times.

These results indicate that if the laser pulsewidth is comparable or larger than the acoustic transient time, the approximation (10) may be used, but it will predict incorrect results otherwise. In order to analyze the range of validity in more detail, we model the closed-aperture Z-scan experiment [35] for thermally induced refraction in the aqueous solution of nigrosine. Figures 6 and 7 show the calculated Z-scan signals using both the full solution of the acoustic Eq. (9) and the approximation (10). To speed up calculations, the thickness of the sample was chosen to be 200 microns so that the «thin-sample» Z-scan formalism commonly used to analyze experimental data is valid.[35]

Care must be exercised in applying the thermal lensing approximation to assure that significant errors do not occur. In particular, one has to be careful conducting experiments defining the thermal optical properties of the materials (Refs. [32-34]) or their absorption characteristics (Ref. [30]). Also we see that when modeling the performance of a liquid based optical limiter on a nanosecond time scale [19,36] we can often apply the thermal lensing approximation if the limiting element is placed near a tightly focused beam. However, if the limiter is located far from focus, where the beam size is large (e.g. as in tandem limiters [36]), the photo-acoustic equation must be solved in order to correctly model the thermal refraction. Figure 8 shows the results of an experiment performed on Nigrosene in water with a focused spot size of 5.9 μm (FWHM) and 7 ns pulses along with modeling. Within uncertainties the thermal lensing approximation is valid for this experiment.

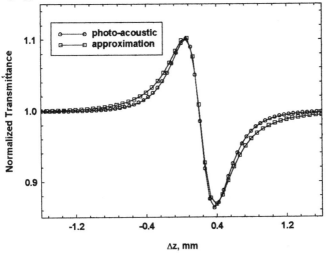

Figure 6. Z-scan of Nigrosene in water modeled with the full acoustic wave solution and making the thermal lens approximation, where $w_0 = 6$ μm and the $\tau_p = 7$ ns.

48

We also display the capabilities of the program to model beam shapes other than Gaussian in Figure 9. Here the spatial beam profile of an initially flat-top beam (made

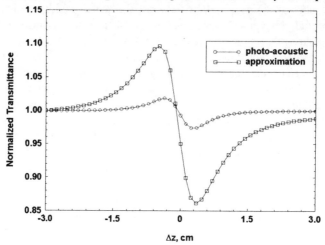

*Figure 7. Z-scan of Nigrosene in water modeled with the full acoustic wave solution and making the thermal lens approximation, where $w_0 = 30 \ \mu m$ and the $\tau_p = 7 \ ns$.*

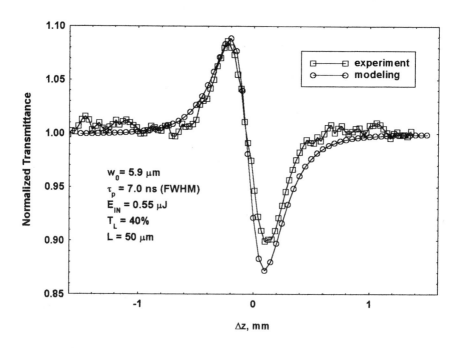

*Figure 8. Z-scan along with the results of modeling (no fitting parameter) of nigrosene dye in water using the parameters given in the figure. $T_L$ refers to the aperture transmittance used in the Z-scan.*

by clipping an expanded Gaussian beam) is shown after the 1 mm of nigrosene dye solution and after propagating through air to the intermediate field along with the results of numerical modeling.

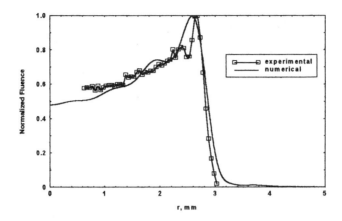

*Figure 9. The beam profile as measured on a CCD camera (line scan- with only half the radially symmetric beam shown, for an initially flat-top beam, along with the results of modeling. The material is again nigrosene dye in water.*

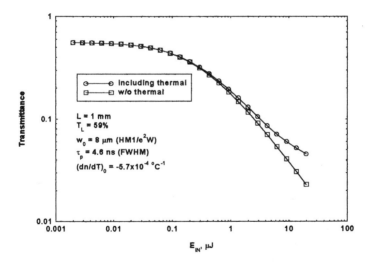

*Figure 10. Results of modeling of optical limiting of RSA (3-level model), with and without including the effects of thermal refraction.*

The consequences of thermal refraction on the performance of optical limiters using liquid dye solutions can be quite significant. Figure 10 shows a simulation of the limiter response of a solution of silicon naphthalocyanine in toluene with and without a thermal nonlinearity. The sample thickness is several Rayleigh ranges. As seen, the thermally induced defocusing reduces the effectiveness of the RSA given by this dye. This reduction is primarily the result of propagating through a material longer than the depth of focus of the input lens, such that the defocusing near the front of the cell widens the beam toward the back of the cell, reducing the fluence there, and therefore reducing the nonlinear response.

## 3. CONCLUSIONS

We have briefly discussed a very versatile method for this rapid spectral characterization of the nonlinear response of optical materials that also gives the temporal response of these nonlinearities. The spectral and temporal response along with the magnitude of the nonlinearity greatly helps in determining the underlying nonlinear mechanisms. Given a characterization of the material response, this response can then be used as the input to a nonlinear propagation code that models the pulse propagation through thick and/or multiple nonlinear elements in an optical limiter geometry. The codes we have developed allow for all the usual nonlinearities encountered in optical limiting with the exception of nonlinear scattering. We are currently using these codes to optimize the optical limiting performance of materials where we have fully characterized the nonlinear response. One particularly difficult nonlinear response to characterize and model is thermal refraction in the nanosecond range where the response can be transient. This has required us to simultaneously solve the electromagnetic and acoustic wave equations.

## 4. ACKNOWLEDGMENTS

This work was carried out under the support of the National Science Foundation (ECS #9970078), the Office of Naval Research (grant number N00014-97-1-0936), the Naval Air Warfare Center Joint Service Agile Program (contract number N00421-98-C-1327), and the Air Force Office of Scientific Research (grant number F49620-93-C-0063).

## 5. REFERENCES

1. «Nonlinear spectrometry of chromophores for optical limiting» D. J. Hagan, E. Miesak, R. Negres, S. Ross, J. Lim and E. W. Van Stryland, Proc. SPIE-3472, 80-90 (1998).
2. K. D. Belfield, B. A. Reinhardt, L. L. Brott, S. J. Clarson, O. Najjar, S. M. Pius, E. W. Van Stryland and Raluca Negres, «Synthesis and characterization of new two-photon absorbing polymers», *Polymer Prep.* **40**, no. 1, 127-128 (1999).
3. M. Sheik-Bahae, D.C. Hutchings, D.J. Hagan and E.W. Van Stryland, "Dispersion of Bound Electronic Nonlinear Refraction in Solids", IEEE J. Quantum Electron. QE-27, 1296-1309 (1991).
4. T.H. Wei, D.J. Hagan, M.J. Sence, E.W. Van Stryland, J.W. Perry and D.R. Coulter, "Direct Measurements of Nonlinear Absorption and Refraction in Solutions of Phthalocyanines», Appl. Phys. B 54, 46 (1991).

5. J. P. Gordon, R. C. C. Leite, R. S. Moore, S. P. S. Porto and J. R. Whinnery, «Long-transient effects in lasers with inserted liquid samples,» J. Appl. Phys. 36, 3-8 (1965).

6. S. A. Akhmanov, D. P. Krindach, A. V. Migulin, A. P. Sukhorukov and R. V. Khokhlov, «Thermal self-action of laser beams,» IEEE J. Quantum Electron. QE-4, 568-575 (1968).

7. C. K. N. Patel and A. C. Tam, «Pulsed optoacoustic spectroscopy of condensed matter,» Rev. Mod. Phys. 53, 517-550 (1981).

8. J. N. Hayes, «Thermal blooming of laser beams in fluids,» Appl. Opt. 11, 455-461 (1972).

9. A. J. Twarowski and D. S. Kliger, «Multiphoton absorption spectra using thermal blooming. I. Theory,» Chem. Phys. 20, 251-258 (1977).

10. S. J. Sheldon, L. V. Knight and J. M. Thorne, «Laser-induced thermal lens effect: a new theoretical model,» Appl. Opt. 21, 1663-1669 (1982).

11. P. R. Longaker and M. M. Litvak, «Perturbation of the refractive index of absorbing media by a pulsed laser beam,» J. Appl. Phys. 40, 4033-4041 (1969)

12. Gu Liu, «Theory of the photoacoustic effect in condensed matter,» Appl. Opt. 21, 955-960 (1982)

13. C. A. Carter and J. M. Harris, «Comparison of models describing the thermal lens effect,» Appl. Opt. 23, 476-481 (1984).

14. A. M. Olaizola, G. Da Costa and J. A. Castillo, «Geometrical interpretation of a laser-induced thermal lens,» Opt. Eng. 32, 1125-1130 (1993).

15. F. Jurgensen and W. Schroer, «Studies on the diffraction image of a thermal lens,» Appl. Opt. 34, 41-50 (1995).

16. S. Wu and N. J. Dovichi, «Fresnel diffraction theory for steady-state thermal lens measurements in thin films,» J. Appl. Phys. 67, 1170-1182 (1990).

17. P. Brochard, V. Grolier-Mazza and R. Cabanel, «Thermal nonlinear refraction in dye solutions: a study of the transient regime,» J. Opt. Soc. Am. B 14, 405-414 (1997)

18. D.J. Hagan, E.W. Van Stryland, M.J. Soileau, Y.Y. Wu and S. Guha, "Self-Protecting Semiconductor Optical Limiters", Opt. Lett. 13, 315 (1988).

19. P. Miles, «Bottleneck optical limiters: the optimal use of excited-state absorbers,» Appl. Opt. 33, 6965-6979 (1994).

20. T. Xia, D. J. Hagan, A. Dogariu, A. A. Said and E. W. Van Stryland, «Optimization of optical limiting devices based on excited-state absorption,» Appl. Opt. 36, 4110-4122 (1997).

21. R. W. Boyd, Nonlinear optics, (Academic Press, Inc. 1992).

22. M. D. Feit and J. A. Fleck, Jr., «Simple method for solving propagation problems in cylindrical geometry with fast Fourier transforms,» Opt. Lett. 14, 662 (1989).

23. W. H. Press, B. P. Flannery, S. A. Teukolsky and W. T. Vetterling, Numerical recipes. The art of scientific computing, (Cambridge University Press, 1986).

24. D. Kovsh, S Yang, D. J. Hagan and E. W. Van Stryland; «Software for computer modeling of laser pulse propagation through the optical system with nonlinear optical elements,» Proc. SPIE 3472, 163-177 (1998).

25. D. Kovsh, S. Yang, D. Hagan and E. Van Stryland, «Nonlinear optical beam propagation for optical limiting,» submitted to Applied Optics.
26. G.R. Hadley, «Wide-angle beam propagation using Padé approximant operators», Opt. Lett. 17, 1426 (1992).
27. M. Sheik-Bahae, A. A. Said and E. W. Van Stryland, «High-sensitivity, single-beam n2 measurements,» Opt. Lett. 14, 955-957 (1989).
28. D. Landau and E. M. Lifshitz, Course of theoretical physics. Volume 6. Fluid mechanics, (Pergamon Press).
29. Dmitriy I. Kovsh, David J. Hagan, Eric W. Van Stryland, «Numerical modeling of thermal refraction in liquids in the transient regime.» Optics Express, 4, 315 (1999)
30. S. R. J. Brueck, H. Kildal and L. J. Belanger, «Photo-acoustic and photo-refractive detection of small absorptions in liquids,» Opt. Comm. 34, 199-204 (1980).
31. J. -M. Heritier, «Electrostrictive limit and focusing effects in pulsed photoacoustic detection,» Opt. Comm. 44, 267-272 (1983).
32. Jian-Gio Tian et al, «Position dispersion and optical limiting resulting from thermally induced nonlinearities in Chinese tea,» Appl. Opt. 32, (1993).
33. Y. M. Cheung and S. K. Gayen, «Optical nonlinearities of tea studied by Z-scan and four-wave mixing techniques,» J. Opt. Soc. Am. B 11, 636-643 (1994).
34. J. Castillo, V. P. Kozich et al, «Thermal lensing resulting from one- and two-photon absorption studied with a two-color time-resolved Z-scan.» Opt. Lett. 19, 171-173 (1994).
35. M. Sheik-Bahae, A. A. Said and E. W. Van Stryland, «High-sensitivity, single-beam $n_2$ measurements.» Opt. Lett. 14, 955-957 (1989).
36. D. J. Hagan, T. Xia, A. A. Said, T. H. Wei and E. W. Van Stryland, «High Dynamic Range Passive Optical Limiters,» Int. J. Nonlinear Opt. Phys. 2, 483-501 (1993).

# THEORETICAL DESIGN OF ORGANIC CHROMOPHORES WITH LARGE TWO-PHOTON ABSORPTION CROSS-SECTIONS

Marius Albota,* David Beljonne,[1] Jean-Luc Brédas,[1†] Jeffrey E. Ehrlich,[#] Jia-Ying Fu,[3] Ahmed A. Heikal,[3] Samuel Hess,* Thierry Kogej,[1] Michael D. Levin,[2] Seth R. Marder,[2†] Dianne McCord-Maughon,[2] Joseph W. Perry,[†] Harald Röckel,[2] Mariacristina Rumi,[†] Girija Subramaniam,[+] Watt W. Webb,* Xiang-Li Wu,[3] Chris Xu*

*School of Applied Physics and Engineering, and Developmental Resource for Biophysical Imaging Opto-Electronics, Cornell University, Ithaca, NY 14853.
[1]Center for Research on Molecular Electronics and Photonics, Université de Mons-Hainaut, Place du Parc 20, B-7000 Mons, Belgium.
[2]The Beckman Institute, California Institute of Technology, Pasadena, California 91125
[3]The Molecular Materials Resource Center, The Beckman Institute, 139-74, California Institute of Technology, Pasadena, California 91125, and Jet Propulsion Laboratory, 67-119, California Institute of Technology, Pasadena, California 91109
[‡]Department of Chemistry, The Pennsylvannia State University, Hazleton, PA 18201

[†]New permanent address: Department of Chemistry, The University of Arizona, Tucson, AZ 85721

*F. Kajzar and M.V. Agranovich (eds.), Multiphoton and Light Driven Multielectron Processes in Organics: New Phenomena, Materials and Applications, 53–65.*

54

## Abstract

Design strategies and structure-property relationships for two-photon absorption in conjugated molecules are described on the basis of correlated quantum-chemical calculations. We first focus on stilbene derivatives with centrosymmetric structures. We found that derivatization of the conjugated molecule with electroactive groups in a quadrupolar-like arrangement leads to a large increase in the two-photon absorption cross section, $\delta$. Quantum-chemical description provides rich insight into the mechanisms for the two-photon absorption phenomenon.

## I. INTRODUCTION

Two-photon absorption is a process by which, in the presence of intense laser pulses, a compound simultaneously absorbs two photons of light that can be of the same energy or of different energies. As a result, the compound reaches a (two-photon symmetry-allowed) excited state which is higher than the ground state by the sum of energies of the two absorbed photons. The process we will consider here is absorption by two photons of the same energy, $\omega$. After having been promoted to a two-photon excited state, the compound can undergo a number of photophysical phenomena, such as internal conversion down to the lowest excited state, intersystem crossing to the triplet manifold, excited-state absorption, or fluorescence emission. A major aspect to point out is that, while the transition probability for the usual one-photon absorption is linearly proportional to light intensity, that for the simultaneous absorption of two identical photons depends on the square of light intensity, $I^2$. This is the reason why, although two-photon absorption was predicted as early as 1931, it was experimentally observed only after the advent of lasers in 1961.

Molecules with a large two-photon absorption cross-section, $\delta$, are in great demand for variety of applications including, two-photon excited fluorescence microscopy (*1- 4*), optical limiting (*5 - 7*), three-dimensional optical data storage (*8, 9*), and two-photon induced biological caging studies (*10*). These applications utilize two key features of two-photon absorption, namely, the ability to create excited states using photons of half the nominal excitation energy, which can provide improved penetration in absorbing or scattering media, and the $I^2$ dependence of the process which allows for excitation of chromophores with a high degree of spatial selectivity in three dimensions using a tightly focused laser beam. Unfortunately, most known organic molecules have relatively small $\delta$, and criteria for the design of molecules with large $\delta$ have not been well developed (*11, 12*). As a result, the full utility of two-photon absorbing materials has not been realized. Here we review our work on design strategies and structure-property studies for two-photon absorption that have lead to the synthesis of fluorescent molecules with unprecedented $\delta$ values.

## 2. DESIGN OF CENTROSYMMETRIC PHENYLENE-VINYLENE CHROMOPHORES WITH LARGE TWO-PHOTON ABSORPTION CROSS-SECTIONS

We have found that 4,4'-bis(di-*n*-butyl)amino-*E*-stilbene, **2** (Fig. 1), in toluene solution, exhibited a strong blue fluorescence that depended on $I^2$ when exposed to 5-ns laser pulses at 605 nm *(13)*. Compound **2** has a linear absorption peak at 385 nm, an emission maximum at 410 nm, and a fluorescence quantum yield, $\Phi_f$, of 0.90. The two-photon excited fluorescence spectrum for **2** was essentially identical to that excited by one-photon absorption into $S_1$, suggesting that there was rapid relaxation of the state reached by two-photon absorption (taken to be $S_2$) to the $S_1$ state, and subsequent fluorescence from that state. Measurement of the two-photon excitation cross section for **2** gave a maximum $\delta$ of 210 x $10^{-50}$ cm$^4$s/photon at an excitation wavelength of 605 nm, which is almost 20 times greater than that of *trans*-stilbene, **1**, (*14*) and is among the largest values of δ reported for organic compounds. We conjectured that the large increase in the two-photon absorption for **2** relative to **1** was related to a symmetrical charge transfer from the amino nitrogens to the conjugated bridge of the molecule.

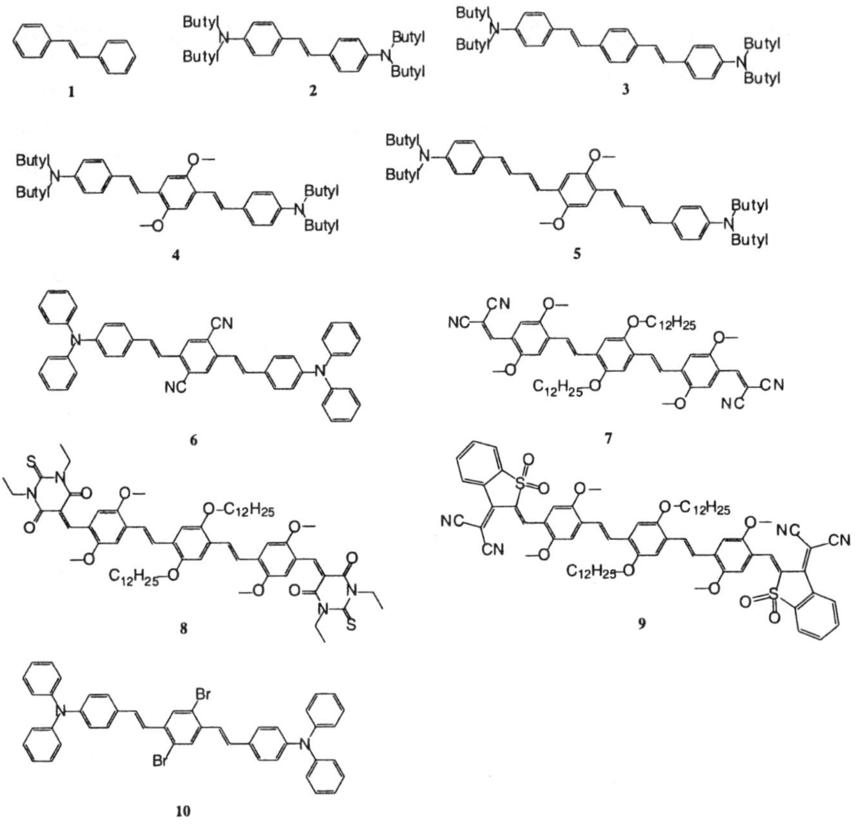

Figure 1: Structures and numbering scheme for compounds studied in this paper.

In order to gain insight into the origin of the large δ value for **2** relative to **1**, we performed quantum-chemical calculations on **1** and 4,4'-bis(dimethyl)amino-*E*-stilbene. Using AM1 (*15*) optimized geometries, the energies (E) and transition dipole moments (M) for the singlet excited states of both compounds were calculated by combining the intermediate neglect of differential overlap (INDO) (16) Hamiltonian with the multi-reference double configuration interaction (MRD-CI) (17) scheme. The frequency dependence of δ, δ(ω), is related to the imaginary part of the second hyperpolarizability, Im γ(-ω;ω,ω,−ω) by:

$$\delta(\omega) = \frac{8\pi^2 \hbar \omega^2}{n^2 c^2} \, L^4 \, \mathrm{Im}\, \gamma(-\omega;\, \omega,\omega,-\omega), \qquad (1)$$

where $\hbar$ is Planck's constant divided by $2\pi$, n is the index of refraction of the medium (vacuum assumed for the calculations), L is a local field factor (equal to 1 for vacuum), and c is the speed of light (18). We calculated Imγ(-ω;ω,ω,−ω) using the sum-over-states (SOS) (19) expression (the damping factor, Γ, has been set to 0.1 eV in all cases in this study).

As can be seen in Table 1, the calculations predict roughly an order of magnitude enhancement in δ upon substitution of *trans*-stilbene with terminal dimethylamino groups, consistent with the experimental data presented above. This two-photon transition is from the ground-state ($S_0$, $1A_g$) to the lowest excited state with $A_g$ symmetry ($S_2$, $2A_g$). For both molecules, the $S_2$ state is located at about 0.8 eV above the lowest one-photon allowed excited state ($S_1$, $1B_u$) (see Fig. 2). A simplified form of the SOS expression for the peak two-photon resonance value of δ(ω) for the $S_0 \rightarrow S_2$ transition, $\delta_{S_0 \rightarrow S_2}$, is

$$\delta_{S_0 \rightarrow S_2} \propto \frac{M_{01}^2 \, M_{12}^2}{(E_1 - E_0 - \hbar\omega)^2 \, \Gamma} \qquad (2)$$

where the subscripts 0, 1, and 2 refer to $S_0$, $S_1$, and $S_2$ states, and $\hbar\omega = (E_2 - E_0)/2$. This expression results from taking $S_1$ as the dominant intermediate state and is valid when $(E_1 - E_0 - \hbar\omega)$ is large compared to the damping factor for the $S_0 \rightarrow S_1$ transition.

Based on the results of the calculations, as illustrated in Fig. 2, we can rationalize the increase in $\delta_{S_0 \to S_2}$ on going from *trans*-stilbene to 4,4'-bis(dimethyl)amino-*E*-stilbene on the basis of: (i) an increase in the $S_1$ to $S_2$ transition dipole moment ($M_{12}$) from 3.1 to 7.2 D; (ii) an increase in the $S_0$ to $S_1$ transition dipole moment ($M_{01}$) from 7.1 D to 8.8 D; and (iii) a decrease in the one-photon detuning term, ($E_1 - E_0 - \hbar\omega$), from 1.8 to 1.5 eV. This enhancement results from the electron donating properties of the terminal amino groups. The calculations also show that the electronic excitation from $S_0$ to $S_1$ is accompanied by a substantial charge transfer (~0.14 e) from the amino groups to the central vinylene linkage, as we had hypothesized, leading to a large change in quadrupole moment upon excitation (a similar sense and magnitude of charge transfer is calculated for the in the $S_0 \to S_2$ transition). This pronounced redistribution of the $\pi$-electronic density is correlated with an increase of electron delocalization in the first excited state and results in a significant increase in the $S_1$ to $S_2$ transition dipole moment, which is the major contributor to the enhanced value of 4,4'-bis(dimethyl)amino-*E*-stilbene with respect to that of *trans*-stilbene. Another consequence of the terminal substitution with electron donors is a shift of the position of the two-photon resonance to lower energy.

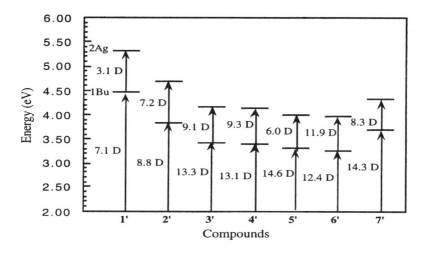

Figure 2: Scheme of the calculated energy levels and transition dipole moments (in Debye, D) for the three lowest singlet states for compounds **1'** to **7'** (which are model compounds for **1** to **7**, wherein alkyl groups on amino or alkoxy groups and phenyl groups on terminal amino groups were replaced by methyl groups).

These results suggested several strategies to enhance $\delta$ and tune the wavelength of the two-photon absorption peak for $\pi$-conjugated organic molecules. Because the symmetric charge transfer and change in quadrupole moment appear to be important, for molecules with small ground-state mesomeric quadrupole moments, we reasoned that structural features that could further enhance the change in quadrupole moment upon excitation could be beneficial for enhancing the corresponding transition dipole moments and the magnitude of $\delta$. We therefore examined both theoretically and experimentally molecules in which: (i) the conjugation length was increased by inserting phenylene-vinylene or phenylene-butadienylene groups (compounds **3** to **5**) to increase the distance over which charge can be transferred; (ii) electron-accepting cyano groups were attached to the central ring of the bis-styrylbenzene backbone (compound **6**), creating a D-A-D motif, to increase the extent of charge transfer from the ends of the molecule to the center; and (iii) the sense of the symmetric charge transfer was reversed by substituting electron donating alkoxy donors on all three rings of the bis-styrylbenzene and attaching relatively strong accepting dicyano vinyl (compound **7**), thiobarbituric acid (compound **8**), or 3-(dicyanomethylidenyl)-2,3-dihydrobenzothiophene-2-ylidenyl-1,1-dioxide (compound **9**) electron accepting groups on both ends, creating A-D-A compounds. Finally, we were interested in introducing heavy atoms (for example, bromine atoms, compound **10**) with a large spin–orbit coupling to facilitate intersystem crossing from the $S_1$ state to the lowest excited triplet state, $T_1$, aiming to create a two-photon absorbing molecule that could act as an efficient triplet sensitizer.

We performed the same INDO-MRD-CI calculations described above on a series of model compounds (**3'** to **7'**) for **3** to **7** wherein alkyl groups on amino or alkoxy groups and phenyl groups on terminal amino groups were replaced by methyl groups. The results which are shown in Table 1 support our proposed design strategies. Increasing the conjugation length of the molecule or increasing the extent of symmetrical charge transfer from the ends of the molecule to the middle or *vice versa* leads to a large increase of δ and a shift of the two-photon absorption peak to longer wavelength relative to that of *trans*-stilbene.

We synthesized **3** to **10** by standard techniques and characterized them by nuclear magnetic resonance, electronic absorption, fluorescence and mass spectroscopies, as well as elemental analysis. The two-photon absorption cross sections of these molecules were measured using the two-photon fluorescence excitation method with nanosecond (*20*) and femtosecond (*11*) laser pulses. In both cases, measurements were performed using fluorophores with well characterized δ-values as reference standards (*11*). The positions and magnitudes of the two-photon resonances, the fluorescence quantum yields, and positions of the one-photon absorption bands are shown in Table 1.

Several important conclusions can be drawn from the data in Table 1: (i) there is good agreement between the peak values of $\delta$ measured using femtosecond and nanosecond pulses, and calculated with the INDO-MRD-CI method; (ii) the INDO-MRD-CI calculations reproduce the trends in the evolution of the position of the two-photon absorption peak (although, as expected, the absolute excitation energies are systematically overestimated by theory, due to over-correlation of the ground state with the MRD-CI scheme); (iii) Increasing the length of the molecule leads to a significant increase in $\delta$, as can be seen by comparing results for **3**, **4**, and **5** with **2**; (iv) our hypothesis that D-A-D and A-D-A compounds should have enhanced $\delta$ is borne out by the observation of large $\delta$-values, in the range of $620 \times 10^{-50}$ to $4400 \times 10^{-50}$ cm$^4$s/photon, for **6** to **9** relative to bis-1,4-(2-methylstyryl)benzene for which $\delta$ is $55 \times 10^{-50}$ cm$^4$s/photon (*21*); (v) there are significant shifts of the peak position of the two-photon absorption to longer wavelength upon increasing both the conjugation length and the extent of symmetric charge transfer; (vi) with the exceptions of **5**, **8**, **9**, and **10**, the compounds have very high fluorescence quantum yields indicating that they could be of interest as fluorescent probes for two-photon microscopy; and (vii) the dibromo-substituted compound **10** has a reasonably large $\delta$-value, and its fluorescence quantum yield is low in comparison to **3**, consistent with efficient intersystem crossing. Preliminary results indicate that **10** is a singlet $O_2$ sensitizer, which makes it a good candidate for cytotoxicity and photodynamic therapy studies in biological tissues (*22*).

We suggest that $\pi$-conjugated molecules with large changes of quadrupole moment upon excitation are worthy of examination as molecules with large two-photon absorption cross sections. Molecules

Table 1. Calculated and experimental two-photon excitation cross sections (δ) and peak positions (TPA $\lambda_{max}$) for compounds in this work. Experimental δ values determined using nanosecond pulses and femtosecond pulses (given in parentheses) are reported. The uncertainty in the experimental δ values is estimated to be ± 15%

| | Theoretical results | | | Experimental results | | |
|---|---|---|---|---|---|---|
| Compound | TPA $\lambda_{max}$ (nm) | δ (10$^{-50}$cm$^4$s/photon) | Compound | TPA $\lambda_{max}$ (nm) | δ (10$^{-50}$cm$^4$s/photon) | $\Phi_f$ |
| 1' | 466 | 27.3 | 1 | 514 [ref. 13] | 12 [ref. 13] | |
| 2' | 529 | 202.4 | 2 | 605 (<620) | 210 (110 @ 620 nm) | 0.90 |
| 3' | 595 | 680.5 | 3 | 730 (~725) | 995 (635) | 0.88 |
| 4' | 599 | 670.3 | 4 | 730 (~725) | 900 (680) | 0.88 |
| 5' | 620 | 712.5 | 5 | 775 (~750) | 1250 (1270) | 0.12 |
| 6' | 625 | 950.0 | 6 | 835 (810) | 1940 (3670) | 0.86 |
| 7' | 666 | 570.4 | 7 | 825; 940 (815) (910) | 480 620 (650) (470) | 0.82 |
| - | | | 8 | 970 | 1750 | 0.06 |
| - | | | 9 | 975 (945) | 4400$^a$ (3700) | 0.0085 |
| - | | | 10 | ~800 | 450 | 0.41 |

$^a$ There is a large uncertainty in this δ value due to a large uncertainty in the rather low value of $\Phi_f$ which is 0.0055(± 45%).

derived from the design strategies described should greatly facilitate a variety of applications of two-photon excitation in biology, medicine, three-dimensional optical memory, photonics (*13*), optical limiting (*5*), and materials science (*13*).

## Acknowledgements

The work in Mons was carried out within the framework of the Belgium Prime Minister Office of Science Policy "Pôle d'Attraction Interuniversitaire en Chimie Supramoléculaire et Catalyse (PAI 4/11)" and is partly supported by the Belgium National Fund for Scientific Research (FNRS-FRFC) and an IBM-Belgium Academic Joint Study; DB is a Chercheur Qualifié FNRS; TK is a Ph.D. grant holder of FRIA (Fund for Research in Industry and Agriculture). Support from the National Science Foundation (Chemistry Division), Office of Naval Research, Air Force Office of Scientific Research (AFOSR) and its Defense University Research Instrumentation Program at Caltech is gratefully acknowledged. The research described in this paper was performed in part by the Jet Propulsion Laboratory (JPL) California Institute of Technology, as part of its Center for Space Microelectronics Technology and was supported by the Ballistic Missile Defense Initiative Organization, Innovative Science and Technology Office and AFOSR through an agreement with the National Aeronautics and Space.

## References and footnotes

1. Denk, W., Strickler, J.H., and Webb, W.W. (1990) *Science*, **248**, 73.

2. William, R.M., Piston, D.W., and Webb, W.W. (1994) *FASEB Journal*, **8**, 804.

3. Denk, W. and Svoboda, K. (1997) *Neuron*, **18**, 351.

4. Kohler, R.H. *et al.*, (1997) *Science*, **276**, 2039.

5. Ehrlich, J.E., *et al.*, (1997) *Opt. Lett.*, **22**, 1843.

6. Said, A.A., *et al.*, (1994) *Chem. Phys. Lett.*, **228**, 646.

7.   He, G.S., Gvishi, R., Prasad, P.N., and Reinhardt, B. (1995) *Opt. Commun.*, **117**, 133.

8.   Parthenopoulos, D.A. and Rentzepis, P.M. (1989) *Science*, **245**, 843 (1989).

9.   Strickler, J.H. and Webb, W.W. (1991) *Opt. Lett.*, **16**, 1780 (1991).

10.  Denk, W. (1997) *Proc. Nat. Acad. Sci.*, **91**, 6629.

11.  Birge R.R. (1986) *Acc. Chem. Res.* **19**, 138.

12.  Xu, C. and Webb, W.W. (1996) *J. Opt. Soc. Am. B*, **13**, 481 (1996); Xu, C., *et al.,* (1996) *Proc. Nat. Acad. Sci.*, **93**, 10763.

13.  Cumpston, B.M., Ananthavel, S.P., Barlow, S., Dyer, D.L., Ehrlich, J.E., Erskine, L.L., Heikal, A.A., Kuebler, S.M., Lee, I.-Y.S., McCord-Maughon, D., Qin, J., Röckel, H., Rumi, M., Wu, X.-L., Marder, S.R., and Perry, J.W. (1999) *Nature* **398**, 51.

14.  Anderson, R.J.M., Holtom, G.R. and. McClain, W.M. (1979) *J. Chem. Phys.*, **70**, 4310.

15.  Dewar, M.J.S., Zoebisch, E.G., Healy, E.F., and Stewart, J.J.P. (1985) *J. Am. Chem. Soc.*, **107**, 3902.

16.  Ridley, J. and Zerner, M. (1973) *Theoret. Chim. Acta*, **32**, 111.

17.  Buenker, R.J. and Peyerimhoff, S.D. (1974) *Theoret. Chim. Acta*, **35**, 33.

18.  Dick, B., Hochstrasser, R.M., and Trommsdorff, H.P., in *Nonlinear Optical Properties of Organic Molecules and Crystals*, Vol. 2, D. S. Chemla and J. Zyss, Eds. (Academic Press, Orlando, 1987) pp. 167-170.

19.  Orr, B.J. and Ward, J.F. (1971) *Mol. Phys.*, **20**, 513.

20.  The two-photon absorption cross-sections were determined via two-photon excited fluorescence measurements, using 5ns, 10Hz pulses from a Nd:YAG-pumped optical parametric oscillator/-amplifier system. The set-up consisted of two arms to allow relative measurements with respect to bis-1,4-(2-methylstyryl)benzene (in cyclohexane), fluorescein (in pH 11

water) and rhodamine B (in methanol). The sample chromophores were dissolved in toluene ($10^{-4}$M concentration); an approximately collimated beam was used to excite the samples over the 1cm length of the cuvettes. The fluorescence emitted at right angles to the excitation was collected and focused onto photomultiplier tube detectors. The intensity of the incident beam was adjusted to assure excitation in the intensity-squared regime.

21.    Kennedy, S.M. and Lytle, F.E. (1986) *Anal. Chem.*, **58**, 2643; We have scaled the peak δ value reported in this paper by a factor chosen to match the long wavelength data reported therein with that of ref. 11.

22.    Lipson, M., Levin, M., Marder, S.R., and Perry, J. W. unpublished results.

# MULTIPHOTON ABSORPTION AND OPTICAL LIMITING

R. L. SUTHERLAND*, D. G. McLEAN*, S. M. KIRKPATRICK*, P. A. FLEITZ, S. CHANDRA*, AND M. C. BRANT*

*Air Force Research Laboratory, Materials and Manufacturing Directorate
3001 P St, Ste 1, Wright-Patterson Air Force Base, Ohio 45433-7750, USA*

## Abstract

Optical limiting requires the development of materials exhibiting strong nonlinear absorption. Multiphoton absorption in materials can take the form of instantaneous two-photon absorption or time-integrating excited state absorption. We examine both types of mechanisms for optical limiting. Several experimental methods are used to study these phenomena and characterize materials. Often new or unusual effects arise in these experiments that require careful analysis to elucidate desired parameters. We describe several of these and give results for different materials relevant to optical limiting.

## 1. Introduction

Optical limiters display a decreasing transmittance as a function of intensity or fluence. Such devices have been explored for applications involving pulse shaping, smoothing, and compression as well as laser power stabilization. In recent years, there has been considerable interest in optical limiters for suppressing high levels of irradiance. Two basic types of optical limiter are refractive (energy redistribution) and absorptive (energy dissipation), although both mechanisms may be at work simultaneously in a single device. Nonlinear absorption is attractive from the standpoint of energy management, with nonlinear refraction then playing a secondary role in modifying beam propagation and possibly beam size, both in the nonlinear medium and in the final focal plane.

Several reviews of optical limiter materials and devices have been published since 1989, including those by Wood *et al.* [1-3], Tutt and Boggess [4], and Hermann and Staromlynski [5]. There have been two published symposium proceedings on materials for optical limiting [6,7], and three symposium proceedings on engineering aspects of optical limiting [8-10]. A recent literature search of optical limiting revealed 380 sources since 1990. Although several of these relate to optical limiting in telecommunications, most have a more general flavor. Clearly, this is an active area of research. Several physical mechanisms have been exploited. Our focus is on nonlinear absorption (NLA).

The principal nonlinear absorption mechanisms that can be exploited for broadband optical limiting are two-photon absorption (TPA) and excited state absorption (ESA). These are illustrated in Fig. 1. TPA has an obvious advantage in the absence of a one-

---

* Science Applications International Corporation, Dayton, OH, USA

*F. Kajzar and M.V. Agranovich (eds.), Multiphoton and Light Driven Multielectron Processes in Organics: New Phenomena, Materials and Applications, 67–81.*
© 2000 *Kluwer Academic Publishers. Printed in the Netherlands.*

photon resonant state, so the material can be colorless yet provide absorption at high intensities. Since there is no real intermediate state, the material absorbs two photons simultaneously. This makes the process sensitive to the instantaneous intensity, which illustrates one of the disadvantages of TPA: the need for high intensity. Once the material makes a transition to a real two-photon allowed state, it can exhibit additional transitions to one-photon allowed states. This is two-photon assisted ESA. The other principal mechanism for ESA is also illustrated in Fig. 1. In this case, there is a one-photon allowed state to which the molecule makes a transition. After relaxation to a nearby state, which may be a lower vibrational level of the excited singlet manifold, or a lower lying singlet manifold, or a triplet manifold reached by intersystem crossing, the molecule can absorb an additional photon and make a transition to another one-photon allowed state. If the higher lying states have a larger absorption cross section than the ground state, the molecular system will have lower transmittance at high fluence levels. Since the process only involves transitions to real states, each with a characteristic lifetime (i.e., the state of the system is determined by the integrated pulse energy), this mechanism is fluence dependent. Such a process is called reverse saturable absorption (RSA) since it is the opposite of the saturable absorption phenomenon wherein light bleaches a high transmission path through the material.

A diversity of nonlinear absorption spectra with wide bandwidths can be found in many organic materials. For this reason, we and other researchers have examined several classes of organic materials, including diphenyl polyenes, dithienyl polyenes, thiophenes, benzithiazoles, carbocyanines, porphyrins, phthalocyanines, and fullerenes among others. A variety of techniques are used to characterize these materials in different time regimes. Interesting phenomena often arise in these experiments, which require careful analysis to extract meaningful nonlinear absorption coefficients. In this paper we present measurements of several organic materials and elucidate the fundamental nonlinear optical properties. A variety of phenomena are considered and observed. We illustrate different effects that may arise in such characterization experiments, and how the data can be analyzed to extract the desired material parameters.

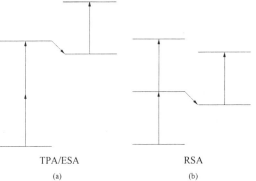

TPA/ESA

(a)

RSA

(b)

Figure 1 Nonlinear absorption mechanisms for optical limiting: (a) two-photon and excited state absorption, (b) reverse saturable absorption.

## 2. Figures of Merit

The operation of an ideal optical limiter is illustrated in Fig. 2a. For low incident intensity or fluence, the device transmittance is linear. For certain materials this may be near 100%. Others will have residual absorption (in some cases by design), and the input-output curve would have a slope <45°. Above some critical input, the transmittance changes abruptly and exhibits an inverse dependence on intensity or fluence. Thus the output is clamped at some desired value. This critical point is called the threshold of the device, while the clamped output is called the limiting value.

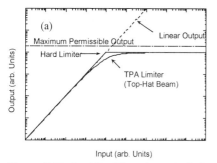 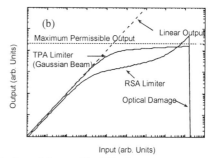

Figure 2 Optical limiting characteristics: (a) ideal limiter, (b) non-ideal limiter.

In an actual device the transition described above is not so abrupt. The definition, or even the concept, of a threshold is thus not quite so precise. However, to retain the terminology of the ideal limiter, a threshold is often defined. One approach is to define it as the input at which the transmittance decreases by some factor from its nominal value. Some authors have defined this as a noticeable departure from linearity, e.g., 5-10%. Others have stipulated a factor of 2 change. Another common definition is given in terms of the point at which the asymptotes of the transmittance in the linear and nonlinear regimes intersect. Also, in a real device the output is not always clamped at a constant value. In some cases the input-output slope merely decreases in the nonlinear regime. This could be due simply to the fact that the optical beam has a Gaussian, or other more complex, profile rather than a top-hat profile. In other cases, molecular phenomena may dictate such dependence. In these instances the output can still rise to unacceptable levels. We desire the optical limiter to provide suppression over a wide range of incident intensity or fluence. Thus if the input-output slope is nonzero, at some level above threshold the device will fail. In some cases the device itself may be damaged. Either of these cases will define a maximum input for which the device provides effective limiting. The ratio of this value to the threshold is called the dynamic range of the limiter.

Simple figures of merit can describe the material properties required to achieve low threshold, wide dynamic range optical limiting. For TPA the figure of merit is the TPA coefficient $\beta$. An order of magnitude estimate of desired values can be obtained with a formula considering the limiting value of pulse energy $W_c$ allowed, the beam spot size $w$ and confocal parameter $b$ in the TPA material, and the pulse width $\tau$, which in the limit of large dynamic range ($>>1$) takes the form

$$\beta \sim \frac{\pi w^2 \tau}{b W_c} \tag{1}$$

If the material exhibits two-photon assisted ESA, then it will have an effective two-photon absorption coefficient $\beta_{eff}$, which depends on the excited state absorption cross section $\sigma_e$ as well as the pulse width.

In RSA materials, the figures of merit are the ground state absorption cross section $\sigma_g$ and the effective excited state cross section. The latter will in general be time dependent. For very short time (~fs-ps) it will be just that of the allowed excited singlet state $\sigma_S$.

For long time (~ns-μs) it will be the appropriate triplet state cross section, modified by the triplet yield ($\phi_T\sigma_T$). For simplicity, we will call this parameter the effective excited state absorption cross section $\sigma_{eff}$. The ground state cross section, as well as the concentration of the chromophore and pathlength of the device, determines the linear transmittance of the sample. The threshold depends on $\Delta\sigma = \sigma_{eff} - \sigma_g$, while the dynamic range will be determined by the ratio $\sigma_{eff}/\sigma_g$.

In addition to low threshold and wide dynamic range, the material must have a broadband response. This will be determined by the spectral property of the above figures of merit. Other properties will generally be necessary as well, such as optical clarity (low haze), color neutrality, and robustness (i.e., resistance to damage and photochemical degradation due to high intensity, humidity, temperature excursions, etc.). These place added constraints on the material. However, the above figures of merit are sufficient for initial screening of materials.

## 3. Measurement Methods

Several methods have been developed to measure the nonlinear optical coefficients needed to assess materials for optical limiting [11]. We have explored several of these for measuring $\beta$, including:

- Transmission measurements
- Three-wave mixing
- Two-photon fluorescence
- Degenerate four-wave mixing
- Z-scan
- Laser calorimetry
- Thermal lensing
- Heterodyned Kerr effect

Each of these has advantages and disadvantages depending on the material under study. Often they can also yield the nonlinear index coefficient $n_2$ as well. Although this is not the major parameter of interest to us for optical limiting, it is often of importance in beam propagation codes used for designing optical limiters, as it may affect beam propagation through the limiter. Hence it is often included in the testing protocol. In some cases, the method may be extended to characterize two-photon assisted ESA, yielding the excited state absorption cross section $\sigma_e$. To measure excited states in general, along with decay rates, pump-probe methods such as laser flash photolysis and transient white-light absorption spectroscopy are employed. We discuss some of these various methods below.

## 4. Investigations of NLA Materials for Optical Limiting

We describe here the various methods and results we have used to characterize NLA materials. The first set has generally been applied to measuring the third-order susceptibility $\chi^{(3)}$ in centrosymmetric materials, and can be used to measure $\beta$ either directly or indirectly. These techniques are degenerate four-wave mixing, Z-scan, and nonlinear fluorescence. Pump-probe methods have been used to measure $n_2$, $\beta$, and excited state absorption cross sections, in some cases all from the same experiment. Techniques designed to specifically characterize the transient states of materials, including decay rates, are laser flash photolysis and transient absorption spectroscopy. The former addresses triplet states exclusively, while the latter monitors the transition of singlet states to triplet states. Finally, the ultimate use of the material in an optical limiting device may employ several mechanisms, and in this case we are interested in just the intensity or fluence dependent transmission. For this instance we employ straightforward nonlinear absorption (i.e., transmission) measurements in the nanosecond regime.

## 4.1 TWO-PHOTON ABSORPTION

The following techniques have been employed by several investigators to characterize third-order nonlinear optical materials. We describe the pertinent aspects of these methods for measuring two-photon coefficients. We have found that subtle effects can often mask the desired phenomenon or profoundly modify the expected signal. An understanding of these effects can sometimes lead to new methods of measurement.

### 4.1.1   Degenerate Four-Wave Mixing (DFWM)

This technique, based on real-time holography, has been utilized for years to characterize nonlinear materials. In a standard experiment, one usually measures the modulus of $\chi^{(3)}$. However, both the real and imaginary components of $\chi^{(3)}$ contribute to the conjugate signal detected in such an experiment, and it is the imaginary component that is related to TPA. In some early work we did studying diphenyl polyenes in the nanosecond regime, we discovered that TPA can produce some profound effects on the temporal behavior of the conjugate signal [12]. Upon investigating this phenomenon, we theoretically determined that such temporal behavior could be analyzed to isolate the imaginary component of $\chi^{(3)}$ and hence determine $\beta$ [13].

There are two aspects of TPA to consider in this problem. First, TPA is related to the formation of an amplitude grating in an ordinary DFWM experiment. Such a grating is normally 90° out of phase with the phase grating that is created due to the real part of $\chi^{(3)}$, or equivalently $n_2$. Usually, the effects of these two gratings cannot be separated, and as stated above, the experimentally measured parameter is the modulus of $\chi^{(3)}$. Secondly, however, TPA produces other effects, such as heating the material or producing excited state species. Either of these will produce further refractive effects, which then give rise to extra phase gratings in the experiment. Two additional aspects of these phenomena are important. First, they depend on the temporal integration of the optical pulse and hence exhibit different time dependence than the phase grating due to $n_2$. Second, their contribution may be 180° out of phase with the $n_2$ contribution (i.e., have opposite sign). This leads to the possibility that at some point within the pulse duration, as the TPA-induced phase grating grows, the two (or more) contributions to the phase grating cancel one another, leaving just the amplitude grating to generate the conjugate signal. Measuring the conjugate signal at this particular point in time will yield the purely imaginary component of $\chi^{(3)}$ and hence $\beta$.

This technique yields an easily measured signal against a weak background, and hence could be applied to measuring small TPA coefficients. In addition, the real part of the susceptibility can be obtained from the same data. Its chief disadvantage lies in the pulse detection and monitoring using a streak camera. We applied this method, calibrated against $CS_2$, to measure the imaginary part of $\chi^{(3)}$ for a solution of dipheynlbutadiene (DPB) in chloroform at 532 nm. The TPA coefficient is related to $\chi^{(3)}$ by

$$\beta = \frac{3\pi \, \text{Im}\left(\chi^{(3)}\right)}{\varepsilon_0 n^2 c \lambda}, \tag{2}$$

where $n$ is the refractive index of the solution, $c$ is the speed of light, $\lambda$ is the laser wavelength, and $\varepsilon_0 = 8.85 \times 10^{-12}$ F/meter. A TPA cross section $\sigma_2$ is defined by

$$\sigma_2 = \frac{hc\beta}{\lambda N}, \tag{3}$$

where $h$ is Planck's constant, and $N$ is the number density of molecules. For DPB, we obtained $\sigma_2 = (40\pm8) \times 10^{-50}$ cm$^4$-s/photon-molecule.

### 4.1.2    Z-Scan

The Z-scan experiment is performed by moving a thin sample along the axis of a focused beam [14]. Two detectors in the far field on the transmission side give a measure of the transmittance through a lens (open) and through an aperture (closed). The open aperture data give a measure of transmittance changes due to nonlinear absorption only. These data can be fit to give a value for the two-photon absorption coefficient at the pump wavelength. Varying the pump frequency (using an OPA) allows for the measurement of the dispersion. The open aperture data can also be used to measure degenerate excited state absorption by varying the pump power and observing the transmittance change. In such a case, the change in the transmittance would remain constant. The closed aperture detector sees changes in transmittance due to both nonlinear absorption and nonlinear refraction. Using the data from the open aperture scan, a subsequent fit of the closed aperture scan yields $n_2$.

Highly conjugated linear chain and resonant ring structures have attracted considerable attention because of their nonlinear properties. We have examined a family of dithienyl polyenes, which are similar in structure to the diphenyl polyenes, with thiophene terminal rings replacing the phenyl units. Both classes of polyenes are transparent in the visible and yield good optical quality in solution or melt. Analyzing open aperture Z-scan data, we can extract a TPA coefficient. Examples of data for molten DPB

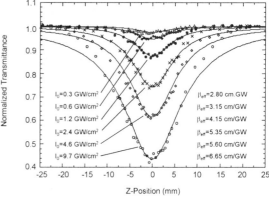

Figure 3 Open aperture Z-scans in molten diphenyl butadiene, exhibiting an intensity dependent TPA coefficient.

are given in Fig. 3. The laser was at 532 nm with 24 ps pulses. It is obvious from these data that $\beta$ is not a constant. An effective intensity dependent TPA coefficient $\beta_{\text{eff}}$ indicates that the system is also exhibiting linear absorption from a two-photon excited state. This also yields an intensity dependent nonlinear refractive index. The data for molten DPB are consistent with the following parameters: $\beta = 2.5$ cm/GW, $\sigma_2 = 36 \times 10^{-50}$ cm$^4$-

s/photon-molecule, $\sigma_e=2.4\times10^{-17}$ cm$^2$, $n_2=4.1\times10^{-14}$ cm$^2$/W, and $\sigma_r=-2.0\times10^{-22}$ cm$^2$, where $\sigma_r$ is the excited state refractive cross section.

We have discovered that this two-photon assisted ESA is more the norm than the exception in organic materials. Figure 4 shows an example of the nonlinear absorption coefficient as a function of irradiance in a solution of 3,3'-dithienyloctatetraene in chloroform. Extrapolation of the linear fit to the data to zero intensity yields $\beta$, while the slope of the line allows an estimation of $\sigma_e$. For 3,3'-dithienyloctatetraene, this yields $\sigma_2=910\times10^{-50}$ cm$^4$-

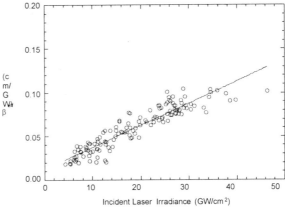

Figure 4 Effective two-photon absorption coefficient in 3,3'-dithienyloctatetraene.

s/photon-molecule and $\sigma_e=15.6\times10^{-18}$ cm$^2$. We have measured $\sigma_2$ as large as $20{,}000\times 10^{-50}$ cm$^4$-s/photon-molecule in 2,2'-dithienyloctatetraene using the Z-scan technique.

### 4.1.3 *Nonlinear Fluorescence*

The tunable nonlinear fluorescence experiment allows us to measure the two-photon absorption cross section as well as investigate any photochemical reactions that may occur at higher pump intensities. The experiment is performed by collecting all of the emitted fluorescence as a function of pump intensity. Any chemical reactions that occur due to an intensity dependent process will give rise to new chemical species having new fluorescence features with different decay rates. In the absence of any photochemical reactions, the total integrated fluorescence can be plotted against pump intensity and will exhibit a quadratic intensity dependence if due to TPA. Using a rate equation analysis describing the molecular levels, we can fit the integrated fluorescence data to extract both $\sigma_2$ and fluorescence yields. Shown in Fig. 5 is a typical set of data for the AF-50 chromophore. The data obviously follow a quadratic intensity dependence indicating TPA. From this fit, $\sigma_2$ can be obtained under some simplifying

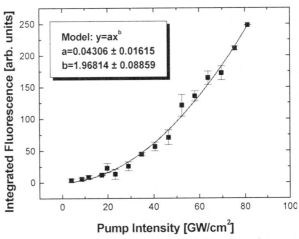

Figure 5 Fluorescence intensity as a function of pump intensity in AF-50 in toluene.

74

assumptions by

$$\sigma_2 = \frac{2A\hbar\omega_0}{\phi_f N\tau\sqrt{\pi/2}},$$  (4)

where $A$ is the fit parameter, $\omega_0$ is the center frequency of excitation, $N$ is the concentration, $\tau$ is the pulse width, and $\phi_f$ is the fluorescence yield. The data for AF50 yield $\sigma_2 = 13.6 \times 10^{-50}$ cm$^4$-s/photon-molecule.

A change in the two-photon fluorescence spectrum signals a photochemical change. This is dependent on intensity and solvent interactions. We have noted such behavior specifically in halogenated solutions of porphyrins.

### 4.1.3    Pump-Probe Measurements

Excited state cross sections measured in two-photon assisted ESA are moderately large, $\sim 10^{-18}$-$10^{-17}$ cm$^2$. We are interested in the properties of these two-photon excited states. Time-delayed pump-probe transmission measurements can provide information about the evolution and strength of these states. However, performing these experiments with coherent laser beams presents complications that must be considered.

Although the long time delay regime yields the evolution of the excited state, transmission data in the vicinity of the two-pulse overlap is more complex. An example of probe transmittance as a function of pump-probe delay for molten DPB using 100 ps pulses is given in Fig. 6. (The three oscillations in these data are due to the overlap of double pulses emitted by the mode-locked laser used in these experiments. The center feature is the one of interest, while the satellite pulses are simply artifacts of the double pulsing.) Several researchers have addressed the coherent artifact (the so-called coherent spike) in nonlinear absorption pump-probe experiments, but the feature in Fig. 6 is different from this. It is due to the linear chirp in the laser pulses.

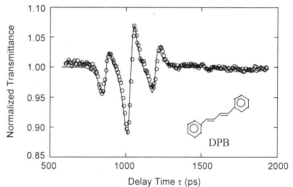

Figure 6 Normalized transmittance as a function of delay time in a pump-probe experiment in molten diphenyl butadiene.

The oscillation in the probe transmission can be explained in terms of energy coupling between the pump and probe mediated by a third-order (Kerr-like) refractive nonlinearity, since for different delay times, the frequencies are non-degenerate because of the chirp. Energy transfer is to the lower frequency beam, and it changes sign as the delay time goes from negative to positive. A symmetric depression of the transmission curve is due to the absorption nonlinearity (TPA). By careful analysis of the data, it is possible to extract the real and imaginary components of $\chi^{(3)}$ as well as the finite response time (4.1 ps) of the refractive nonlinearity [15], which is presumably a molecular reorientation or bond vibration/rotation in DPB. The real part of $\chi^{(3)}$ was determined to be

$5.2 \times 10^{-12}$ esu, while the imaginary part yields values of $\beta$ and $\sigma_2$, by Eqs. (3) and (4), that agree with Z-scan and DFWM measurements within experimental error.

Interestingly, a response time shorter that the laser pulse can be resolved in this type of experiment. The coupling is a maximum, and hence the transmittance a maximum or minimum, when the instantaneous beat period of the two pulses is of the order of the decay time of the nonlinearity. As long as the pulse is long compared to this, the decay time is readily obtained by measuring the delay time between transmission maximum and minimum in the pump-probe data.

Some molecules with larger absorption exhibit a strong absorption from a state coupled to the ground state by a single photon. Pump-probe measurements of these systems generally exhibit long time decays indicative of the excited electronic singlet state. However, we have observed a coherent effect in chlorinated diethylthiadicarbocyanine, similar to that in DPB, mediated by a nonlinearity about $10^4 \times$ that of $CS_2$. The measured decay time of this fast nonlinearity is only 3.1 ps, and is probably related to relaxation by photoisomerization. A longer decay of the transmittance envelope of 1.2 ns is indicative of a singlet-triplet relaxation.

## 4.2 REVERSE SATURABLE ABSORPTION

The following techniques have been employed by several investigators to characterize transient optical properties of materials. We describe the pertinent aspects of these methods for measuring excited state cross sections and decay rates.

### 4.2.1 *Laser Flash Photolysis*

The laser flash photolysis (LFP) experiment is used to characterize excited state properties and chemical interactions that affect these properties. A Q-switched laser pumps molecules to the excited state, which are then probed with a CW Xenon flashlamp to measure the sample transmission. The system provides transient absorption and emission measurements from 5 ns to >1ms.

This experiment can measure the triplet-triplet absorption spectrum $\sigma_T(\lambda)$, the triplet lifetime $\tau_T$, and the triplet formation quantum efficiency $\phi_T$. The measured transient waveforms may be directly converted to a differential absorbance. This is calibrated to molar extinction using the singlet depletion method. The lifetime is extracted from the transient decay using Levenberg-Marquardt nonlinear curve fitting. The pump energy or sample concentration is reduced to diminish triplet-triplet quenching and isolate the first order exponential decay. The triplet formation quantum efficiency is measured using energy transfer in the comparison method [16].

The time-dependent differential absorbance spectrum of the molecule G9 (a platinum poly-yne) was obtained, and Fig. 7 shows the resulting spectrum. In the transition state, fluorescence emission is observed at 400 nm and 800 nm. This fluorescence decays at the same rate as the transition state, indicating that the transition state is the first excited singlet. The sharp peaks in the transition state spectrum are striking. These may be due to the population of higher vibrational states during the laser pulse. The energy relaxation of this material is shown in Fig. 8 at two wavelengths. The discrepancy in extracted decay times is due the presence of the transition state.

We show the calibrated triplet and ground state absorption spectra of zinc-octobromo tetraphenyl porphyrin (ZnOBP) in Fig. 9, using the singlet depletion method

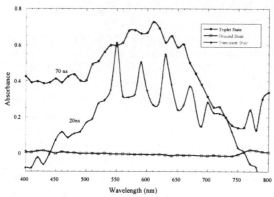

Figure 7 Excited state absorption spectra in G9.

for calibration. In the region around 530-540 nm, we find $\sigma_T/\sigma_g \geq 20$. The decay time of the ZnOBP triplet state is much shorter than that of zinc-tetraphenyl porphyrin ($\tau_T = 20$ μs) due to bromination.

The LFP experiment is also useful to monitor real-time chemical changes in solutions as they are pumped by UV or visible light. These have important implications for the use of materials in optical limiters.

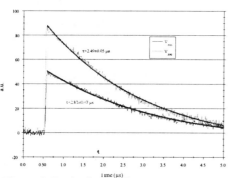

Figure 8 Excited state decay rates at two wavelengths in G9.

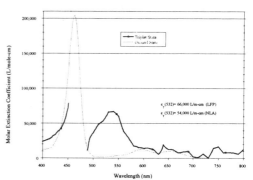

Figure 9 Calibrated singlet and triplet absorption spectra in ZnOBP.

### 4.2.2  *Transient White Light Absorption Spectroscopy*

One method for accurately probing and measuring the excited state dynamics of any material is transient white light absorption spectroscopy [17-20]. The pulse from a femtosecond laser is split in two by an ultrafast beam splitter. The higher energy pulse is used as a pump beam and is propagated down a variable delay line. The lower energy pulse is propagated along a fixed delay line (designated seed pulse) and is used to generate a white light supercontinuum in the single filament regime. Note that this setup is contrary to a typical two-color pump-probe experiment where the probe pulse is delayed relative to a fixed pump. This is because the generation of the supercontinuum is highly nonlinear both in intensity and in bandwidth, requiring a stable input pulse. The probe beam is split into a signal and reference beam, and both are imaged into a spectrometer. A liquid nitrogen CCD array collects the signal.

It is desirable to collect as many wavelengths simultaneously as possible in order to get a good picture of the dynamics occurring within the sample. However, care must be taken to avoid ground state saturation from the probe pulse, and multiple order signals on a low groove density grating. Some authors elect to use a tunable étalon to select single probe wavelengths from the supercontinuum and then to detect them on two cooled PMTs to normalize out background electrical noise, shot noise, fluorescence, and scattering [19,20]. In order to simplify this dual detector process, we split our probe beam into two spots, one of which makes an angle of 20° with respect to the pump beam and the sample cell (designated signal pulse). The other beam is used as a reference pulse and is propagated around the cell. Both spots are imaged onto two separate regions of the CCD detector and collected simultaneously to give single pulse normalization. A 600 g/mm grating and appropriate long and short pass filters are used to probe a given wavelength region (typically 150 nm at a time). This also allows us to adjust probe intensity to match the dynamic range of the detection setup as the supercontinuum peaks in intensity at the generation wavelength of 800 nm and spans from 400 nm to 1.1 μm. Selection of a smaller wavelength region also prevents higher order diffraction images within the spectrometer from obscuring the signal. The zero-time delay of the experiment is set using a KDP crystal and observing second harmonic generation by varying the optical delay line.

In order to obtain accurate spectra, four configurations are used to collect data. First, a spectrum is taken with no sample in the path (designated $I_c$). This gives the calibration curve of the reference pulse relative to the signal pulse and should correspond to the optical density of the beam splitter. Second, a spectrum is taken of the sample with no pump beam (designated $I_g$). These two spectra combined give the ground state absorbance and correspond to that taken by standard absorption techniques. Third, a spectrum is taken with the pump beam overlapping the signal pulse at the sample cell (designated $I_p$). Finally, a spectrum is taken with the pump incident on the sample, but with both probes blocked (designated $I_s$). This last spectrum gives the amount of fluorescence and scatter caused by the pump pulse on both regions of the CCD. The ground state and excited state absorbance of the sample are then given by

$$A_{GSA} = -\log_{10}\left[\frac{\left(I_g^{sig}/I_g^{ref}\right)}{\left(I_c^{sig}/I_c^{ref}\right)}\right] \tag{5}$$

$$A_{ESA} = -\log_{10}\left[\frac{\left[\left(I_p^{sig}-I_s^{sig}\right)/\left(I_p^{ref}-I_s^{ref}\right)\right]}{\left(I_g^{sig}/I_g^{ref}\right)}\right] \tag{6}$$

where the superscript indicates either the signal or reference region of the detector. Using this method we can eliminate most sources of noise by using the same detector and optical path for single pulse normalization, giving an excellent signal-to-noise ratio throughout most of the probe spectrum.

Figure 10 shows the ground state absorbance for a 13 μm film of lead phthalocyanine as measured by Shirk et al. Note that at 800 nm the material has an optical density of 1.2, while beyond 900 nm it drops to zero. It has been postulated that excitation at 800 nm will result in a charge-transfer state and a subsequent increase in the excited state absorbance in the near IR. Single wavelength measurements at 1064 nm have

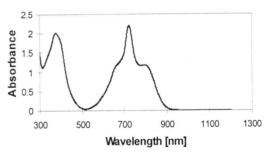

Figure 10 Linear absorption spectrum of PbPC(CP)$_4$ in CHCl$_3$.

78

shown an increase in the excited state absorption of 0.1 at a time delay of 1 ns from excitation [21].

Shown in Fig. 11 are two absorption spectra taken as described above, illustrating the ground state absorption (dashed) and the excited state absorption (solid). The excited state absorption spectrum was taken at a delay of 400 fs relative to the zero overlap of the fundamental pulses. This spectrum is uncorrected for chirp. As can be seen, an increase in the optical density is observed in the infrared following excitation by 800-nm light.

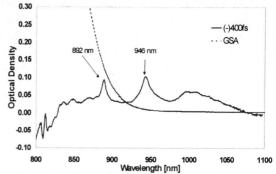

Figure 11 Ground and excited state absorption spectra of PbPC(CP)$_4$.

## 4.3 NONLINEAR TRANSMISSION

Previous research has focused on the excited state absorption properties of RSA materials in solution. Recently we have been investigating the excited state absorption cross section of ZnOBP doped in an elastic host, EPO-TEK epoxy. For our application, we are interested in the excited state absorption cross section of the material when exposed to Q-switched, nanosecond laser pulses. A simple model is used to calculate the effective excited state absorption cross section from nonlinear absorption measurements [22].

The experiment is designed to measure nonlinear absorption in samples as thick as 1 cm. A 532 nm, 6.8 ns (FWHM) pulse is focused to ~100 μm radius in the sample. We use a long focal length lens (500 mm) to provide a collimated beam throughout the sample. The incident and transmitted energies are measured for 10 laser pulses at a 10-Hz repetition rate. This is repeated at increasing laser energies up to $10^5$ times greater than the lowest energy. Linear operation of the silicon energy meters over this large energy range is maintained by inserting calibrated neutral density filters as needed. The transmission of the system with no sample is also measured over this energy range to ensure the linearity of the experiment to within 1%. In addition, an integrating sphere is used to collect forward-scattered light within a f/2 cone. The sample does not scatter light outside this cone angle. Thus any nonlinearities in transmission are attributed solely to the nonlinear absorption of the sample.

The nonlinear absorption of ZnOBP/EPO-TEK was measured for samples ranging from 0.04 mM to 1 mM. At concentrations higher than this the ZnOBP begins to agglomerate, which invalidates the measurements. Figure 12 is representative of data taken for these concentrations.

We assume that the molecules are either in the ground or a single excited state. For the pulse width used here, we expect that this excited state will be the first excited triplet state, but the model only calculates an effective excited state. We further assume that data are obtained at fluence levels below strong saturation. This is true for this data set, but must me checked for each data fit. Thus the population of excited states higher than the first is assumed negligible. The model predicts a transmittance of the form,

$$T = T_0 \frac{\ln\left(1 + F_p/F_c\right)}{F_p/F_c}, \qquad (7)$$

where $F_p$ is the peak on-axis fluence, $T_0$ is the linear internal absorption driven transmission,

$$F_c = \frac{2\hbar\omega}{\Delta\sigma\left(1 - T_0\right)}, \qquad (8)$$

$\hbar\omega$ is the photon energy, and $\Delta\sigma$ is the difference between the effective excited and ground state cross sections, the molecular property that determines the nonlinear transmission of an RSA material at low fluence. Obviously, this does not include relaxation effects, but it will reflect the behavior of the material in many practical applications.

The model was fit to the data and $\Delta\sigma$ calculated for four tests on the same sample. The average $\Delta\sigma$ is $1.87\times10^{-16}$ cm$^2$, with a standard deviation of $7.4\times10^{-18}$ cm$^2$. The molar

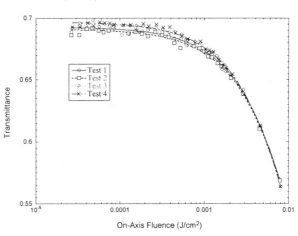

Figure 12 Nonlinear transmittance as a function of on-axis fluence in ZnOBP/EPO-TEK.

extinction of ZnOBP in solution was previously measured and the ground state cross section calculated to be $2.1\times10^{-17}$ cm$^2$. Thus the excited state cross section is $2.08\times10^{-16}$ cm$^2$, which is ten times greater than the ground state. These data imply that $F_c=13$ mJ/cm$^2$, and the assumptions described above are satisfied for the most part in this experiment.

The ZnOBP/EPO-TEK material has demonstrated a large excited to ground state ratio. This provides strong nonlinear absorption in a solid host, which is useful for our application.

## 5. Conclusions

Optical limiting requires the reduction of transmission as a function of increasing intensity or fluence. NLA provides a useful means of accomplishing this. Two phenomena that show promise for exploiting NLA in this regard are TPA and RSA. Important figures of merit for optical limiting materials involve the parameters of TPA coefficients and excited state absorption cross sections. We have investigated several materials using a variety of methods to measure these parameters. Such measurements are complex and often involve a combination of effects. Careful analysis is required to extract meaningful material parameters.

A wide variety of $\beta$ and $\sigma_{eff}$ values have been measured in these experiments and by others. Some are obviously too small to be of significance for optical limiting. However, they are useful for elucidating structure-property relations of molecules and guiding design paradigms for classes of molecular structures. A few cases are of definite interest for optical limiting, and prototype development of devices is underway. However, the full character of these materials over all interesting spectral regions is still largely unknown. This and the characterization of newly synthesized materials will continue to occupy the time of experimental investigators. Most likely, new or unusual phenomena will frequently arise in these experiments that will require careful interpretation for meaningful analysis of pertinent parameters. In the meantime, researchers will continue to press the capabilities of currently available materials in unique designs to extend the state of the art of optical limiting devices.

## References

1. Wood, G. L., Said, A. A., Hagan, D. J., Soileau, M. J., and Van Stryland, E. W. (1989) Evaluation of passive optical limiters and switches, *Proc. SPIE* **1105**, 154-180.

2. Wood, G. L., Clark III, W. W., and Sharp, E. J. (1990) Evaluation of thermal defocusing, nonlinear scattering and nonlinear quarter-wave stack switches, *Proc. SPIE* **1307**, 376-394.

3. Wood, G. L., Mott, A. G., and Sharp, E. J. (1992) Material requirements for optical limiting, *Proc. SPIE* **1692**, 2-14.

4. Tutt, L. W. and Boggess, T. F. (1993) A review of optical limiting mechanisms and devices using organics, fullerenes, semiconductors, and other materials, *Prog. Quant. Electr.* **17**, 299-338.

5. Hermann, J. A. and Staromlynska, J. (1993) Trends in optical switches, limiters, and discriminators, *Int. J. Nonlin. Opt. Phys.* **2**, 271-337.

6. Crane, R., Lewis, K., Van Stryland, E., and Khoshnevisan, M. (eds., 1995), *Materials for Optical Limiting*, Proc. Mater. Res. Soc. Vol. 374, Materials Research Society, Warrendale.

7. Sutherland, R., Pachter, R., Hood, P., Hagan, D., Lewis, K., and Perry, J. (eds., 1997), *Materials for Optical Limiting II*, Proc. Mater. Res. Soc. Vol. 479, Materials Research Society, Warrendale.

8. Lawson, C. M. (ed., 1997) *Nonlinear Optical Liquids and Power Limiters*, Proc. SPIE Vol. 3146, SPIE – The International Society for Optical Engineering, Bellingham.

9. Lawson, C. M. (ed., 1998) *Nonlinear Optical Liquids for Power Limiting and Imaging*, Proc. SPIE Vol. 3472, SPIE – The International Society for Optical Engineering, Bellingham.

10. Lawson, C. M. (ed., 1999) *Power Limiting Materials and Devices*, Proc. SPIE Vol. 3978, SPIE – The International Society for Optical Engineering, Bellingham.

11. Sutherland, R. L. (1996) *Handbook of Nonlinear Optics*, Marcel Dekker, Inc., New York.

12. Fleitz, P. A., Sutherland, R. L., Natarajan, L. V., Pottenger, P., and N. C. Fernelius (1992) Effects of two-photon absorption on degenerate four-wave mixing in solutions of diphenyl polyenes, *Opt. Lett.* **17**, 716-718.

13. Sutherland, R. L., Rea, E., Natarajan, L. V., Pottenger, T., and Fleitz, P. A. (1993) Two-photon absorption and second hyperspolarizability measurements in diphenylbutadiene by degenerate four-wave mixing, *J. Chem. Phys.* **98**, 2593-2603.

14. Sheik-Bahae, M.; Said, A. A.; Wei, T.-H.; Hagan, D. J., and Van Stryland, E. W. (1990) Sensitive measurement of optical nonlinearities using a single beam, *IEEE J. Quantum Electron.* **26**, 760-769.

15. Tang, N. and Sutherland, R. L. (1997), Time-domain theory for pump-probe experiments with chirped pulses, *J. Opt. Soc. Am. B* **14**, 3412-3423.

16. Bensasson, R. V., Land, E. J., and Truscott, T. G., (1983) *Flash Photolysis and Pulse Radiolysis*, Pergamon Press, New York.

17. Martin, M. M., Nesa, F., Breheret, E., and Meyer, Y. H. (1988) Supercontinuum spectroscopy of ethyl violet using a simple pulse compression technique, in Yajima, T., Yoshihara, K., Harris, C. B., and Shionoya, S. (eds.), *Ultrafast Phenomena VI* Vol. 48, Springer-Verlag, Berlin, pp. 473-476.

18. Alfano, R. R. (1989) *The Supercontinuum Laser Source*, Springer-Verlag, New York.

19. Ohta, K., Naitoh, Y., Saitow, K., Tominaga, K., Hirota, N., and Yoshihara, K. (1996) Ultrafast dynamics of photoexcited trans-1,3,5-hexatriene in solution by femtosecond transient absorption spectroscopy, *Chem. Phys. Lett.* **256**, 629-634.

20. Meyer, Y. H., Pittman, M., and Plaza, P. (1998) Transient absorption of symmetrical carbocyanines, *J. Photochem. Photobio. A* **114**, 1-21.

21. Shirk, J. (1999) private communication.

22. McLean, D. G., Brandelik, D. M., Brant, M. C., Sutherland, R. L., and Frock, L. (1995) Solvent effects on $C_{60}$ excited state cross sections, in Crane, R., Lewis, K., Van Stryland, E., and Khoshnevisan, M. (eds.) *Materials for Optical Limiting*, Proc. Mater. Res. Soc. Vol. 374, Materials Research Society, Warrendale, pp. 293-298.

# MOLECULAR AND MATERIAL ENGINEERING FOR OPTICAL LIMITING WITH FULLERENE BASED SOL-GEL MATERIALS

R. SIGNORINI, M. MENEGHETTI, R. BOZIO
*Department of Physical Chemistry, University of Padova, 2 Via Loredan, I-35131 Padova, Italy; E-mail: R.Bozio@pdchfi.unipd.it*

M. MAGGINI, G. SCORRANO
*Centro Meccanismi di Reazioni Organiche del CNR, Department of Organic Chemistry, University of Padova, Italy*

M. PRATO
*Department of Pharmaceutical Sciences, University of Trieste, Italy*

G. BRUSATIN, P. INNOCENZI AND M. GUGLIELMI
*Department of Mechanical Engineering, University of Padova, Italy*

## 1. Introduction

A material whose transmittance strongly and quickly drops as the intensity of a laser pulse traversing it increases beyond a threshold level accomplishes optical limiting (OL)[1]. Materials of this kind can be used to implement passive devices for the protection of optical sensors, including the eye, against damage or injuries due to exposure to intense laser pulses. For pulse duration in the picosecond to microsecond domain, the physical quantity that needs to be put under control for the protection of the eye is fluence (energy per unit area) rather than irradiance (power per unit area). In this case, the reverse saturable absorption (RSA) phenomenon is a natural candidate for the microscopic mechanism of OL. In fact, RSA is active when an excited state that can be efficiently populated by optical pumping has an absorption cross-section ($\sigma_E$) larger than that of the ground state ($\sigma_G$) at the pump wavelength. Thus, the RSA acts to actually reduce the total pulse energy rather than simply reducing the fluence or irradiance by way of scattering or beam deflection.

Carbon materials occupy a special position among the many material classes that are currently under investigation for their OL properties and their potential use in passive protection devices. In fact, different substances belong to this class and their OL behavior stems from different microscopic mechanisms. Carbon black suspensions (CBS) [2,3], fullerenes [4], functionalised fullerenes[5,6], single-walled nanotubes (SWN) [7] and multi-walled nanotubes (MWN) [8] are representative of this variety. Nonlinear scattering induced by optical breakdown and RSA are believed to be the

*F. Kajzar and M.V. Agranovich (eds.), Multiphoton and Light Driven Multielectron Processes in Organics: New Phenomena, Materials and Applications, 83–98.*

dominating mechanisms for OL in CBS and the fullerenes, respectively. However, other nonlinear (NL) processes such as instantaneous (coherent) and/or sequential (incoherent) multiphoton absorption and refractive index changes are likely to contribute to a different extent in different materials.

The discovery of the optical limiting properties of [$C_{60}$]fullerene[4] spurred much research activity aiming at (i) understanding the microscopic mechanism(s) underlying the phenomenon and (ii) exploiting the OL properties in practical protection devices. There is general consensus on the fact that RSA accounts, at least qualitatively, for most of the observed features of the NL transmission properties of $C_{60}$ in solution (toluene is generally used as the solvent). From the viewpoint of an RSA mechanism, $C_{60}$ offers some advantages over other RSA chromophores. Two features of the spectral[9] and photophysical[10] properties of $C_{60}$ appear advantageous: (i) the weak, almost flat ground state absorption extending over most of the visible range and attributed to symmetry-forbidden (due to the high molecular symmetry) electronic transitions; (ii) the very fast inter-system crossing (ISC) to a long lived lowest excited state ($\tau_{ISC} \approx 1$ ns) with almost unitary quantum efficiency ($\Phi_{ISC}$). These two features make $C_{60}$ particularly appealing as an RSA material for broadband optical limiters. On closer examination, one actually sees that the subtle balance between spectral and photophysical parameters required to optimize the OL properties actually restricts these potentialities to the red side of the visible spectrum[11].

Regarding the use of $C_{60}$, as well as of other RSA materials, for the production of practical protection devices, several issues must be addressed and solved concerning materials properties and device design. The foremost requirement is that RSA materials must be made available in solid forms that are easily processible and have great mechanical, thermal and photochemical stability. For optical limiting applications, additional requirements must be met: (i) incorporation of the NL chromophores into the solid matrix must not affect their RSA properties; (ii) the resulting composite material must exhibit as high as possible resistance to laser damage in view of its use in devices with high optical gain.

Optical polymers, like PMMA, have been considered[12] and are still under investigation as solid matrices for RSA materials. A possible advantageous alternative has been made available by the rapid progress in the field of glassy materials produced by the sol-gel technique. However, $C_{60}$ presents the drawback of being practically insoluble in polar solvents used for sol-gel processing. First attempts[13] at incorporating $C_{60}$ in conventional sol-gel matrices have, in fact, produced materials that are either very lightly doped or that present phase separation and structural inhomogeneities. Fortunately, the flexible chemistry[14] of fullerenes is a great advantage that allows one to overcome this problem by way of chemical modification and functionalization of $C_{60}$.

Thus, fullerene chemistry sol-gel processing provide good opportunities to exploit molecular and materials engineering for the tailoring of OL properties, for investigation of microscopic mechanisms of OL and of their possible subtle interplay and for development of practical protection devices.

## 2. Pyrrolidino-fullerenes for sol-gel processing of OL materials

In the last few years, work at the University of Padova[5] has been aiming at synthesizing, investigating and exploiting sol-gel materials for optical limiting based on functionalized $C_{60}$ derivatives.

$C_{60}$ derivatives have been synthesized in order to increase the solubility of $C_{60}$ in polar solvents and to endow the fullerenes with a silicon alkoxide group able to

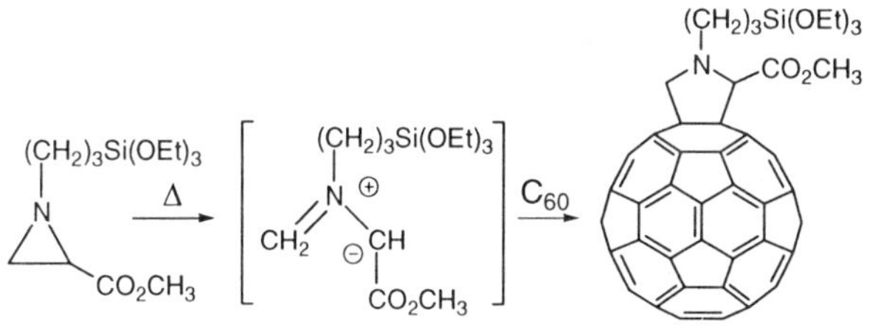

Figure 1. Synthetic scheme and chemical structure of FULP.

covalently link the silica network of sol-gel glasses. Here we focus on one of these derivatives: a functionalized pyrrolidino-fullerene, hereafter denoted as FULP, whose structure is shown in Fig. 1. FULP was generated by thermal ring-opening of an aziridine bearing both an electron-attracting -COOCH₃ group and the silicon alkoxide functionality in the presence of $C_{60}$[15]. The solubility of FULP in THF, a polar solvent suitable for the processing of sol-gel glasses, is 43 mg/ml.

The electronic absorption spectrum of FULP is similar to that of other 1,2-dihydrofullerens[16,17]. The main differences with respect to that of the parent $C_{60}$ are most likely related to the broken symmetry and the reduced number of π electrons in FULP (58 electrons instead of 60). They consist of a typical narrow peak around 430 nm, a slow decrease of absorption intensity in the visible and, most notable, a weak band around 700 nm, that is at wavelengths where $C_{60}$ practically does not absorb[9]. Combined with a shift of the main peak of the triplet absorption spectrum from 750 nm in $C_{60}$ to about 700 nm in FULP and with a still high (almost unitary) quantum efficiency for intersystem crossing[18,19] the change in the ground state absorption leads one to expect a higher OL efficiency for FULP than for $C_{60}$ in the deep red region of the spectrum. This is ascribed to an RSA process in which the higher ground state absorptivity of FULP around the peak wavelength of the triplet absorption spectrum (700 nm) makes it easier to pump the derivative into the strongly absorbing triplet state.

This expectation has been confirmed experimentally by measuring and comparing the NL transmittance of toluene solutions of pyrrolidino-fullerenes and of $C_{60}$ at various wavelengths[15,17]. These data, as well as those reported here for solution samples, have been measured using ca. 15 ns pulses delivered by an excimer pumped dye laser focused with a 200 mm focal lenght lens into 10 mm liquid cells.

The OL behavior observed for FULP in solution agrees with all the expectations based on the assumption that RSA is the microscopic mechanism of the observed phenomenon. Direct comparison of the data obtained for toluene solutions of both FULP and $C_{60}$ is made possible by the fact they have all been obtained under the same, well controlled experimental conditions. Both $C_{60}$ and the FULP derivative exhibit improved performances in the red compared to the green part of the spectrum. By virtue of the spectral changes following the chemical modification, FULP proves to be superior to $C_{60}$ at wavelengths around the maximum of the triplet excited state absorption around 700 nm.

Using literature data for absorption cross sections in assessing the potential usefulness of pyrrolidino-fullerenes as OL materials for practical protection devices, if other mechanisms such as NL scattering, multi-photon absorption, etc. do not add to or interfere with the RSA one, the pyrrolidino-fullerenes can be considered as good candidates for practical devices in a spectral region ca. $100 \div 150$ nm wide centered around the triplet absorption peak at 700 nm[11]. Note that this does not hold for $C_{60}$ itself since it cannot be efficiently pumped at wavelengths longer than 700 nm whereas the triplet absorption peaks at 750 nm.

Exploitation of the favourable properties of the FULP derivative allowed us to obtain good quality sol-gel glasses with a high fullerene content (molar ratio $FULP/SiO_2 = 4 \times 10^{-3}$). The first organically modified silica matrices were based on methyl-triethoxy-silane (MTEOS) and tetraethoxy-silane (TEOS) as silica precursors. However, they showed some deficiencies in view of their use as OL materials. The films obtained were too thin (1 - 2 µm) to observe NL transmission and bulk matrices about 2 mm thick had to be prepared for this purpose. Unfortunately, these bulk matrices showed scattering problems probably due to microstructural inhomogeneities.

These aspects have stimulated the development of new hybrid organic-inorganic matrices that appear to fulfil all the requirements for the construction of OL devices in various configurations including the multilayer structures required by the *bottleneck* optical limiter design[20]. Materials for this purpose should present: (i) easy processibility in the form of very thick films (up to a few hundreds microns), (ii) high homogeneity to avoid scattering, (iii) high resistance against laser damage. The new matrices (Fig. 2) are based on glycidoxypropyl-trimethoxy-silane (GPTMS) and TEOS and, beside the above mentioned features, they appear to possess other favourable properties such as low porosity, high scratch resistance and low optical losses.

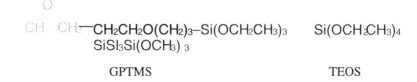

| GPTMS | TEOS |

*Figure 2.* Precursors of the hybrid organic-inorganic sol-gel matrices used in this work. GPTMTS = glycidoxypropyl-trimethoxy-silane; TEOS = Tetratethoxy-silane.

### 3. Optical limiting properties of FULP solutions and FULP-doped sol-gel materials

Only recently has the issue of comparing the OL performance of RSA materials in glassy matrices with the same materials in solution been addressed. In their work on $C_{60}$-doped PMMA, Kost and collaborators[12] noted that the OL performance of the polymer was poorer than that of solution samples, a fact that they attributed to a less efficient NL scattering in PMMA. More recently, McBranch and coworkers[18]carried out an investigation of the excited state dynamics of fullerene by comparing solutions, evaporated thin films and doped sol-gel glasses. The decay of the excited state absorption was much faster in the evaporated films than in solution and the sol-gel glasses exhibited a behavior intermediate between the two. A faster decay of the excited state absorption would mean a less efficient OL and this was in fact verified by comparing the fluence dependent NL transmittance of their sol-gel samples with that of solution samples. The microscopic origin of the faster decay in the solid samples was ascribed to exciton-exciton annihilation in the microcrystalline $C_{60}$ composing the evaporated films or in $C_{60}$ aggregates which were supposed to be present in the sol-gel samples.

The tendency of $C_{60}$ to form clusters and eventually to phase-separate in sol-gel glasses is well known. The addition of a reactive, easily hydrolizable silicon alkoxide group in our pyrrolidino-fullerenes is just meant to favor a molecular dispersion of the NL chromophores by way of their stable linking to the silica network in the early stage of densification of the sol-gel materials. Preliminary results[19] of investigations of the excited state dynamics of our sol-gel samples have shown a long lived component of the transient absorption with decay time in the microsecond range, comparable to that of the triplet state in solution samples.

Another factor that may cause a lower OL efficiency in glassy matrices than in solution is related to the structural disorder of these materials. Disorder causes inhomogeneous broadening of optical transitions. In accounting for its possible effects on an RSA process one must consider the possibility that the inhomogeneous distributions of the transition from ground to singlet excited state and that of the transition from lowest triplet to higher triplet states are uncorrelated.

Model calculations[21] show that there is a decrease in the RSA efficiency as the ratio $\sigma_x/\gamma_S$ between the inhomogeneous broadening and the homogenous width of the ground state absorption increases and as the correlation coefficient $r$ between the inhomogeneous distributions of the ground $\rightarrow$ singlet and triplet $\rightarrow$ triplet transitions decreases. However, for completely uncorrelated inhomogeneous distributions ( $r = 0$ ) and for a value of the ratio $\sigma_x/\gamma_S = 10$, the ratio of the triplet state to ground state absorbance, $\alpha_T(\omega_L)/\alpha_G(\omega_L)$, is still $\approx 65$ % of the corresponding value for a homogeneous system. This suggests that the matrix disorder should not have a very dramatic effect on the RSA properties of chromophoric systems at room temperature. In fact, not too large values of $\sigma_x/\gamma_S$ are expected to apply to these systems.

As mentioned above, the tendency of fullerenes to form clusters may have a negative effect on the OL properties. It has been reported[22] that the formation of clusters in polar solvents is signaled by a change in the absorption spectrum of fullerene into a broad or even structureless shape. The great majority of our sol-gel samples exhibited spectra quite similar to those of pyrrolidino-fullerenes in solution. However, depending on the fullerene concentration and on details of the sol-gel processing (hydrolysis and condensation time, thermal treatment, etc.), a few of our samples showed an absorption spectrum where all the structures characterizing the $C_{60}$ derivatives, including the sharp band at 430 nm, were smoothed out. Despite this, the measured absorbance levels were those expected from Beer's law based on the concentration and on the thickness of the samples and there was no clear sign of scattering from inhomogeneities. Comparison of the NL transmission properties of the two types of samples supports the idea that the formation of clusters of nanoscopic size was the origin of the anomalous absorption spectrum. In fact, two samples having the same linear transmittance of 77 %, one with a structureless spectrum the other with a solution-like spectrum, exhibited quite different NL transmittance values. The figure of merit, $T_{lin}/T_{NL}$ (see below), at the same moderate input energy was 1.3 and 2.2,

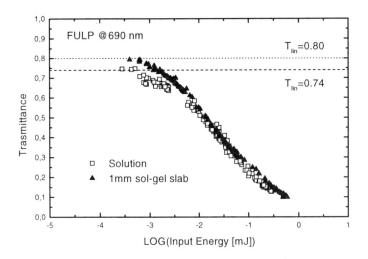

*Figure 3.* Non-linear transmission data at 690 nm for a toluene solution of FULP in a 10 mm liquid cell with 74 % linear transmittance (open squares) and for a 1.0 mm sol-gel slab with 80 % linear transmittance (full triangles).

respectively.

Figure 3 shows the NL transmission data (full triangles) measured at 690 nm with an f/66 focusing optics for a FULP-doped sol-gel slab 1.0 mm thick having an 80 % linear transmittance. The data are compared with analogous data (open squares) from a toluene solution of FULP contained in a 10 mm liquid cell with 74 % linear transmittance. This result shows that the inclusion in the sol-gel glassy matrix does not affect appreciably the OL properties of pyrrolidino-fullerene. In particular, our sol-gel materials should fulfill the requirements for achieving the desired attenuation factor based on the RSA mechanism according to the conclusion drawn in the previous section.

A summary of our results in the characterization of the OL properties of $C_{60}$ and FULP solutions is given in Table I. Table II provides the corresponding data for FULP in sol-gel samples. In assessing the OL efficiency of RSA materials a commonly used figure of merit (FOM) is the ratio of the linear transmittance $T_{lin}$ to the minimum trasmittance, $T_{NL}$, that is obtained at maximum input fluence. Strictly, the maximum input fluence should correspond to the threshold for laser damage of the OL material or of the liquid cell. Furthermore, the FOM as defined above should be used to compare measurements carried out under strictly identical conditions, on account of the dependence of $T_{NL}$ on the sample concentration and thickness, on the focusing optics and on the temporal and spatial intensity profile of the laser pulse.

TABLE I. Figures of merit (FOM) for optical limiting of $C_{60}$ and FULP solutions in toluene.[a]

| $T_{lin}$ | $T_{NL}$ | FOM | Max. input energy (mJ) | Beam waist diam. (μm) |
|---|---|---|---|---|
| | | $C_{60}$ @ 532 nm | | |
| 0.77 | 0.30 | 2.6 | 0.32 | 70 |
| 0.66 | 0.17 | 3.9 | 0.63 | 90 |
| 0.62 | 0.15 | 4.1 | 0.56 | 90 |
| | | $C_{60}$ @ 690 nm | | |
| 0.87 | 0.15 | 5.8 | 1.00 | 200 |
| | | FULP @ 532 nm | | |
| 0.78 | 0.27 | 2.9 | 0.40 | 200 |
| 0.74 | 0.25 | 3.0 | 0.63 | 200 |
| 0.74 | 0.15 | 4.9 | 0.40 | 100 |
| | | FULP @ 690 nm | | |
| 0.92 | 0.20 | 4.6 | 0.56 | > 200 |
| 0.74 | 0.10 | 7.4 | 0.32 | > 200 |
| 0.74 | 0.05 | 14.8 | 0.32 | 200 |
| 0.40 | 0.025 | 16.0 | 0.79 | > 200 |

[a] Liquid cells with 10 mm path length were used. The focusing optics had f/66.

TABLE II. Figures of merit (FOM) for optical limiting of FULP-doped sol-gel samples[a].

| $T_{lin}$ | $T_{NL}$ | FOM | Max. input energy (mJ) | Beam waist diam. ($\mu m$) |
|---|---|---|---|---|
| Two-layer sample, each layer 100 $\mu m$ thick, C = $2.0 \times 10^{-3}$ M | | | | |
| 0.90 | 0.13 | 6.9 | 1.00 | 100 - 200 |
| Four-layer sample, each layer 200 $\mu m$ thick, C = $9.7 \times 10^{-4}$ M | | | | |
| 0.90 | 0.20 | 4.5 | 1.00 | > 200 |
| Six-layer sample, each layer 600 $\mu m$ thick, C = $3.9 \times 10^{-4}$ M | | | | |
| 0.90 | 0.27 | 3.3 | 1.00 | > 200 |
| Ten-layer sample, each layer 500 $\mu m$ thick, C = $8.1 \times 10^{-3}$ M | | | | |
| 0.77 | 0.10 | 7.7 | 0.79 | 200 |
| Bulk sample, 0.6 mm thick, C = $7.43 \times 10^{-3}$ M | | | | |
| 0.63 | 0.025 | 25 | 0.56 | ~ 100 |

[a] For all the reported samples, the total thickness was 2150 $\mu m$. The focusing optics had f/66.

The data in Table I demonstrate the enhanced OL efficiency in the red spectral region (690 nm) compared to the green one (532 nm) for both $C_{60}$ and FULP in solution. Notice that all the reported NL transmission data have been measured for input energies well below the damage threshold and with a not too tightly focused laser beam propagating into 10 mm path length liquid cells. As a result, the corresponding laser fluences are well below the values required to reach full saturation of the strongly absorbing triplet level. The differences in input energies and estimated beam waist diameter should be taken into consideration when comparing the FOM's of different substances and at different wavelengths. For instance, the larger beam waist at 690 nm partially spoils the enhancement of the OL efficiency in the red region in the case of $C_{60}$.

Regarding the comparison with FULP at 690 nm, notice that FOM's larger than those of $C_{60}$ at the same wavelength are obtained for lower input energy and comparable or larger beam waist.

The OL data of sol gel samples collected in Table II confirm that the limiting properties of FULP are substantially maintained upon inclusion in the solid matrix. The first three samples reported in Table II have been prepared in order to investigate the possible role of the fullerene concentration in sol-gel materials. The linear transmittance of these samples has been kept constant by varying the number of FULP doped layers and their thickness while their concentration was decreased from $2.0 \times 10^{-3}$ M to $3.2 \times 10^{-4}$ M, that is almost one order of magnitude. The absence of cluster formation was checked for the more concentrated sample by recording their linear absorption spectra. Also, the total number of layers, doped and undoped, was kept constant in order to have the same number of interfaces with sodalime glass layers. In all samples the total thickness was 2.15 mm. It appears that samples with a higher FULP concentration in the active layers exhibit improved OL efficiency. A similar tendency has been recently reported also for solution samples [23]. This phenomenon and its microscopic origin deserve further investigation. Table II also reports data for a bulk sample, 0.62 mm thick, containing FULP at a concentration of $7.43 \times 10^{-3}$ M. A value of FOM = 25 could be easily obtained for this sample, showing that high performance OL can indeed be obtained from our sol-gel samples.

According to a simple RSA model, the maximum achievable FOM for a thin sample where all the NLO units have been brought to the strongly absorbing excited state is set by the product $(\sigma_E - \sigma_G)NL$, with $N$ the number density of molecules and $L$ the length of the sample. Comparing the available spectral and photophysical data of FULP with those of RSA materials such as metallo-phthalocyanines (M-Pc), considered as benchmarks, one sees that the expected RSA performance of FULP around the maximum of the triplet absorption at 700 nm is at least as good as that of Sn-Pc at 532 nm.[24]

With the molecular RSA properties maintained in the sol-gel matrix and with a high threshold for laser damage of our sol-gel composite glasses[25], it becomes sensible to use these materials to implement the *tandem*[26] or *bottleneck* [20]schemes in order to reach full saturation of the triplet state while preserving the RSA material from laser damage. We have recently reported results of our initial tests on a multilayer structure. For the ease of fabrication, this sample was not fully optimized according to the bottleneck design principle. A NL attenuation factor of 250 was estimated assuming that the transfer of population of the FULP molecules to the triplet state was complete. In contrast, after identifying the location for maximum performance of the limiter by a Z-scan measurement, the attenuation factor measured with f/5 optics turned out to be lower that the predicted value by more than one order of magnitude. Similar discrepancies have been reported in previous attempts reported in the literature[27]. One of the factors that limit the predicted performance of tandem or bottleneck limiters has been already recognized by Van Stryland and collaborators[28]. They have shown that it is essential to account for the dynamics of the population transfer and NL absorption. One sees that, under these circumstances, the fluence required for the transfer of population to approach completeness is remarkably higher than the previously considered[20] value of $2.5 \times F_S$ (saturation fluence). This is a consequence of the fact that the population transfer rate must be integrated over the pulse envelope and this, at any time, reduces the accumulated excited state population.

Obviously, a deeper understanding of the OL phenomenon in our materials can be gained through:
(i) comparison between experimental results and model calculations of the NL transmission performed according to an RSA model and identification of qualitative and quantitative discrepancies;
(i) detailed experimental investigation of possible concurrent phenomena adding to or interfering with the RSA one.

We have worked out a numerical code for calculating NL transmission in thick samples (considerably thicker than the Rayleigh range for the optics employed in our measurements) due to an RSA process. The NL absorbers are modeled as six-level systems [29] and the propagation equation, coupled with the rate equations for the level populations, are integrated over space and over time to obtain the energy of the transmitted pulse.

Representative results are shown in Fig. 4. The NL transmittance calculated at 532

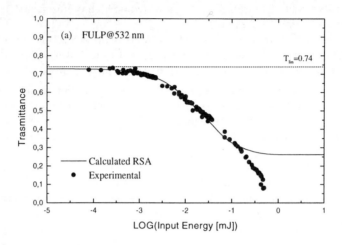

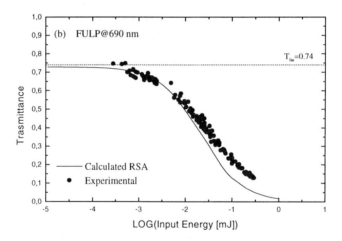

*Figure 4.* Calculated (line) and experimental (circles) non-linear transmission data for a toluene solution of FULP in a 10 mm cell with 74% linear transmittance at 532 nm (panel a) and at 690 nm (panel b)

nm, shown as the full line in Fig, 4(a), underestimates the OL performance for input energies greater than about 0.15 mJ. Note that the calculation has been performed with almost no adjustable parameter.

Absorption cross sections and excited state dynamical parameters have been taken from the literature and attempts at adjusting them to account for uncertainties in the reported experimental values did not produce significant improvements. Adjustment of the beam waist parameters indicates a best fit value close to that directly measured by a CCD camera (see below). At 690 nm the discrepancy between calculated and measured NL transmittance is in the opposite sense compared to the 532 nm data. With increasing input energy, the calculated transmittance decreases faster than the measured one.

We have undertaken an experimental investigation of the possible interfering factors that cause the observed OL behavior to deviate from the predictions of the RSA model. The possible effects of NL scattering have been investigated for solution samples of both $C_{60}$ and FULP. NL transmission measurements have been performed in the closed aperture and open aperture modes at 532 and 690 nm. In the closed aperture mode, only the transmitted laser beam reaches the detector, so that the apparent transmittance results from absorption as well as from scattering processes. In the open aperture mode the transmitted radiation propagating in a wide solid angle is collected and sent to the detector. Comparison of the results obtained in the two measurements allows an estimate of the NL scattering phenomena. Results of these measurements are shown in Fig. 5.

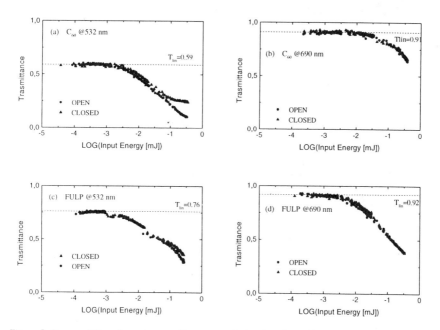

*Figure 5.* Open and Closed aperture non–linear transmission data for a toluene solution (panels a and b) and FULP (panels c and d) in a 10 mm liquid cell .The linear transmittance of $C_{60}$ is 59% at 532 nm (a) and 91% at 690 nm (b), the linear transmittance of FULP is 72% at 532 nm (c) and 96% at 690 nm (d).

94

The NL transmission data for 10 mm solution samples of $C_{60}$ are shown in panels (a) and (b) of Fig. 5 for laser wavelengths at 532 and 690 nm, respectively. At 532 nm there is a clear discrepancy between the closed aperture and the open aperture data showing that a remarkable contribution to the OL behavior at input energies exceeding 0.1 mJ comes from NL scattering rather than from RSA. This was already recognized by Nashold and Walters[3] in their comparison of the OL properties of $C_{60}$ and CBS. A similar effect is much less evident, if at all present, in the data at 690 nm.

(a)                                                    (b)

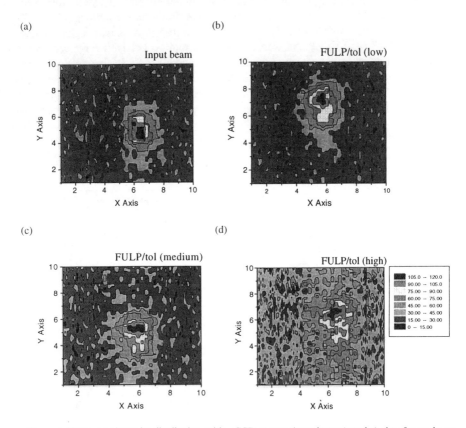

(c)                                                    (d)

*Figure 6.* Transverse intensity distribution, with a CCD camera: input beam (panel a), data for a toluene solution of FULP, in a 10 mm cell, at 532 nm, at increasing pump energy (panels b-d).

It should be noted however that measurements at the two wavelengths have been performed for constant concentration but markedly different linear transmission of the samples. It is quite reasonable to assume that the NL scattering phenomenon increases with increasing concentration of the NL absorber. Panels (c) and (d) show the analogous data for FULP solutions. At 532 some signature of NL scattering is detectable, though much less pronounced than for $C_{60}$. It appears that the NL scattering contribution to OL is less important for FULP than for $C_{60}$, although some care is suggested by the different linear transmission of the samples. Finally, the closed aperture and open aperture data for FULP solution at 690 nm practically overlap each other and NL scattering does not appear to contribute. However, a more systematic investigation of the concentration dependence is needed and is currently under way.

Evidence of the effects of the refractive index changes with increasing input energy has been collected by imaging with a CCD camera the pump beam at the focal plane inside 10 mm solution samples. Typical results are shown in Fig. 6: panels (a) through (d) are maps of the transverse intensity distribution at increasing pump energy. Analysis of these images shows that the beam waist dimension progressively increases to about 150 % of its value in air. This is likely due to a thermal lensing effect that causes the beam focus to shift downstream along the beam z-axis. Such a shift has been estimated to be 7.5 mm at the highest pump energy reported in Fig. 6 (d).

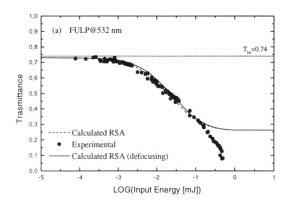

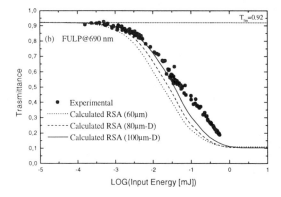

*Figure 7.* Calculated (lines) and experimental (circles) non-linear transmission data for a toluene solution of FULP in a 10 mm cell with 74% linear transmittance at 532 nm (panel a) and 92% at 690 nm (panel b). The calculated data take into account the defocusing effects.

Attempts at accounting, in a phenomenological way, for the thermal lensing effect in our RSA model were performed by using different focal plane positions and beam waist dimensions at different input energies according to the indications from the CCD camera imaging. Figure 7 (a) shows that the NL transmission at 532 nm calculated for a FULP solution in a 10 mm cell is practically insensitive to changes in the beam propagation parameters and, in all cases, the high energy transmittance could not be reproduced.

This is consistent with the suggestion that NL scattering, that cannot be accounted for by any RSA model, adds a significant contribution to NL absorption at high energies. Similar calculations performed at 690 nm (Fig. 7(b)) show that neither in this case a satisfactory agreement with the experimental data is obtained. Nonetheless, the calculated curves for increasing beam waist dimension progressively approach the high energy transmission data, suggesting that an intensity dependent change of the beam properties may account for the smaller slope of the experimental NL transmission curves compared with those calculated by a simple RSA model.

## 4. Conclusions

The reported results show that our FULP-doped sol-gel materials possess the requisites (RSA properties, mechanical, thermal and photochemical stability) for the preparation of solid state OL devices operating in the red spectral range. Molecular and materials engineering have both contributed to this achievement. Chemical modification of $C_{60}$, yielding high solubility in polar solvents, has allowed inclusion into stable sol-gel matrices at a comparatively high concentration. Molecular dispersion of the RSA-active units, overcoming their strong tendency to aggregate into clusters of nanometric size, has been obtained by linking the molecules to the silica matrix through their silicon alkoxide functionality. The spectral changes occurring upon chemical modification of $C_{60}$ all combine to enhance the RSA properties of FULP compared to those of the parent compound.

The development of hybrid organic-inorganic sol-gel matrices has provided a suitable solid environment for the fullerene molecules that allows one to preserve their favorable RSA properties. Processing methodologies have been worked out for the production of thick layers and bulk samples of high optical quality and good resistance to laser damage. These are, in fact, essential requirements for the development of practical solid state protection devices that embody the limiter into a high gain optical systems.

It is worth emphasizing, however, that some difficulties still must be overcome before reaching this goal. In fact, implementation into practical devices of the FULP species, as well as of any other RSA material presently available, requires that excited state absorption be exploited to its maximum limit. Concurrent phenomena, such as thermal effects, NL refraction and scattering, two-photon absorption, must be known and carefully considered in designing devices such as tandem or bottleneck limiters. First results in this direction have been reported here. They indicate that NL scattering and NL refraction, likely through thermal lensing, are dominant among all possible concurrent phenomena. For solution samples, they appear to exhibit a concentration and wavelength dependence that deserves further investigation. Preliminary results suggest that the relative importance of these phenomena in sol-gel samples differs from solution ones.

Theoretical models that account for these phenomena and for their interplay are much needed tools for device design as well as for materials selection.[30] .

### References

1. Crane R. (editor), Lewis K, Van Stryland SV, Khoshnevisan M, *Materials for Optical Limiting*, 1994, MRS Fall Meeting Boston MA, Materials Research Society, Aug. 1995.
2. Mansour, K., Soileau M.J., and Van Stryland, E.W. (1992) *J. Opt. Soc. Am. B* **9**, 1100.
3. Nashold, K., and Walter, D.P. (1995) *J. Opt. Soc. Am B* **12**, 1228.
4. Tutt, L.W., and Kost, A. (1992) *Nature* **326**, 225.
5. Maggini M., Scorrano G., Prato M., Brusatin G., Innocenzi P., Guglielmi M., Renier A., Signorini R., Meneghetti M., Bozio R., (1995) *Adv. Mat.* **7**, 4, 404.
6. Sun Y-P., Riggs J.E. and Liu B., (1997) *Chem Mater.*, **9**, 1268.
7. Vivien, L., Anglaret, E., Riehl, D., Bacou, F., Journet, C., Goze, C., Andrieux, M., Brunet, M., Lafonta, F., Bernier, P., and Hache, F. (1999) *Chem. Phys. Lett.* **307**, 317.
8. Chen, P., Wu, X., Lin, J., Li, W., and Tan, K.L. (1999) *Phys. Rev. Lett.* **82**, 2548.
9. Leach S., Vervloet M., Desprès A., Bréheret E., Hare J.P., Dennis T.J., Kroto S.W.H., Taylor R., and Walton D.R.M., (1992) *Chem. Phys.* **160**, 451.
10. Arbogast et al., (1991) *J.Phys.Chem.*, **95**,11.
11. Signorini R. et al., (1996) *Chem. Comm*, 1891; Signorini R et al., (1999) *NLO* section B, **21**, 1-4, 143.
12. Kost A et al., (1993) *Opt. Lett.*, **18**, 5, 334.
13. Dai S., Compton R.N., Young J.P., Mamantov G., (1992) *J.Am.Ceram.Soc.*,**75** 2865.
14. Hirsch A., (1994) *The Chemistry of the Fullerenes*, Thieme, Stuttgart; Prato M., (1997) *J.Mat.Chem.*, **7**,1097.
15. Maggini M., De Faveri C., Scorrano G., Prato M., Brusatin G., Guglielmi M., Meneghetti M., Signorini R. and Bozio R., (1999), *Chem.Eur.J*, **5**, 9, 2501; Prato M., Maggini M., Giacometti G., Scorrano G., Sandonà G., and Farnia G., (1996) *Tetrahedron* **52**, 5221.
16. Bensasson R.V., Bienvenue E., Janot J.-M., Leach S., Seta P., Schuster D.I., Wilson S.R., and Zhao H., (1995) *Chem. Phys. Lett.* **245**, 566.
17. Williams R. M., Zwier J.M., and Verhoeven J.W. , (1995) *J. Am. Chem. Soc.* **117**, 4093.
18. Klimov V., Smilowitz L., Wang H., Grigorova M., Robinson J.M., Koskelo A., Mattes B.R., Wudl F., and McBranch D.W., (1997) *Res. Chemical Intermediates* **23**, 587.

19. Guldi D. et al., private communication.
20. Miles P.A., (1994) *Appl. Opt.* **33**, 6965.
21. Fantinel F. and Bozio R., to be published.
22. Sun Y.-P., Ma B., Bunker Ch.E., and Liu B., (1995) *J. Am. Chem. Soc.* **117**, 12705.
23. Riggs J.E., Sun Y.P., (1999) *J.Phys.Chem.*, **103**, 485.
24. Signorini R. et al, (1999) *Carbon*, in press.
25. Innocenzi P., Brusatin G.et al. ,(1999) *SPIE procedings*, in press.
26. Hagan et al., (1995) *Mat.Res.Soc.Symp.Proc.*, 374, 161.
27. Perry J.W. et al., (1996), *Science*, **273**, 1533.
28. Xia T. et al., (1997) *Appl.Opt*, **36**, 18, 4110.
29. With the photophysical parameters that apply to $C_{60}$ and FULP some degree of approximation is tolerated, this model can be reduced to an effective three-level system.
30. Kovsh DI, Yang S, Hagan DJ, Van Stryland EW, (1999) *Appl. Opt.*, **38**, 24, 5168.

# A NEW APPROACH FOR OPTICAL LIMITING IN THE IR

**C. Taliani, C. Fontanini, M. Murgia and G. Ruani**

*Istituto di Spettroscopia Molecolare, C.N.R.*
*Via P. Gobetti 101, I 40126, Bologna, Italy*

## 1. Abstract

In this paper we explore a new approach to Reverse Saturable Absorption (RSA) which circumvent the limitation given by the optical gap of *closed-shell* organics. An optical limiting RSA device in the near IR is obtained by designing a solid state system that takes advantage of the reduced energy gap between the LUMO and LUMO+1 molecular orbitals in conjugated carbon *open-shell* systems with a weak charge transfer ground state absorption. A solid solution of ZnPc and $C_{60}$ deposed by UHV molecular beam shows indeed a CT state with an absorption maximum at 880 nm (1.4 eV). Further evidence is obtained from the photovoltaic effect in the IR as well as the combination of PL and differential absorption spectra. The degree of CT (0.18 e⁻) is established by means of the Raman scattering shift of the pitch mode of $C_{60}$. Finally and most importantly a preliminary evidence of optical limiting at 1.06 $\mu$m (1.16 eV) is given.

## 2. Introduction

Reverse Saturable Absorption (RSA) is one of the most promising routes for obtaining Optical Limiting (OL) [1]. The principle of RSA is based essentially on the possibility that the excited state absorption cross section, in certain organic materials, is much larger than the ground state cross section. The instantaneous preparation of excited states by the pumping beam result in the population of relatively long lived excited states. These states show large excited state cross section in the same spectral range of the ground state absorption acting therefore as an efficient passive filter at that wavelength.

Several organic materials based on phthalocyanines [2] and fullerenes [3] have been shown to be efficient RSA systems at 530-550 and 650-700 nm respectively offering efficient passive protection to the eye and sensors in the visible spectral range.

Passive OL protection in the near infrared (NIR) transparency window of the atmosphere at 1-1.4 $\mu$m is also very appealing for protecting sensors. The lower limit of the damage threshold of a solid state NIR detector is estimated to be in the range of 1 $mJ/cm^2$ for ns pulses [4].

The extension of the RSA activity in the NIR in *closed-shell* organics is restricted by the fact that the lower limit in optical absorption is approximately at 700-800 nm. Within a one-electron picture (Hückel), the optical energy gap, in the first approximation, is equal to the separation between the highest occupied (HOMO) and the lowest unoccupied (LUMO) molecular orbitals, is given by twice the resonant

*F. Kajzar and M.V. Agranovich (eds.), Multiphoton and Light Driven Multielectron Processes in Organics: New Phenomena, Materials and Applications, 99–108.*

integral. Since the resonance integral, which measures the electron interaction of the carbon conjugated chains, reaches a saturation for large delocalized conjugated electron systems, the optical gap of *closed-shell* organic systems is spectrally limited to about 800nm. On the other hand, within the same approximation, the separation between the LUMO and LUMO+1 molecular orbitals in conjugated carbon based systems is half the energy of the optical gap.

We have chosen fullerene $C_{60}$ radical anion as a possible candidate for the excited state absorption at about 1μm, which is shown to be intense both in solution [5] and in the solid phase [6]. In this paper we will explore the possibility of generating a stable charge transfer in the ground state (CT) with a suitable donor (D), which may give rise to a weak CT absorption band that would act as the primary step for the preparation of $C_{60}^-$. We make the hypothesis that instantaneous absorption of the CT ground state to the excited CT state (CT*), should provide the excess energy for the system to dissociate in free e-h pairs ($C_{60}^- - D^+$). $C_{60}^-$ species, as a result, would give rise to a strong absorption at about 1 μm.

There are however strong limitations for the choice of D: 1) the HOMO level of the donor should be just at higher energy than the LUMO of the acceptor, 2) the CT absorption should be at approximately the same energy as that of $C_{60}^-$ and, 3) the CT absorption should be weak compared to the $C_{60}^-$ absorption at 1 μm. A schematic representation of the energetics of the system that we propose is shown in Fig. 1.

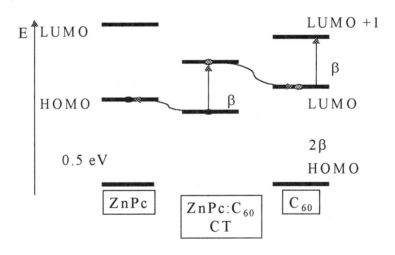

Fig. 1.Energy level scheme of the occupied and unoccupied molecular orbitals of ZnPc (left) and $C_{60}$ (right) leading to the renormalization of the ground state with the formation of a stable CT state (middle). Excitation of the CT into CT* is followed by charge separation and formation of $C_{60}$ radical -anion which absorbs strongly at the same energy.

In this paper, we will show that Zn-phthalocyanine (ZnPc) (Fig. 2a) is a suitable choice as a D for RSA in the NIR since it forms a stable CT state with $C_{60}$ (Fig. 2b). Intimate mixing of D and A are obtained by molecular beam deposition in ultra high vacuum (UHV).

We will demonstrate first that there is a new state at about 1 μm observed by fluorescence spectroscopy in ZnPc/$C_{60}$ thin films when the evaporation condition is such to generate an intimate intermixing of the two components as opposed to the layer by layer growth. This state may be a metastable or a stable state. Secondly, we will show that the CT is indeed a stable ground state by comparing the linear absorption of the layer-by-layer grown film with the intimately intermixed film. The I/V characteristics indeed show that even by IR excitation, there is a rectifying action and charges are formed at energies below than the energies of the individual components. The degree of CT is established by Raman spectroscopy. Finally, we will demonstrate optical limiting at 1.06 μm (1.16 eV).

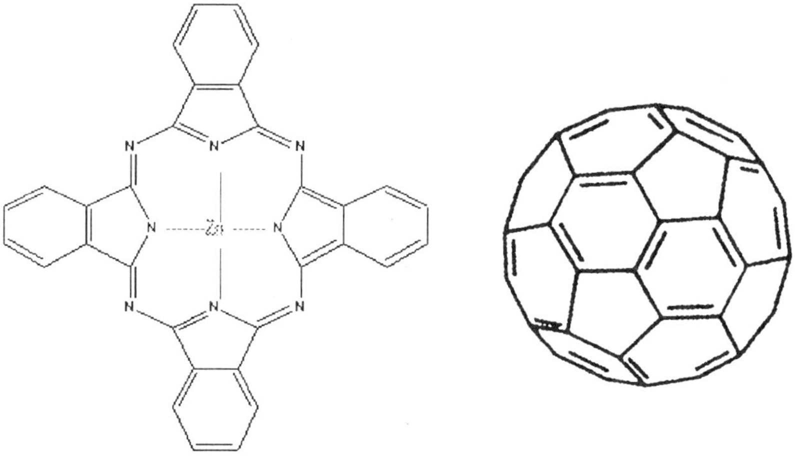

Fig. 2. Molecular structures of Zn-phthalocyanine (ZnPc) (left) and fullerene $C_{60}$.

## 3. Experimental

Zn-phthalocyanine and fullerene $C_{60}$ were purified by chromatography and vacuum sublimation. Molecular beam growth has been accomplished by means of a three UHV chambers system especially designed for growth and characterization of organic thin films (RIAL). Special home made Knudsen cells were used for achieving accurate control of the evaporation conditions. Base pressure during evaporation was better than $10^{-9}$ mbar. Contemporary evaporation was made possible by controlling separately the growth rate of the individual Zn-phthalocyanine and fullerene $C_{60}$ Knudsen cells by means of separate calibrated quartz balances. Morphology of the thin films was routinely controlled by ex situ AFM microscopy. Photoluminescence and Raman scattering were measured by means of a modified Fourier transform interferometer

(Bruker RFT 100) equipped with a liquid nitrogen cooled Ge detector. Absorption spectra were recorded by a Fourier-transform interferometer (Bruker IFS 88) equipped with a Si detector for the region from 1.2 μm to 590 nm (1 to 2.1 eV) and a grating monochromator spectrometer (JASCO 500) for the VIS spectral range. The low scattering cross-section of thin films require very long time of accumulation of the order of 1-2 hours. I-V measurements have been performed by using a computer controlled Keithley source-measuring unit mod. 236. Optical limiting at 1.06 μm (1.16 eV) was checked by using a set-up based on a 20 Hz, 10 ns Nd-YAG laser (Quantel).

## 4. Results and Discussion

### 4.1 THIN FILM GROWTH

A well defined solid state topology offering intimate mixing at D and A, which is a prerequisite for the formation of a stable CT state, is not easy to obtain. One of the most serious obstacles is the tendency towards aggregation of the individual D and A entities in the case of spin coating or other solution based techniques. We have explored the possibility offered by molecular beam growth among the several approaches. This is possible if D and A have sufficiently high melting points. This is indeed the case of ZnPc and $C_{60}$. In fact, ZnPc [7] and $C_{60}$ [8] have been widely explored as materials for the formation of thin films. Since the molecular mean free path is larger than the flight of the molecules leaving the source and impinging onto the target, the growth proceeds by minimizing the aggregation of the individual ZnPc and $C_{60}$ by this technique. Moreover, it is in principle possible to generate well-defined interfaces between ZnPc and $C_{60}$.

We have followed three different approaches:

- the layer by layer growth of ZnPc and $C_{60}$ multilayers on an optical substrate.
- the growth of complex nanostructured graded multilayers on metal electrodes.
- the co-deposition of intimately intermixed ZnPc and $C_{60}$ on an optical substrate.

By the first method, we aimed to investigate the properties of the D/A interface.

The second approach consisted in the growth of a structure of the type: ITO/ZnPc/graded interpenetrated network of ZnPc:$C_{60}$/ $C_{60}$ /Al. The interpenetrated network of ZnPc:$C_{60}$ is obtained by co-evaporation of the two constituents by gradually increasing the rate of evaporation of one and reducing the other component. This was done in order to generate a large interface at which the charge transfer could dissociate and allow at the same time the efficient withdrawal of photogenerated charges, towards the metal electrodes, by means of the percolating networks provided by the pristine D and A.

The last approach consisted in the simultaneous co-evaporation of the two constituents on a substrate to probe the optical properties of the intimate mixture of D and A.

### 4.2 PHOTOCURRENT MEASUREMENTS

By using the growing technique described in the previous paragraph we were able to measure the photocurrent produced by different photons with different photon energies spanning from the VIS into the IR on a *p-n* junction built using ZnPc as p-type material and $C_{60}$ as the n-type, when a large interface of 30 nm in depth among the two molecular layers was generated by co-evaporation. I-V characteristics obtained exciting with He-Ne (632 nm) and Ar+ ion (514 nm) lasers indicate a substantial photovoltaic effect. To our surprise the response extended into the IR much further than the onset of optical absorption. A linear direct photogeneration of carriers have been indeed observed by exciting at 1.06 μm (1.16 eV) by means of a c.w. Nd-YAG laser at different power densities. While either one or the other of the two constituent molecules absorb directly at 632 and 514 nm, there is no clear evidence of absorption at 1.06 μm (1.16 eV) unless a new renormalized ground state is formed by the interaction of ZnPc and $C_{60}$. The intensity dependence of the $I_{sc}$ is linear indicating that the carrier formation cannot be ascribed neither to triplet-triplet annihilation, that eventually can be excited in the near IR with an almost vanishing cross section, nor to two-photon absorption mechanisms. Moreover, the large $I_{sc}$ and the very low optical density of the device at the same energy compared with those observed for the other two excitations reveals a large efficiency of a direct photogeneration of carriers at 1.06 μm (1.16 eV).

## 4.3. PHOTOLUMINESCENCE

In the near IR region, the photoluminescence spectrum of a two-layer ZnPc and $C_{60}$ recorded in air shows a long tail that goes to zero at about 0.75 eV with a relatively narrow peak superimposed at 1.28 μm (0.97 eV) (see **Fig.3**).

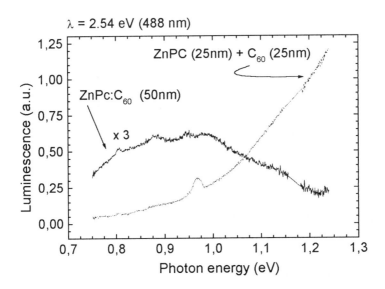

Fig. 3. Photoluminescence spectra of a ZnPc:$C_{60}$ co-evaporated (50 nm thickness) (full line), compared to the ZnPc(25 nm), $C_{60}$ (25 nm) double layer (dotted line).

Measurements were performed in air using both the 488 and 514 nm laser lines of the Ar+. The peak at 1.28 μm is assigned to the emission of singlet oxygen that is populated directly via energy transfer from the $C_{60}$ triplet state. The same effect, which is quite common in organic systems, was observed for the first time in solid $C_{60}$ thin films by Denisov et al. [9] (Fig 4a). Despite the very short lifetime of triplet in the aggregated state of $C_{60}$, the energy transfer from the triplet to the oxygen singlet is so efficient to be observed in photoluminescence; moreover, it has been shown that singlet oxygen is the main factor responsible of photodegradation of $C_{60}$, inducing oxygen-incision of the carbon cage [10]. The photoluminescence detected under the same experimental condition in an equivalent co-evaporated sample show a complete different pattern: the $C_{60}$ luminescence is strongly suppressed and only a broad peak with maximum at 1.38 μm (0.9 eV) with a line-width (fwhm) of about 1200 cm$^{-1}$ (0.15 eV) is observed (Fig. 3). The new emission is assigned to a new in-gap state that appears in the solid state due to the interaction among the two different molecules. The strong reduction of the PL integral intensity reveals the presence of a new non-radiative pathsway that is not present in the two-layer film (Fig. 4b). Moreover, the absence of the oxygen singlet emission reveals that also the $C_{60}$ triplet formation is inhibited by the interaction with Zn-Pc. This indicates that intersystem crossing between the new excited electronic state and the $C_{60}$ triplet manifold is negligible.

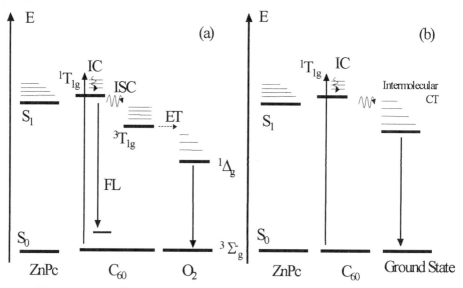

Fig. 4. Schematics of radiative and non-radiative excited state relaxation.
a) double layer thin film. Excitation into the $C_{60}$ manifold eventually relax by intersystem crossing (ISC) into the triplet manifold. Efficient triplet energy transfer into the ground state oxygen triplet results in the generation of excited state singlet oxygen and subsequent radiative emission.
b) co-evaporated thin film: The ISC to the $C_{60}$ triplet is suppressed and the new renormalized ground state is made easily accessible to Förster transfer by the intimate mixing of ZnPc and $C_{60}$.

## 4.4 FURTHER EVIDENCE OF A STABLE CHARGE TRANSFER STATE

### 4.4.1. *Differential optical absorption*

A direct evidence of charge transfer in the ground state is given in the first place by the previous observation of a photovoltaic effect by illuminating at 1.06 $\mu$m (1.16 eV). More evidence of this state may be given by optical absorption. At first instance, the absorption of the double layer film is just the sum of the individual absorption spectra of the separate components. This is not surprising since the number of charge transfer states would be limited to the interface and therefore would give a negligible absorption.

This is not so in the case of the co-evaporated thin film. In order to extract this information we have compared the two-layer film of ZnPc and $C_{60}$ and an equivalent thickness of a co-evaporated material. By doing so we were able to avoid all unwanted sources of optical losses and to observe a new absorption band in the near region infrared with a maximum at 880 nm (1.4 eV), which posses a mirror image compared to the photoluminescence. This gives us confidence that the absorption and the photoluminescence originate from the same state. Namely, the stable ground state CT.

### 4.4.2. *Raman spectra*

Raman spectroscopy is a method of choice to derive the degree of CT. In particular it has been shown that Raman scattering measurement is a powerful method for controlling the number of electrons transferred to the $C_{60}$. In fact, the Raman active pinch mode of $C_{60}$ reveal a softening of approximately 8 cm$^{-1}$ per transferred electron. [11].

In order to check the degree of charge transfer that occurs among $C_{60}$ and Zn-Pc in co-evaporated films we have performed Raman scattering characterization by exciting at 1.16 eV. The deconvolution of the observed Raman peak around 1468 cm$^{-1}$ (the energy of the pinch mode in pristine $C_{60}$) reveals the presence of a second red-shifted component by 1.8 cm$^{-1}$. This is indicative of a net charge transfer of 0.18 electrons.

The presence of both the peaks indicates that not all the $C_{60}$ molecules participate to the charge transfer process. This may be due to some tendency of the individual components to segregate. Unfortunately, Raman scattering, being a resonant process cannot be used to evaluate the percentage of the charge transfer phase formed with respect to the total amount of material deposited. In fact, the energy of the charge transfer state is of the order of 1.4 eV which is quite close to the excitation energy, while the lower energy optical transition associated to the pristine $C_{60}$ is much higher. At this preliminary stage we observe that there is still a lot of room for improvement of the co-evaporation technique for achieving a better and more intimate intermixing of the constituents especially considering the diffusion controlled character of the growth process which indicates that the substrate temperature variance is a very critical parameter to play with [12].

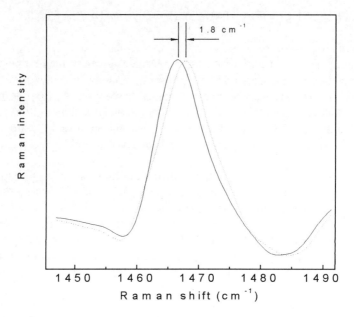

Fig. 5. Raman scattering spectra of the ZnPc:$C_{60}$ co-evaporated thin film (full line) compared with the pristine $C_{60}$ (dotted line).

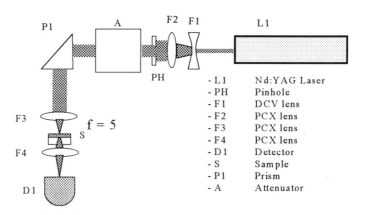

Fig. 6. Experimental set-up used for the measurement of the optical power limiting at 1.06 μm (1.16 eV). Neutral density filter have been used as attenuator.

## 4.5 OPTICAL LIMITING

Preliminary measurements of optical limiting were performed on a 60 nm thick co-evaporated film at 1.06 μm (1.16 eV) using a set-up illustrated in Fig. 6.

The plot of power in vs power out is shown in Fig. 7. The response is linear in the range up to 8 mJ while at higher fluence it tends to saturate. The large threshold for saturation is due to the limited spectral optical density of the actual film at 1.06 μm (1.16 eV). In principle this is not a major limitation. The generation of thicker films with good optical quality is among the future planned developments. At higher fluence, the film tends to undergo to permanent damage due to removal of material. We plan to explore the growth of multilayer hetero-structures with the intercalation of transparent and (ZnPc:$C_{60}$) co-evaporated thin films with the optimal optical density for achieving a good optical limiting effect.

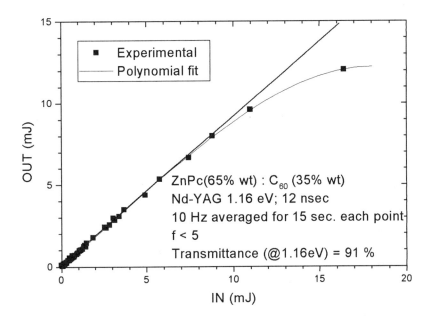

Fig. 7. Optical power limiting of a thin co-evaporated film of ZnPc:$C_{60}$ with a transmittance at 1.06 μm (1.16 eV) of 91%.

## 5. Conclusion

In conclusion, we have shown that a new approach to optical limiting in the near IR based on the formation of appropriate CT ground state absorption is possible. The combination of ZnPc and $C_{60}$ in an intimately intermixed solid by molecular beam growth shows indeed a CT state with an absorption maximum at 880 nm (1.4 eV). This evidence is proved by the photovoltaic effect in the IR as well as the combination of PL and differential absorption spectra. The degree of CT (0.18 e⁻) is established by means of the Raman scattering shift of the pitch mode of $C_{60}$. The large cross section of $C_{60}^-$ absorption on the low energy tail of the CT absorption provides the proper requisites for RSA: namely weak ground state versus large excited state absorption cross sections.

Indeed, we show preliminary evidence of the tendency to reach saturation of output power at large input power at 1.06 μm (1.16 eV) in excess of 7 mJ.

## 6. Acknowledgements

We would like to acknowledge Mr. Paolo Mei for technical support and the financial support of the E.U. Brite-Euram III project n° BRPR-CT97-0564 (MOLALC) and the CNR funding (legge 95/95) as well as the E.U. TMR-Network project n°EBRFMRX-CT97-0155 (FULPROP).

## 7. References

1.  Kajzar, F., (1999) *Proceedings of First International Workshop on Opitcal Power Limiting*. Vol. 21, Cannes: Gordon and Breach Science Publishers. 560.
2.  Miles P.A., (1994) Bottleneck optical limiters: the optimal use of excited-state absorbers. Applied Optics, 33(30): p. 6965-6970.
3.  Bozio R., *et al.*, (1999) Optical limiting of multilayer sol-gel structures containing fullerenes. Synthetic Metals, 103(1-3): p. 2474-2475.
4.  Groulier-Mazza V., (1999) *The specification of laser protection*, in *Proceedings of First International Workshop on Opitcal Power Limiting*, F. Kajzar, Editor., Gordon and Breach Science Publishers: Cannes. p. 73-83.
5.  Baumgarten M., (1993) Epr and optical absorption spectra of reduced buckminsterfullerene. Advanced materials, 5(6): p. 458-461.
6.  Pichler T., (1992) *et al.*, Electronic-transitions in KxC60 (0 less-than-or-equal-to x less-than-or-equal-to 6) from in situ absorption-spectroscopy. Solid State Commun., 81: p. 859.; *In-situ UV/VIS and infrared spectroscopy of potassium-doped C60. Solid State Science*, H. Kuzmany, M. Mehring, J. Fink, Editors., Springer-Verlag 113: p. 497
7.  Schmidt A., *et al.*, (1999) *Epitaxial phthalocyanine ultrathin films grown by organic molecular beam epitaxy (OMBE)*, in *Phthalocyanines: Properties and Applications*, Leznoff C.C. and Lever A.B.P., Editors. VCH. p. 307-342.
8.  Yase K., *et al.*, (1998) Aggregation mechanism in fullerene thin films on several substrates. Thin Solid Films, 331(1-2): p. 131-140.
9.  Denisov V.N., *et al.*, (1993) Oxygen effect on photoluminescence of fullerite C60 thin films. Synthetic Metals, (55-57): p. 3119-3124.
10. Taliani C., *et al.*, (1993) Light -induced oxygen incision of C60. Chemical Communications, (3): p. 220-222.
11. Kuzmany H. and Winter J., (1993) in *Electronic properties of fullerenes*, Kuzmany H., *et al.*, Editors. p. 273.
12. Biscarini F., *et al.*, (1997) Scaling behavior of anisotropic organic thin films grown in high vacuum., Physical Review Letters, 78(12): p. 2389-2392.

# QUANTUM CONFINEMENT AND SUPERRADIANCE OF SELF–TRAPPED EXCITONS FROM 1D J-AGGREGATES

V.M. AGRANOVICH and A.M. KAMCHATNOV

*Institute of Spectroscopy*
*Russian Academy of Sciences*
*Troitsk, Moscow obl. 142092 Russia*

## Abstract

In one-dimensional (1D) molecular crystals with finite length $2L \ll \lambda$ ($\lambda$ is the optical wavelength) an overwhelming part of the total oscillator strength is concentrated in the lowest excitonic state and is equal to $F_1 \cong 0.85 f_0(2L/a)$, where $f_0$ is the oscillator strength of a monomer and $a$ is the lattice constant. This leads to the superradiance from the lowest excitonic state and its domination in the absorption spectrum of the crystal. We show that self-trapping of excitons destroys this simple picture so that it takes place only for short chains with length $2L$ small as compared to the length $2l_0$ of self-trapping. For long enough chains the value of $F_1$ does not increase with growth of $L$, as it occurs in linear case, but tends to the saturation limit $F_1 \cong 5 f_0(l_0/a)$. The oscillator strength of the next bright state also tends to the same limit with growth of $L$, but it takes place only at the length $L > 9l_0$, and analogous relations are true for the following bright states. We consider also the influence of quantum confinement and self-trapping on the superradiance of 1D molecular crystals.

# 1  Introduction

In semiconductor crystals the characteristic length, which determines the properties of the quantum confinement phenomenon, is the Bohr radius $a_B = \hbar^2 \epsilon / M e^2$ ($M$ is the exciton mass, $\epsilon$ the dielectric constant of the crystal). As a rule, this radius is equal to 30–100Å and even more. The very thin crystalline films of high quality with the width 50–100Å can be fabricated by the molecular beam epitaxy method, and therefore in these semiconductor crystals the quantum confinement is studied very well and widely used in investigations of different electrical and optical (linear and nonlinear) properties of quantum wells and superlattices [1]. In

*F. Kajzar and M.V. Agranovich (eds.), Multiphoton and Light Driven Multielectron Processes in Organics: New Phenomena, Materials and Applications, 109–122.*

organic crystals with one-dimensional Wannier-Mott excitons we can expect that the quantum confinement effects appear also at the crystal length of the order of magnitude of Bohr radius of the exciton. A natural question arises about possibility of quantum confinement in those molecular crystals where the electroneutrality of molecules is not violated and the Wannier-Mott excitons do not appear. The present paper (see also [2]) is devoted to consideration of this problem. We shall show here that even in molecular crystals in which the lowest electronic excitations are the Frenkel excitons, i.e. where the Bohr radius is of order of the molecular size, the quantum confinement is possible, if, for example, one takes into account the electron-phonon interaction and the possibility of self-trapping of excitons. It is clear that in this case a new characteristic length, a size of the self-trapped state, appears, whence the phenomenon of quantum confinement becomes possible, if this length is of order of magnitude of the film thickness (or the molecular chain length).

If the exciton-phonon interaction is weak, i.e. the width of exciton optical transition line is much smaller than the exciton band width, then at small film thickness we shall get the usual space quantization of exciton states inside the band. It will be demonstrated below for a simplest and well-known example of 1D molecular chain. For ideal linear chains of finite length without exciton-phonon interaction and in approximation of nearest neighbors interaction (this model is often used for analysis of spectra of $J$-aggregates, see, e.g., [3]), the energy of exciton states is equal to

$$E_j = E_0 - 2J \cos\left(\frac{\pi j}{N+1}\right), \tag{1}$$

where $N$ is the number of molecules in the chain, $J$ the matrix element of the excitation transfer to the neighboring site, $j = 1, \ldots, N$ enumerates the states, and $E_0$ is the energy of molecular excitation including the gas-condensed matter shift. In what follows we neglect the variation of this quantity at the ends of the chain. The exciton band (1) extends from $E = E_0 - 2J$ to $E = E_0 + 2J$, and is symmetric around $E = E_0$. For $J > 0$, the $j = 1$ state lies at the bottom of the band (direct band edge), whereas for $J < 0$ this state is located at the top of the band. For large $N \gg 1$ we have

$$E_j \cong E_0 - 2J \cos\left(\frac{\pi j a}{2L}\right), \tag{2}$$

where $a$ is the 1D lattice constant and $2L$ denoted the length of the chain. At the bottom of the band (i.e. for small $j$ and supposing $J > 0$) the expansion of this expression defines an effective mass of the exciton:

$$M = \frac{\hbar^2}{2J a^2}. \tag{3}$$

It can be shown [3, 4] that only eigenstates with $j$ odd have nonzero oscillator strength, $F_j \propto |\mu_j|^2$,

$$
\begin{aligned}
\mu_j^2 &= \tfrac{2\mu_{\mathrm{mon}}^2}{N+1} \cot^2\left(\tfrac{\pi j}{2(N+1)}\right), & j &= \text{odd}, \\
\mu_j^2 &= 0, & j &= \text{even},
\end{aligned}
\tag{4}
$$

where $\mu_j$ is the transition dipole moment to the state $j$, $\mu_{\mathrm{mon}}$ the dipole moment of a monomer. Elementary analysis of these expressions shows that the state $j = 1$ contains an overwhelming part of the total oscillator strength up to 81% for $N \gg 1$, what is expressed by the formula

$$
F_{j=1} \cong \frac{8}{\pi^2} N f_0,
\tag{5}
$$

where $f_0$ is the oscillator strength of a monomer. This leads to superradiant emission with radiative lifetime $\tau \cong \tau_0/N$, where $\tau_0$ is the radiative lifetime of a monomer, and to domination of this state in the absorption spectrum (for the first time the superradiance from $J$-aggregates was observed in [5]). The oscillator strengths of higher states drop off as $1/j^2$ for $j \ll N$. The exciton-phonon interaction leads to widening of lines corresponding to the states with different $j$. In the really observed absorption spectra these transition usually overlap with each other. Such an overlap, however, does not shift considerably the position of the maximum of the absorption line. Therefore we can accept with sufficient accuracy that the maximum of the absorption line corresponds to the transition to the state $j = 1$. It follows from Eq. (1) that for $N \gg 1$ the frequency corresponding to the maximum of the absorption band has to vary with $N$ according to the law

$$
\Delta \propto \frac{1}{N^2}.
\tag{6}
$$

We neglect here the influence of disorder. In this approximation the shift occurs in the direction of higher frequencies (blue shift) for $J > 0$ and in the direction of lower frequencies (red shift) for $J < 0$.

Now let us turn to the discussion of the systems with exciton-phonon interaction. As is known, in this case the self-trapping of excitons becomes possible. Two limiting cases can be distinguished in this phenomenon (see, e.g. [6]). The first limiting case, which was firstly discussed by Peierls [7] and Frenkel [8], corresponds to the excitons with a narrow exciton band $J \ll \hbar\omega_0$, $\omega_0$ being the characteristic frequency of phonons strongly interacting with excitons. In this case the exciton is the slow subsystem and the lattice deformation around the exciton is able to follow its position. As it was metaphorically described by Frenkel, the exciton, when traveling in the lattice, "drags with itself the entire load of atomic displacement".

In this paper we shall consider another limiting case which corresponds to the excitonic bands being much wider than the typical energy of phonons, $J \gg \hbar\omega_0$ (the case of "light" excitons [6]). Such a limiting case takes place, as is known, for many $J$-aggregates [3]. For the first time this model was applied to excitons by Rashba [9]. Just for this case of light excitons and weak exciton-phonon coupling one can expect a large size of the self-trapped excitonic states and quantum confinement becomes the most interesting. In what follows we consider the phenomenon of quantum confinement in the model of 1D molecular crystal. This model is often used for analysis of optical properties of $J$-aggregates. These properties have been investigated for a long time, but recently this interest has risen considerably in connection with new experimental results on the superradiance of excitonic states and its temperature dependence at low temperatures [10]. These both problems (i.e. the problem of quantum confinement and that of the superradiance) are closely connected to each other, although such a connection has not been discussed yet. Taking into account the quantum confinement in 1D crystals and at low temperatures is very essential, because according to the well-known results of Rashba [9, 6] the self-trapped states in 1D crystals (contrary to those in 3D crystals) arise at arbitrarily weak exciton-phonon coupling without any energy barrier. This feature of self-trapping in 1D crystals is taken into account in the present paper and used in the discussion as of quantum confinement phenomenon so of the excitonic superradiance.

## 2 Self-trapping of light excitons in 1D nanostructures

We shall start following [6, 9, 12] with a short introduction into the main concepts of the self-trapping theory which is based on the supposition that excitons interact with optical phonons via non-polar interaction and with longitudinal acoustical phonons via deformation potential. For definiteness we shall consider the case of optical phonons. Then the exciton-phonon Hamiltonian can be written in the form

$$H = -\frac{\hbar^2}{2M}\frac{\partial^2}{\partial x^2} + \sum_k \hbar\omega_0 a_k^+ a_k + \frac{\gamma}{\sqrt{N}}\sum_k \exp(ikx)\left(a_{-k}^* + a_k\right), \qquad (7)$$

where $M$ is the exciton effective mass, $a_k^+$ and $a_k$ the creation and annihilation operators of the phonon with wave number $k$, $\hbar\omega_0$ the characteristic phonon frequency, $\gamma$ the exciton-phonon coupling constant, and $N$ is the number of sites in the crystal lattice. The Fourier components $q_k$ of displacement coordinates and their conjugate momenta $p_k$ are related to the operators $a_k^+$ and $a_k$ by the well-known formulas

$$q_k = \frac{1}{\sqrt{2}}\left(a_k + a_{-k}^+\right), \qquad p_k = \frac{1}{\sqrt{2i}}\left(a_{-k} - a_k^+\right), \qquad (8)$$

with

$$q_k p_{k'} - p_{k'} q_k = i\delta_{kk'}. \qquad (9)$$

The basic feature of the self-trapping phenomenon is that the exciton creates a static lattice deformation and then becomes self-trapped in the potential well that results from this deformation. Therefore it is natural to define the displacement $q_k$ as a sum of two terms,

$$q_k = d_k + \tilde{q}_k, \qquad (10)$$

where the variables $d_k$ refer to the static potential well deformation ($d_k = d_k^*$, $d_k = d_{-k}$) and $\tilde{q}_k$ to small vibrations around this displaced field. On substituting Eqs. (8,10) into (7), we obtain the exciton-phonon Hamiltonian in terms of new variables

$$H = -\frac{\hbar^2}{2M}\frac{\partial^2}{\partial x^2} + \frac{\hbar\omega_0}{2}\sum_k (p_k p_{-k} + \tilde{q}_k \tilde{q}_{-k} - 1) + \hbar\omega_0 \sum_k d_k^2$$
$$+ \frac{\hbar\omega_0}{2}\sum_k d_k \tilde{q}_k + \frac{\gamma}{\sqrt{N}}\sum_k \exp(ikx)(d_k + \tilde{q}_{-k}). \tag{11}$$

In obtaining Eq. (11) we have used Eq. (9) and the fact that $d_k$ is an even function of wave vector $k$. The displacements $d_k$ caused by the existence of the exciton with wave function $\psi(x)$ can be estimated by means of the variational principle. We average the Hamiltonian (11) over trial function $\psi(x)\phi_0$, where $\phi_0$ is the phonon ground state function, to obtain the functional

$$H[\psi, d_k] = -\frac{\hbar^2}{2M}\int \psi^*(x)\frac{\partial^2 \psi(x)}{\partial x^2}dx + \hbar\omega_0 \sum_k d_k^2$$
$$+ \frac{\gamma}{\sqrt{N}}\sum_k d_k \int \exp(ikx)\psi^*(x)\psi(x)dx, \tag{12}$$

where the usual normalization condition

$$\int \psi^*(x)\psi(x)dx = 1 \tag{13}$$

was used. The first term in Eq. (12) corresponds to the kinetic energy of the exciton which after integration by parts and with taking into account vanishing of $\psi(x)$ at the boundaries of the system becomes

$$E_{\text{kin}} = \int \frac{\hbar^2}{2M}|\psi_x|^2 dx, \tag{14}$$

where $\psi_x = \partial\psi/\partial x$. The other terms in the right-hand side of Eq. (12) represent the deformation energy of the lattice. We determine the displacements $d_k$ by minimizing $H[\psi, d_k]$ with respect to the set $\{d_k\}$. Then the condition

$$\frac{\partial H[\psi, d_k]}{\partial d_k} = 0$$

gives

$$d_k = -\frac{\gamma}{2\hbar\omega_0\sqrt{N}}\int \exp(ikx)\psi^*(x)\psi(x)dx. \tag{15}$$

Substitution of Eq. (15) into (12) yields

$$H[\psi] = \int \frac{\hbar^2}{2M}|\psi_x|^2 dx - \frac{\gamma^2}{4\hbar\omega_0 N}\int\int\sum_k \exp[ik(x+x')]\psi^*(x)\psi(x)\psi^*(x')\psi(x')dxdx'. \tag{16}$$

In the lattice site representation we have

$$\sum_k \exp[ik(x + x')] = N\delta(x + x').$$

Taking into account the rule $\sum_n \longleftrightarrow \int dx/a$ of transition from summation over lattice sites to integration over the coordinate along the chain and also that $\psi^*(x)$ and $\psi(x)$ have the same parity with respect to inversion transformation $x \to -x$, we finally arrive at the Hamiltonian

$$H[\psi] = \int \left[ \frac{\hbar^2}{2M} |\psi_x|^2 - g|\psi|^4 \right] dx, \tag{17}$$

where

$$g = \frac{\gamma^2 a}{4\hbar\omega_0} \tag{18}$$

is an effective nonlinear coupling constant and

$$E_g = g \int |\psi(x)|^4 dx \tag{19}$$

is the deformation energy of the lattice. In case of interaction of exciton with acoustical phonons we obtain the same Hamiltonian with different definition of the effective coupling constant $g$.

The equation for the exciton wave function $\psi(x)$ is obtained by minimizing the Hamiltonian (17) with respect to $\psi(x)$. We introduce a Lagrange multiplier $E$ in order to maintain the normalization condition (13) and get the nonlinear Schrödinger (NLS) equation

$$-\frac{\hbar^2}{2M} \frac{d^2\psi}{dx^2} - 2g|\psi|^2\psi = E\psi. \tag{20}$$

The nonlinear Schrödinger equation for exciton (what is equivalent to the polaron case) for the first time was obtained in [13].

The time dependent wave-function $\psi(t, x)$ obeys the Schrödinger equation which can be written in the Hamiltonian form

$$i\hbar \frac{\partial \psi}{\partial t} = \frac{\delta H}{\delta \psi^*}, \tag{21}$$

where $\delta/\delta\psi^*$ is the variational derivative. Then the stationary states have the wave functions in the form

$$\psi(t, x) = \exp\left(-iEt/\hbar\right) \psi(x), \tag{22}$$

and $\psi(x)$ satisfies Eq. (20). In an infinite 1D system Eq. (20) has the well-known solution [9, 14]

$$\psi(x) = \frac{1}{\hbar} \sqrt{\frac{Mg}{2}} \frac{1}{\cosh\left(Mgx/\hbar^2\right)} \tag{23}$$

with

$$E = -Mg^2/2\hbar^2. \qquad (24)$$

This solution is normalized according to the condition (13). As is clear from the solution (23), the characteristic length of localization is equal to

$$l_0 = \frac{\hbar^2}{Mg}, \qquad (25)$$

and can be large for light exciton and small $g$.

Now let us turn to the study of properties of the self-trapped states in finite nanostructures. An important feature of the exciton theory in nanostructures is the condition of vanishing of the wave function at the boundaries of the structure[1]. In the case under consideration of the layer of the width $2L$ with plain boundaries (or 1D chain of the length $2L$) this condition can be written in the form

$$\psi(L) = \psi(-L) = 0, \qquad (26)$$

where $\psi(x)$ depends only on the coordinate $x$ perpendicular to the boundaries (or along the chain), and the points $x = \pm L$ correspond to the boundaries positions. By analogy with usual linear Schrödinger equation, the wave functions can be constructed from the periodic solution of the NLS equation. The necessary particular periodic solution was found long ago [16, 17]. In accordance with the symmetry of the system with respect to inversion $x \to -x$, the eigenfunctions must be even or odd functions of $x$ and, correspondingly, have the form

$$\psi(x,t) = A \exp\left[-\frac{i}{\hbar}\left(\frac{\hbar q^2}{2M} - gA^2\right)t\right]\left\{\begin{array}{c} \text{cn} \\ \text{sn} \end{array}\right\}\left(\sqrt{q^2 + \frac{2MA^2g}{\hbar^2}}\,x, m\right), \qquad (27)$$

where $\text{cn}(u,m)$ and $\text{sn}(u,m)$ are the Jacoby elliptic functions [18] with the parameter

$$m = \frac{gA^2}{\hbar^2 q^2/2M + gA^2}. \qquad (28)$$

The amplitude $A$ and the "wave number" $q$ are determined by the boundary condition (26),

$$\sqrt{q^2 + \frac{2MA^2g}{\hbar^2}}\cdot L = j\,\text{K}(m), \qquad (29)$$

and the normalization condition (13),

$$\int_{-L}^{L} |\psi|^2 dx = \frac{2jA^2}{\sqrt{q^2 + 2MA^2g/\hbar^2}}\,\frac{\text{E}(m) - (1-m)\,\text{K}(m)}{m} = 1, \qquad (30)$$

---

[1]Note that in the paper [15] Rashba imposed the periodicity condition on the wave function and investigated the threshold behavior of the self-trapping phenomenon as a function of the coupling constant. The wave function which he used did not vanish at $x = \pm L$ (in fact, it could be expressed in terms of elliptic "dn" function instead of our "cn" or "sn" functions; see below) and only the ground one-exciton state was studied.

where $K(m)$ and $E(m)$ are the complete elliptic integrals of the first and the second kind, respectively, and $j$ is equal to $(1, 3, 5, \ldots)$ for even states and $(2, 4, 6 \ldots)$ for odd states. The equations (28–30) permit one to express the length $L$ and the energy (17) as functions of the parameter $m$. Simple calculation yields

$$L(m) = l_0 j^2 \, K(m) \left[ E(m) - (1 - m) \, K(m) \right], \tag{31}$$

$$E(m) = \frac{E_s}{3j^2} \frac{(1 - 2m) \, E(m) - (1 - 3m)(1 - m) \, K(m)}{\left[ E(m) - (1 - m) \, K(m) \right]^3}, \tag{32}$$

where the localization length $l_0$ is given by Eq. (25) and $E_s$ is the absolute value of the self-trapped state energy in the infinite chain (see (24))

$$E_s = \frac{Mg^2}{2\hbar^2} = \frac{\hbar^2}{2Ml_0^2}. \tag{33}$$

The formulas (31) and (32) give the dependence of the $j$-th exciton energy level $E_j$ on the chain length $2L$ in a parametric form.

In the limit of strong quantum confinement for $j$-th level, $L \ll j^2 l_0$ (i.e. $m \ll 1$), this dependence can be expressed explicitly. The series expansions of $L(m)$ and $E(m)$ in powers of the small parameter $m$ have the forms

$$L(m) \simeq \frac{l_0 j^2 \pi^2}{8} \, m \left( 1 + \frac{3}{8} m \right), \qquad E(m) \simeq \frac{16 E_s}{j^2 \pi^2 m^2} \left( 1 - \frac{3}{2} m \right).$$

Elimination of $m$ from these expressions gives the $j$-th energy level of the exciton as a function of $L$:

$$E_j(L) \simeq \frac{\hbar^2}{2M} \left( \frac{j\pi}{2L} \right)^2 - \frac{3}{4} \frac{g}{L}, \qquad L \ll j^2 l_0, \tag{34}$$

where two terms of series expansion in powers of the small parameter $(L/j^2 l_0)$ are taken into account. The first term in (34) represents the usual energy levels of a quantum particle with mass $M$ moving inside the potential well of width $2L$. The second term describes the correction to this energy due to the self-trapping interaction.

In the opposite limit $L \gg j^2 l_0$ we obtain in a similar way

$$E_j \simeq -\frac{E_s}{3j^2}, \qquad L \gg j^2 l_0, \tag{35}$$

and in this limit the energy levels with $j$ satisfying the above inequality do not depend on $L$. All these states beginning from the state $j = 1$ have negative energy and thus are self-trapped. It is clear that the number of these states depends on $L$. The dependence of the first lowest three levels of the exciton on the length $L$ is shown in figure 1. From this figure and above formulas we conclude that for a given $L$ the quantum confinement influences considerably the levels with $j \leq \sqrt{L/l_0}$,

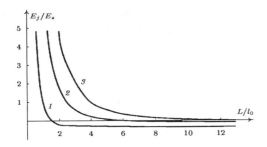

Figure 1: The dependence of energies $E_j$ (in units $E_s$) for three lowest energy excitonic states ($j = 1, 2, 3$) on the length $L$ (measured in units $l_0$). The states $j = 1$ and $j = 3$ are bright ($F_{1,3} \neq 0$), whereas the state $j = 2$ is dark ($F_2 = 0$).

while the other levels with $j \gg \sqrt{L/l_0}$ remain actually the same as those of a free exciton in a linear chain.

It is important to note that one exciton spectrum in the finite 1D system of length $2L$ differs drastically from the spectrum of the infinite 1D system with self-trapping interaction. In the infinite chain the spectrum consists of a continuous spectrum of freely moving particle and of single discrete level corresponding to the self-trapped state of the exciton. In the finite chain all spectrum is discrete and the number of self-trapped states (with $E_j < 0$; see (35)) depends on $L$. In the limit $L \to \infty$ the lowest level $j = 1$ goes to the self-trapped state described by the formulas (22–24) with the wave function vanishing at infinity. Other states (35) ($j \neq 1$) do not correspond to eigenfunctions vanishing at infinity and therefore they exist only in finite systems.

## 3 Oscillator strengths and superradiance of excitons from a finite molecular chain

Now let us turn to the analysis of the oscillator strengths of transitions to the self-trapped excitonic states. The matrix element of a dipole moment transition consists of two factors. The first one is determined by the excitonic wave function and can be calculated with the use of the results of the preceding section (see below). The second factor is determined by the overlap of the lattice vibrational wave functions of the crystal in the ground and excited states. In the expression for the oscillator strength it results in the so called Franck-Condon factor. This factor depends crucially on the properties of vibrations with which the excitonic transition under consideration interacts most intensively. In many organic crystals, as, e.g., in $J$-aggregates of pseudoisocyanine (PIC), it is supposed [10] that the lowest electronic transitions interact most intensively with the lattice optical vibrations with the frequencies of the order of magnitude 100–300 cm$^{-1}$. The dispersion of such vibrations is usually small ($\sim 10$ cm$^{-1}$). If this dispersion is neglected, then the above mentioned Franck-Condon factor is equal to (see, e.g. [9])

$$\exp\left(-\frac{E_g}{\hbar\omega_0}\right), \qquad (36)$$

where $E_g$ is the deformation energy of the crystal (see (19)). If the energy $E_g \ll \hbar\omega_0$, this factor is very close to unity. Therefore in this case (the case of weak exciton-phonon coupling) we can concentrate our attention on the calculation of the part of the matrix element which is determined by the excitonic wave function.

The coupling of the radiation field of wave vector $\mathbf{k}$ to the eigenstate $j$ is determined by the value of the sum of the monomer dipole moments $\mu_{\text{mon}}$ multipled by the amplitudes $\psi_j(n)$ that the exciton is situated at the site $n$, and the phase factor $\exp(i\mathbf{k}\mathbf{r}_n)$, $\mathbf{r}_n$ being the radius-vector of the site $n$,

$$\sum_{n=1}^{N} \mu_{\text{mon}}\psi_j(n)\exp(i\mathbf{k}\mathbf{r}_n).$$

We consider aggregates with the length $2L$ small compared to an optical wavelength ($|\mathbf{k}|L \ll 1$), so the transition dipole moment of the aggregate in the continuous approximation is equal to

$$\mu_j = \frac{\mu_{\text{mon}}}{a}\int_{-L}^{L}\psi_j(x)dx, \qquad (37)$$

where $a$ is the lattice constant. The oscillator strength of the aggregate, which determines the one-photon absorption and the radiative lifetime of the $j$-th state is proportional to $\mu_j^2$, and it can be easily calculated with the use of already known eigenfunctions (27).

The integral in (37) vanishes for odd eigenfunctions $\psi(x)$, so that only the states with $j = 1, 3, 5 \ldots$ contribute to the radiative transitions. Substitution of the "cn" wave function (27) into (37) leads to the table integral (see formula 16.24.2 in [18]) and after simple calculation yields

$$\int_{-L}^{L}\psi(x)dx = (-1)^{(j-1)/2}\frac{2A}{\sqrt{q^2 + 2MgA^2/\hbar^2}}\cdot\frac{\arcsin\sqrt{m}}{\sqrt{m}}, \qquad j = 1, 3, 5, \ldots \quad (38)$$

where $A, q$, and $m$ are related with each other by the Eqs. (28–30). Taking them into account, we obtain the oscillator strength of the chain as a function of the parameter $m$:

$$F_j = f_0\frac{2l_0}{a}\arcsin^2\sqrt{m}, \qquad j = 1, 3, 5\ldots, \qquad (39)$$

where $m$ depends on the length $L$ and the number of the state $j$ according to eq. (31).

The formulas obtained permit us to make the analysis of some features of the superradiance from such 1D molecular crystals as, e.g., $J$-aggregates, as a function of their length and exciton-phonon interaction constant. As is clear from the above formulas, the oscillator strengths of radiation are determined by the ratio of the chain's length $L$ to the localization length $l_0$. At given $L$ there are states $j$ satisfying the inequality $L \ll j^2 l_0$ ($j$ odd). For these states the corresponding values of the parameter $m$ are much less than unity and according to eq. (31) are equal to

$$m \cong \frac{8}{\pi^2} \frac{1}{l_0 j^2}, \tag{40}$$

so the corresponding limit of eq. (39) takes the form

$$F_j \cong f_0 \cdot \frac{8}{\pi^2} \frac{2L}{a} \cdot \frac{1}{j^2}. \tag{41}$$

If $L \ll l_0$, the formula (41) holds for all odd $j = 1, 3, 5, \ldots$. But if $9l_0 > L > l_0$, then this formula is valid only for $j = 3, 5, \ldots$. Of course, $F_j = 0$ for all dark states $j = 2, 4, 6, \ldots$. Let us return to the limit of strong confinement when $L \ll l_0$ and hence equation (41) is valid for all $j$. In this limit we reproduce from (41) the equation (5) for the state $j = 1$. Summation of (41) over all odd values of $j$ gives an obvious sum rule

$$\sum_{j=1}^{\infty} F_{2j-1} = f_0 \cdot \frac{2L}{a} = f_0 \cdot N. \tag{42}$$

Note, that this sum rule takes place for any values of $L$, even if the condition of strong confinement is not satisfied for all states.

Let us consider this case in more detail. We assume that $L \gg j^2 l_0$ ($j$ odd). Then there are a few states with values of $j$ satisfying this inequality. For these states $m \cong 1$, and for corresponding value of $F_j$ we obtain

$$F_j \cong f_0 \cdot \left(\frac{\pi}{2}\right)^2 \cdot \frac{2l_0}{a}. \tag{43}$$

which already does not depend on $L$, i.e. there is saturation of the oscillator strength. It is remarkable that all these states (with $j \ll \sqrt{L/l_0}$) have equal oscillator strengths what contrasts drastically to the theory without self-trapping. The dependence of the oscillator strength $F_j$ on $L/l_0$ for levels $j = 1, 3, 5$ is shown in figure 2.

## 4    Discussion

As we see from figure 2, for all energy levels with odd $j = 1, 3, 5 \ldots$ the oscillator strengths tend to the same saturation limit but this limit is achieved at different lengths of the chain. This feature of the self-trapping can play an important role

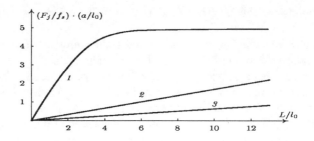

Figure 2: The dependence of oscillator strengths $F_j$ (in units $f_0 l_0/a$) for $j = 1, 3, 5$ on the length $L$ (measured in units $l_0$).

in discussion of the radiative properties of molecular nanostructures, for example, of $J$-aggregates. Indeed, if $L$ is greater than $l_0$, the oscillator strength of the lowest energy state $j = 1$ reduces considerably compared to the case of system without self-trapping. In this case only about $(\pi/2)^2(2l_0/a)$ molecules radiate coherently. Using the common terminology, we can say that the coherence length is equal to $(\pi/2)^2 2l_0$ and does not depend on the length of the chain. This effect appears due to self-trapping and is not connected with scattering of excitons by phonons or lattice defects.

Let us make now some estimations. For example, at $l_0/a \sim 10$ the oscillator strength, and hence the radiation width of the state, is about 50 times as large as the value for a monomer and this leads to the lifetime 50 times as small as the lifetime of the excited state in a monomer. For the state $j = 1$ such a decrease of the radiative lifetime takes place for the chain length $2L \geq 2l_0$. At this $L$ the oscillator strength $F_j$ for states $j = 3, 5, \ldots$ is much less than the value of $F_1$. Therefore in this case only the lowest energy exciton state ($j = 1$) has a considerable oscillator strength and therefore this situation is similar to that for linear (without self-trapping) chain with the same length. However, with increase of the length $L$ up to the values in the interval $9l_0 < L < 25l_0$ the oscillator strength $F_1$ does not change, whereas $F_3 \to F_1$ and for $j = 5, 7, \ldots$ we have $F_j \ll F_1$. Thus, already for this chain length the relationship between oscillator strengths of the two low energy levels differs considerably from that for linear case. We can say that now the states with small oscillator strengths are shifted to the region $j \geq 5$, for longer chains to $j \geq 7$, and so on.

Note also that in the theory under consideration the main characteristic parameters are $l_0/a$ and $L/l_0$, and the relationship between radiative excitonic lifetime and position of its energy levels arises very naturally. For example, let the exciton-phonon coupling constant be such that $l_0/a \cong 10$. Then, according to figure 2, at length $L \cong 2l_0$ we have $F_1 \cong 30f_0$ so the radiative lifetime of this state is $\tau_1 \cong \tau_0/30$. On the other hand, according to figure 1 at the same value of $L$ the energy of the lowest level is equal to $E_1 \cong -E_s = -\hbar^2/(6Ml_0^2)$, or with account of (2) $E_1 \cong -(J/3)(a/l_0)^2$. Thus, if as in [10] we take the value $J = 630\,\mathrm{cm}^{-1}$, then at $l_0/a \cong 10$ and $L \cong 2l_0$ we obtain $E_1 \cong -2\,\mathrm{cm}^{-1}$. The next state with $j = 2$ is dark and from figure 1 we find that at $L \cong 2l_0$ its energy is equal to $5E_s$. Thus the gap between bright (lowest energy $j = 1$ state) and the nearest dark state ($j = 2$) is equal to $\cong 6E_s \cong 12°\mathrm{K}$. Therefore from this estimate it follows that at temperature $T < 10°\mathrm{K}$ the radiative lifetime will be equal to $\tau \cong \tau_1$ and does not depend on temperature $T$. Only at higher temperature $T > 10°\mathrm{K}$, when an occupation of the first dark state begins, we can expect that the radiative lifetime will start to arise going gradually to the radiative lifetime of a monomer. It is clear that this estimate of the temperature is correct only up to the order of magnitude. For more accurate treatment it is necessary to perform detailed calculations of the temperature dependence of the superradiation with the use of kinetic equations (see, e.g. [10]) and taking into account of the self-trapped states. Recently, the new data on the temperature dependence of radiative lifetime of $J$-aggregates at low temperature were discussed in the literature (see [10]). As follows from our illustrative estimates, even in the case of weak exciton-phonon interaction (i.e. when $E_g \ll \hbar\omega_0$), the effects of quantum confinement and self-trapping should be taken into account in the discussion of this interesting phenomenon. More thorough treatment of temperature effects and of influence of disorder on the superradiance require considerable numerical calculations and will be discussed elsewhere.

## Acknowledgements

We acknowlege partial support from Grant 97–1074 of Russian Ministry of Science and Technology and Grant 99–03–32178 of Russian Foundation of Basic Researches.

## References

[1] Burstein, E. and Weisbuch, C. (eds) (1995) *Confined Electrons and Photons. New Physics and Applications.*, Plenum, N.Y.

[2] Agranovich, V.M. and Kamchatnov, A.M. (1999) *Chem. Phys.* **245**, 173.

[3] Fidder, H., Knoester, J., and Wiersma, D.A. (1991) *J. Chem. Phys.* **95**, 7880.

[4] Hochstrasser, R.M. and Whiteman, J.D. (1972) *J. Chem. Phys.* **56**, 5945.

[5] De Boer, S. and Wiersma, D.A. (1990) *Chem. Phys. Lett.* **165**, 45.

[6] Rashba, E.I. (1982) Self-Trapping of Excitons, in E.I. Rashba and M.D. Sturge (eds), *Excitons*, North-Holland, p.543.

[7]  Peierls, R. (1932) *Ann. Phys.* **13**, 905.

[8]  Frenkel, J. (1931) *Phys. Rev.* **37**, 17, 1276.

[9]  Rashba, E.I. (1957) *Opt. Spektrosk.* **2**, 75, 88; **3**, 568.

[10]  Potma, E.O. and Wiersma, D.A. (1990) *J. Chem. Phys.* **108**, 4894.

[11]  Feynman, R.P. (1972) *Statistical Mechanics,* W.A. Benjamin, Inc., Reading, Chapter 8.

[12]  Shaw, P.B. and Whitfield, G. (1978) *Phys. Rev.* **B17**, 1495.

[13]  Deigen, E.I. and Pekar, S.I. (1951) *Zh. Exp. Teor. Fiz.* **21**, 568.

[14]  Holstein, T. (1959) *Ann. Phys.* **8**, 325, 343.

[15]  Rashba, E.I. (1994) *Synth. Metals* **64**, 255.

[16]  Talanov, V.I. (1964) *Izv. Vuzov Radiofizika* **7**, 264.

[17]  Ostrovsky, L.A. (1966) *Zh. Exp. Teor. Fiz.* **51**, 1189.

[18]  Abramowitz, M. and Stegun, I. (1968) *Handbook of Mathematical Functions,* Dover Publications, Inc., New York.

# The Mixing of Frenkel- and Charge-Transfer Excitons in 1D-Structures: Application to PTCDA and MePTCDI

M. Hoffmann, K. Schmidt, T. Fritz, T. Hasche,
V.M. Agranovich*, K. Leo
*Institut für Angewandte Photophysik (www.iapp.de)*
*TU Dresden, 01062 Dresden, Germany*

*permanent address:*
*Institute of Spectroscopy*
*Russian Academy of Sciences, Troitsk,*
*Moscow Reg. 142092, Russia*

**Abstract:** The exciton structure of crystalline PTCDA (3,4,9,10-perylenetetra-carboxylic dianhydride) and MePTCDI (N-N'-dimethylperylene-3,4,9,10-dicarbox-imide) is modeled by a one-dimensional Hamiltonian, which includes interactions between Frenkel excitons with several vibronic levels and charge transfer excitons. With appropriate fitting parameters, which are verified by quantum chemical cal-culations, this model can explain the main features of the low temperature absorp-tion spectra. Polarized absorption spectra of MePTCDI show different polarization ratios for the various peaks. This polarization behavior is explained by the varying contribution of the CT transition dipole, which has a direction different from the Frenkel transition dipole.

# 1 Introduction

Perylene derivatives like PTCDA (3,4,9,10-perylenetetracarboxylic dianhydride) and MePTCDI (N-N'-dimethylperylene-3,4,9,10-dicarboximide) are widely inve-stigated due to their potential applications in thin film optoelectronic devices. In vapor-deposited films, they typically form microcrystals or epitaxial layers with well defined crystal structure [1].

The excited states of the isolated molecules are well understood. The strong absorption band in the visible range is due to a $\pi$-$\pi^*$-transition. This electronic transition strongly couples to C-C and C=C stretching modes of the carbon back-bone, which leads to one single vibronic progression [2] as can be seen in Fig. 1. In the crystalline phase, these excited states of the monomer are strongly influenced by intermolecular interactions, which leads to dramatic changes in the absorption spectra (Fig. 1). Models capable to describe significant aspects of the solid-state absorption spectra are just emerging [2].

The crystal structure of PTCDA, MePTCDI and many related perylenes is characterized by quasi one-dimensional stacks. The small distance between the molecular planes within the stacks causes strong interactions of the $\pi$-electron systems. The interaction between different stacks is considerably smaller due to

F. Kajzar and M.V. Agranovich (eds.), Multiphoton and Light Driven Multielectron Processes in Organics: New Phenomena, Materials and Applications, 123–134.

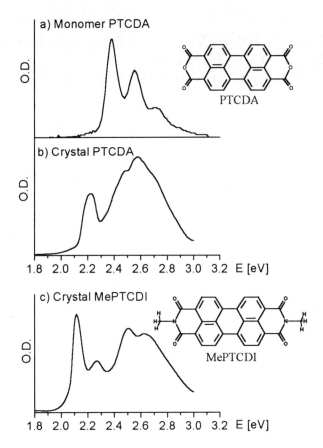

Figure 1: Normalized absorbance spectra: a) PTCDA dissolved in DMSO at room temperature, b) PTCDA thin film at 5 K, c) MePTCDI highly oriented domain in thin film at 5 K

their larger separation. Therefore, we use a one-dimensional model Hamiltonian to describe the excited states of the crystal.

Within such a one-dimensional model, we try to give the simplest description that can explain the main features of the absorption spectra. We concentrate on the low temperature spectrum of MePTCDI, which is roughly characterized by four peaks. For PTCDA, the peak structure is not as pronounced but seems similar to MePTCDI. Since the monomer absorption of both dyes is dominated by three peaks, it can be well described by an electronic transition with three vibronic levels. These three levels become three mixed exciton bands if energy transfer between excited molecules is included in the 1D Hamiltonian for the crystal. However, such a Frenkel exciton model can not explain the four significant peaks in the crystal spectrum. To achieve that, we include one nearest neighbor charge transfer state

(CT) into the model. The importance of CT states in PTCDA has already been demonstrated [2]. Furthermore, our quantum chemical calculations also indicate strong mixing of Frenkel excitons with CT excitons lying in the same spectral region.

A CT state can have a transition dipole moment with a direction different from the molecular transition dipole. Then, the direction of the total transition dipole for a mixed state depends on the relative contribution of Frenkel and CT states. We consider this effect to explain the experimentally observed polarization ratio, which varies for the different bands. Such a varying polarization ratio is a new phenomenon entirely due to the CT mixing - in the framework of pure Frenkel exciton theory it can not occur if one molecular electronic transition is considered.

The actually investigated crystals contain two molecules per unit cell which leads to a small but observable Davydov splitting. However, this resonant interaction between two stacks can be clearly analyzed as a pertubative correction of the predictions of the one dimensional model.

## 2 Model Hamiltonian

For the description of the excited states in a one-dimensional molecular crystal with one molecule per unit cell we use the following Hamiltonian:

$$
\begin{aligned}
\mathcal{H} &= \mathcal{H}^{\mathrm{F}} + \mathcal{H}^{\mathrm{FF}} + \mathcal{H}^{\mathrm{C}} + \mathcal{H}^{\mathrm{FC}} &\qquad (1)\\
\mathcal{H}^{\mathrm{F}} &= \sum_{n\nu} \Delta_{\mathrm{F}}^{\nu} \mathcal{B}_{n\nu}^{\dagger} \mathcal{B}_{n\nu}\\
\mathcal{H}^{\mathrm{FF}} &= {\sum_{\substack{n\nu \\ m\mu}}}' M_{nm}^{\nu\mu} \mathcal{B}_{n\nu}^{\dagger} \mathcal{B}_{m\mu}\\
\mathcal{H}^{\mathrm{C}} &= \sum_{n\sigma} \Delta_{\mathrm{CT}} \mathcal{C}_{n\sigma}^{\dagger} \mathcal{C}_{n\sigma}\\
\mathcal{H}^{\mathrm{FC}} &= \sum_{n\nu} \{ \epsilon_{\mathrm{e}}^{\nu} (\mathcal{B}_{n\nu}^{\dagger} \mathcal{C}_{n,+1} + \mathcal{B}_{n\nu}^{\dagger} \mathcal{C}_{n,-1}) + \epsilon_{\mathrm{h}}^{\nu} (\mathcal{B}_{n\nu}^{\dagger} \mathcal{C}_{n+1,-1} + \mathcal{B}_{n\nu}^{\dagger} \mathcal{C}_{n-1,+1}) + \mathrm{h.c.} \}
\end{aligned}
$$

Here the operator $\mathcal{B}_{n\nu}^{\dagger}$ ($\mathcal{B}_{n\nu}$) describes the creation (annihilation) of a neutral local excitation (Frenkel exciton) at lattice site $n$. Only one electronically excited local state is considered, and the index $\nu$ specifies the excited vibrational level ($\nu = 0, 1, .., r$). Then $\Delta_{\mathrm{F}}^{\nu}$ is the on-site energy of a Frenkel exciton and $M_{nm}^{\nu\mu}$ the hopping integral for excitation transfer from level $\nu$ at site $n$ to level $\mu$ at site $m$. In the summation in $\mathcal{H}^{\mathrm{FF}}$ the terms for $n = m$ are omitted. The Hamiltonian $\mathcal{H}^{\mathrm{F}} + \mathcal{H}^{\mathrm{FF}}$ can be used to describe the well-studied case of mixing of molecular configurations for Frenkel excitons with several excited states in Heitler-London approximation [3, 4, 5].

In addition to these Frenkel excitons we now include nearest neighbor charge transfer excitons. A localized CT exciton with the hole at lattice site $n$ and the electron at lattice site $n + \sigma$ ($\sigma = -1, +1$) is created (annihilated) by the operator $\mathcal{C}_{n\sigma}^\dagger$ ($\mathcal{C}_{n\sigma}$). For simplicity, only the vibrational ground state is considered for the CT excitons with $\Delta_{\text{CT}}$ as their on-site energy. Hopping of CT states will not be considered.

The mixing between Frenkel and CT excitons is expressed in the last part $\mathcal{H}^{\text{FC}}$ of the Hamiltonian. Here, the transformation of a CT state into any Frenkel state at the lattice site of either hole or electron is allowed. The relevant transfer integrals $\epsilon_{\text{e}}^\nu$ ($\epsilon_{\text{h}}^\nu$) can be visualized as transfer of an electron (hole) from the excited molecule $n$ to its nearest neighbor.

In order to diagonalize the Hamiltonian (1), we first transform all operators into their momentum space representation:

$$\mathcal{B}_{k\nu} := \frac{1}{\sqrt{N}} \sum_n e^{-ikn} \mathcal{B}_{n\nu} \tag{2}$$

$$\mathcal{C}_{k\sigma} := \frac{1}{\sqrt{N}} \sum_n e^{-ikn} \mathcal{C}_{n\sigma} \tag{3}$$

Then the Hamiltonian takes the form:

$$\mathcal{H} = \sum_k ( \mathcal{H}_k^{\text{F}} + H_k^{\text{FF}} + \mathcal{H}_k^{\text{C}} + \mathcal{H}_k^{\text{FC}} ) \tag{4}$$

$$\mathcal{H}_k^{\text{F}} = \sum_\nu \Delta_{\text{F}}^\nu \mathcal{B}_{k\nu}^\dagger \mathcal{B}_{k\nu}$$

$$\mathcal{H}_k^{\text{FF}} = \sum_{\nu\mu} L_k^{\nu\mu} \mathcal{B}_{k\nu}^\dagger \mathcal{B}_{k\mu}$$

$$\mathcal{H}_k^{\text{C}} = \sum_\sigma \Delta_{\text{CT}} \mathcal{C}_{k\sigma}^\dagger \mathcal{C}_{k\sigma}$$

$$\mathcal{H}_k^{\text{FC}} = \sum_\nu \mathcal{B}_{k\nu}^\dagger \{ (\epsilon_{\text{e}}^\nu + \epsilon_{\text{h}}^\nu e^{ik}) \mathcal{C}_{k,+1} + (\epsilon_{\text{e}}^\nu + \epsilon_{\text{h}}^\nu e^{-ik}) \mathcal{C}_{k,-1} \} + \text{h.c.}$$

Here, the symbol $L_k^{\nu\mu}$ is used to abbreviate the lattice sum

$$L_k^{\nu\mu} := {\sum_m}' e^{ikm} M_{0m}^{\nu\mu}. \tag{5}$$

The Hamiltonian (4) is already diagonal with respect to $k$. It still contains mixed terms of the $0..r$ operators for the Frenkel excitons and the two operators for CT excitons. These are altogether $r + 3$ molecular configurations, which would yield $r + 3$ mixed exciton bands.

With several simplifying assumptions we now reduce the number of parameters in the Hamiltonian, which also allows the separation of one non-mixing exciton band. The parameters are given by the transition matrix elements with the considered states. Using product wavefunctions in Born-Oppenheimer approximation we can write for the ground state

$$|0\rangle = |\prod_n \varphi_n^0 \chi_n^{00}\rangle^{(-)} \tag{6}$$

and for the Frenkel excitons

$$\mathcal{B}_{n\nu}^{\dagger}|0\rangle = |\varphi_n^1 \chi_n^{1\nu} \prod_{n' \neq n} \varphi_{n'}^0 \chi_{n'}^{00}\rangle^{(-)}. \tag{7}$$

Here $\varphi_n^0$ and $\varphi_n^1$ denote the electronic part of the wavefunction of molecule $n$ in the ground and first electronically excited state. $\chi_n^{\mu\nu}$ is the vibrational wavefunction of molecule $n$ in its $\mu$th electronic and $\nu$th vibrational state. The upper index $(-)$ at the Dirac bracket indicates that the product wavefunction is antisymmetrized with respect to all electrons. The CT states are represented by:

$$\mathcal{C}_{n,+1}^{\dagger}|0\rangle = |\varphi_n^+ \chi_n^{+0} \varphi_{n+1}^- \chi_{n+1}^{-0} \prod_{\substack{n' \neq n \\ n' \neq n+1}} \varphi_{n'}^0 \chi_{n'}^{00}\rangle^{(-)}, \tag{8}$$

$$\mathcal{C}_{n,-1}^{\dagger}|0\rangle = |\varphi_{n-1}^- \chi_{n-1}^{-0} \varphi_n^+ \chi_n^{+0} \prod_{\substack{n' \neq n \\ n' \neq n-1}} \varphi_{n'}^0 \chi_{n'}^{00}\rangle^{(-)}, \tag{9}$$

where $\varphi_n^{\pm}$ and $\chi_n^{\pm}$ refer to the ionized molecules.

Using these representations the Frenkel exciton hopping integral can be split into an electronic and a vibronic part:

$$\begin{aligned} M_{nm}^{\nu\mu} &= \langle \mathcal{B}_{n\nu}^{\dagger} 0 | \mathcal{H} | \mathcal{B}_{m\mu}^{\dagger} 0 \rangle \\ &= M_{nm} S_\nu S_\mu \end{aligned} \tag{10}$$

The vibronic overlap factors $S_\nu = \langle \chi_n^{1\nu} | \chi_n^{00} \rangle$ defined here are directly related to the Franck-Condon factors $F_{0\nu} = |\langle \chi^{1\nu} | \chi^{00} \rangle|^2$. Since the Frenkel exciton hopping integrals $M_{nm}$ in dipole approximation are decreasing with distance by $(n-m)^{-3}$, in the one-dimensional case we can neglect all but nearest neighbor transfer interactions

$$M_{nm} = \delta_{n\pm 1,m} M \tag{11}$$

with $M$ being the nearest neighbor exciton hopping integral. With this approximation $L_k^{\nu\mu}$ from (5) is reduced to

$$L_k^{\nu\mu} = 2M S_\nu S_\mu \cos k. \tag{12}$$

The vibrational overlap factors for the Frenkel states can be easily derived from the absorption spectrum of the molecule in solution. Since in our case the coupling to *intra*molecular vibrations is strong compared to *inter*molecular phonon coupling, the overlap factors from isolated molecules can be used for the crystalline phase as well.

For CT excitons the situation is different. There is no direct access to the vibrational overlap factors connected with $\chi^{+0}$ and $\chi^{-0}$. We therefore make the strongly simplifying assumption that $\chi^{+0} = \chi^{-0} = \chi^{00}$, which means that the coupling of CT excitons to the considered vibrations is neglected. Then we obtain

$$\epsilon_e^{\nu} = \langle \mathcal{B}_{n\nu}^{\dagger} 0 | \mathcal{H} | \mathcal{C}_{n,+1}^{\dagger} 0 \rangle$$

$$= \quad {}^{(-)}\langle\varphi_n^1 \prod_{n'\neq n} \varphi_{n'}^0 |\mathcal{H}|\varphi_n^+\varphi_{n+1}^- \prod_{\substack{n'\neq n \\ n'\neq n+1}} \varphi_{n'}^0\rangle^{(-)} \underbrace{\langle\chi_n^{1\nu}|\chi_n^{00}\rangle}_{S_\nu}$$

$$\underbrace{\phantom{{}^{(-)}\langle\varphi_n^1 \prod_{n'\neq n} \varphi_{n'}^0 |\mathcal{H}|\varphi_n^+\varphi_{n+1}^- \prod \varphi_{n'}^0\rangle^{(-)}}}_{\epsilon_e}$$

$$= \quad \epsilon_e S_\nu \tag{13}$$

$$\epsilon_h^\nu = \langle \mathcal{B}_{n\nu}^\dagger 0 |\mathcal{H}| \mathcal{C}_{n+1,-1}^\dagger 0\rangle$$

$$= \quad {}^{(-)}\langle\varphi_n^1 \prod_{n'\neq n} \varphi_{n'}^0 |\mathcal{H}|\varphi_n^-\varphi_{n+1}^+ \prod_{\substack{n'\neq n \\ n'\neq n+1}} \varphi_{n'}^0\rangle^{(-)} \underbrace{\langle\chi_n^{1\nu}|\chi_n^{00}\rangle}_{S_\nu}$$

$$\underbrace{\phantom{{}^{(-)}\langle\varphi_n^1 \prod_{n'\neq n} \varphi_{n'}^0 |\mathcal{H}|\varphi_n^-\varphi_{n+1}^+ \prod \varphi_{n'}^0\rangle^{(-)}}}_{\epsilon_h}$$

$$= \quad \epsilon_h S_\nu \tag{14}$$

With these simplifications the Hamiltonian for the Frenkel-CT-mixing becomes:

$$\mathcal{H}_k^{FC} = \sum_\nu S_\nu \mathcal{B}_{k\nu}^\dagger \left\{ (\epsilon_e + \epsilon_h e^{ik})\mathcal{C}_{k,+1} + (\epsilon_e + \epsilon_h e^{-ik})\mathcal{C}_{k,-1} \right\} + \text{h.c.} \tag{15}$$

We now introduce two new operators with even and odd symmetry with respect to change of the direction of the charge transfer:

$$\tilde{\mathcal{C}}_{kg} := \frac{1}{\sqrt{2\epsilon_k}} \cdot \left\{ (\epsilon_e + \epsilon_h e^{ik})\mathcal{C}_{k,+1} + (\epsilon_e + \epsilon_h e^{-ik})\mathcal{C}_{k,-1} \right\} \tag{16}$$

$$\tilde{\mathcal{C}}_{ku} := \frac{1}{\sqrt{2\epsilon_k}} \cdot \left\{ (\epsilon_e + \epsilon_h e^{ik})\mathcal{C}_{k,+1} - (\epsilon_e + \epsilon_h e^{-ik})\mathcal{C}_{k,-1} \right\}, \tag{17}$$

where

$$\epsilon_k := \sqrt{\epsilon_+^2 \cos^2 \frac{k}{2} + \epsilon_-^2 \sin^2 \frac{k}{2}} \tag{18}$$

with

$$\epsilon_\pm := \epsilon_e \pm \epsilon_h \tag{19}$$

The Hamiltonian for the CT states then simplifies to:

$$\mathcal{H}_k^C = \Delta_{CT} \{\tilde{\mathcal{C}}_{kg}^\dagger \tilde{\mathcal{C}}_{kg} + \tilde{\mathcal{C}}_{ku}^\dagger \tilde{\mathcal{C}}_{ku}\} \tag{20}$$

$$\mathcal{H}_k^{FC} = \sum_\nu \sqrt{2}\epsilon_k S_\nu \cdot \mathcal{B}_{k\nu}^\dagger \tilde{\mathcal{C}}_{kg} + \text{h.c.} \tag{21}$$

The odd operator $\tilde{\mathcal{C}}_{ku}$ does not mix with the Frenkel operators anymore, which reduces the number of mixed exciton bands by one. The remaining even part $\mathcal{H}_k^g$ of the Hamiltonian can be formally diagonalized by transformation to new operators $\xi_{k\beta}$

$$\xi_{k\beta} := \sum_{\nu=0}^r (u_{\nu\beta}^*(k)\mathcal{B}_{k\nu}) + c_\beta(k)\tilde{\mathcal{C}}_{kg} \tag{22}$$

where the transformation matrix $(u_{\nu\beta}, c_\beta)$ is the solution of the Eigenvalue problem for the matrix $H_{\alpha\beta}$ of the coefficients in the Hamiltonian:

$$H_{\alpha\beta} = \left( \begin{array}{c|c} \left( \begin{array}{ccc} & \vdots & \\ \cdots & \delta_{\nu\mu}\Delta_F^\nu + 2MS_\nu S_\mu \cos k & \cdots \\ & \vdots & \end{array} \right)_{\nu\mu} & \left( \begin{array}{c} \vdots \\ \sqrt{2}\epsilon_k S_\nu \\ \vdots \end{array} \right)_\nu \\ \hline \left( \begin{array}{ccc} \cdots & \sqrt{2}\epsilon_k S_\mu & \cdots \end{array} \right)_\mu & \Delta_{CT} \end{array} \right)$$

The diagonalization of this matrix was always carried out numerically.

With the knowledge of the excited states we can also calculate the transition dipole moments for optical excitation. From (18) it follows that for $k=0$ $\mathcal{H}_k^{FC}$ is only determined by $\epsilon_{k=0} = |\epsilon_+|$. If $\vec{P}$ is the total transition dipole operator, the transition dipole moment for the state $\xi_{0\beta}^\dagger|0\rangle$ can be expressed as a sum of Frenkel and CT transition dipoles:

$$\begin{aligned} \vec{P}^\beta &= \langle\xi_{0\beta}^\dagger 0|\vec{P}|0\rangle \\ &= \underbrace{\sum_{\nu=0}^r u_{\beta\nu}^*(k=0) \cdot \langle\mathcal{B}_{0\nu}^\dagger 0|\vec{P}|0\rangle}_{FE} + \underbrace{c_\beta(k=0) \cdot \langle\tilde{\mathcal{C}}_{0g}^\dagger 0|\vec{P}|0\rangle}_{CT} \\ &= \vec{P}_{FE}^\beta + \vec{P}_{CT}^\beta \end{aligned} \tag{23}$$

These total Frenkel and CT transition dipoles can be related to molecular properties as follows: For the Frenkel exciton we obtain with (2) and (7):

$$\vec{P}_{FE}^\beta = \sum_{\nu=0}^r u_{\beta\nu}^*(k=0) S_\nu \cdot \frac{1}{\sqrt{N}} \sum_n {}^{(-)}\langle\varphi_n^1 \prod_{n'\neq n}\varphi_{n'}^0|\vec{P}|\prod_{n''}\varphi_{n''}^0\rangle^{(-)} \tag{24}$$

If *inter*molecular exchange effects are neglected in this expression, $\vec{P}$ can be split into a sum of molecular transition dipole operators $\vec{P}_n$ with $\vec{p}_{FE} = \langle\varphi_n^1|\vec{P}_n|\varphi_n^0\rangle$ being the molecular transition dipole moment:

$$\vec{P}_{FE}^\beta = \sqrt{N}\vec{p}_{FE} \cdot \sum_{\nu=0}^r u_{\beta\nu}^*(k=0) S_\nu \tag{25}$$

In the same way we obtain from (8), (9) and with $\chi^{+0} = \chi^{-0} = \chi^{00}$ as used in (13) and (14):

$$\vec{P}_{CT}^\beta = c_\beta(k=0) \frac{1}{\sqrt{N}} \sum_n {}^{(-)}\langle\underbrace{\frac{\varphi_n^+ \varphi_{n+1}^- + \varphi_n^- \varphi_{n+1}^+}{\sqrt{2}}}_{|CT_n^g\rangle} \prod_{\substack{n'\neq n \\ n'\neq n+1}}\varphi_{n'}^0|\vec{P}|\prod_{n''}\varphi_{n''}^0\rangle^{(-)} \tag{26}$$

Here, we can separate an even CT state for a dimer $|CT_n^g\rangle = |1/\sqrt{2}(\varphi_n^+ \varphi_{n+1}^- + \varphi_n^- \varphi_{n+1}^+)\rangle^{(-)}$ with a transition dipole $\vec{p}_{CT}$. If we again neglect exchange of electrons between the CT dimer and the other molecules in the ground state we get:

$$\vec{P}_{CT}^{\beta} = \sqrt{N}\vec{p}_{CT} \cdot c_{\beta}(k=0) \tag{27}$$

Even if the CT transition dipole is vanishingly small, the oscillator strength of the Frenkel exciton is distributed over all mixed exciton bands. This is expressed by the notion that the CT exciton 'borrows' oscillator strength from the Frenkel states.

In summary, there are two main features of the model Hamiltonian (1): (i) If the coupling constant $\epsilon_+$ is in the same order as the energetic separation between the CT and the Frenkel excitons, the model predicts a strong mixing with $r+2$ significantly absorbing bands. (ii) With a finite CT transition dipole the polarization direction of the bands will vary due to the varying composition of the bands.

# 3 Application to Polarized Absorption Spectra

Because of the monomer absorption spectra (for PTCDA see Fig. 1, MePTCDI looks almost identical) with three characteristic peaks, we consider $r+1 = 3$ vibronic levels for the Frenkel exciton. From the peak positions we obtain the vibronic energy $\hbar\omega$ and by that $\Delta_F^{\nu} = \Delta_F^0 + \nu\hbar\omega$, and from the relative intensities we derive the vibronic overlap factors $S_{\nu}$. With appropriate parameters $\Delta_F^0, \Delta_{CT}, M, \epsilon_+$ the model Hamiltonian (1) can predict 4 absorption peaks as observed in the crystalline thin film (compare Fig. 1).

For MePTCDI it was possible to prepare films with macroscopic orientation. The dye was vapor-deposited onto a stretched polymer substrate at room temperature. We could then observe domains of a size up to about $10\mu m \times 100\mu m$, which show a very high degree of polarization. The unit cell of MePTCDI contains 2 molecules per unit cell with both molecular planes approximately in the (102) plane of the crystal lattice. From the polarization behavior of the film we conclude that in the highly oriented domains the (102) plane is parallel to the substrate plane and the crystal axes are uniformly oriented.

We measured the absorbance spectra of such domains for perpendicular incidence and two orthogonal polarization directions ($s$ and $p$) at low temperature (5 K). The $p$ direction corresponds to the direction with highest absorption for the peak at about 2.7 eV. The spectra are shown in Fig. 2 together with the peak positions from a peak fitting analysis using a four band model. Remarkably, the peak positions are slightly different for both polarization directions. We interpret this peak shifts as Davydov splitting since their values show the expected order of magnitude. From the occurrence of Davydov splitting it follows that the exciton states of the two non-equivalent stacks are coherently coupled. Also shown is the polarization ratio $R^{\beta}$ (ratio of the peak areas) for the $s$ and $p$ polarized spectra. $R^{\beta}$ varies for the different bands, which is an immediate consequence of a varying polarization direction within the 1D stacks according to the 1D model.

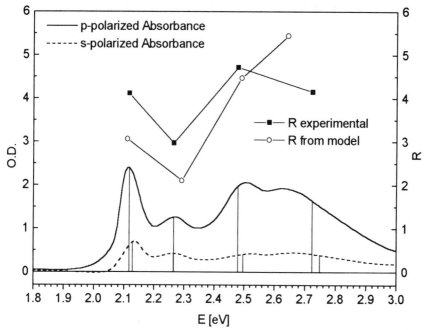

Figure 2: Polarized absorbance spectra and polarization ratio $R^\beta$ of MePTCDI, highly oriented domain in thin film at 5 K

We carried out the following quantitative analysis of the MePTCDI spectra. From quantum chemical calculations (see below), we derive the direction of the CT transition dipole $\vec{p}_{CT}$ of two neighbored molecules in the stack. $\vec{p}_{CT}$ has a large component (94%) within the molecular plane. This is surprising if the CT state is considered as a dipole of two point charges located at the centers of the molecules. However, in the case of MePTCDI molecules the carriers are delocalized over a $\pi$ electron system which is much larger than the separation between the molecules. Therefore, the actual CT transition dipole is mainly governed by the offset of the two molecules along the molecular planes and by the subtle structure of the overlap between the $\pi$ orbitals. We now only consider the component of $\vec{p}_{CT}$ in the molecular plane, which forms an angle of $\gamma = 68°$ with the molecular transition dipole. Then the direction of the total transition dipole for a mixed exciton band $\xi^{\dagger\beta}|0\rangle$ in the 1D stack is given according to (23), (25) and (27) by the relative CT transition dipole strength $p_{CT}^{rel} := |\vec{p}_{CT}|/|\vec{p}_{FE}|$.

For weak coherent coupling between the two non-equivalent stacks the two Davydov components will have total transition dipoles of $\vec{P}^\beta(A)\pm\vec{P}^\beta(B)$. Then the absorption cross section and the polarization ratio for each band in the polarized spectra from Fig. 2 can be expressed by

1. the model parameters $\Delta_F^0$, $\Delta_{CT}$, $M$ and $\epsilon_+$

2. the relative CT transition dipole $p_{CT}^{rel}$

3. the angle $\gamma$ of the CT transition dipole

4. the angle $\phi$ between the nonequivalent molecules.

For MePTCDI $\phi = 36°$ is known from the crystal structure [6], which leaves five unknown parameters.

These five parameters we used to fit the experimental peak positions and intensities. A least square optimization gave one and the same solution independently on the initial parameter choice, and the found parameters do not sensibly depend on the weighting factors. We found the following values: $\Delta_F^0 = 2.23$ eV, $\Delta_{CT} = 2.17$ eV, $M = 0.10$ eV, $\epsilon_+ = 0.10$ eV and $p_{CT}^{rel} = 0.31$.

The model predictions for peak positions and polarization ratios according to these fitting parameters are shown in Fig. 2. The first three peaks are in good agreement with the experimental spectra. The relatively small polarization ratio of the second peak at 2.26 eV is particularly well described. This peak is most strongly affected by the CT exciton. Since the transition dipole of the CT exciton is roughly perpendicular to the molecular transition dipole, the CT- and Frenkel excitons influence the polarization ratio in the opposite way.

Only for the fourth peak at 2.72 eV, the model and experiment deviate considerably. This discrepancy is not surprising since already the peak analysis of the spectra obviously shows that several peaks contribute to this part of the spectrum. Here we reach the limits of our simple four-band model. In reality, there are several vibronic modes coupled to the electronic transition, which is clearly revealed in resonant Raman spectroscopy [7]. In the isolated molecule, these vibronic modes are almost degenerate. In the crystal, excitation transfer will remove this degeneracy and spread the modes over a wider energy range.

For PTCDA, the application of our model is much less reliable. We could not obtain samples that show polarized absorption spectra. This difficulty is not surprising since in PTCDA the two non-equivalent molecules are almost perpendicular to each other [2], so that the crystal itself is less anisotropic than in the case of MePTCDI. Furthermore, the bands in the absorption spectra (Fig. 1) are not as well separated. If we neglect the CT transition dipole and apply the same analysis as for MePTCDI we find model parameters in the same order of magnitude. Especially the Frenkel-CT-coupling constant $\epsilon_+ = 0.11$ eV and the general structure of the exciton bands is very similar to the situation in MePTCDI.

In order to obtain an independent confirmation for the proposed model we carried out a quantum chemical analysis. Using the semi-empirical ZINDO/S method as provided by HyperChem 5.01 (Hypercube, Inc., Waterloo, Canada), we calculated the excited states for a co-facial dimer of two MePTCDI or PTCDA molecules. We took the geometry from the geometry in the actual crystal structure [6]. The Hartree-Fock calculation of such a dimer gives molecular orbitals which are delocalized over both molecules. In order to reconstruct the Frenkel and CT states, which are defined by transitions between orbitals localized at one molecule, we projected the dimer orbitals onto monomeric orbitals. We then applied

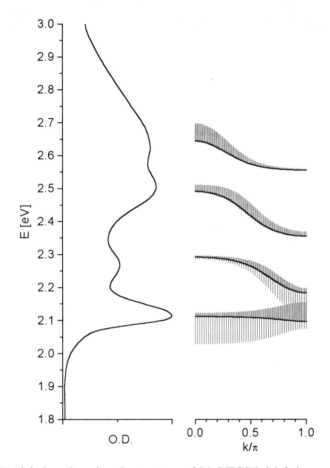

Figure 3: Modeled exciton band structure of MePTCDI (right) compared to the absorption spectrum from Fig. 1 (left). The upper shaded stripe at the bands visualizes the Frenkel part of the oscillator strength $|\vec{P}_{FE}^{\beta}|^2/p_{FE}^2$ from (25), the lower shaded stripe gives the CT part $|\vec{P}_{CT}^{\beta}|^2/p_{CT}^2$ from (27)

this projection to the excited dimer states calculated by configuration interaction. Thus, the excited states can be expressed as a superposition of Frenkel and CT excitons and the electronic interaction parameters can be obtained.

For both MePTCDI and PTCDA this dimer analysis shows the same scenario as predicted by our crystal model: The Frenkel and the CT excitons are energetically resonant and mix strongly. Furthermore, the CT exciton contributes a considerable transition dipole. For MePTCDI this analysis gives $M=0.20$ eV, $\epsilon_+=0.16$ eV and $p_{CT}^{rel}=0.27$. The order of magnitude of these parameters agrees with the fitting parameters from the model Hamiltonian, which confirms the plausibility of the model. From quantum chemical overlap calculations, Hennessy et al. [2] found similar values of $\epsilon_+ = 0.09$ eV for PTCDA.

Quantum chemistry also provides the parameter $\epsilon_-$, which is not accessible from absorption experiments. For MePTCDI we get $|\epsilon_-| = 0.05$ eV. With $\epsilon_-$ the full momentum dependent band structure for the excitons can be calculated. In order to give a qualitative picture, we scaled $\epsilon_\pm$ from the quantum chemical calculations so that $\epsilon_+$ corresponds to the model fit. The resulting band-structure is shown in Fig. 3. There the composition of the bands is also indicated by a schematic visualization of the $k$-dependent values $|\vec{P}_{FE}^{\beta}|^2$ from (25) and $|\vec{P}_{CT}^{\beta}|^2$ from (27).

In conclusion, we presented a simple model Hamiltonian which is capable to describe energetic positions, peak intensities and a varying polarization ratio for absorption spectra of quasi one-dimensional crystalline perylene derivatives. The exciton structure is essentially determined by a strong mixing of Frenkel and CT excitations. For the first time, a mechanism is introduced that leads to polarization dependent spectra due to the contribution of a CT transition dipole. The model can be semi-quantitatively confirmed by a quantum chemical analysis of the *inter*molecular interactions.

**Acknowledgements:** M.H. thanks the Deutsche Forschungsgemeinschaft for financial support. V.M.A. thanks the Technische Universität Dresden for hospitality and support. He also acknowledges partial support from Grant 97-1074 of the Russian Ministry of Science and Technology.

# References

[1] Forrest, S.R. (1997) Ultrathin organic films grown by organic molecular beam deposition and related techniques, *Chem. Rev.* **97**, 1793-1896.

[2] Hennessy, M.H., Soos, Z.G., Pascal Jr., R.A., Girlando, A. (1999) Vibronic structure of PTCDA stacks: the exciton-phonon-charge-transfer dimer, *Chem. Phys.* **245**, 199-212.

[3] Craig, D.P. (1955) The polarized spectrum of anthracene. Part II. Weak transitions and second-order crystal field perturbations. *J. Chem. Soc.*, 2302-2308.

[4] Agranovich, V.M. (1961) On the theory of excitons in molecular crystals, *Sov. Phys. Sol. State* **3**, 592 [*Fiz. Tverd. Tela* **3**, 811].

[5] Craig, D.P., Walmsley, S.H. (1968) *Excitons in Molecular Crystals. Theory and Applications*, Benjamin, New York.

[6] Hädicke, E., Graser, F. (1986) Structure of eleven perylene-3,4:9,10-bis(dicarboximide) pigments. *Acta Cryst.* **C42**, 189-195.

[7] Akers, K., Aroca, R., Hor, A.M., Loutfy, R.O. (1988) Molecular organization in perylene tetracarboxylic di-imide solid films, *Spectrochim. Acta* **44A**, 1129-1135.

# MODELLING OF BITHIOPHENE ULTRAFAST PHOTOPHYSICS: ELECTRONIC OSCILLATOR AND MOLECULAR GEOMETRY EVOLUTION

T. PÁLSZEGI[1], V. SZŐCS[2], M. BREZA[1] AND V. LUKEŠ[1]

[1] *Faculty of Chemical Technology, Slovak Technical University, Radlinského 9, SK-812 37 Bratislava, Slovakia*

[2] *Institute of Chemistry, Comenius University, Faculty of Natural Sciences, Mlynská dolina CH2, SK-842 15 Bratislava, Slovakia*

**Abstract.** Theoretical analysis of the ultrafast electronic-nuclear evolution of electronically excited *bithiophene* ($2T$) molecules, generalizable to longer oligothiophenes ($nT$) is presented. A molecular equation of motion (EOM) is derived, approximating the system of coupled Heisenberg equations of the one-electron density operators of $\pi$ electrons and of the normal mode phonon (vibrational) displacement operators. The initial electronic state of the molecular evolution is formed by the sum of the ground state one-electron density-matrix (DM) and of a vertically excited electronic oscillator (EO) DM. The $2T$ Hamiltonian used is the extended Hubbard one for electrons, harmonic for the nuclear part, and contains linear, non-local one- and two-electron-phonon couplings. The fs time-scale evolution of the $2T$'s geometry, charge densities, bond orders and EO energies, influenced by carbon-carbon bond stretchings and thiophene-thiophene inter-ring torsional angle deformation, is numerically investigated.

## 1. Introduction

At present the ultrashort (fs and early ps) multipulse optical spectroscopy techniques give the most detailed information on the photophysics and

e-mails: palszegi@cvt.stuba.sk, szocs@fns.uniba.sk, breza@cvt.stuba.sk and lukes@theochem.chtf.stuba.sk

*F. Kajzar and M.V. Agranovich (eds.), Multiphoton and Light Driven Multielectron Processes in Organics: New Phenomena, Materials and Applications, 135–150.*
© 2000 *Kluwer Academic Publishers. Printed in the Netherlands.*

non-linear optical (NLO) properties of $\pi$-conjugated molecules. Thiophene oligomers belong to the most popular conjugated systems investigated by means of ultrafast methods, being excellent model compounds for understanding $\pi$-electronic optical responses. With respect to oligothiophene, $nT$, and analogous systems, pump-probe (PP) studies have discovered, e.g. fs-scale vibrational- and torsional relaxation $(4T)$ [1]; fs planarization of an $S_1$ state (hexamethyl-$6T$ ) [2]; ultrafast Frenkel and charge-transfer exciton formation $(\alpha - 6T$ ) [3]; fs and ps dynamics of $S_1$ state ($\alpha$-$5T$) [4]. Recently, phonon modes of $\alpha - 6T$, strongly coupled to the lowest optical transition, were directly probed on fs time-scale by PP (coherent vibrational) technique, determining beating in the differential transmission on tens of fs time-scale [5]. These phenomena are, clearly, consequences of (molecular) structural and dynamical grounds, which can be revealed by theoretical modelling of spectroscopic information, providing deep insight into the optical structure-properties relationships.

The standard optical signal modelling techniques are based on the knowledge of many-electronic eigenstates or PES [6]-[11], and are cumbersome for larger molecules, e.g. for typical $\pi$-conjugated molecules. For this class of compounds new methods have been developed recently, based on one-electron real DM [12] or on the coupled electronic oscillator (CEO) [13] descriptions. Our modelling strategy, concentrated on the search of an effective method, follows the same direction. It represents a combination of approaches (i) used by *Mukamel and co.* for the CEO study of conjugated systems NLO [12], [13],[14], (ii) of one-electron and phonon DM evolution descriptions of *Lindenberg, Brown and West* [15] and (iii) of coupled el-ph pure state evolutions by *Förner* [16]. The photophysics in our attention corresponds to the sequence of (a) instantaneous laser pulse optical excitation and (b) ultrafast free electronic-nuclear evolution (i.e. electronic charge and bond density redistribution, molecular geometry changes). In the present paper the process (a) is represented by excited electronic oscillators (EO, see [12]) in a *fictitious* concept. The process (b) is modelled and calculated *explicitly* by means of coupled Heisenberg equations of one-electron density operators and of the normal phonon (vibrational) modes (cf.[15] and [16]).

The Introduction is followed by Section 2, where the model bithiophene ($2T$) Hamiltonian is specified. The derivation of the electronic-nuclear Equation Of Motion (EOM) of the $2T$ model is reported in Section 3. On the basis of $2T$ geometrical, vibrational and electronical information the 4-th order Runge-Kutta [17] numerical solution of the electronic-nuclear EOM is used in Section 4 to study the ultrafast $2T$ evolution. The extent of the molecular geometry relaxation, electronic charge density and bond order evolution and the changes of EO energies in time are analysed. Section 5 contains the conclusions.

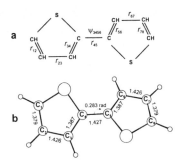

*Figure 1.* (a) Idealized ground state geometry of $2T$, bond lengths ($r_{nm}$) and inter-ring torsion ($\psi_{3456}$); (b) AM1 method ground state geometry, bond lengths (Å) and torsional angle (rad).

## 2. The bithiophene Hamiltonian

Before modelling ultrafast photophysical processes the Hamiltonian of the system under study is to be specified. Let us define the geometry of the $2T$ molecule. The distances $r_{nn+1} \equiv q_n^{str}$ between neighbouring ion cores (C-C stretching, str= stretch) for $n = 1, ..., 7$, and the torsional angle between the thiophene rings $\psi_{3456} \equiv q^{trs}$ (inter-ring torsion, trs= torsion), see Fig.1(a), represent the dominant geometry information. They contribute to the internal coordinates $\vec{R} = (q_1^{str} q_2^{str}, ..., q_7^{str}, q^{trs}, ...)$ and to the displacement $\vec{u} = \vec{R} - \vec{R}_0$ from the minimum energy geometry $\vec{R}_0$, i.e. $\vec{u} = (u_1^{str}, u_2^{str}, ..., u_7^{str}, u^{trs}, ...) \equiv (u_1, u_2, ..., u_8, ...)$. The 7 stretching and 1 torsional coordinates play active role in the el-ph coupling, but in the vibrational analysis all 48 Cartesian coordinates of the atoms forming the $2T$ molecule are involved. The chosen $\pi$-electron *semiempirical* Hamiltonian, $H_{2T}$, is the extended Hubbard one (see e.g. [18]) for the electron-electron Coulomb terms, Su-Schrieffer-Heeger (SSH, [19]) type for one-electron terms and it contains also Coulomb integral dependent (two-particle) non-local linear el-ph couplings [16], and can be written in a compact form

$$
\begin{aligned}
H_{2T} &= \sum_{n,m(8),\sigma} t_{nm}^0 \hat{c}_{n,\sigma}^+ \hat{c}_{m,\sigma} + \sum_{n(8)} U_n^0 \hat{n}_{n\uparrow}^+ \hat{n}_{n\downarrow} + \sum_{n(7),\sigma,\sigma'} V_{1n}^0 \hat{n}_{n,\sigma}^+ \hat{n}_{n+1,\sigma'} \\
&+ \sum_{n(7),\sigma,\sigma'} \left( 2a_n^{str} u_n^{str} \hat{P}_{nn+1,\sigma,\sigma'} + b_n^{str} u_n^{str} \hat{n}_{n,\sigma}^+ \hat{n}_{n+1,\sigma'} \right) \\
&+ \sum_{\sigma,\sigma'} \left( 2a^{trs} u^{trs} \hat{P}_{45,\sigma,\sigma'} + b^{trs} u^{trs} \hat{n}_{4,\sigma}^+ \hat{n}_{5,\sigma'} \right) + H_{ph}.
\end{aligned} \tag{1}
$$

Here $n(i) \equiv n = 1, ..., i$, further $t_{nm}^0$ are the one-electron resonance integrals, non-zero for the nearest neighbours. $U_n^0$ and $V_{1n}^0$ are one-center and two-center two-electron Coulomb integrals. $\{a_n^{str}, a^{trs}, b_n^{str}, b^{trs}\}$ is the set of the linear non-local el-ph coupling parameters. $\widehat{c}_{m,\sigma}^+ (\widehat{c}_{n,\sigma})$ creates (annihilates) a $\pi$ electron with spin $\sigma$, $\widehat{n}_{n,\sigma} = \widehat{c}_{n,\sigma}^+ \widehat{c}_{n,\sigma}$ is the electron charge density operator and $\widehat{p}_{nn+1,\sigma,\sigma'} = \frac{1}{2}(\widehat{c}_{n,\sigma}^+ \widehat{c}_{n+1,\sigma'} + \widehat{c}_{n+1,\sigma}^+ \widehat{c}_{n,\sigma'})$ is the bond density operator. The definitions and approximations leading to the $H_{2T}$ Hamiltonian may be found e.g. in [18]. The electron spin variables will be ignored in our treatment (the same simplification see in [12]).

The $H_{2T}$ Hamiltonian can be rewritten into fully second quantized form $\widehat{H}_{2T}$. To this purpose, $H_{ph}$ is replaced by second quantized phonon Hamiltonian $\widehat{H}_{ph}$ and the Cartesian atomic displacements from equilibrium position, $r_n$, substituted by displacement operators $\widehat{r}_n$ (for details see e.g. [16]). In particular, for $\widehat{r}_n$ and the normal phonon (vibrational) mode displacement operators $\widehat{b}_k + \widehat{b}_k^+$ the relation follows: $\widehat{r}_n = \sum_k \left[\frac{\hbar}{2M_n \omega_k}\right]^{0.5} U_{nk}(\widehat{b}_k + \widehat{b}_k^+)$.

Here $\widehat{b}_k$ $(\widehat{b}_k^+)$ creates (annihilates) the $k$th normal phonon (vibrational) mode, $\hbar$ is the Planck constant, $M_n$ denotes the $n$th diagonal element of the mass matrix $\mathbf{M}$, $\omega_k$ is the $k$th normal mode frequency, $U_{nk}$ is the $n$th element of the dimensionless normal mode eigenvector $\overrightarrow{U}_k = (U_{1k}, U_{2k}, ...)$. Using $\overrightarrow{U}_k$ the set $\left\{\widetilde{U}_{nk}\right\}_{n=1,...,7}$ can be expressed, representing the $k$th normal mode C-C bond stretching amplitudes, and $\widetilde{U}_{8k}$, giving the inter-ring torsional amplitude. More precisely, $\widetilde{U}_{ik}$ correspond to the differences of bond lengths (for i=1,...,7) and to the inter-ring torsion (for $i = 8$) in maximum normal mode amplitude displacement and in the ground state equilibrium molecular geometry. Formally, $\widetilde{U}_{ik} = d_k^{max}(C_i C_{i+1}) - d^{eq}(C_i C_{i+1})$, for $i = 1, ..., 7$ and $\widetilde{U}_{8k} = \psi_{3456,k}^{max} - \psi_{3456}^{eq}$. Here $d_k^{max}(C_i C_{i+1})$ is the $i$-th maximal dimensionless C-C bondlength in the k-th normal mode and $\psi_{3456,k}^{max}$ is the maximal inter-ring torsion. In stretching and inter-ring torsional coordinates one arrives the operator form $\widehat{u}_n = \sum_k \left[\frac{\hbar}{2M\omega_k}\right]^{0.5} \widetilde{U}_{nk}(\widehat{b}_k + \widehat{b}_k^+)$, where

$M$ is the mass of the ion-core, equal for all carbon ion-cores in present work. Then the $\widehat{H}_{2T}$ Hamiltonian reads

$$\widehat{H}_{2T} = \widehat{H}_\pi^0 + \sum_{n(7),k} \hbar\omega_k [2A_{nn+1k}^{str}(\widehat{b}_k + \widehat{b}_k^+)\widehat{p}_{nn+1} + B_{nn+1k}^{str}(\widehat{b}_k + \widehat{b}_k^+)\widehat{n}_n^+ \widehat{n}_{n+1}$$

$$+ 2A_k^{trs}(\widehat{b}_k + \widehat{b}_k^+)\widehat{p}_{45} + B_k^{trs}(\widehat{b}_k + \widehat{b}_k^+)\widehat{n}_4^+ \widehat{n}_5] + \widehat{H}_{ph} . \qquad (2)$$

Here $\widehat{H}_\pi^0$ is the one- and two electron part, described above in $H_{2T}$, containing the non-zero nearest-neighbour resonance and the extended Hubbard Coulomb terms. The set $\left\{A_{yk}^x, B_{yk}^x\right\}$ of k-th normal mode el-ph coupling parameters are given as $A_{nn+1k}^{str} = \frac{a_n^{str}}{\omega_k} \left[\frac{\hbar}{2M\omega_k}\right]^{0.5} \tilde{U}_{nk}$, $A_k^{trs} = \frac{a^{trs}}{\omega_k} \left[\frac{\hbar}{2M\omega_k}\right]^{0.5} \tilde{U}_{8k}$,

$B_{nn+1k}^{str} = \frac{b_n^{str}}{\omega_k} \left[\frac{\hbar}{2M\omega_k}\right]^{0.5} \tilde{U}_{nk}$ and $B_k^{trs} = \frac{b^{trs}}{\omega_k} \left[\frac{\hbar}{2M\omega_k}\right]^{0.5} \tilde{U}_{8k}$.

Let us continue with the identities $\Delta \hat{t}_{nn+1}^x = \sum_k \hbar\omega_k A_{nn+1k}^x (\hat{b}_k + \hat{b}_k^+)$

and $\Delta \hat{\gamma}_{nn+1}^x = \sum_k \hbar\omega_k B_{nn+1k}^x (\hat{b}_k + \hat{b}_k^+)$. The substitution of the sets $\left\{\Delta \hat{t}_{nn+1}^x\right\}$

and $\{\Delta \hat{\gamma}_{nn+1}^x\}$ into $\hat{H}_{2T}$ (Eq.(2)) leads to the Hamiltonian

$$
\begin{aligned}
\hat{H}_{2T} = {} & \sum_{n(7)} 2t_{nn+1}^0 \hat{p}_{nn+1} + \sum_{n(8)} U_n^0 \hat{n}_n^+ \hat{n}_n + \sum_{n(7)} V_{1n}^0 \hat{n}_n^+ \hat{n}_{n+1} + \hat{H}_{ph} \qquad (3) \\
& \sum_{n(7)} \left(2\Delta \hat{t}_{nn+1}^{str} \hat{p}_{nn+1} + \Delta \hat{\gamma}_{nn+1}^{str} \hat{n}_n^+ \hat{n}_{n+1}\right) + 2\Delta \hat{t}_{45}^{trs} \hat{p}_{45} + \Delta \hat{\gamma}_{45}^{trs} \hat{n}_4^+ \hat{n}_5 ,
\end{aligned}
$$

employed in the next Section for the derivation of electronic-nuclear EOM. The first line of this form of the $2T$ Hamiltonian represents the one- and two $\pi$-electronic part, $\hat{H}_\pi^0$. In the second line the el-ph part, denoting it by $\hat{H}_{el-ph}$, can be expressed as

$$
\begin{aligned}
\hat{H}_{el-ph} = {} & \sum_k \sum_{n(7)} \hbar\omega_k (2A_{nn+1k}^{str} \hat{p}_{nn+1} + \delta_{n,4} 2A_{45k}^{trs} \hat{p}_{45} \qquad (4) \\
& + B_{nn+1k}^{str} \hat{n}_n^+ \hat{n}_{n+1} + \delta_{n,4} B_{45k}^{trs} \hat{n}_4^+ \hat{n}_5)(\hat{b}_k + \hat{b}_k^+).
\end{aligned}
$$

Here the sum of the first two terms in the round bracket, multiplied by $\hbar\omega_k$, will be denoted by $\Delta \hat{t}_{nn+1}^k$. Similarly is $\Delta \hat{\gamma}_{nn+1}^k$ defined on the basis of the third and forth terms in $\hat{H}_{el-ph}$. Then the operator $\hat{G}_k$, playing important role in the next section is introduced as follows

$$
\hat{G}_k = \sum_{n(7)} \left(\Delta \hat{t}_{nn+1}^k + \Delta \hat{\gamma}_{nn+1}^k\right). \qquad (5)
$$

## 3. Derivation of the electronic-nuclear EOM

The derivation of the $2T$ one-electron DM $\rho(t)$ and molecular geometry evolution equations (i.e. electronic-nuclear EOM) influenced by an semiclassically described external electromagnetic field is based on the approximative solution of the following set of Heisenberg equations:

$$
i\hbar \,\dot{\hat{\rho}}_{nm} = \left[\hat{\rho}_{nm}, \hat{H}_{2T} + \hat{H}_{mol-f}\right], \qquad n,m = 1,...,8 \qquad (6)
$$

and

$$\dot{\hat{b}}_k = \frac{i}{\hbar}\left[\hat{H}_{2T},\hat{b}_k\right] , \qquad \dot{\hat{b}}_k^+ = \frac{i}{\hbar}\left[\hat{H}_{2T},\hat{b}_k^+\right] , \qquad k=1,...,42 \qquad (7)$$

where $\hat{\rho}_{nm} = \hat{c}_m^+\hat{c}_n$ is the ($\pi$-electron) density operator and Eq.(7) is defined for the 42 ground state normal (not translational and rotational) vibrational (phonon) modes of the all atoms forming the $2T$ molecule. The external field is described in dipole approximation and the corresponding molecule-field interaction Hamiltonian is given as follows (see [12]) $\hat{H}_{mol-f} = -E(t)\hat{P}$. Here $E(t)$ is the electric field intensity, $\hat{P}$ is the molecular polarization operator equal to $\sum\limits_{n,m,\sigma} \mu_{mn}\hat{c}_{m,\sigma}^+\hat{c}_{n,\sigma}$, where $\mu_{mn}$ is the transition dipole matrix element.

Several restrictions and approximations have to be performed to obtain *numerically tractable* electron DM and molecular geometry EOMs. The steps of the derivation of the EOMs, which is in the present paper very informative (a more detailed one will be presented in our forthcoming article) are taken and combined from [12],[15] and [16]. First, let us suppose that the density operator of our model $2T$ el-ph system is approximated by a *factored* el-ph density operator $\hat{\rho}_{el-ph}(t)$, i.e. by $\hat{\rho}_{el-ph}(t) \simeq \hat{\rho}_{el}(t)\cdot\hat{\rho}_{ph}(t)$. In the present work, $\hat{\rho}_{el}(t) \cdot \hat{\rho}_{ph}(t)$ is expressed by pure states: by the total many-electron state $\mid \Psi(t)\rangle$, i.e. $\hat{\rho}_{el}(t) = \mid \Psi(t)\rangle \langle\Psi(t)\mid$, and by the phonon part $\hat{\rho}_{ph}(t)$ by $\hat{\rho}_{ph}(t) = \mid \Phi(t)\rangle \langle\Phi(t)\mid$, where $\mid \Phi(t)\rangle$ is a phonon state. We take $\hat{\rho}_{el}(t)$ in single Slater determinant representation ($\mid \Psi_{Sl}(t)\rangle$) and for $\mid \Phi(t)\rangle$ a coherent phonon state [15], i.e. $\mid \Phi_{coh}(t)\rangle = \exp\left[-b_k(t)\hat{b}_k^+ + b_k^*(t)\hat{b}_k\right]\mid 0\rangle_{ph}$ (here $\mid 0\rangle_{ph}$ is the phonon vacuum state and $b_k(t)$ represents the $k$-th normal mode displacement).

The next step to derive the EOM is the one-electron DM representation of the above electronic density operator Heisenberg equation (Eq.(6)) in correspondence with *TDHF approximation*, analogously to [12]. First, the operation $\langle\Psi_{Sl}(t) \mid \cdot \mid \Psi_{Sl}(t)\rangle = Tr_{el}[\cdot \mid \Psi_{Sl}(t)\rangle\langle\Psi_{Sl}(t)\mid]$ is to be applied on the Eq.(6) for the $\hat{H}_{2T}$ Hamiltonian given by the Eq.(3), denoting the matrix element $\langle\Psi_{Sl}(t) \mid \cdot \mid \Psi_{Sl}(t)\rangle$ by $\langle\cdot\rangle$. Introducing $\rho_{nm}(t) = \langle\Psi_{Sl}(t) \mid \hat{\rho}_{nm} \mid \Psi_{Sl}(t)\rangle$ one obtains

$$i\hbar \dot{\rho}_{nm}(t) = \sum_i \left[(t_{ni}^0 + \Delta\hat{t}_{ni})\rho_{im}(t) - (t_{im}^0 + \Delta\hat{t}_{im})\rho_{ni}(t)\right] \qquad (8)$$

$$+U_n^0\langle\hat{\rho}_{nn}\hat{\rho}_{nm}\rangle - U_m^0\langle\hat{\rho}_{mm}\hat{\rho}_{nm}\rangle + \frac{1}{2}\sum_{i\neq n}(\gamma_{ni}^0 + \Delta\hat{\gamma}_{ni})\left[\langle\hat{\rho}_{ii}\hat{\rho}_{nm}\rangle - \langle\hat{\rho}_{nm}\hat{\rho}_{ii}\rangle\right]$$

$$-\frac{1}{2}\sum_{i\neq m}(\gamma_{mi}^0 + \Delta\hat{\gamma}_{mi})\left[\langle\hat{\rho}_{ii}\hat{\rho}_{nm}\rangle - \langle\hat{\rho}_{nm}\hat{\rho}_{ii}\rangle\right] + E(t)\sum_i\left[\mu_{ni}\rho_{im}(t) - \mu_{im}\rho_{ni}(t)\right] ,$$

where $\Delta \hat{t}_{nn+1} = \Delta \hat{t}_{nn+1}^{str}$ for $n \neq 4$ and $\Delta \hat{t}_{nn+1} = \Delta \hat{t}_{nn+1}^{str} + \Delta \hat{t}_{45}^{trs}$ for $n = 4$ (for $\Delta \hat{\gamma}_{nn+1}$ analogously). Then in Eq.(8) the two-electron densities $\langle \hat{\rho}_{nm} \hat{\rho}_{ij} \rangle$ are factorized into products of single electron densities, i.e. $\langle \hat{\rho}_{nm} \hat{\rho}_{ij} \rangle = \rho_{nm}(t) \rho_{ij}(t) - \rho_{im}(t) \rho_{nj}(t)$ to obtain the TDHF approximation of EOM. The one-electron DM evolution equation has partially second quantized form and we write it down for the case of zero external field giving the *Random Phase Approximation (RPA)*. The corresponding EOM reads

$$i\hbar \, \dot{\rho} \, (t) = \left[ \hat{h}(t), \rho(t) \right] , \quad (9)$$

where the elements of the matrix operator $\hat{h}(t)$ are given by $\hat{h}_{nm}(t) = t_{nm}^0 + \Delta \hat{t}_{nm} + \delta_{n,m} \sum_l (\gamma_{nl}^0 + \Delta \hat{\gamma}_{nl}) \rho_{ll}(t) - (\gamma_{nm}^0 + \Delta \hat{\gamma}_{nm}) \rho_{nm}(t)$.

As the *last step* of the derivation of our EOM the $Tr_{ph}[ \cdot \mid \Phi_{coh}(t) \rangle \langle \Phi_{coh}(t) \mid ]$ operation has to be performed on the Eq.(9) to replace the operators $\Delta \hat{t}_{nn+1}^{str} = \sum_k \alpha_{nn+1}^{str,k} (\hat{b}_k + \hat{b}_k^+)$, etc by their expectation values. Analogously to [15] (i) the matrix elements $\langle \Phi_{coh}(t) \mid \hat{b}_k + \hat{b}_k^+ \mid \Phi_{coh}(t) \rangle$ are substituted by $\langle \Phi_{coh}(0) \mid \hat{b}_k(t) + \hat{b}_k^+(t) \mid \Phi_{coh}(0) \rangle$ on the basis of the relationship between Schrödinger and Heisenberg representations. Further, (ii) the explicit normal mode displacement operators are replaced by their integral form, integrating formally the Eqs.(7), similarly to [15], which gives $\hat{b}_k(t) + \hat{b}_k^+(t) = e^{-i\omega_k t} \hat{b}_k(0) + e^{i\omega_k t} \hat{b}_k^+(0) - \frac{2}{\hbar} \int_0^t d\tau \sin \omega_k(t-\tau) \langle \hat{G}_k^+(\tau) \rangle$. Finally, (iii) using that for coherent phonon states the normal mode displacement $b_k(t) + b_k^*(t)$ is equal to the expectation value of the operator $\hat{b}_k(t) + \hat{b}_k^+(t)$, gives the expression

$$b_k(t) + b_k^*(t) = \quad (10)$$

$$e^{-i\omega_k t} b_k(0) + e^{i\omega_k t} b_k^*(0) - \frac{2}{\hbar} \int_0^t d\tau \sin \omega_k(t - \tau) \langle \hat{G}_k^+(\tau) \rangle_{RPA} ,$$

where the RPA has been also applied to $\langle \hat{G}_k^+(\tau) \rangle$, i.e.

$$\langle \hat{G}_k^+(t) \rangle_{RPA} = \sum_{n(7)} \{ \hbar \omega_k \left( A_{nn+1k}^{str} + \delta_{n,4} A_{45k}^{trs} \right) (\rho_{nn+1}(t) + \rho_{n+1n}(t)) +$$

$$\hbar \omega_k \left( B_{nn+1k}^{str} + \delta_{n,4} B_{45k}^{trs} \right) (\rho_{nn}(t) \rho_{n+1n+1}(t) - \rho_{n+1n}(t) \rho_{nn+1}(t)) \} \quad (11)$$

Eqs.(10) and (11) indicate a quadratic dependence of the $b_k(t) + b_k^*(t)$ on the one-electron density matrix elements $\{\rho_{nm}(t)\}$. The integral in Eq.(10) represents the time-dependent shift in the average value of the normal mode displacements. The actual internal coordinate displacement, i.e. the persistent shift plus oscillations, is given using the result of [16] as follows

$$u_n(t) = \sum_k \left[ \frac{2\hbar}{M\omega_k} \right]^{0.5} \tilde{U}_{nk} \operatorname{Re} b_k(t) \quad . \tag{12}$$

The final one-electron DM EOM is represented by the set of 1st order integro-differential equations with variables in the 4th power, i.e.

$$i\hbar \, \rho \, (t) = [\mathbf{h}(t), \rho(t)] \quad , \tag{13}$$

where the matrix elements of $h(t)$ are given by the formula $h_{nm}(t) = t_{nm}^0 + \Delta t_{nm} + \delta_{n,m} \sum_l (\gamma_{nl}^0 + \Delta\gamma_{nl})\rho_{ll}(t) - (\gamma_{nm}^0 + \Delta\gamma_{nm})\rho_{nm}(t)$. Here only the $\Delta t_{nn+1}$ and $\Delta t_{n+1n}$ one-electron el-ph coupling terms are non-zero, as implied by the equation $\Delta t_{nn+1} = \sum_k \hbar\omega_k \alpha_{nn+1k}(b_k(t) + b_k^*(t))$ , where $\alpha_{nn+1k} = A_{nn+1k}^{str}$, for $n \neq 4$ and $\alpha_{nn+1k} = A_{nn+1k}^{str} + A_{45k}^{trs}$ for $n = 4$. The form of the two-electron non-zero $\Delta\gamma_{nn+1}$ and $\Delta\gamma_{n+1n}$ el-ph coupling contributions is analogous, expressed on the basis of the formulas $\beta_{nn+1k}$ (defined using $\left\{ B_{nn+1k}^x \right\}$) and $b_k(t) + b_k^*(t)$.

Comparing Eqs.(10,11,13) it implies, that the solution of the electronic-nuclear EOM (Eq.(13)) is dependent on the calculation of normal mode displacements (Eq.(10)). Then, on the basis of the formula given by Eq.(12), it is possible to get the time-evolution of the C-C stretching and inter-ring torsional coordinates of the $2T$ molecule. The solution of Eq.(13) gives us also the charge density and bond order redistributions during the material evolution between the light pulses. In every time moment it is available to calculate the EO energies (see Sec.4), in correspondence with the actual geometry, which are important for the optical transitions caused by a next light pulse. An important comment: the influence of semiclassical electromagnetic field can be included substituting the last term of Eq.(9) into the one-electron DM evolution (Eq.(13)). In this way could be pulsed optical excitation processes numerically modelled with little difficulty.

## 4. Numerical analysis of the 2T ultrafast photophysics

Let us begin our analysis with the parameters involved in the $H_{2T}$ Hamiltonian (Eq.(1)). They are dependent on the ground state molecular geometry obtained by AM1 [21] method. The corresponding values of C-C bond lengths and of the thiophene-thiophene torsional angle (given by the angle between the normal vectors to the thiophene rings) are in Fig.1(b). The basic value, $t_0$, for the resonance integrals $t_{nm}^0$ has been set to 3.4 eV, multiplied by factors 0.7, 0.8, 1 and 0.75, going from the peripheral

C-C bonds to the bond connecting the thiophene rings. The multiplicative factors have been chosen on the basis of thiophene resonance parameters in [22]. $t_0$ is larger (by 1 eV) as the usual value of resonance integrals for polyene systems (e.g. in [12] and [23]), but the stability of the used Runge Kutta method called for this shift. The diagonal Coulomb term, $U_i^0$, is given by $\frac{U_0}{\epsilon}$ ($U_0 = 11.13$ eV and the semiempirical screening parameter $\epsilon = 1.5$, see [12]The value of $U_0$ has been reduced by factor 0.8 for the more diffuse first, 4th, 5th and 8th carbon atoms. The nearest-neighbour two-particle Coulomb terms, $V_{1i}^0$, is defined by the Ohno formula $\gamma_{nm}^0 = \frac{U_0}{(1+(\frac{r_{nm}}{a_0})^2)^{0.5}}$ [12] will be used, $a_0 = 1.2935$ Å [12] and $r_{nm}$ is the distance of the carbons in equilibrium geometry. The 4-th two-particle Coulomb term, $V_{14}^0$, has been multiplied by $\cos(\psi_{3456})$ to include the torsional effect. The stretching determined one particle el-ph couplings have been chosen linearly dependent on the C-C bond length displacement (by factors $a_n^{str} \in \langle 0.5; 4 \text{ eV/Å}\rangle$); and sinusoidally on $\psi_{3456}$ (factor $100a^{trs}$ for $a^{trs} \in \langle 0.5; 4 \text{ eV/Å}\rangle$). The two-particle, stretching dependent el-ph couplings $\{b_n^{str}\}$ have been defined by the first derivative of the Ohno formula, while their inter-ring torsion counterpart, $b_n^{trs}$, by the expression $100\sin(\psi_{3456})V_{14}^0$. The values of $\{b_n^{str}\}$ and $b_n^{trs}$ have been under 10 eV/Å and 10 eV/rad, respectively. The optimization of the parameter set will be performed in our future study e.g. on the basis of approaches in [24] or [23].

The limitations of the standard theoretical modelling of molecular photophysics are caused by time and hardware consuming calculations of the excited electronic PES. From this reason the simple and fast CEO method based adiabatic electronic energy calculation [25] is desirable, that is still *enough exact* (see [14]) in comparison with the etabled quantum chemical methods. Very recent article [26] illustrates well the potential of the CEO approach, reporting a vibrational dynamics study on the excited state PES of a protonated Schiff base, determined by the geometry dependent EO energies. To apply the EOM to $\pi$-conjugated systems, derived in the previous section, EOs are not inevitable. But, arguing with the effectiveness of EO representation for larger systems, we follow this strategy, at least in the present and in a next papers.

The CEO approach [13] is a reduced description of the molecular electronic structure which uses only the relevant excited state information contained in the one-electron density matrix $\rho$. An EO DM, $\xi_\nu$, is a linear stationary correction to the ground state one-electron DM, $\bar{\rho}$. This means, $\bar{\rho} + \xi_n$ is an excited stationary solution of the RPA approximated and linearized one-electron DM Heisenberg equation, following form Eq.(13), omitting the el-ph interaction. The set $\{\xi_\nu\}$, with eigenvalues (energies) $\{\Omega_\nu\}$, has a counterpart $\{\xi_{-\nu}\}$, having opposite (negative) energies, $\{-\Omega_\nu\}$. The EOs represent an approximation to the exact transition DMs, i.e. for their

144

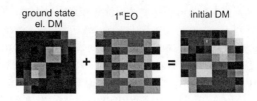

*Figure 2.* 2T initial one-electronic DM equal to the sum of ground state DM and $1^{st}$ EO DM.

matrix elements:

$$(\xi_v)_{nm} \doteq \langle g \mid \widehat{c}_n^+ \widehat{c}_m \mid \nu \rangle \quad . \tag{14}$$

Here $\mid g \rangle$ and $\mid \nu \rangle$ are the exact ground and excited many-electronic eigenstates, respectively. The molecular optical response is possible to express in terms of coupled EOs, using TDHF solution of the molecule-field Heisenberg equation (see e.g. [12]).

It is very advantageous to calculate the HF molecular orbitals (MO) first and then the EOs [12]. From this reason, using the electronic part, $\widehat{H}_\pi^0$, of the Hamiltonian $\widehat{H}_{2T}$ (Eq.(2)), iterative procedure [12] has given the set of MOs of the 2T molecule. The 16 positive energies of EOs, $\{\xi_v\}$, have been obtained by solving the linear approximation of the RPA (TDHF) DM equation of motion, given for the one-electron DM in MO basis, corresponding to electron-hole transition DM elements. Collecting the EOs, the structural information basis for the forthcoming numerical solution of Eq.(13) is complete. Accepting the results of the previous ultrafast dynamics studies of $\pi$-conjugated systems [1]-[5], it is possible to formulate a qualitative photophysical model. One can suppose, that in the time $0^+$ the 2T molecule is yet vertically excited, i.e. saving the ground state geometry the molecule is excited into a new one-electron DM state. The DM of this state, in general, can be represented by a linear combination of EOs. The geometry evolution starts and is coupled to the one-electron DM dynamics. The initial values of the normal mode displacements are from the set $\{b_k(0)\}$, together with their complex conjugates. In the time $t$ the molecule achieves a geometry, corresponding to a new set of normal mode displacements.

The numerical solution of the EOM, Eq.(13), has been performed by the 4-th order Runge-Kutta method [17]. The initial normal mode displacements given by Eq.(10) were taken static. The initial one-electron DM,

$\rho(0^+)$ has been chosen in the form $\bar{\rho} + \xi_n$. An initial DM, represented by the linear combination $\bar{\rho} + C_n\xi_n + C_{-n}\xi_{-n}$, where $C_n$ and $C_{-n}$ are coefficients, would correspond to the vertical excitation into the electronic state $|g\rangle + C_n|n\rangle$. Unfortunately, our numerical procedure after some femtoseconds of the time evolution become unstable for such an initial condition. In Fig.2 the initial DM of the $2T$ molecule is displayed for the case: $\rho(0^+) = \bar{\rho} + \xi_1$. The diagonal elements of $\bar{\rho}$ represent the $\pi$ electron charge densities on the carbons $(C_1, ..., C_8)$ and the non-diagonal elements are the coherences $\{\rho_{nm}, n \neq m\}$ (the values are increasing with the luminosity of the squares.

Fig.3 demonstrates the $2T$ geometry evolution (i) for initial electronic DM, formed by the first EO perturbed ground state (vertical excitation); (ii) for initial geometry, corresponding to the maximum amplitude of the C-C stretchings and inter-ring torsion in the first vibrational normal mode and (iii) for one-electron el-ph coupling, given by $\{a_n^{str}\} = a^{trs} = 4$ eV/Å. The frequency of the (ground state) AM1 normal mode chosen is $\omega_1 = 16$ cm$^{-1}$ $\equiv$ cca 330 fs and it is the torsion dominated one, i.e. the maximum amplitudes of the stretchings are near zero. Fig.3 shows that the bonds, which are the shorter ones in the initial (ground state) geometry ($1^{st}, 3^{rd}, 5^{th}$ and $7^{th}$ bond) are shortened after the evolution starts, while the longer bonds ($2^{nd}, 4^{th}$ and $6^{st}$) are elongated. The stretchings are oscillating in time with cca 20 fs period, modulated with a slower, cca 150 fs mode. These frequencies represent the collective behaviour of the normal modes. The average length of shorter bonds are shortened in time, while of the longer ones elongated. The initial, torsion corresponding to the ground state, i.e. cca 0.3 rad, is lowered, shaping $2T$ more planar as in the initial, $\bar{\rho} + \xi_1$, state. The evolution of charge densities and bond orders allow an other view into the ultrafast $2T$ photophysics. Fig.4 demonstrates the effect of one-electron el-ph coupling strengths on the charge densities of $C_5$ and $C_6$ carbons for initial condition, represented by vertical excitation of the $1^{st}$ EO and by $1^{st}$ normal mode determined (maximal amplitude) outgoing displacement. Here, the $\pi$ electron charge density on the i-th carbon is defined by $1 - 2\rho_{ii}$. The time dependence of the charge densities exhibit a little influence of the el-ph coupling, only the phase shifts are growing for longer time (see the insert in Fig.4). The beating in the charge density (Fig.4) has a 0.5-1 fs long period, corresponding to oscillations of cca 10000-5000 cm$^{-1}$. The calculations of the charge density dynamics for different initial EO have shown no systematic qualitative changes in their character. The situation is quite different for the bond order evolution. Although the influence of the el-ph coupling was the same as in the case of charge densities (i.e. slow phase shift for longer time) on the other side, the bond order dynamics dramatically depends on the initially excited EO. Figs.5(a) and 5(b) display

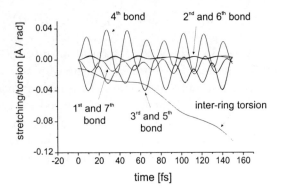

*Figure 3.* 2T geometry (bond lengths and torsional angle) evolution for $1^{st}$ EO excited initial DM (see in Fig.2) and for initial geometry, given by the $1^{st}$ vibrational normal mode maximum amplitude.

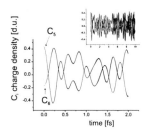

*Figure 4.* $\pi$-electron charge density evolution of $C_5$ and $C_6$ atoms in charge density units (d.u.). One-electron el-ph couplings: 0.5 eV, 1 eV, 2 eV and 4 eV, superimposed solid lines. Insert: the same for longer time.

the time dependences of the bond orders, defined by the sum $\rho_{ii+1} + \rho_{i+1i}$, for the bonds $C_4$–$C_5$ (interring bond, top of the panels) and $C_5$–$C_6$ (bottom of the panels), respectively. The EOs are denoted as follows: the $1^{st}$ and $3^{rd}$ by solid lines; $2^{nd}$ and $4^{th}$ by dotted lines. The pulsation of the bond orders can be seen, the widest one for the $2^{nd}$ EO initial excitation, and the pulsation frequency is increased for higher outgoing EOs. We have yet no explanation for this strange behavior of the 2T bond orders. The last figure (Fig.6) demonstrates the effect of the molecular geometry evolution on the EO energies for different one-electron el-ph couplings and for the initial $1^{st}$ EO excitation and the $1^{st}$ normal mode maximal amplitude geometry. With increasing coupling strength the EO energies are changing more and more dramatically in time. This should be the common consequence of

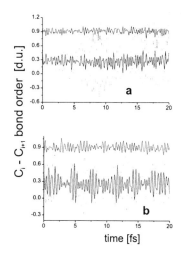

*Figure 5.* C-C $\pi$-electron bond order time-dependence; $C_4$-$C_5$ (on the top of panels) and $C_5$-$C_6$ (on the bottom of panels). Initially excited EOs: (a) $1^{st}$ and $2^{nd}$ (solid and dotted line,respectively), (b) $3^{rd}$ and $4^{th}$ (solid and dotted line, respectively).

bond length oscillations and torsional relaxation, mediated by the geometry dependence of resonance and Coulomb integrals in the Hamiltonian of $2T$ molecule. Comparing the positions of the energy minima and maxima in Figs.6(a), (b) and (c) with the extreme of stretching coordinates in Fig.3 one can see, that the time moments of extreme energies and stretchings are the same. The situation is more complicated for EO energy evolution in the case of el-ph coupling equal to 4 eV/Å (Fig.6(d)). The positions of EO energy minima and maxima are shifted and their number is larger as in the case of weaker el-ph coupling strengths. Let we conclude, that while the EO energies, in principle, well approximate the electronic excited state energies (see e.g. in [13]), the time evolution of distances and intersections of EO potential curves could play an important role to judge the chance vertically excite the $2T$ molecule in a time moment by a photon of definite wave length. The calculation of the EO energy dynamics could also give information on the geometries, corresponding to high probabilities of non-radiative transitions (e.g. internal conversion, intersystem crossing).

## 5. Summary and conclusions

A preliminary theoretical study of the ultrafast photophysics of a $2T$ molecule has been presented on the basis of an one-electron DM EOM and $2T$ model

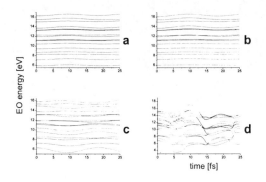

EO energy [eV]

time [fs]

*Figure 6.* 2*T* geometry evolution determined time-dependence of EO energies. One-electron el-ph couplings: (a) 0.5 eV/Å, (b) 1 eV/Å, (c) 2 eV/Åand (d) 4 eV/Å.

Hamiltonian. The derivation of the EOM, founded on a combination and modification of methods reported in papers [12], [15] and [16], is resulting in a set of 1st order integro-differential equations with one-electron DM elements, $\rho(t)$, in the 4th power (Eq.(13)). The EOM has been solved numerically by the 4-th order Runge-Kutta procedure for different initial excited EOs, geometries of the 2*T* and el-ph coupling strengths, giving $\rho(t)$ on the femtosecond time-scale. Using $\rho(t)$, the molecular geometry (bond lengths and inter-ring torsion) evolution of the model 2*T* molecule has been determined by means of Eqs.(10) and (12).

Our approach is built on several *dynamical* and *structural* model assumptions as well as on *approximations* in the solution of the model derived. The 2*T* Hamiltonian defining the molecular structure is semiempirical $\pi$-electronic one of extended Hubbard type for two-electronic terms, it contains linear, non-local one- and two-el-ph couplings, and is based on the set of ground state vibrational modes. The time evolution is restricted to the factorized one-electron DM and coherent phonon normal mode amplitude dynamics, i.e. two-electron and el-ph correlations are neglected, no thermal effects are involved. The approximations in the model calculations have structural and dynamical aspects. The el-ph coupling is restricted to the dominant stretching and torsional nuclear coordinates, only the nearest-neighbour Coulomb interaction is considered in the Hamiltonian, the parameters of which have not been optimized. The vibration spectrum and modes have been calculated by an all-valence electron Hamiltonian of the semiempirical AM1 method. The approximations in the solution of dynamics are connected with the definition of the initial state for the time evolution, represented by static coherent phonon state and by single EO.

The restriction to single EO excitation is caused by stability problems of the Runge-Kutta method used. The limitations in the parameter set chosen in the $2T$ Hamiltonian (the value of $t_0$) are connected with the numerical stability.

Despite these restrictions and approximations the analysis of the one-electron DM and molecular geometry evolutions gives interesting view into the $2T$ photophysics. Oscillations, formed by cca 20 and 150 fs components, have been registered in the C-C bond lengths, representing collective (wave-packet) evolution of the vibrational normal modes of the $2T$ molecule (Fig.3). The largest amplitudes of this collective mode correspond to the maxima of the stretching dominated normal modes and the average length of shorter C-C bonds are shortened in time, while these of the longer ones are elongated. While the $\pi$-electronic charge density time dependence is not sensitive on the el-ph coupling strength (small phase shifts in time, see Fig.4) or on the initially excited EO DM (not qualitative effect), the C-C bond orders of the $2T$ molecule show interesting behaviour. They are displaying amplitude pulsations dependent on the excited EO (Fig.5), representing some hidden behaviour of the one-electron DM evolution. On the other hand, the EO energy time dependence is significantly determined by the geometry (see Fig.6). The possibility to find conical intersections of the EO curves in some time moments represent a chance for new type of non-radiative transition studies, following from the computationally modest calculation of excited PES by the CEO method [14].

The directions to develop more *realistic* versions of our study are represented (i) by including the el-ph correlations using the memory function technique of [16], (ii) optimizing the parameters of the Hamiltonian presented, (iii) implementing a more exact Hamiltonian. Further, (iv) better description of the initial phonon and electronic DM is desirable, including also thermal effects and the explicit excitation by optical frequency field; (v) the influence of phonon-phonon coupling and phonon mode time-dependence should be included. Finally, (vi) the numerical procedure used for the solution of EOM has to be improved for better numerical stability of the calculations. After such improvements has the presented modelling method a great chance to help in the interpretation of ultrafast optical experiments, in particular of the [5] type. Our very approximative *preliminary* study, devoted to a model $2T$ molecule, shows clearly that this goal could be achieved.

## 6. Acknowledgements

The work has been supported by the Grant Agency VEGA of the Slovak Republic (projects No.1/6268/99, 1/4199/97, 1/4205/97 and 1/4012/99)

150

and by the Fonds zur Förderung der wissenschaftlichen Forschung, Wien (project P12566-PHY). V.S. thanks to the Österreichische Nationalbank (project No. 7318).

## References

1. Lanzani, G., Nisoli, M., De Silvestri, S. and Tubino, R. (1996) *Chem.Phys.Lett.* **251**, 339.
2. Lanzani, G., Nisoli, M., De Silvestri, S. and Tubino, R. (1996) *Synth.Met.* **76**, 39.
3. Lanzani, G., Frolov, S., Nisoli, M., Lane, P.A., De Silvestri, S., Tubino, R., Abbate A. and Vardeny, Z.V. (1997) *Synth.Met.* **84**, 517.
4. Sugita, A., Shiraishi, Y. and Kobayashi, T. (1998) *Chem.Phys.Lett.* **296**, 365.
5. Cerullo, G., Lanzani, G., Muccini, M., Taliani, C. and De Silvestri, S. (1999) *Phys.Rev.Lett.* **83**, 231.
6. Chen, G., Mukamel, S., Beljonne, D. and Bredas, J.-L. (1996) *J.Chem.Phys.* **104**, 5406.
7. Mukamel, S. (1995) *Principles of Nonlinear Optical Spectroscopy*, Oxford University Press, New York.
8. Yan, Y.J. and Mukamel, S. (1988) *J.Chem.Phys.* **88**, 5735.
9. Schen, Y-C. and Cina, J.A. (1999) *J.Chem.Phys.* **110**, 9793.
10. Raab, A., Worth, G.A., Meyer, H.-D. and Cederbaum, L.S. (1999) *J.Chem.Phys.* **110**, 936.
11. Kühn, O., Renger, T., May, V., Voigt, J., Pullerits, T. and Sundström, V. (1997) *Trends in Photochem. and Photobiol.* **4**, 213.
12. Takahashi, A. and Mukamel, S. (1994) *J.Chem.Phys.* **100**, 2366.
13. Tretiak, S., Chernyak, V. and Mukamel, S. (1998) *Int.J.Quant.Chem.* **70**, 711.
14. Tretiak, S., Chernyak, V. and Mukamel, S. (1997) *J.Am.Chem.Soc.* **119**, 11408.
15. Brown, D.W., Lindenberg, K. and West, B.J. (1987) *Phys.Rev. B* **35**, 6169.
16. Förner, W. (1998) *J.Phys.:Condens.Matter* **10**, 2631.
17. Press,W.H., Teukolsky, S.A., Vetterling, W.T. and Flannery, S.P. (1992) *Numerical Recipes in Fortran (The Art of Scientific Computing)*, University Press, Cambridge.
18. Baeriswyl, P., Campbell, D.K. and Mozumdar, S. (1992) in *Conjugated Conducting Polymers*, ed. Kiess, H.G. *Springer Series in Solid State Sciences* **102**, 7.
19. Heeger, A.J., Kivelson, S., Schiffer, J.R. and Su, W.P. (1987) *Adv.Mod.Phys.* **60**, 781.
20. Davydov, A.S. and Kislukha, N.I. (1973) *Phys.Status.Solidi b* **59**, 465.
21. Dewar, M.J.S. and Thiel, W. (1986) *AMPAC, Austin Model 1 Package*, Austin, TX: University of Texas.
22. Viallat, A. and Rossi,G. (1990) *J.Chem.Phys.* **92**, 4548.
23. Girlando, A., Painelli, A. and Soos, Z.G. (1993) *J.Chem.Phys.* **98**, 7459.
24. Utz, W. and Förner, W. (1998) *Phys.Rev. B* **57**, 10512.
25. Tsiper, E.V., Chernyak, V., Tretiak, S. and Mukamel, S. (1999) *J.Chem.Phys.* **110**, 8328.
26. Tsiper, E.V., Chernyak, V., Tretiak, S. and Mukamel, S. (1999) *Chem.Phys.Lett.* **302**, 77.

# "HOLE BURNING SPECTROSCOPY OF ORGANIC GLASSES"

## Application for the investigation of low temperature glass dynamics

B.M.KHARLAMOV

*Institute of spectroscopy, RAS*

*142092, Troitsk, Moscow region, Russia*

## Abstract

The recent results of low temperature organic glass relaxation on broad temperature and time scales, investigated via optical hole burning spectroscopy, are reviewed. A short introduction to the persistent hole burning effect and the burning mechanisms is given. The manifestation of glass relaxation in optical spectra in form of spectral diffusion (SD) is interpreted in terms of the two level system (TLS) model. The main characteristics of equilibrium SD in organic glasses are investigated and quantitatively described by the TLS model. Nonequilibrium effects, caused by various reasons such as sample aging or external perturbations are discussed. The last section presents the new observations of SD on PMMA samples at relatively high temperature, which can not be explained in a framework of TLS model.

## 1. Introduction

Spectral hole burning is a common way for elimination of inhomogeneous broadening in all kinds of spectra. Depending on the investigated systems, different techniques are used for hole burning. For example, the most used method in gas spectroscopy is the saturation of an optical transition – population hole burning. In the present article a particular region of solid state spectroscopy - the persistent hole burning (PHB) and its application to investigation of low temperature organic glass relaxations – is presented. Therefore, the discussion will be limited by organic systems only, although the hole burning spectroscopy of organic and inorganic glasses have many common points.

PHB spectroscopy arose from the spectroscopy of intramolecular optical transitions of organic molecules embedded in a solid matrix. In the course of the investigation of the nature of broad-band optical spectra of complex organic molecules [1-3] the PHB effect was discovered in amorphous [4] and crystalline [5] matrices. Shortly after, PHB turned

*F. Kajzar and M.V. Agranovich (eds.), Multiphoton and Light Driven Multielectron Processes in Organics: New Phenomena, Materials and Applications, 151–166.*

into a powerful technique for high resolution spectroscopy of complex molecules [6-11] in a solid surrounding. A dramatic increase of spectral resolution (up to 4 – 5 orders of magnitude) made the PHB technique very convenient for investigation of external field effects: electric [12,13], magnetic [14-16] and pressure [17]. High resolution spectra of impurity molecules are also very sensitive to intimate details of impurity-host interaction. Numerous investigations of the homogeneous line width temperature dependence shed light on many aspects of the electron-phonon interaction in crystalline and amorphous solids [7-10,18-21]. Investigations of optical dephasing in glasses proved that anomalies in low temperature thermodynamical, acoustical and dielectric properties of glasses (see, for references, reviews [22-24]), persist also in the optical region. It was found, in particular, that holes in chromophore spectra in glasses are time dependent [25]. That gave rise to extensive examinations of glass dynamics via optical hole burning [26] and photon echo spectroscopy [27-29]. Both methods proved to be very efficient in this field. The main subject of the present review is the discussion of the results on the long-time organic glass relaxations, obtained in the last years via optical hole burning spectroscopy.

## 2. Principles of hole burning spectroscopy

### 2.1 HOLE PROFILE AND BURNING KINETICS

The hole burning spectroscopy of impurities in solids at low temperatures is based on two fundamental facts. The first one is a large inhomogeneous broadening, which exceeds the homogeneous spectral line width at helium temperatures by 3 – 5 orders of magnitude in organic glasses and up to 2 – 3 orders of magnitude in crystals. The inhomogeneous broadening smoothes the fine structure of optical spectra of impurity molecules and makes the techniques, such as hole burning, very attractive for fine structure measurements of impurity optical spectra. The second fact is the existence of a number of low temperature photoreactions, providing a technical possibility to burn persistent holes. Both facts were revealed firstly in [4,5]. A wealth of new information about the nature of inhomogeneous and homogeneous broadening of impurity spectra along with the types of photoreactions was obtained in the course of later investigations (see, reviews [9-11,18-21,29-31]).

The homogeneous optical band of an impurity in a solid at low temperatures consists of a narrow zero phonon line (ZPL), which has a lorentzian shape, and a broad phonon wing (PW). Their relative intensities are expressed via Debye-Waller factor:

$$\alpha(T) = \frac{I_{ZPL}}{I_{ZPL} + I_{PW}} \tag{1}$$

where $I_X$ is the integral intensity of ZPL or PW correspondingly.

$\alpha$ is temperature dependent. It reaches a finite value at $T \to 0$ and tends to zero at increasing temperature. That means, at temperatures above 70 - 100 K the ZPL disappears in organic systems and all intensity of the optical band is concentrated in the structureless PW. This limits the application range of the site selection spectroscopy in general, and the persistent hole burning method in particular.

The burning of a persistent hole is technically very simple: The sample should be irradiated by a narrow-band excitation source with a moderate intensity (the burning fluences, depending on the temperature and the burning efficiency, are usually in the region between $\mu J/cm^2$ and $mJ/cm^2$). The photoreaction causes a frequency selective absorption bleaching, in other words, the occurrence of a spectral hole. The hole growth in a simplified case of an absence of back reactions and a very broad inhomogeneous band can be described as:

$$\Gamma(v - v_b, t) = \int_{v_0} dv_0 \cdot \varepsilon(v - v_0) \cdot n(v_0) \cdot [1 - \exp(-P\phi t \cdot \varepsilon(v_0 - v_b))] \qquad (2)$$

where: $\varepsilon(v)$ is the homogeneous absorption band of the impurity molecule; $n(v)$ is the inhomogeneous frequency distribution function; $P$ is the burning power density; $\phi$ is the burning quantum efficiency; $v_b$ is the burning frequency; $t$ is the burning time.

If we are interested only in ZPL, we can neglect the part of the hole profile, connected with PW, which is usually well separated from ZPL[1]. In that case the only part of the hole, connected with a resonance burning through ZPL in the impurity absorption band, can be considered. For the low burning fluence limit we can expand the exponent in a series and keep only the first term. In this case the hole profile will be a Lorentzian with the double homogeneous line width:

$$\Gamma_{ZPL}(v - v_b, t) \approx n(v_b) \cdot \sigma \cdot P \cdot \phi \cdot t \frac{(2\Gamma_h)^2}{(v - v_b)^2 + (2\Gamma_h)^2} \qquad (3)$$

where: $\sigma$ is the peak ZPL absorption cross section and $\Gamma_h$ is the halfwidth of ZPL.

Equation (3) gives the basis for the measurement of the homogeneous line width of impurities via optical hole burning. There are many complications in real experiments. At increasing burning fluences the saturation effect will broaden the hole (see, for details [7,32,33]). Many other effects can broaden a hole and relation (3) will be broken: sample heating by the burning laser, power saturation etc. So the measurement of a nonsaturated hole requires a careful elimination of all above disturbing factors, which can be rather complicated in definite cases (see, for example [34]).

The above analysis is correct in the case of crystals, where a hole profile does not change spontaneously in time. But this is not true for glasses. Due to so called spectral diffusion, the hole broadens with time. This problem will be analyzed in detail further.

## 2.2 BURNING MECHANISMS

For the most of applications of the persistent hole burning technique the particular burning mechanism is irrelevant. However a few words should be said about the especially often used photoreactions. They can be roughly divided in two groups - intramolecular and intermolecular.

The broadly used burning mechanism, belonging to the first group is a photoinduced proton tautomerization of free-base porphyrins and phthalocyanines (see, for details and further references [26]). Formally this tautomerization is equal to a rotation of the molecule by 90°. In a low-symmetry matrix it induces a small absorption band shift (up to a few tens of wavenumbers). The dark tautomerization at He temperatures is very weak, that means, burned holes are very stable. Some porphyrins are enough stable with respect to dark tautomerization even at temperatures near 100 K. The quantum effi-

---

[1] The detailed analysis of a hole profile taking into account both PW and ZPL in impurity absorption band can be found in [7,9].

154

ciency of proton tautomerization depends on particular molecule structure and is usually in the region $10^{-3} \div 10^{-4}$. Porphyrins are very convenient for the use as optical chromophore probes in the low temperature investigation of organic systems. First, their absorption bands, situated in the red, are very suitable for the single-frequency laser excitation. Second, due to the intramolecular burning mechanism, any organic system, where porphyrins can be put into, is accessible for the hole burning. Third, with variations of molecular structure the absorption band of porphyrin molecules can be moved in a broad spectral range, roughly from 550 nm to 720 nm.

Another broadly used type of photoreaction is so called "nonphotochemical hole burning" (NPHB). This type of photoreactions belongs to intermolecular. In the most cases the existence of NPHB depends mainly on the properties of matrix, and, to a much lesser degree, on the impurity molecule. Strong NPHB exists in polar glasses and some polymers. In fact, NPHB exists almost in all organic glasses. The main difference between various glasses is the ratio of photoactive and photostable chromophore molecules. In alcohol glasses practically every chromophore molecule is photoactive, in some polymers, like polimethylmethacrylate (PMMA) the amount of burnable molecules does not exceed 15 – 20 %. NPHB burning quantum efficiency lies in the region $10^{-3} - 10^{-5}$. The substantial difference between NPHB and proton tautomerization is the temperature stability. The barrier heights in NPHB have a very broad distribution, starting from very low values [35,36]. Therefore, a dark back reaction in the case of NPHB takes place at any temperature. But since the hole filling is approximately a logarithmic function of time ([37] for references, see also [26]), at given temperature nonphotochemic holes also persist practically infinitely long.

NPHB was discovered in the very first experiment on persistent hole burning [4], but its nature is non-clear yet. The first model of NPHB was proposed in [38]. It was supposed, that impurity in amorphous matrix creates a specific defect state with double well potential. It forms an "extrinsic" TLS, strongly interacting with the impurity molecule. Photoinduced transitions between these TLS sublevels are responsible for NPHB. This model explains many properties of NPHB, such as its rather high universality for glasses and absence of NPHB in crystals. But, there are some characteristics of NPHB, which can not be quantitatively explained by above model (see, for more detail, [10, 21,26,34,39,40]).

As was pointed out, NPHB is also a relatively universal burning mechanism for glasses. It is broadly used in the cases, when a particular molecule, without internal phototransformation mechanisms should be used.

## 3. Spectral diffusion

Spectral holes are very sensitive to small changes in the environment. Numerous experiments with external electric and magnetic fields and external pressure have proven it. This sensitivity and the very long lifetime of spectral holes appeared to be very suitable for investigation of low temperature glass dynamics. This section will illustrate the application of the hole burning technique for investigation of low temperature glass dynamics on a very broad time scale.

## 3.1. GENERAL DESCRIPTION

More than twenty years ago the first experimental measurements of specific heat of amorphous quartz below 1 K showed the strong anomaly in its temperature dependence as compared with crystalline samples [41]. Very soon the anomalies were found also in heat conductivity, ultrasound attenuation etc. These anomalies appeared to be very typical for glasses, both inorganic and organic. Very new effects, such as time dependence of specific heat, phonon echo and heat release were discovered in course of extensive research of low temperature glass dynamics. All these phenomena were partially explained and partially forecasted by the so called TLS model, invented independently by Anderson at al. [42] and Philips [43].

The model postulated the existence of specific low energy local states with double well potentials, which fluctuate between two lowest energy states due to interaction with phonons. The structure of the TLS is presented in figure 1. The main properties of TLS can be described by two parameters. The commonly used are the TLS asymmetry $\Delta$ and the tunneling parameter $\lambda$. It is expressed in terms of other characteristic potential parameters:

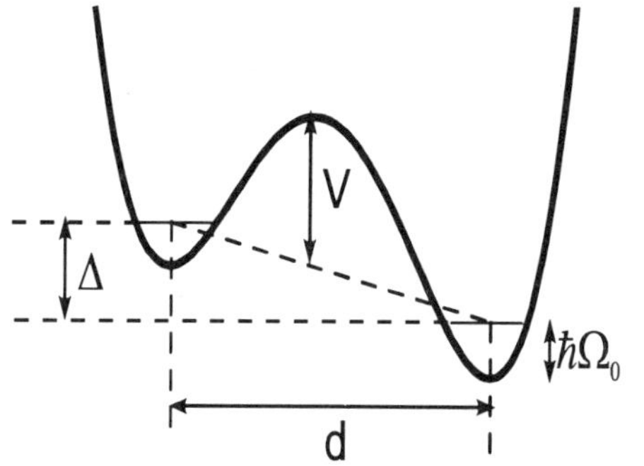

$$\lambda = \frac{d}{\hbar}\sqrt{2mV} \qquad 4)$$

$\lambda$ is related to tunneling matrix element $\Delta_0$:

$$\Delta_0 = \hbar\omega_0 \exp(-\lambda) \qquad (5)$$

The complete TLS energy splitting is:

$$E = \sqrt{\Delta^2 + \Delta_0^2} \qquad (6)$$

The very simple broad distribution of TLS in $\Delta$ and $\lambda$ was introduced in the first publications [42,43], which we will call the standard model distribution function:

$$P(\Delta,\lambda) = const \qquad (7)$$

Figure 1. The structure of TLS.

This distribution function gives a very good qualitative and in many cases even quantitative agreement with experimental data. Minor modifications of this function were used for better experimental data fit in particular publications.

Very specific for low temperature glass effect is spectral diffusion (SD). Spectral diffusion was observed firstly in spin glasses and theoretically described by Klauder and Anderson [44]. Later their formalism was modified by Black and Halperin for the description of phonon echo in glasses [45]. In optical spectra SD manifests itself in spontaneous broadening of any preselected monochromatic ensemble of chromophore transition frequencies. The interpretation of this effect in frames of TLS model is based on the interaction of chromophore molecules with the ensemble of spontaneously flipping TLS [25,46].

## 3.2. THEORETICAL DESCRIPTION OF EQUILIBRIUM SD.

We will follow in description of SD to Reinecke [46]. TLS flips cause the spontaneous diffusional frequency jumps of chromophore transition frequency. For the dipole-dipole chromophore-TLS interaction and low spatial density of TLS a Lorentzian diffusion kernel was obtained:

$$D(\omega - \omega_0) = \frac{1}{\pi} \frac{\Gamma(t,T)}{(\omega - \omega_0)^2 + (\Gamma(t,T))^2} \tag{8}$$

The time and temperature dependence of this Lorentzian kernel is given by:

$$\Gamma(t,T) = \frac{2\pi^2}{3\hbar} C \cdot \left\langle \frac{\Delta}{E} \cdot n(t,T) \right\rangle_{\lambda,\Delta} \tag{9}$$

Here C is the averaged value of the chromophore-TLS interaction constant and $n(t,T)$ is a probability for a single TLS to leave its initial state after a time $t$ at given temperature $T$. The calculation of $n(t,T)$ for thermally equilibrium ensemble of TLS is straightforward (see, for example, [26]) and gives:

$$n\left(\frac{E}{k_B T}\right) = \frac{1}{2} \sec h^2 \left(\frac{E}{2k_B T}\right) \tag{10}$$

After the ensemble averaging we come to :

$$\Gamma(t,T) = \frac{2\pi^2}{3\hbar} C \cdot k_B T \int dx \cdot \sec h^2(x) \int d\lambda \cdot \frac{\Delta}{E} P(\Delta, \lambda) \cdot \left[1 - \exp(-r(E, \Delta_0, T)t)\right] \tag{11}$$

For the standard model distribution function a simple approximate dependence follows:

$$\Gamma(t,T) \propto T \int dx \cdot \sec h^2(x) \cdot \ln(r_{max}(x,T)t) \tag{12}$$

where, $r_{max}$ is the maximum TLS relaxation rate, which corresponds to E = $\Delta_0$, and for the case of one-phonon assisted TLS relaxations is:

$$r_{max} = r(E = \Delta_0) = A\Delta_0^3 \cdot \coth\left(\frac{E}{2k_B T}\right) \tag{13}$$

So, in this case the spectral line broadening is a logarithmic function of time and approximately linear in temperature.

In real experiments the sum of homogeneous and diffusional broadening is always measured. For example, in hole burning experiments the measured hole width is:

$$\Gamma = 2\Gamma_H + \Gamma_{SD} \tag{14}$$

were $\Gamma_H$ is the homogeneous linewidth of the optical transition, the second term represents the diffusional part.

Due to limited time resolution the measurable part of the diffusional line broadening is the difference:

$$\Delta\Gamma_{SD}(t) = \Gamma(t) - \Gamma(t_0) \tag{15}$$

For this quantity the expression is even more simple:

$$\Delta\Gamma_{SD}(t) \propto T \cdot \ln\left(\frac{t}{t_0}\right) \tag{16}$$

So, SD behavior within the framework of TLS model with standard distribution function is very simple: about linear in temperature and logarithmic in time. It should be noted, that such simple dependence is valid only for the equilibrium TLS relaxation, for nonequilibrium situations the SD behavior is more complicated. Nonequilibrium TLS dynamics will be discussed later, in the next section the experimental data on equilibrium SD are presented.

## 3.3. EQUILIBRIUM SD: EXPERIMENTAL DATA

The first experimental observation of SD in optical spectra was realized in [47]. Observation of a persistent holes in chromophore absorption spectrum on a time scale covering 4 orders of magnitude proved qualitatively the logarithmic time dependence in SD line broadening. For a long time measurements of SD in optical spectra were rather fragmentary and could not serve as a real test of the TLS model. The first high-precision experiments on a long time scale ($10 - 10^6$ s) were fulfilled on PMMA at 0.5 and 1 K in [48] and later at 2 K in [49]. A substantial deviation from the logarithmic law was found. The equilibrium conditions were specially controlled in above experiments: the sample was kept at the experimental temperature for about 2 months before the measurements of the equilibrium SD. The whole SD time dependence in this experiment could be roughly separated into two regions. SD line broadening in the short time scale could be treated as logarithmic, the long time part looked like a square root asymptotic. Two theoretical models were offered for the explanation of nonlogarithmic SD behavior.

The first one takes into account the TLS-TLS interaction [50,51]. Such interaction causes the modification of the TLS distribution function, increasing the relative amount of slowly relaxing TLS clusters. The estimation of the upper temperature boundary of existence of such clusters, made in [50] was too low for the temperature range of our experiments. The later modification of the model, taking into account strong TLS-phonon interaction improved the theoretical estimation [51].

The second model supposed the existence of an additional group of TLS in PMMA, related to some specific molecular groups or impurities [52]. As a possible source of such specific TLS water molecules were suggested, their presence is almost inevitable in PMMA.

Further measurements were carried out at higher temperatures. They contradicted both models, even more, the doubts in a validity of TLS model for these data explanation appeared [53]. These experimental results will be discussed separately further.

The detection of nonlogarithmic long-time SD dynamics in PMMA provoked the question about its universality. Up to now very few objects are investigated in adequate time region, so the data are very fragmentary. Similar to PMMA behavior was found in protein [54]. The resemblance of SD in PMMA and protein includes not only the time dependence, but the absence of the aging effect at temperatures near 4 K. Another system, investigated on a broad time scale is an alcohol glass -glycerol/dimethylformamide. It shows a logarithmic SD time dependence on a time scale $10 - 10^6$ s at temperatures $0.1 - 4$ K as well as an aging effect. The possible connection between the absence of aging effect and nonlogarithmic SD we will discuss below.

## 3.4. NONEQUILIBRIUM EFFECTS

In this section we limit our discussion only by the scope of the TLS model. That means, the essentially nonequilibrium nature of a glassy state will not be taken into consideration and only the influence of nonequilibrium population of TLS ensemble on SD time dependence will be analyzed. It is known that TLS relaxation rates at He temperatures are distributed in a huge time interval, from picoseconds till, probably, months or years. In that case, a TLS ensemble after the sample cooling comes in the equilibrium with phonons very slow. Practically, a part of the TLS always remains in a nonequilibrium state on a real experimental time scale, measured by hours or days. Therefore it is important to know, what is the influence of such partial nonequilibrium on SD, how it depends on sample thermal history, etc.

The first attempt to take into account the influence of the sample cycle heating on SD in framework of TLS model was done in [55,56]. Later a similar approach was used in [57,58,26] for general analysis of nonequilibrium effects in SD line broadening. Here we will follow [26].

Let us suppose that initial ($T_i$) and final ($T_f$) temperatures of the sample during the experiment are not the same. In that case, the probability for TLS to change its state after the temperature change from $T_i$ to $T_f$ can be written in the form:

$$n\left[\left(\frac{E}{k_B T_f}\right),\left(\frac{E}{k_B T_i}\right)\right] = \frac{1}{2}\left[1 - \tanh\left(\frac{E}{2k_B T_i}\right) \cdot \tanh\left(\frac{E}{2k_B T_f}\right)\right] \quad (17)$$

This expression converts into eq.(10), if $T_i = T_f$. It is convenient to transform it into the sum of two terms. We denote the first one as "equilibrium":

$$n_e\left(\frac{E}{k_B T_f}\right) = \frac{1}{2}\left[1 - \tanh^2\left(\frac{E}{2k_B T_f}\right)\right] = \frac{1}{2}\sec h^2\left(\frac{E}{2k_B T_f}\right) \quad (18)$$

and the second one as "nonequilibrium":

$$n_{ne}\left[\left(\frac{E}{k_B T_f}\right),\left(\frac{E}{k_B T_i}\right)\right] = \frac{1}{2}\tanh\left(\frac{E}{2k_B T_f}\right) \cdot \left[\tanh\left(\frac{E}{2k_B T_f}\right) - \tanh\left(\frac{E}{2k_B T_i}\right)\right] \quad (19)$$

The physical sense of such designation is evident: the first term coincides with the probability of TLS flip in equilibrium case, the second one vanishes at equilibrium conditions.

The time dependence of SD line broadening depends not only on $T_i$ and $T_f$, but on the thermal history of the sample. Let us analyze some typical scenarios.

### 3.4.1. Thermal relaxation (aging effect).

a) **Predictions of the TLS model.** The sample is cooled down very fast from the temperature $T_i$ to the final temperature. Then the hole is burned with the time delay $t_d$. We can separate the SD line broadening into two parts, equilibrium ($\Gamma_e$) and nonequilibrium ($\Gamma_{ne}$) ones.

The equilibrium part is the same, as for the case when the sample would be kept infinitely long at $T = T_f$:

$$\Delta\Gamma_e(t,T_f) = Ak_B T_f \cdot \ln\left(\frac{t}{t_0}\right) \cdot \int dx \cdot n_e(x) \quad (20)$$

The nonequilibrium part depends on the sample history (see, for details [26,58]):

$$\Gamma_{ne}(t,T_i,T_f) = Ak_B T_i \cdot \ln\left(\frac{t+t_d}{t_d}\right) \cdot \int dx \cdot n_{ne}\left[\left(x\frac{T_f}{T_i}\right),x\right] \quad (21)$$

Also, the nonequilibrium part is also a logarithmic function of $t$, but its slope depends both on $T_f$ and $T_i$. It is easy to show that for the case $T_i \gg T_f$ $\Gamma_{ne} \propto T_i$. There are three time intervals with substantially different ratio of equilibrium and nonequilibrium parts of SD.

a) $t \ll t_d$. In this case $\Gamma_e \gg \Gamma_{ne}$ and equilibrium SD dominates.

b) $t \gg t_d$. In this situation $\Gamma_e \ll \Gamma_{ne}$, SD line broadening is governed by nonequilibrium TLS relaxation.

c) $t \approx t_d$. Both components are comparable.

The start of strong nonequilibrium SD is roughly a mirror reflection of the sample cooling time relative the hole burning time. Therefore, if we assume the infinite time distribution of TLS relaxation rates, the glass is never in equilibrium even within the framework of the TLS model. In that case the approximately equilibrium conditions exist for SD dynamics on a time scale $t \ll t_d$. If this condition is not fulfilled, strongly nonlogarithmic SD can be observed even in the case of standard model TLS distribution function.

**b) Experimental observations.** The first observation of aging effect in the optical region was carried out in [59]. Later a more detailed investigation of the main peculiarities of this effect were performed in [60]. It was pointed out in this publication, in particular, the independence of nonequilibrium SD on the experimental temperature on a broad scale. The systematic investigation of nonequilibrium SD and the explicit model treatment of the effect was done in [58] on alcohol glass samples. Figure 2 represent the results of these measurements. Strongly nonlogarithmic behavior is clearly seen due to a transition from equilibrium to nonequilibrium SD regime. The data are in a very good agreement with the above model calculation. The initial TLS ensemble temperature ($T_i$) was used as fitting parameter at the data approximation. Because it does not coincide with the initial temperature of the sample before cooling, it was named the "effective temperature" of the TLS ensemble. The physical sense of this parameter is rather simple. At this temperature the relaxation rates of the main part of TLS ensemble become very slow as compared with the sample cooling time. It proved to be very reliable to the variation of experimental conditions: time interval between sample cooling and hole burning $t_d$ and experimental temperature $T_f$. The value of $T_i$, obtained from the data fitting, reflects the average population in TLS ensemble immediately after the sample cooling. The value of $T_i$ for alcohol glass for experimental temperatures $T_f$ in the region $0.1 - 0.8$ K is about 12 K. For $T_f = 4.2$ K $T_i \approx 18$ K was

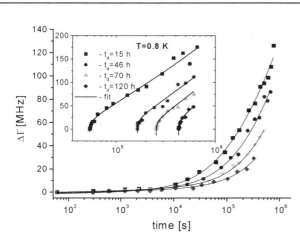

Figure 2 Aging effect at T=100 mK. Time dependence of hole width for 4 holes burned at different delay times $t_d$ are 19 h, 66 h, 114 h and 140 h from left to right respectively. The insert present the 800 mK data for the same system. The data are plotted here over a common log(t) axis with origin at the time when T = 800 mK was reached. (The data are taken from 58)

obtained [54]. The values of $T_i$, obtained for PMMA at $T_f = 0.1 - 1$ K, are about 4 K. The same value of $T_i$ was obtained for protein samples [61]. As follows from the above definition of effective temperature, the aging effect should not exist at temperatures above $T_i$. And that was the case in the experiments on PMMA [62] and protein [54] samples at temperatures around 4 K.

### 3.4.2. Thermal cycling.

a) **Model predictions.** A cyclic change of the sample temperature is a powerful tool for glass dynamics investigation on a broad temperature region. Direct SD measurement at temperatures above 4 - 6 K is difficult, because the homogeneous line width grows with increasing temperature much faster then diffusional part. But homogeneous component is completely reversible on a time scale of the sample temperature change. The behavior of diffusional component is more complicated. Let us analyze the following scenario.

a) time interval between the hole burning (at $t=0$) and the beginning of the thermal cycle: $0 < t < t_1$. The sample temperature is $T_f$.

SD line broadening in this time is equilibrium and is described by eq.(20) with $T = T_f$.

b) time interval, when the sample temperature was fast increased to $T = T_c$: $t_1 < t < t_2$. Equilibrium part of SD is described by the same equation, but, due to temperature dependence of TLS relaxation rates, all relaxations accelerate. This effect is rather small as compared with nonequilibrium SD which appears due to the temperature change.

c) time interval after the end of thermal cycle: $t > t_2$.

The equilibrium component is, as before, described by eq.(20). The nonequilibrium part of SD has now the negative slope:

$$\Gamma_{ne}(t,T_f,T_c) = A k_B T_f \int dx \cdot n_{ne} \left[ x, \left( x \frac{T_f}{T_c} \right) \right] \cdot \ln \left[ 1 + \frac{r_{max}(x,T_c) \cdot (t_2 - t_1)}{r_{max}(x,T_f) \cdot (t - t_2)} \right] \quad (22)$$

The equation (22) makes evident the following statement: SD line broadening due to thermal cycle should be almost completely reversible. As was firstly pointed out in [55], the very good test of the temperature boundaries of TLS model validity should be investigation of spectral hole width reversibility in thermal cycling experiments. But in practice it is not so simple. The relaxation to unperturbed equilibrium SD takes time, proportional to thermal cycle duration and dependent on $T_c$, and what is especially important, on the TLS relaxation rate temperature dependence. This "memory time" can be qualitatively estimated as:

$$t_m \approx (t_2 - t_1) \frac{r(x,T_c)}{r(x,T_f)} \quad (23)$$

For one-phonon-assisted TLS relaxations the temperature dependence of relaxation rates is rather weak, therefore at very low temperatures $t_m$ should be comparable with the thermal cycle duration. But in the activation regime it is an exponential function of temperature and $t_m$ in that case can be very long. In that case it would be hard to distinguish between a very long reversible relaxation and an irreversible hole broadening, which would designate the deviation from TLS-like SD behavior.

**Experiment.** Many experiments with thermal cycles were performed in the last decade [63-67]. But in these experiments only a hole broadening as a function of cycle temperature was measured without monitoring its time evolution. Therefore, the very interesting question about reversibility of SD line broadening in thermal cycles was not investigated. The first observation of diffusional hole narrowing regime after the sample heating was made in [57]. But experimental conditions in this experiment were not enough pure for precise quantitative comparison with the model: hole burning and sample heating were made by one laser pulse. The measurements of a hole narrowing after the "well-defined" thermal cycle were fulfilled in [58] on alcohol glass samples. After one hour long thermal cycle (0.1 K $\rightarrow$ 4 K $\rightarrow$ 0.1 K) the hole narrowing was monitored during one week, and even after this time the hole width did not reach the asymptotic value. The detailed investigation of the irreversible SD hole broadening in experiments with short heat pulses were carried out on polivinylbutyral (PVB) samples in [68]. These measurements showed that SD in PVB can be satisfactory described by TLS model up to temperatures near 20 K, at higher temperatures SD becomes much stronger as compared with TLS model predictions.

### 3.4.3. Electric field cycle induced SD.

a) **Theory.** The nonequilibrium can be created in TLS ensemble not only by the temperature change, but also via an external perturbation, such as electric field. Indeed, TLS in many glasses have substantial dipole moment. In external electric field the energy $E$ of TLS with electric dipole moment $\mu$ will be changed to:

$$E_f = E - \vec{\mu}\vec{F} \tag{24}$$

Due to the field-induced energy change the TLS ensemble comes out of equilibrium. If an electric field cycle would be performed, we become electric-field induced nonequilibrium SD. It will cause the hole broadening when the field is on and the nonequilibruim part will cease after the end of a field cycle. The probability for TLS to change its state when electric field is on can be written in the form (see, for details [26,69,70]):

$$\tilde{n}_{ne}(T,E,F) = \tfrac{1}{2}\tanh\left(\frac{E}{2k_BT}\right)\cdot\left[\tanh\left(\frac{E}{2k_BT}\right) - \frac{k_BT}{\mu F}\ln\left(\frac{\cosh\left(\dfrac{E+\mu F}{2k_BT}\right)}{\cosh\left(\dfrac{E-\mu F}{2k_BT}\right)}\right)\right] \tag{25}$$

The nonequilibrium part of SD hole width after the end of the electric field pulse can be written for standard model distribution function as:

$$\Gamma_{ne}(t,F) = AT\ln\left[\frac{t-t_1}{t-t_2}\right]\int dx \cdot \tilde{n}_{ne}(T,E,F) \tag{26}$$

b) **Experiment.** Experiments with electric field cycles provide the unique opportunity for purely optical measurements of an average value of TLS electric dipole moment. This opportunity was realized for the first time in [71]. Later the detailed investigations of the electric field induced SD on temperature, field strength and field cycle duration were fulfilled in [69,70]. Figure 3 shows the dependence of hole width evolution on the electric field cycle duration. The hole narrowing is clearly seen, the time dependence and dependence on the cycle duration are in very good accordance with the model predictions. In general, the experimental data verify the predictions of TLS model very good. Some interesting new effects were found. For example, the correlation between TLS electric dipole moment and relaxation rate was found (see insert in figure 3.)

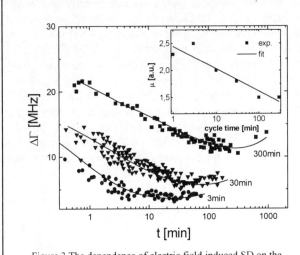

Figure 3 The dependence of electric field induced SD on the duration of the electric field cycle. The insert shows dependence of the average value of TLS electric dipole moment on the cycle duration. (From 69).

Electric field can be switched very fast, that creates the opportunity to increase the time resolution in SD measurements via persistent hole burning. The time resolution up to microseconds is realized [72] and it extends the whole scale of SD investigation via persistent hole burning up to about 12 decades. It should be noted, that the above correlation between TLS electric dipole moment and relaxation rate, firstly revealed on the time scale $10 - 10^4$ s, manifests itself even more clear on a shorter time scale.

## 3.5. LONG-TIME SD IN PMMA AT HIGH TEMPERATURES (3 – 4 K)

As was pointed out earlier, the experiments at temperatures above 1 K were necessary

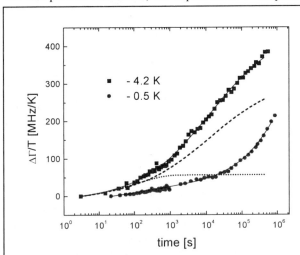

Figure 4. Long-time SD hole broadening in PMMA. Solid lines are fit curves with modified TLS distribution function. Corresponding model does not take into account the absence of an aging effect. Dashed curve is the fit with cutoff in TLS energy distribution; the dotted curve is based on cutofff in TLS tunneling parameter distribution. (see, for details 53,62).

for understanding the nature of nonlogarithmic long time SD in PMMA. Such experiments are now in progress, the first results of the measurements at 4.2 K were presented in [53]. The experimental data are shown in figure 4. Old 0.5 K data are also shown for comparison. The experimental data do not confirm the preliminary supposition about square root SD long time asymptotic. The nonlogarithmic part in a hole diffusional broadening does not cease, as should be in the case of TLS-TLS interaction. As it was expected on the basis of effective temperature estimation (see section 3.4.1.) there is no aging effect in SD in PMMA at 4.2 K.

Let us analyze in more detail, what signifies the absence of an aging effect in terms of TLS distribution function. There are two opportunities: TLS distribution function has cutoff or in energy near 4 K, or in relaxation times near 10 minutes (the order of magnitude of sample cooling).

Suppose, there is cutoff in energy distribution. In that case, due to absence of TLS with energy far above 4 K the aging effect would be observable only at temperatures substantially below 4 K, what is just the case in PMMA. But this hypothesis is inconsistent with other experimental data. First, there are no indications of such cutoff in the experimental data, measured via non-optical methods [73,74]. Second, in that case we should have the saturation of SD rate as a function of temperature. That is not the case in our experiments.

Suppose now that cutoff in $\lambda$-distribution exists, such that TLS relaxation times at 4 K do not exceed tens of minutes. In that case the effective temperature of TLS ensemble would be about 4 K and it would be no aging effect. But again we come to the contradiction with experimental data: it this case SD time dependence should saturate at times around tens of minutes.

Hence, we come to contradiction between the absence of aging effect and an absence of saturation in time or temperature dependence of SD. This contradiction can not be solved within the framework of TLS model. It looks like nonlogarithmic part of SD hole broadening has non-TLS origin. We can speculate that this part of SD is caused by glass structure relaxation, which are strongly pronounced at temperatures near glass transition and are not observed by conventional technique at helium temperatures.

## 4. Conclusion

As was shown, the hole burning technique is a very powerful tool for investigation of low temperature glass dynamics on a very broad time scale. The main characteristics of equilibrium and nonequilibrium spectral diffusion are recently investigated on a time scale up to $10^6$ s and quantitatively interpreted in terms of TLS model. At the same time the distinct limitations of TLS model both in time and temperature scale are found. As usual, more new question arose than answers obtained.

For example, only one object is investigated in detail – PMMA. The question is, how universal are the results, obtained on this system? Definitely, there is difference in SD behavior for different glasses (see previous section). Another very important question: What is the influence of impurity molecules on the low energy excitation spectrum of matrix. We know, that at very high concentration chromophore molecules can modify this spectrum substantially [75-77] and we have now the only indirect evidences that at low concentrations they do not disturb it.

Very important and interesting direction of feature investigations is the extension of SD measuremnts on the higher temperature region, were TLS model fails in SD description. The experiments in this temperature region will supply information for the glass description in the intermediate region between the vicinity of glass transition region, where the models of structural relaxation work, and very low temperatures, where the TLS model proved it's correctness.

164

## 5. Acknowledgements

The part of the work made in Russia was supported by RFBR (Gr. No.99-02-18056), the part of the work made in Germany was supported by the Volkswage Foundation (Grant AZ I/70526), the author's participation in the NATO Advanced Research Workshop was sponsored by NATO grant. The author is grateful to all his collaborators in Russia and Germany for their contribution to the experiments reviewed in the manuscript.

## 6. References

1. Personov R.I., Al'shits E.I., Bykovskaja L.A. (1972) The effect of fine structure appearance in laser-excited fluorescence spectra of organic compounds in solid solutions, *Optics Comm.*6, 169.
2. Personov R.I., Kharlamov B.M. (1973) Extreme narrowing of bands in the fluorescence excitation spectra of organic molecules in solid solutions, *Optics Comm.* 7, 417.
3. Personov R.I., Al'shits E.I., Bykovskaja L.A., Kharlamov B.M. (1973) Fine structure of luminescence spectra of organic molecules at laser excitation and the nature of broad-band spectra, *JETF* 65, 1825.
4. Kharlamov B.M., Personov R.I., Bykovskaja L.A. (1974) Stable "gap" in absorption spectra of solid solutions of organic molecules by laser irradiation, *Optics Comm.,* 12, 191.
5. Gorokhovskii A.A., Kaarli R.K., Rebane L.A. (1974) Hole burning in the contour of purely electronic line in the Shpol'skii system, *Pisma JETF* 20, 474.
6. Personov R.I. (1983) Site selection spectroscopy of complex molecules in solution and its applications, in Agranovich V.M. and Maraduduin A.A. (eds.), *Modern problems in condensed matter sciences,* North-Holland, Amsterdam, N.Y., Oxford.
7. Rebane L.A., Gorokhovskii A.A., Kikas J.V. (1982) Low-Temperature Spectroscopy of Organic Molecules in Solids by Photochemical Hole Burning, *Appl.Phys.B., 29,* 235-250.
8. Small G.J. (1983) Persistent nonphotochemical holeburning and the dephasing of impurity electronic transitions in organic glasses, in: Agranovich V.M. and Maraduduin A.A. (eds.), *Modern problems in condensed matter sciences,* North-Holland, Amsterdam.
9. Friedrich J., Haarer D. (1984) Photochemical Hole Burning: A Spectroscopic Study of Relaxation Processes in Polymers and Glasses, *Angew.Chem.Int.Ed.Engl., 23,* 113-140.
10. Personov R.I., Kharlamov B.M. (1986) Photochemical and photophysical hole burning in electronic spectra of complex organic molecules, *Laser Chemistry,* 6, 181.
11. Moerner W.E. ed. (1988) *Persistent spectral hole burning: science and applications,* (Topics in current physics, v.44) Springer, Berlin.
12. Maier M. (1986) Persistent spectral holes in external fields, *Appl.Phys.B.*41, 73.
13. Kohler B.E., Personov R.I., Woehl J.C. (1995) Electric field effects in molecular systems studied via persistent hole burning, in: Myers A.B. and Rizzo T.R. (eds.), *Laser Techniques in Chemistry* (Techniques of Chemistry Series, Vol.XXIII), John Willey & Sons,Inc., Ch.8.
14. Van den Berg R., Van der Laan V., V"lker S. (1987) Zeeman effects in amorphous solids: A hole-burning study of free-base porphin in polyethylene, *Chem.Phys.Lett.,* 142, 535 (and references therein).
15. Ulitsky N.I., Kharlamov B.M., Personov R.I. (1990) Effect of high magnetic field on the degenerate S-S transitions of organic molecules in amorphous matrices via hole burning: Zn-phthalocyanine in polyvinylbutural, *Chem.Phys.,* 141, 441-445 (and references threin).
16. Krausz E.R., Riesen H., Schatz P., Gasyna Z., Duford C.L., Williamson B.E. (1996) Magneto-optical hole burning studies of matrix isolated phthalocyanines, *J.Luminescence* 66&67, 19-24.
17. Laisaar A., Kikas J.V. (1997) Persistent spectral hole-burning in doped organic crystasls and polymers at high hydrostatic pressures, *J.Luminescence* 72-74, 515-516.
18. Macfarlane R.M., Shelby R.M. (1987) Homogeneous line broadening of optical transitions of ions and molecules in glasses, *J.Luminescence* 36, 179-207.
19. Völker S., (1989) Hole-burning spectroscopy, *Ann.Rev.Phys.Chem.,* 40, 499-530.
20. Osad'ko I.S. (1991) Optical dephasing and homogeneous optical bands in crystals and amorphous solids:dynamic and stochastic approaches, *Phys.Rep.* 206, 45-97.
21. Jankowiak R., Hayes J.M., Small G.J. (1993) Spectral Hole-Burning Spectroscopy in Amorphous Molecular Solids and Proteins, *Chem.Rev.* 93, 1471.

22. Hunklinger S., Raychandhuri A.K. (1986) Termal and elastic anomalies in glasses at low temperatures, in: Brewer D.F. (ed.), *Progressin Low Temperature Physics*, **Vol.9**, Elsevier Science Publishers B.V., Ch.3.

23. Phillips W.A. (1987) Two-level states in glasses, *Rep.Prog.Phys.* **50**, 1657.

24. Esquinazi P., (ed.), (1998) *Tunneling systems in amorphous and crystalline solids*, Springer-Verlag, Berlin-Heidelberg-New York-London-Paris-Tokio-Hong Kong-Barcelona-Budapest).

25. Friedrich J., Haarer D. (1986) Structural relaxation processes in polymers and glasses as studied by high resolution optical spectroscopy, in: Zschokke-Granacher I. (ed.), *Optical spectroscopy of glasses*, Reidel, Dordrecht., pp.149-198.

26. Maier H., Kharlamov B.M., Haarer D. (1998) Investigation of tunneling dynamics by optical hole-burning spectroscopy in: Esquinazi P. (ed), *Tunneling systems in amorphous and crystalline solids*, Springer-Verlag, Berlin-Heidelberg-New York-London-Paris-Tokio-Hong Kong-Barcelona-Budapest, Ch.6.

27. Meijers H.C., Wiersma D.A. (1994) Low temperature dynamics in amorphous solids: A photon echo study, *J.Chem.Phys.* **101**, 6927-6943.

28. Thorn Leeson D., Berg O., Wiersma D.A. (1994) Low-tempeature protein dynamics studied by the long-lived stimulated photon echo, *J.Phys.Chem.*, **98**, 3913-3916.

29. Narasimhan L.R., Littau K.A., Pack D.W., Elschner A., Bai Y.S., Fayer M.D. (1990) Probing organic glasses at low temperature with variable time scale optical dephasing measurements, *Chem.Rev.* **90**, 439-457.

30. Orrit M., Bernard J., Personov R.I. (1993) High-resolution spectroscopy of organic molecules in solids: from fluorescence line narrowing and hole burning to single molecule spectroscopy, *J.Phys.Chem.* **97**, 10256-10268.

31. Skinner J.L., Moerner W.E. (1996) Structure and Dynamics in Solids As Probed by Optical Spectroscopy, *J.Phys.Chem.* **100**, p.13251-13262.

32. Köhler W., Breinl W., Friedrich J. (1985) Laser photochemistry with polarised light in low-temperature glasses, *J.Phys.Chem.* **89**, 2473-2477.

33. Kador L., Schulte G., Haarer D. (1986) Relation between hole-burning parameters and molecular parameters: free-base phtalocyanine in polymer hosts, *J.Phys.Chem.* **90**, 1264-1270.

34. Müller J., Khodykin O.V., Haarer D., Kharlamov B.M. (1999) Nonphotochemical hole burning in Zn-TBP/PMMA. Unusual burning kinetics and very narrow holewidth limit, *Chem.Phys.* **243**, 201-208

35. Köhler W., Friedrich J. (1987) Distribution of barrier heights in amorphous organic materials, *Phys.Rev.Lett.* **59**, 2199.

36. Köhler W., Friedrich J. (1988) Probing of extremely low configurational barriers in an organic glass, *J.Chem.Phys.* **88**, 6655.

37. Köhler W., Meiler J., Friedrich J. (1987) Tunneling dynamics of doped organic low-temperature glasses as probed by a photophysical hole-burning system, *Phys.Rev.B*, **35**, 4031.

38. Hayes J.M., Small G.J. (1978) Non-photochemical hole burning and impurity site relaxation processes in organic glasses, *Chem.Phys.* **27**, 151.

39. Shu L., Small G.J. (1990) On the mechanism of nonphotochemical hole burning of optical transitions in amorphous solids, *Chem.Phys.* **141**, 447-455.

40. Shu L., Small G.J. (1992) Mechanism of nonphotochemical holeburning: Cresyl violet in polyvinyl alcohol films, *JOSA B* **9**, 724-732.

41. Zeller R.C., Pohl R.O. (1971) Thermal conductivity and specific heat of noncrystalline solids, *Phys.Rev.B* **4**, 2029.

42. Anderson P.W., Halperin B.I., Varma C.M. (1972) Anomalous low-temperature thermal properties of glasses and spin glasses,. *Phil.Mag.* **25**, 1-9.

43. Phillips W.A. (1972) Tunneling states in amorphous solids, *J.Low Temp.Phys.* **7**, 351-360.

44. Klauder J.R., Anderson P.W. (1962) Spectral diffusion decay in spin resonance experiments, *Phys.Rev.* **125**, 912.

45. Black J.L., Halperin B.I. (1977) Spectral diffusion, phonon echoes, and saturation recovery in glasses at low temperatures, *Phys.Rev.B* **16**, 2879.

46. Reinecke T.L. (1979) Fluorescence linewidth in glasses, *Sol.St.Comm.* **32**, 1103-1106.

47. Breinl W., Friedrich J., Haarer D. (1984) Spectral diffusion of a photochemical proton transfer system in an amorphous organic host: Quinisarin in alkohol glass, *J.Chem.Phys.* **81**, 3915.

48. Maier H., Kharlamov B.M., Haarer D. (1996) Two-level system dynamics in the long-time limit: a power-law time dependence, *Phys.Rev.Lett.* **76**, 2085.

49. Hannig G., Maier H., Haarer D., Kharlamov B.M. (1996) Interacting tunneling States: A hole-burning study of spectral diffusion, *Mol.Cryst.Liq.Cryst.* **291**, 11-16.

50. Burin A.L., Kagan Yu. (1994) Low energy collective excitations in glasses. New relaxation mechanism for ultralow temperatures, *JETF* **106**, 633-647.

51. Neu P., Reichman D.R., Silbey R.J. (1997) Spectral diffusion on ultralong time scales in low-temperature glasses , *Phys.Rev.B* **56**, 5250.

52. Heuer A., Neu P. (1997) Tunneling dynamics of side chains and defects in proteins, polymer glasses, and OH-doped network glasses, *J.Chem.Phys.* **107**, 8686.

53. Müller J., Haarer. D, Khodykin O.V., Kharlamov,B.M. (1999) Long-time scale spectral diffusion in PMMA: Beyond the TLS model?, *Physics Letters A* **255**, 331-335

54. Fritsch K., Eicker A., Friedrich J., Kharlamov B.M., Vanderkooi J.M. (1998) Spectral diffusion in proteins, *Europh.Lett.* **41**, 339-344.

55. Bai Y.S., Littau K.A., Fayer M.D. (1989) The nature of glass dynamics: thermal reversibility of spectral diffusion in a low temperature glass, *Chem.Phys.Lett.* **162**, 449-454.

56. Littau K.A., Bai Y.S., Fayer M.D. (1990) Two-lewel systems and low-temperature glass dynamics: Spectral diffusion and thermal reversebility of hole-burning linewidths, *J.Chem.Phys.* **92**, 4145.

57. Kharlamov B.M., Haarer D., Jahn S. (1994) Spectral diffusion in organic glasses. Studies in a millisecond range, *Optics and Spectroscopy* **76**, 302-313

58. Fritsch K., Friedrich J., Kharlamov B.M. (1996) Non-equilibrium phenomena in spectral diffusion physics of organic glasses, *J.Chem.Phys.* **105**, 1798-1806.

59. Jahn S., Müller K.-P., Haarer D. (1992) Two-level-system dynamics in doped polymers glasses below 1 K: hole burning as an optical analog to heat-release experiments, *JOSA B* **9**, 925-930.

60. Maier H., Haarer D. (1995) Equilibrium and nonequilibrium tunneling dynamics and spectral diffusion in the millikelvin regime, *J.Luminescence* **64**, 87-93.

61. Fritsch K., Friedrich J., Kharlamov B.M., unpublished results.

62. Maier H., Hannig G., Müller J., Haarer. D, Khodykin O.V., Kharlamov,B.M (1999), *J.Chem.Phys.*, to be published

63. Schulte G., Grond W., Haarer D., Silbey R.J. (1988) Photochemical hole burning of phtalocyanine in polymer glasses: Thermal cycling and spectral diffusion, *J.Chem.Phys.* **88**, 679-685.

64. Köhler W., Friedrich J. (1988) Irreversible features of thermal line broadening in glasses as probed by persistent optical holes, *Europh.Lett.* **7**, 517.

65. Köhler W., Zollfrank J., Friedrich J. (1989) Thermal irrevesibility in optically labelled low-temperature glasses, *Phys.Rev.B* **39**, 5414.

66. Al'shits E.I., Ulitsky N.I., Kharlamov B.M. (1991) Investigation of light- and thermo-induced spectral diffusion in impurity amorphous systems via Stark hole burning spectroscopy, *Isv.AN ESSR, phys.-mat.* **40**, 161-171.

67. Schellenberg P., Friedrich J. (1994) Optical spectroscopy and disorder phenomena in polymers, proteins and glasses, in: Richert R. And Blumen A. (eds.), *Disorder effects on relaxation processes*, Springer-Verlag, Berlin, Heidelberg, pp.407-424.

68. Khodykin O.V., Ulitsky N.I., Kharlamov B.M. (1996) Thermal irreversibility in low temperature polymers: investigation of spectral diffusion, *Opt.Spektr.* **80**, 489-496.

69. Kharlamov B.M., Wunderlich R., Maier H., Haarer D. (1998) Optical investigation of electric-field-induced relaxations in amorphous solids at low temperatures, *J.Luminescence*, **76&77**, 283-287.

70. Wunderlich R., Maier H., Haarer D., Kharlamov B.M. (1998) Optical investigation of low-temperature electric-field-induced relaxations in amorphous solids, *J.Phys.Chem.B* **102**, 10150-10157.

71. Maier H., Wunderlich R., Haarer D., Kharlamov B.M., Kulikov S.G. (1995) Optical Detection of Electric Two Lewel System Dipoles in a Polymeric Glass, *Phys.Rev.Lett.* **74**, 5252-5255.

72. Wunderlich R., Haarer D., Kharlamov B.M. (1999) Optical investigation of electric field induced spectral diffusion in polymers on a time scale $10^{-5} - 10^5$ s, to be published.

73. Federle G., Hunklinger S. (1982) Ultrasonic attenuation in polymethylmethacrylate at low temperatures, in: *Nonmetalic materials and composites at low temperatures*, v.2, Plenum press, NY&London.

74. Nittke A., Scherl M., Esquinazi P., Lorenz W., Li Junyun, Pobell F. (1995) Low temperature heat release, sound velosity and attenuation, specific heat and thermal conductivity in polymers, *J.Low Temp.Phys.* **98**, 517-547.

75. Wunderlich R., Maier H., Haarer D., Kharlamov B.M. (1998) Light-induced spectral diffusion in heavily doped polymers, *Phys.Rev.B* **57**, R5567

76. Den Hartog F.T.H., Bakker M.P., Silbey R.J., Völker S. (1998) Long-time spectral diffusion induced by short-time energy transfer in doped glasses: concentration-, wavelength- and temperature dependennce of spectral holes, *Chem.Phys.Lett.* **297**, 314-320.

77. Den Hartog F.T.H., van Papendrecht C., Silbey R.J., Völker S (1999) Spectral diffusion induced by energy transfer in doped organic glasses: delay-time dependence of spectral holes, *J.Chem.Phys.* **110**, 1010.

# LEAKY MODES IN NONLINEAR OPTICAL RESONATORS

R. Reinisch*, G. Vitrant*, E. Popov*[1], M. Nevière**

* Laboratoire d'Electromagnétisme, Microondes et Optoélectronique, URA CNRS 833, ENSERG, 23 ave des Martyrs, BP 257, 38016 Grenoble Cedex, France.
** Laboratoire d'Optique Electromagnétique, URA CNRS 843, Faculté des Sciences et Techniques de St Jérome, 13397 Marseille Cedex 20, France.

## 1. Introduction

The aim of this lecture is to show that leaky modes provide a convenient mean for the study of nonlinear optical interactions in "open" resonators such as Fabry-Perot's, multilayer structures, prism or grating couplers, etc.

The starting point is the coupled mode formalism where the electromagnetic field is expanded on the *guided* modes (i. e. originating from *proper* poles) and the radiation fields. Since open guiding structures, such as prism couplers, have *no* proper poles and, consequently, propagate *no* guided modes, the expansion of the coupled mode approach *only* includes the radiation fields. In other words, one is left with an *integral representation* of the solution.

Due to the open characteristic of these resonators, the solution of the associated homogeneous problem yields improper poles and, as a result, leaky modes. The difficulty with these modes comes from the fact that they diverge at infinity in the cross-section plane. We show that a leaky mode constitutes a good approximation of the radiation fields excited at resonance. This is an important result which serves as a basis to derive the spatio-temporal equation describing nonlinear optical interactions in leaky resonators. In this equation, the leaky mode normalization constant *does not* diverge. The case of grating couplers is considered with a special emphasize. Several examples are given which concern nonlinear optical interactions in leaky resonators: optical Kerr effect (optical bistability, self-pulsing, cavity solitons, chaos, short pulse regime of the prism coupler), second harmonic generation (different phase-matching schemes, subwavelength gratings).

## 2. The coupled mode formalism

It is appropriate to consider first the longitudinal, or Fourier, representation of the electromagnetic (EM) field from which two other expansions are derived : the coupled mode representation (also called the transverse expansion) which involves the guided modes and the radiation field(s) supported by the structure and the leaky mode representation which is convenient when considering "open" or leaky resonators.

To begin with the structure of interest is a linear planar multilayer configuration illuminated by a light beam (time dependence $e^{-i\omega t}$) under incidence $\theta_i$. Figure 1 shows an example corresponding to a system involving three media 1, 2, 3 with relative permittivity $\varepsilon_1$, $\varepsilon_2$ and $\varepsilon_3$ with $\varepsilon_2 > (\varepsilon_1, \varepsilon_3)$ in order to allow for guided modes.

For simplicity, we assume a two-dimensional situation where $\dfrac{\partial}{\partial z} = 0$. Thus the EM field is either TE or TM polarized (specified by the z-component $\Phi_z(x, y)$ of the electric or magnetic field respectively). The analysis closely follows refs. [1-5].

---

[1] On leave from Institute of Solid State Physics, 72 Tzarigradsko Chaussee, 1784 Sofia, Bulgaria.

F. Kajzar and M.V. Agranovich (eds.), Multiphoton and Light Driven Multielectron Processes in Organics: New Phenomena, Materials and Applications, 167–182.

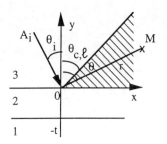

*Figure 1.* The structure of interest. A leaky mode only contributes to the radiation field within the shaded region (see section 3).

## 2.1 THE LONGITUDINAL REPRESENTATION

The solution $\Phi_z(x,y)$ is expressed as a Fourier integral :

$$\Phi_z(x,y) = \int_{-\infty}^{+\infty} \varphi_z(\gamma,y)e^{i\gamma x}d\gamma. \tag{1}$$

Let $\alpha_q$ denote the y-component of the wavevector in the outside media q (q=1,3). The quantities $\alpha_q$ and $\gamma$ are related by :

$$\alpha_q^2 + \gamma^2 = k_q^2, \tag{2a}$$

or

$$\alpha_q = \pm\sqrt{k_q^2 - \gamma^2}, \tag{2b}$$

with :

$$k_q^2 = \frac{\omega^2}{c^2}\varepsilon_q \qquad \text{(c : speed of light in vacuum).} \tag{3}$$

The advantage of eq. 1 is its relative simplicity. The drawback comes from the fact that this expression does not exhibit explicitely possible guided modes which, as is known, may be present in the structure fig. 1.

## 2.2 THE TRANSVERSE REPRESENTATION

This representation is obtained from eq. 1 by performing a contour deformation where the path P of integration along the real $\gamma$-axis is transformed into a path $P_\gamma$ in the complex $\gamma$-plane. Such a procedure implies a search of the singularities of $\varphi_z(\gamma,y)e^{i\gamma x}$[6]. These singularities are of two types: i) poles which contribute to the integral in the complex $\gamma$-plane through the residue theorem, ii) branch points, due to the square root in eq. 2b, which require cuts in the complex $\gamma$-plane in order to avoid multivalued functions.

The cuts are chosen in such a way that the radiation condition at infinity is fulfilled on the *entire* top sheet of the four-sheeted Riemann $\gamma$-plane. The radiation condition writes :

$$\text{Im}(\alpha_q) > 0 \qquad (q=1,3), \tag{4a}$$

and the cuts correspond to :

$$\text{Im}(\alpha_q) = 0 \qquad (q=1,3). \tag{4b}$$

With these cuts the entire top sheet of the four-sheeted $\gamma$-plane is mapped on the upper half of the complex $\alpha_q$-planes (q=1,3). The resulting contour of integration $P_\gamma$ is illustrated in fig. 2.

All these mathematical considerations lead to the following transverse representation of $\Phi_z(x,y)$:

$$\Phi_z(x,y) = \sum_m c_m \varphi_{m,z}(y)e^{i\gamma_m x} + \sum_{Q=I,II} \int_0^{+\infty} c_Q(\alpha_Q)\varphi_{Q,z}(\alpha_Q,y)e^{i\gamma(\alpha_Q)x}d\alpha_Q. \tag{5}$$

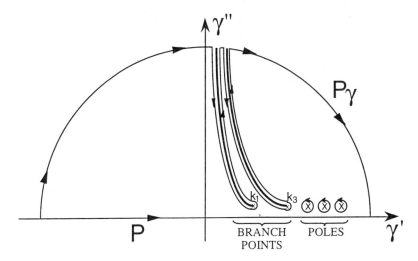

*Figure 2.* Top sheet of the complex γ-plane. The bold lines represent the branch cuts.

Since the *entire* integration is carried out in the top sheet of the complex γ-plane, *all* the poles comply with the radiation condition eq. 4a. These poles, which are called "proper", are associated to non-diverging EM fields obtained solving the homogeneous problem, i. e. without excitation. Thus the corresponding residues account for guided modes described by the summation over m in eq. 5. The two continuous spectra (Q=I,II) come from the paths around the cuts and describe two types of radiation fields. The term $\alpha_Q$ denotes $\alpha_1$ for class Q=I and $\alpha_3$ for class Q=II. The radiation fields correspond to the solution obtained when illuminating the system by a plane wave of unit amplitude incident on the interfaces y=-t and y=0 for Q=I and Q=II respectively.
Therefore the transverse representation corresponds to the expansion of the EM field used in the coupled-mode formalism[7]. It is worth noting that "open" resonators, such as prism couplers, do *not* exhibit proper poles. Thus for these structures the coupled mode representation of the EM field *only* includes the radiation fields (second term on the right-hand side of eq. 5) which lead to an integral representation of the solution. Consequently, when no guided modes are present, eq. 5 is not more helpful than eq. 1 is.

## 3. Leaky modes

Open resonators are conveniently handled introducing another complex plane which we call the complex w-plane. Let us consider the EM field $\Phi_z(x,y)$ in medium q=3. The integration in eq. 1 is carried out in the complex w-plane :

$$w = u + iv, \tag{6a}$$

through the following change of variables :

$$\gamma = k_3 \sin w, \tag{6b}$$
$$\alpha_3 = k_3 \cos w, \tag{6c}$$

together with the use of polar coordinates :

$$x = r \sin \theta, \tag{6d}$$
$$y = r \cos \theta, \tag{6e}$$

where r and θ are shown in fig. 1.
The mapping eqs. 6b,c plots the four quadrants of the top and bottom sheets of the γ-plane into the regions denoted $T_i$ and $B_i$ (i=1,2,3,4) respectively (fig. 3). The path $P_w$ in fig. 3 is the image of the original contour of integration P, used in the integration of eq. 1, through the transformation eqs. 6b, c.

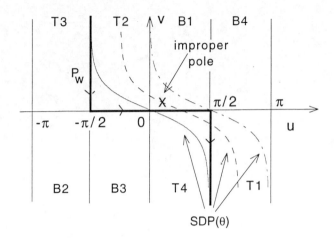

*Figure 3.* The complex w-plane. Solid curve : SDP ($\theta=0$), dashed curve : SDP ($\theta= \pi/4$), dotted-dashed curve : SDP ($\theta= \pi/2$).

In view of an asymptotic evaluation of $\Phi_z(x,y)$ in medium 3, $\Phi_z(x,y)$ is written (according to eqs. 6) :

$$\Phi_z(x,y) = \int_{P_w} \phi_z(w,y)e^{ik_3 r \cos(w-\theta)}dw . \tag{7a}$$

The leaky mode representation involves a contour deformation from path $P_w$ to the steepest descent path (SDP). The calculation shows that there is a saddle point at $w=\theta$ and that the SDP obeys the equation :

$$\cos(u - \theta)\cosh v = 1. \tag{7b}$$

Before going further on some comments are in order.

i) According to eq. 7b, the position of SDP in the complex w-plane *depends on the direction of observation $\theta$* i. e. SDP=SDP($\theta$). The SDP curve crosses the v=0 axis at u= $\theta$.

ii) *Only those poles located between paths $P_w$ and SDP* contribute to the integral eq. 7a. Since SDP generally passes through strips B1 or B3, poles located on the *bottom* sheet of the $\gamma$-plane may be captured. These poles, which do not fulfill the radiation condition eq. 4a, are called "improper", and correspond to leaky modes whose amplitude diverges at infinity. But fig. 3 shows that an improper pole contributes *provided the angle of observation $\theta$ is greater than a critical angle,* $\theta_{c,\ell}$ (the index $\ell$ labels a pole), the value of which depends on the location of the pole in the w-plane (in the example fig. 3 $\theta_{c,\ell} = \pi/4$). Stated differently, the improper poles contribution *only occurs within a finite angular region* corresponding to $\theta > \theta_{c,\ell}$ (shaded region of fig. 1) preventing, as $y \rightarrow +\infty$, any divergence in the EM field representation.

iii) These leaky modes are characteristic of open resonators : the improper feature of pole $\gamma_\ell$ comes from the EM coupling with the outside. These modes are obtained solving the homogeneous problem.

iv) According to section 2.2 improper poles cannot be captured within the transverse expansion. Thus *leaky modes are part of the radiation field.*

Finally it is seen that the EM field eq.7a is the sum of the space wave $\Phi_{sw,z,3}$ resulting from the integration along SDP and of the residues contribution coming from the poles located between $P_w$ and SDP :

$$\Phi_z(x,y) = \sum_\ell \Gamma(\theta - \theta_{c,\ell})c_\ell \phi_{\ell,z}(y)e^{i\gamma_\ell x} + \Phi_{sw,z,3}. \tag{8}$$

In eq. 8 $\Gamma(\theta - \theta_{c,\ell})=1$ if $\theta > \theta_{c,\ell}$, $\Gamma(\theta - \theta_{c,\ell}) = 0$ if $\theta < \theta_{c,\ell}$.

The discrete summation in eq. 8 is carried out considering all the poles located between $P_w$ and SDP : the leaky ones and the "proper" ones within the angular regions where they contribute.

To summarize for large y (x constant), leaky modes do not contribute and the space wave constitutes a good approximation of the radiation field. In, or very close to, the guiding layer (medium 2), it can be shown[1,4] that a leaky mode provides a good description of the radiation fields involved in the resonance process. Thus:

$$c_\ell(x)\varphi_{\ell,z}(y)e^{i\gamma_\ell x} \approx \sum_{Q=I,II} \int_{\Delta_Q} c_Q(\alpha_Q,x)\varphi_{Q,z}(\alpha_Q,y)e^{i\gamma(\alpha_Q)x}d\alpha_Q. \tag{9}$$

The following hypothesis are assumed which correspond to usual situations when interested in "open" resonators: their exists *isolated* improper poles, ii) the working point remains in the vicinity $\Delta_Q$ of such a pole i. e. of the order of the distance from the pole to the real $\gamma$-axis.

Equation 9 is important since it allows to derive the equation of evolution of a leaky mode amplitude. As shown in refs. [8,9] the leaky mode amplitude obeys the following first order differential equation:

$$\frac{dc_\ell}{dx} = -i\omega\frac{\left\langle \vec{E}_\ell^-(y).\vec{\mathcal{P}}^{NL}(x,y)\right\rangle}{N_Q}e^{-i\gamma_\ell x} + \gamma_\ell''T_Q\mathcal{A}_i(x)e^{i(\gamma_i-\gamma_\ell)x}. \tag{10a}$$

In eq. 10a the first term arises from the existence of the nonlinear interaction whereas the second one describes the in-coupling of possible incident beam(s) on the system, $\vec{E}_\ell^-(y)$ is the electric field deduced from the backward leaky mode, $\gamma_\ell''T_Q$ is the in-coupling coefficient. The theoretical study shows that the calculation of the leaky mode normalization constant $N_Q$ has to be done "closing" the resonator[9]:

$$N_Q = \xi\gamma_m \int_{-\infty}^{+\infty}\frac{1}{\eta(y)}\varphi_{m,z}^2(y)dy, \tag{10b}$$

where $\eta(y)=1, \varepsilon(y); \xi = \dfrac{-2}{\omega\mu_0}, \dfrac{2}{\omega\varepsilon_0}$ in the TE and TM cases respectively. In eq. 10b, $\varphi_{m,z}(y)$ corresponds to the transverse field map of the *"closed"* cavity. In other words, considering, for example a prism coupler, $\varphi_{m,z}(y)$ is the transverse field map of the guided mode m *without* the prism. Consequently, despite the fact that leaky modes are used, there is *no problem of divergence at infinity* when calculating $N_Q$.

## 4. Grating couplers

Let us consider first a planar waveguide. We know that it supports guided modes. Concentrating on a guided mode m, the transverse representation (eq. 5) gives the expression of the corresponding EM field:

$$\Phi_{m,z}(x,y) = c_m\varphi_{m,z}(y)e^{i\gamma_m x}. \tag{11a}$$

Now let us see how the homogeneous solution eq. 11a is modified when the groove depth h is different from zero (Fig. 4). The periodicity along x implies that $\varphi_{m,z}$ is no longer a function of y alone: it becomes also x-dependent with periodicity d[10]. Thus $\varphi_{m,z}$ can be expanded in Fourier series. For reasons which will soon become clear, it is convenient to make the substitution: index m $\rightarrow$ index $\ell$.

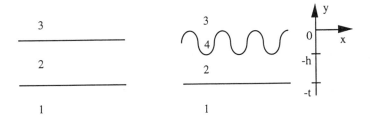

the waveguide : h=0       the grating structure : h≠0

*Figure 4.* A waveguide and a typical grating structure obtained by modulating the interface between media 3 and 2 (periodicity d, groove depth h), the modulated region is labelled 4.

Therefore eq. 11a takes the following form:

$$\Phi_{\ell,z}(x,y) = c_\ell \sum_{n=-\infty}^{+\infty} \psi_{\ell,n,z}(y) e^{i(\gamma_\ell + nK)x},$$ (11b)

with

$$K = \frac{2\pi}{d}.$$ (11c)

It is known that outside the modulated region(s) $\Phi_{\ell,z}(x,y)$ can be expressed as a Rayleigh expansion. Thus in regions q=1 or q=3 $\Phi_{\ell,z}(x,y)$ writes:

$$\Phi_{\ell,z}(x,y) = c_\ell \sum_{n=-\infty}^{+\infty} a_{q,\ell,n} e^{i\left[(\gamma_\ell + nK)x \pm \alpha_{q,\ell,n} y\right]},$$ (12a)

with

$$\alpha_{q,\ell,n}^2 + (\gamma_\ell + nK)^2 = k_q^2.$$ (12b)

In eq. 12a signs +, - correspond to q=3,1 respectively.

Thus the periodicity of the grating structure converts a guided mode m into a mode which is the sum of an infinite number of space harmonics with longitudinal wavevector component:

$$\tilde{\gamma}_{\ell,n} = \gamma_\ell + nK.$$ (13)

The pole $\gamma_\ell$ and the amplitudes $a_{q,\ell,n}$ are obtained looking for the solution of the homogeneous problem. This is achieved using the rigorous theory of diffraction in linear optics[10]. It is found that the pole $\gamma_\ell$ is complex even for grating structures without dielectric losses. Thus the quantities $\tilde{\gamma}_{\ell,n}$ are also complex:

$$\tilde{\gamma}_{\ell,n} = \gamma'_\ell + nK + i\gamma''_\ell = \tilde{\gamma}'_{\ell,n} + i\gamma''_\ell.$$ (14a)

The complex feature of $\tilde{\gamma}_{\ell,n}$ is due to the radiative losses, i. e. EM leakage, coming from the propagating diffracted orders in the outer media.

The labelling of the space harmonics is chosen such that:

$$\lim_{h\to 0} \tilde{\gamma}_{\ell,0} = \gamma_m.$$ (14b)

Practically for almost reasonable groove depth we have:

$$\tilde{\gamma}'_{\ell,0}(h \neq 0) = \gamma'_\ell(h \neq 0) \approx \gamma_m.$$ (14c)

Under these conditions among all the space harmonic, the n=0 one has two peculiarities: its transverse field map remains close to that of the guided mode $\gamma_m$ (see eqs. 14b,c) and it is the only one whose amplitude does not tend to zero in the flat limit case (i. e. $h \to 0$).

According to eqs. 12,13 the diffracted orders can be classified in two groups: i) those which propagate (or are radiated) in one of the outside medium q fulfill: $|\gamma'_\ell + nK| < k_q$, ii) those which are evanescent in one of the outside medium q fulfill: $|\gamma'_\ell + nK| > k_q$. Equation 14a shows that $\tilde{\gamma}'_{\ell,n}$ can be positive or *negative* although $\gamma'_\ell$ and $\gamma''_\ell$ remain positive. Moreover when lossless outside media are considered, eqs. 12b,13 show that for each diffracted order one has:

$$\alpha'_{q,\ell,n}\alpha''_{q,\ell,n} = -\tilde{\gamma}'_{\ell,n}\gamma''_\ell \qquad \text{(q=1 or q=3)}.$$ (15)

On the basis of eq. 15, it is expected that the behaviour of radiated diffracted orders when $y \to \pm\infty$ depends on the sign of $\tilde{\gamma}'_{\ell,n}$. Figure 5 summarizes the results concerning the nature of the diffracted orders.

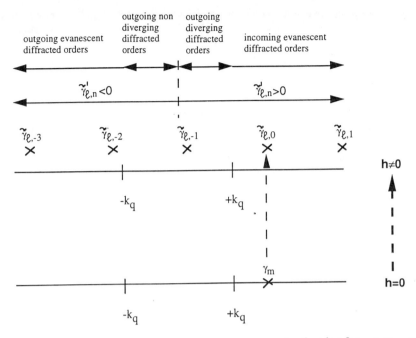

*Figure 5.* Poles in the complex $\gamma$-plane for the two cases h=0 (bottom) and h$\neq$0 (top). When $h = 0 \rightarrow h \neq 0$ the pole $\gamma_m$, which is on the real $\gamma$-axis, becomes a multiplicity of poles in the complex $\gamma$-plane. Top: also reported are the different types of diffracted orders constituting a leaky mode in the outside media q=1,3. In this example there is only one radiated diffracted order in medium q: the order -1.

Since the propagating diffracted orders give rise to an EM leakage, grating structures behave as open, or leaky, resonators.

Remembering that a mode is the solution of the homogeneous problem, it is seen that the EM field associated to the leaky mode is described by eq. 11b. In other words *a* leaky mode of a grating coupler is constituted by the *whole* set of diffracted orders (evanescent or propagating). Thus it should be called a diffracted leaky mode. According to eq. 11b $c_\ell$ denotes the amplitude of such a mode; $\psi_{\ell,n,z}(y)$ (in eq. 11b) and $a_{q,\ell,n} e^{\pm i\alpha_{q,\ell,n} y}$ (in eq. 12a) refer to the transverse field map of a space harmonic n $\forall y$ (especially in the modulated region q=4) and in the outside media q=1,3 respectively.

Let us now consider a grating coupler illuminated (in medium 3) by a plane wave with longitudinal wavevector component $\gamma_i$ and amplitude $A_i$. The expression of the EM field writes

* in medium q=3:

$$\Phi_z(x,y) = A_i e^{i[\gamma_i x - \alpha_i y]} + \sum_{p=-\infty}^{+\infty} A_{3,p}(\gamma_i) e^{i[(\gamma_i + pK)x + \alpha_{3,i,p} y]},$$  (16a)

* in medium q=1:

$$\Phi_z(x,y) = \sum_{p=-\infty}^{+\infty} A_{1,p}(\gamma_i) e^{i[(\gamma_i + pK)x - \alpha_{1,i,p} y]},$$  (16b)

with

$$\alpha_{q,i,p}^2 + (\gamma_i + pK)^2 = k_q^2.$$  (16c)

When:

$$\gamma_i + \hat{p}K = \gamma_\ell \Rightarrow \gamma_i + (\hat{p} + n)K = \tilde{\gamma}_{\ell,n}',$$  (17a)

the rigorous theory of diffraction in linear optics[10] shows that the transmission $\left|\tau_{\hat{p}+n}(\gamma_i)\right|^2$ at y=-t associated to a resonantly excited evanescent diffracted order $\hat{p}+n$ is very well approximated by a lorentzian in the associated resonance domain defined by:

$$\gamma_i \in \left[\tilde{\gamma}'_{\ell,-\hat{p}} - \gamma''_\ell, \tilde{\gamma}'_{\ell,-\hat{p}} + \gamma''_\ell\right]. \tag{17b}$$

Thus we may write:

$$\tau_{\hat{p}+n}(\gamma_i) \equiv \frac{A_{1,\hat{p}+n}(\gamma_i)}{A_i} = \frac{-i\gamma''_\ell T_{\hat{p}+n}}{\gamma_i + (\hat{p}+n)K - (\gamma_\ell + nK)} = \frac{-i\gamma''_\ell T_{\hat{p}+n}}{\gamma_i + \hat{p}K - \gamma_\ell}. \tag{18}$$

When eqs. 17, 18 are fulfilled, the grating structure allows the resonant excitation of the diffracted leaky mode as *a whole*: the space harmonic n=0 by the diffracted order labelled $\hat{p}$, the space harmonics $n \neq 0$ by the evanescent or radiated diffracted orders $\hat{p}+n$ (see eq. 13).

Among all the space harmonics only one has a transverse field map which is close to that of the guided mode m: this space harmonic corresponds to n=0 (see eqs. 14b,c). Moreover it is generally the strongest. In that way the grating structure allows the resonant excitation of the guided mode $\gamma_m$ through the diffracted order $\hat{p}$ (see eqs. 17a, 14c) and acts as a grating coupler and forward or backward resonant excitation is possible.

Considering now the influence of the nonlinear polarization, at a given frequency generally one space harmonic (labelled $\hat{n}$) plays a special role in the nonlinear interaction. For processes independent of phase-matching usually $\hat{n} = 0$. For other types of interactions the leading space harmonic is that for which phase-matching is satisfied. The theory shows that it is possible to generalize eq. 10a to grating couplers leading to the following spatio-temporal equation obeyed by the envelope of the diffracted leaky modes[9]:

$$\frac{\partial \mathcal{A}^{NL}_{1,\hat{p}+\hat{n}}}{\partial x} + \frac{1}{v_{g,\ell}} \frac{\partial \mathcal{A}^{NL}_{1,\hat{p}+\hat{n}}}{\partial t} + i\left[\gamma_i + \hat{p}K - \gamma_\ell\right]\mathcal{A}^{NL}_{1,\hat{p}+\hat{n}} =$$

$$-i\omega \frac{\left\langle \vec{E}^-_{\ell,\hat{n}}(y).\vec{\mathcal{P}}^{NL}(x,y)\right\rangle}{\left(N_{\hat{n}}(\omega)/a_{1,\ell,\hat{n}}\right)} e^{-i\left[\gamma_i + (\hat{p}+\hat{n})K\right]x} + \gamma''_\ell T_{\hat{p}+\hat{n}} \mathcal{A}_i, \tag{19}$$

where:

$$c_\ell(x)a_{1,\ell,n} = \mathcal{A}_{1,\hat{p}+n}(x)e^{i\left(\gamma_i + \hat{p}K - \gamma_\ell\right)x} \text{ and } \mathcal{A}(x,t) = \int_{-\infty}^{+\infty}\mathcal{A}(x,\omega)e^{-i(\omega-\omega_0)t}d\omega. \tag{20}$$

The factor $a_{1,\ell,\hat{n}}$, present in eq. 19 comes from the choice of the origin of the y-axis and from the chosen normalization. Therefore this quantity is not essential since its expression is not characteristic of the problem.

## 5. Kerr-type leaky resonators

For the sake of generality we consider non instantaneous and non local Kerr media for which the nonlinear term obeys a spatio-temporal "material" equation which is written in a phenomenological way as[11-14]:

$$T_d \frac{\partial U}{\partial t_{red}} - L^2 \frac{\partial^2 U}{\partial x^2_{red}} = -U + \left|g_1^{NL}\right|^2. \tag{21a}$$

Moreover as shown in refs. [11, 15-17] the resonant forward-backward excitation of leaky modes has to be taken into account in order to describe properly the diffraction under, or close to, normal incidence. Then it can be shown that the following "electromagnetic" equation holds[18,19]:

$$i\frac{\partial g_1^{NL}}{\partial t_{red}} + \frac{\partial^2 g_1^{NL}}{\partial x^2_{red}} + 2i\gamma_{i,red}\frac{\partial g_1^{NL}}{\partial x_{red}} + \left[i + \eta_K\left(-\Delta_{red} + U\right)\right]g_1^{NL} = ig_i\left(x_{red}, t_{red}\right). \tag{21b}$$

Equations 21 have been written using dimensionless quantities: $x_{red}$ and $t_{red}$ have been rescaled to the coupling length and the cavity built-up time of the cavity, $\gamma_{i,red}$ is also rescaled to the coupling length and denotes the reduced longitudinal wavevector component of the incident beam, $\Delta_{red}$ is the reduced detuning, $T_d$ and $L$ are the reduced Debye time and diffusion length respectively; $g_1^{NL}$ is the total reduced electric field and $\eta_K U$ is the reduced nonlinear term ( $\eta_K = +1, -1$ for focusing, defocusing nonlinearity respectively).

## 5.1 THE CASE OF COMPLETE DIFFUSION

This case corresponds to $L \rightarrow \infty$. Thus eqs. 21 write:

$$i\frac{\partial g_1^{NL}}{\partial t_{red}} + \left[i + \eta_K\left(-\Delta_{red} + U\right)\right]g_1^{NL} = ig_1\left(x_{red}, t_{red}\right), \tag{22a}$$

$$T_d\frac{\partial U}{\partial t_{red}} = -U + \left|g_1^{NL}\right|^2. \tag{22b}$$

The dimension of the phase space being 3, chaos is possible[20]. The stability analysis[21] shows that not only the negative slope part of the bistable loop is unstable but also the upper branch of the bistable cycle which exhibits a Hopf-bifurcation. Moreover a speeding-up[14] phenomena has been predicted where an *increase* of the Debye time leads to a *decrease* of the characteristic response time of the Kerr-type leaky resonator (fig. 6a).

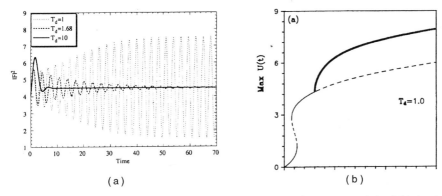

(a)  (b)

*Figure 6.* Nonlinear dynamical response (a) and bifurcation diagrams (b) of a Kerr-type leaky resonator. $\eta_K=+1$, $\Delta_{red}=3$. (a): the output intensity $\left|g_1^{NL}\right|^2$ is plotted versus time for three values of the Debye time. (b): Bifurcation diagrams for the Debye time $T_d=1$ displaying the unstable domain arising from the Hopf bifurcation. After ref. [14].

The sytem is stable for $T_d=1.68$ and $T_d=10$, it is unstable for $T_d=1$: oscillations start and reache a limit cycle which, for these numerical values, corresponds to self-pulsing. Such a behavior is typical of a Hopf bifurcation. The corresponding bifurcation diagram is reported in fig. 6b. It is seen that the Kerr-type cavity returns *faster* to equilibrium for $T_d=10$ than it does with $T_d=1.68$. This is the speeding-up phenomenon.

The dynamical regimes corresponding to self-pulsing, frequency doubling and chaos are represented on fig. 7 together with the corresponding trajectory in the phase space.

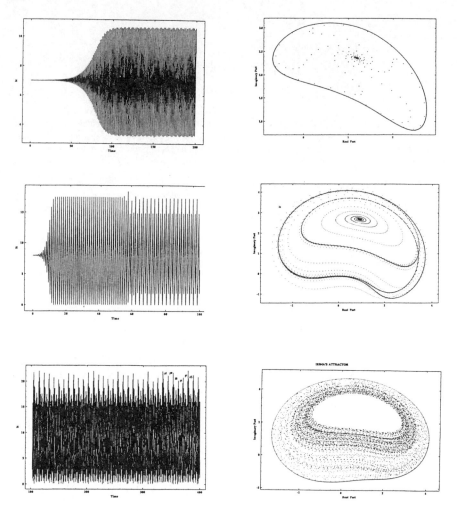

*Figure 7.* Kerr-type leaky resonators. $\eta_K=+1$, $\Delta_{red}=5$, $T_d=1$. Left: temporal evolution of $\left|g_1^{NL}\right|^2$ for different values of the steady-state solution $\left|g_1^{NL*}\right|^2$ $\left(\left[1+\left(\left|g_1^{NL*}\right|^2-\Delta_{red}\right)^2\right]\left|g_1^{NL*}\right|^2 = g_i^2$, $g_i$ : reduced amplitude of the incident plane wave): $\left|g_1^{NL*}\right|^2$ =7 (top, just beyond the Hopf bifurcation: $\left|g_1^{NL*}\right|^2$ =6.82), $\left|g_1^{NL*}\right|^2$ =8 and $\left|g_1^{NL*}\right|^2$ =9 (bottom). Right: associated phase trajectory. After ref. [13].

Chaos is present in the bottom of fig. 7. These examples illustrate a route from periodicity to chaos through successive period-doubling bifurcations.

## 5.2. SPATIO-TEMPORAL INSTABILITIES IN KERR-TYPE RESONATORS[18, 19].

Solving numerically eqs. 21 also allows to study the influence of the angle of incidence on the nonlinear dynamics of Kerr-type cavities. Some results are reported in fig. 8 for not "too" large angles of incidence where all the points at y=0 can be considered as simultaneously illuminated.

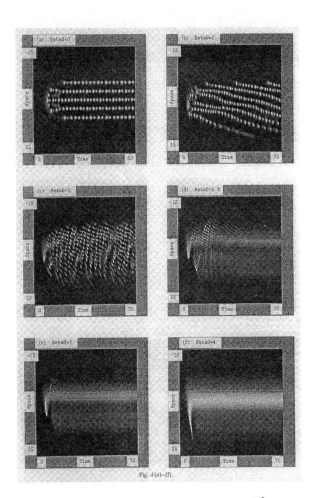

Fig. 4 (a)–(f).

*Figure 8.* Influence of the angle of incidence. The intensity of the transmitted pattern $\left|g_1^{NL}\right|^2$ is shown in the spatio-temporal $(x_{red}, t_{red})$ plane. Excitation: $g_i(x_{red}) = g_0 \exp-\left(x_{red}/X_0\right)^2$, $X_0 = 10$, $\eta_K=+1$, $\Delta_{red}=5$, $L=.1$, $T_d=1$; $g_0^2$ is linearly increased from 0 at $t_{red}=0$ to 40 at $t_{red}=5$ and then kept constant to 40 which corresponds to a steady-state plane wave output $\left|g_1^{NL*}\right|^2 = 7.144$. Normal incidence: (a): $\gamma_{i,red} = 0$. Oblique incidence: (b): $\gamma_{i,red} = 1$, (c): $\gamma_{i,red} = 2$, (d): $\gamma_{i,red} = 2.6$, (e): $\gamma_{i,red} = 3$, (f): $\gamma_{i,red} = 4$. After ref. [19].

The parameters are chosen in such a way that the system is unstable statically and dynamically. Figure 8a corresponds to normal incidence and serves as a reference: period doubling occurs since bright and less bright peaks alternate: the spatial structure corresponds to cavity solitons. Under slightly oblique incidence (fig. 8b), the behavior is completely different from the normal case since it no longer tends towards a regular regime. After up-switching, the soliton-like quasi-periodic self-pulsing peak undergoes a translational motion and disappears. In the meantime switch-on occurs in the other part of the beam and gives a second soliton-like quasi-periodic self-pulsing peak that undergoes the same evolution as the previous one. It is seen that under slightly oblique incidence the pattern exhibits a drift motion which affects the period doubling phenomenon. Under more oblique incidence (fig. 8c), a similar behavior is observed although it is accompanied by some turbulence, influencing the regularity of the pattern which looks like a flow. If the angle of incidence is further increased (figs. 8d,e), a sudden change occurs: self-pulsing disappears and the system spontaneously evolves towards a stationary regime after some hesitations. On fig. 8f, the system directly stabilizes on its stationary solution after a rapid transitory regime. Finally it is important to notice that the angle of incidence, related to $\gamma_{i,red}$, can be very small since $\gamma_{i,red}^2$ has to be compared to the resonance width (in the $\gamma_i^2$ variable) which has been here chosen equal to unity.

All the numerical results (not presented here) demonstrate the strong influence of the Debye time on the stability of the solutions. Dynamic instabilities arising from the Hopf bifurcation can be avoided for Debye time $T_d$ either several orders of magnitude smaller than the cavity built-up time $T_c$ or for $T_d$ only a few times $T_c$. Therefore stable patterns constituted of cavity solitons require either very fast or "slow" Kerr media with not too large diffusion length.

Let us now consider the highly oblique incidence and short pulse regime[22]. The interest is in the influence of the propagation effect on the spatio-temporal evolution of the leaky mode intensity $\left|g_1^{NL}\right|^2$ in the presence of Kerr nonlinearity. So eq. 21a is disregarded and eq. 21b writes:

$$\frac{\partial g_1^{NL}}{\partial t_{red}} + \frac{\partial g_1^{NL}}{\partial x_{red}} - i\left[i + \eta_K\left(-\Delta_{red} + \left|g_1^{NL}\right|^2\right)\right]g_1^{NL} = g_i\left(x_{red}, t_{red}\right). \tag{23}$$

In the highly oblique incidence regime and in the short pulse regime the expression of $g_i\left(x_{red}, t_{red}\right)$ has to include the fact that all the points at y=0 are not illuminated at the same time.

A numerical integration of eq. 23 in the case of a Kerr-type prism coupler leads to the spatio-temporal behaviour of the leaky wave intensity $\left|g_1^{NL}\left(x_{red}, t_{red}\right)\right|^2$ plotted on fig. 9, the incident pulse being gaussian both in space and time. The pulse duration is denoted $T_p$ in cavity build-up time unit. The top-left graph corresponds to a rather long incident pulse ($T_p=10$) at low intensity, i.e. in linear regime. The output pulse is more or less circular, slightly temporally asymmetric due to the fact that the response time of the resonator ($T_c=1$) is not completely negligible in comparison of $T_p$. In the nonlinear regime (top-right plot) switching occur both in the rising and falling edges of the pulse, at the level of the two sharp peaks which are visible on the plot. The sharpness of these peaks demonstrate that switching is very fast. Same graphs are plotted in the bottom line for short pulses ($T_p=1$). The interesting feature is that fast switching remains for such a short pulse. The bottom-left plot is for the linear case. The spot is tilted in the $\left(x_{red}, t_{red}\right)$ plane due to propagation of light: when time increases, the peak output intensity moves along the $x_{red}$-axis. In the nonlinear regime, a behaviour very similar to the case $T_p=10$ is noticed, for the same incident peak intensity. Fast switching still occurs and the output pulse presents the same characteristics as in the top-right plot. The only difference is that the spot is much more tilted in the $\left(x_{red}, t_{red}\right)$ plane because propagation effects are much more pronounced.

Thus such a device is able to exhibit fast switching with a response time even shorter than the cavity build-up time[22]. Such a theoretical prediction comes from the fact that propagation effect are *correctly* handled, both for the incident beam and for the excited guided wave. In particular, when dispersion is negligible, the incident spot on the base of the prism propagates at the same speed as the excited guided wave. These results show that the prism coupler appears to be very similar to nonlinear coherent coupler. In the later case two guided waves are coupled while, in the former one, a guided wave is coupled to a wave propagating in free space. The consequence is that prism coupler is predicted to be almost instantaneous at the time scale of the cavity build-up time[22].

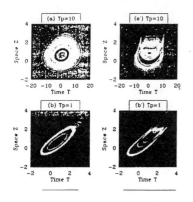

*Figure 9.* Plot, in a two dimensional-representation, of the spatio-temporal output pulse $\left|g_1^{NL}(x_{red}, t_{red})\right|^2$ for different input gaussian pulse durations.

Moreover it is interesting to notice that nonlinear slowing down of switching is predicted when all the points on the base of the prism are assumed to be *simultaneously* illuminated: typical switching times of the order of 100 are calculated where 0.1 is obtained with the correct model, leading to an error of a factor 1000 ! Therefore it is very important to properly include the propagation effects in the theoretical description of distributed couplers.

## 6. Second harmonic generation at grating couplers

Here we concentrate on the different phase-matching schemes when using grating structures. Equation 10a written at $\omega$ (index 1) and at $2\omega$ (index 2) shows that resonance occurs when:

$$\gamma_{\omega,i} + \left(\hat{p}_1 + \hat{n}_1\right)K - \tilde{\gamma}'_{\omega,\ell,\hat{n}_1} = 0 \qquad \text{at } \omega, \tag{24a}$$

$$\gamma_{2\omega,i} + \left(\hat{p}_2 + \hat{n}_2\right)K - \tilde{\gamma}'_{2\omega,\ell,\hat{n}_2} = 0 \qquad \text{at } 2\omega, \tag{24b}$$

and that several phase-matching schemes are possible:
1) direct phase-matching corresponds to:

$$\hat{n}_1 = \hat{n}_2 = p = 0 \Rightarrow 2\gamma'_{\omega,\ell} = \gamma'_{2\omega,\ell}, \tag{25a}$$

2) quasi phase-matching occurs when:

$$\hat{n}_1 = \hat{n}_2 = 0, \, \hat{p} \neq 0 \Rightarrow 2\gamma'_{\omega,\ell} + pK = \gamma'_{2\omega,\ell}, \tag{25b}$$

where a nonzero Fourier component of the nonlinear susceptibility $\left[\chi_p(\omega_v, y)\right]$ is assumed.

3) grating assisted phase-matching takes place when:

$$\hat{n}_v \neq 0 \text{ for } v=1 \text{ or } v=2 \Rightarrow 2\tilde{\gamma}'_{\omega,\ell,\hat{n}_1} = \tilde{\gamma}'_{2\omega,\ell,\hat{n}_2}, \tag{25c}$$

(neglecting the possible periodic modulation of the nonlinear susceptibility at $2\omega$)
Of course eqs. 25b,c can be combined leading to:

$$2\tilde{\gamma}'_{\omega,\ell,\hat{n}_1} + pK = \tilde{\gamma}'_{2\omega,\ell,\hat{n}_2}. \tag{25d}$$

Some different phase-matching schemes are represented in fig. 10 (with p=0, for the sake of simplicity).
Grating-assisted phase-matching allows counterpropagating second harmonic generation (fig. 10c). But due to the large value of $\left(2\gamma'_{\omega,\ell} + \gamma'_{2\omega,\ell}\right)$ small period are required, i. e. $d < \lambda$, in order to avoid situations where $\hat{n}_1, \hat{n}_2 \gg 1$. This determines the need of subwavelength gratings for counterpropagating phase-matching with the interesting consequence that the grating structure may be closed at $\omega$ and even at $2\omega$[23,24].

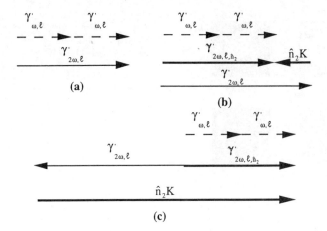

*Figure 10.* Different phase-matching schemes when p=0: (a): forward direct phase-matching between the $\hat{n}_1 = \hat{n}_2 = 0$ space harmonics at $\omega$ and at $2\omega$, (b): forward grating assisted phase-matching between the $\hat{n}_1 = 0$ space harmonic at $\omega$ and the $\hat{n}_2$ space harmonic at $2\omega$ with longitudinal wavevector component $\tilde{\gamma}'_{\ell,2\omega,\hat{n}_2}$, (c): backward grating assisted phase-matching between the $\hat{n}_1 = 0$ space harmonic at $\omega$ and the $\hat{n}_2$ space harmonic at $2\omega$ with longitudinal wavevector component $\tilde{\gamma}'_{\ell,2\omega,\hat{n}_2}$.

The authors of ref. [23, 24] report the first observation of backward grating assisted phase-matching (fig. 11): $\hat{n}_2 = 1$ and the subwavelength grating of period 160nm behaves as a closed resonator. The pair (TM$_{\omega,0}$, TM$_{2\omega,2}$) modes was involved in this counterpropagating phase-matching scheme.

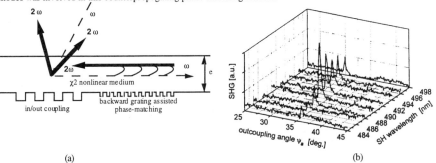

*Figure 11.* a) geometry for SHG at a subwavelength grating, the second harmonic beam is out coupled by the in/out coupling grating, b) second harmonic signal generated through the backward grating assisted phase-matching scheme. The pump wavelength is 980nm and the out-coupled angle is close to 34° (in absolute value). Gratings depth : 100nm, χ[2]-nonlinear medium: PNA-PMMA side chain polymer. After refs. [23,24].

## 7. Conclusion

Equation 19 allows to study nonlinear optical interactions at open resonators. There are as many sets of eq. 19 as frequencies present in the device. These equations apply to a wide class of nonlinear optical effects at leaky resonators: $\chi^2$, $\chi^3$, etc, at prism couplers, Fabry-Perot's, multilayer devices, grating structures, etc. Whatever the geometry and the specificities of the resonator may be, eq. 19 brings into play a set of parameters characterizing the associated *linear* device. In the case of gratings these quantities are derived from the rigorous theory of diffraction in *linear* optics where the groove depth of the grating is *not* considered as a perturbative parameter[10]. For planar systems the indices which do not refer to one of the media have to be dropped. In this sense, the theory presented here constitutes a unified approach of leaky resonators in nonlinear optics (Fig. 12).

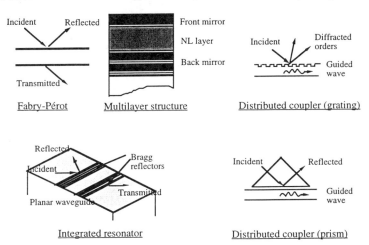

*Figure 12.* Examples of leaky nonlinear optical resonators for which eq. 19 applies.

The main advantage of the leaky mode representation is that *a* leaky mode replaces the *"packet"* of resonantly excited radiation modes. The resulting simplification leads to eqs. 19 which no longer depend on y and allow an easy insight in the physics of nonlinear optics in "open" resonators. Therefore leaky modes provide a convenient framework and a powerful tool for the study of nonlinear optical interactions in such devices.

## References

1. Vassalo, C. (1985) *Théorie des Guides d'Ondes Electromagnétiques*, Tomes 1 and 2, CNET-ENST, Eyrolles, Paris.
2. Vassalo, C. (1979) Radiating normal modes of lossy planar waveguides, *J. Opt. Soc. Am.* **69**, 311.
3. Tamir, T. and Oliner, A. A. (1963) Guided complex waves, fields at an interface, *Proc. IEE* **110**, 310.
4. Tamir, T. and Felsen, L. B. (1965) On lateral waves in slab configurations and their relation to other wave types, *IEEE Trans. Antennas and Propag.* **13**, 410.
5. Shevchenko, V. V. (1971) *Continuous Transitions in Open Waveguides*, The Golem Press, Boulder, Colorado.
6. Arfken, G. (1985) *Mathematical Methods for Physicists*, Third Edition, Academic Press.
7. Kogelnik, H. (1975) Theory of dielectric waveguides, in T. Tamir (ed.), *Integrated Optics*, Springer-Verlag, New-York.
8. Reinisch, R. (1998) Nonlinear waveguiding optics in F. Kajzar and R. Reinisch, (eds.), *Beam shaping and control with nonlinear optics*, NATO ASI Series, Plenum Press, New York.
9. Nevière, M., Popov, E., Reinisch, R., and Vitrant G. to be published *Electromagnetic resonances in nonlinear optics*, Gordon and Breach.
10. Nevière, M. (1980) The homogeneous problem in R. Petit (ed.), *Electromagnetic Theory of Gratings*, Springer-Verlag, New-York.
11. Vitrant, G. (1989) *Effets transverses et bistabilité optique dans les résonateurs optiques non linéaires*, Thèse d'Etat, Institut National Polytechnique de Grenoble (F).
12. Vitrant, G., Reinisch, R., Paumier, J. Cl., Assanto, G., and Stegeman, G. (1989) Nonlinear prism coupler with nonlocality, *Optics Lett.* **14**, 898.
13. Danckaert, J. (1992) *Nonlinear planar resonators for optical processing: simple models, including stratification, time and polarization*, Thèse Vrije Universiteit, Brussel (B.).
14. Danckaert, J., Vitrant, G., Reinisch, R., and Georgiou, M. (1993) Nonlinear Dynamics in Single Mode optical Resonators, *Phys. Rev.* **A48**, 2324.
15. Haelterman, M. (1989) *Contribution à l'étude théorique de l'optique non linéaire dans les guides et les cavités : propagation non linéaire et bistabilité*, Dissertation, Université Libre de Bruxelles, Brussel (B.).
16. Haelterman, M., Vitrant, G., and Reinisch R. (1990) Transverse effects in nonlinear planar resonators I. Modal theory, *J. Opt. Soc. Am* **B7**, 1309.

17. Vitrant, G., Haelterman, M., and Reinisch R. (1990) Transverse effects in nonlinear planar resonators II. Modal analysis for normal and oblique incidence, *J. Opt. Soc. Am* **B7**, 1319.

18. Danckaert, J. and Vitrant, G. (1993) Modulational instabilities in diffusive Kerr-type resonators, *Opt. Comm.* **104**, 196.

19. Vitrant, G. and Danckaert, J. (1994) Suppression of modulational instabilities by optical transport in nonlinear Kerr-type resonators, *Chaos, Solitons and Fractal* **4**, 1369.

20. Milonni, P. W., Shih, M. L., and Ackerhalt, J. R. (1987) *Chaos in laser-matter interactions*, World Scientific Lecture Notes in Physics, Vol. 6, World Scientific Publishing, Singapore.

21. Tabor, M. (1989)*Chaos and integrability in nonlinear dynamics*, John Wiley and Sons.

22. Fick, J. and Vitrant, G. (1995) Fast optical switching in nonlinear prism couplers, *Opt. Lett.* **20**, 1462.

23. Blau, G. (1996) *Doubleurs de fréquences en optique intégrée utilisant des polymères et des nano-réseaux de diffraction* , Thèse Institut National Polytechnique de Grenoble (F.).

24. Blau, G., Hübner, H., and Schnabel, B. (1997) Second harmonic generation using subwavelength gratings in planar waveguides, *Pure Appl. Opt.* **6**, L23.

# TOWARDS STABLE MATERIALS FOR ELECTRO-OPTIC MODULATION AND PHOTOREFRACTIVE APPLICATIONS

*From Molecular Engineering of NLO Chromophores to Elaboration of Hybrid Organic-Inorganic Materials*

M. BLANCHARD-DESCE[a], M. BARZOUKAS[b], F. CHAPUT[c],
B. DARRACQ[d], M. MLADENOVA[a], L. VENTELON[a], K. LAHLIL[c],
J. REYES[d], J.-P. BOILOT[c], Y. LEVY[d]

[a] *Département de Chimie (UMR CNRS 8640), Ecole Normale Supérieure*
*24, rue Lhomond 75231 Paris Cedex 05, France*
[b] *Institut Charles Sadron*
*6, rue Boussingault, 67083 Strasbourg Cedex, France*
[c] *Laboratoire de Physique de la Matière Condensée (UMR CNRS 7643D), Ecole Polytechnique*
*91128 Palaiseau, France*
[d] *Laboratoire Charles Fabry de l'Institut d'Optique (UMR CNRS 8501)*
*Bâtiment 503, B.P. 147, 91403 Orsay Cedex, France*

## 1. Introduction

A large effort have been devoted to the preparation of organic polymeric materials for electro-optic modulation [1-3]. These materials contain push-pull chromophores (i.e. combining electron-releasing and electron-withdrawing groups interacting through a conjugated linker) either incorporated as guest in the polymeric matrix (doped polymers) or grafted onto (or into) the polymeric matrix (functionalized polymers) [4-5]. By heating the polymer above its glass transition temperature ($T_g$), orientation of the dipolar chromophores can be achieved via application of a strong external electric field. After cooling to room temperature, noncentrosymmetrical poled-polymeric materials are obtained. Such materials are interesting in view of both processability and chemical flexibility that allows for "engineering" of the nonlinear responses. When optimized chromophores are incorporated and poled in polymers, materials with large electrooptical coefficient can in principle be obtained. However, besides the optimization of chromophores aiming at obtaining large molecular figure of merit (FOM), a number of additional issues are to be considered if materials with both large and permanent nonlinear responses are to be achieved. For instance, an interesting way to improve the orientational stability consists in using polymers with very high $T_g$ (such as polyimides for instance) [6]. However, this demands organic chromophores with high thermal stability. Also, increasing the chromophore concentration in order to have larger $\chi^{(2)}$ (and electro-optic coefficient) values can be detrimental since dipolar interactions between push-pull chromophores can favor antiparallel association thus impeding the poling process, as pointed out by Dalton and coworkers [7].

*F. Kajzar and M.V. Agranovich (eds.), Multiphoton and Light Driven Multielectron Processes in Organics: New Phenomena, Materials and Applications, 183–198.*

Push-pull chromophores have also been used as molecular components in organic photorefractive polymers [5, 8-10]. Photorefractive polymers (PRP) are composite systems that combine photogenerators, photoconductors, traps and non-linear optical chromophores. In such systems, the mechanism of the light-induced refractive index change is a space-charge field induced photorefractivity. Basically, two classes of photorefractive polymers have been investigated. Guest-host systems are based upon the inclusion of push-pull chromophores at a high concentration in a photoconductive polymer matrix. In these materials that have low $T_g$ (i.e. close to room temperature), a spatially modulated irradiation leads to a space charge-field that induces a spatially periodic orientation of the push-pull chromophores in addition to the uniform poling due to application of an external electric field. As a result, the Pockels electro-optic effect (PEO) is enhanced by a factor of 2 and the periodic orientation induces a spatially modulated birefringence that also generates a modulated refractive index change [10-11]. This orientational birefringence (OB) actually is the dominant contribution to the photorefractivity for the low-$T_g$ PRP that have been designed up to now, leading to guest-host systems with excellent photorefractive performances [5, 12-14]. These materials yet present several drawbacks. For instance, the low glass transition temperature that is responsible for the orientational birefringence phenomenon can also facilitate phase segregation processes and prevents permanent poling as a consequence of the chromophore mobility. Maintaining polar orientation of the push-pull chromophores thus requires the application of a strong electric field for *in situ* poling. These limitations could be circumvented with high-$T_g$ systems where permanent poling could be achieved and chromophore migration impeded. A much lesser number of studies has been devoted to this second class, although a few high-$T_g$ functionalized PRP were shown to have interesting photorefractive properties [15-20]. Such materials hold promise for applications and do not require the application of external field to assist the photo-induced building-up of the periodic space charge since the strong internal field resulting from the permanent poling of the push-pull chromophores can play this role [17]. Thus high-$T_g$ systems present several advantages but require significant improvement both at the molecular level - by designing optimized chromophores with very large molecular PEO figure of merits and suitable thermal stability and solubility - and at the supramolecular level - by designing functionalized polymers with large orientational stability and photoconductive properties - to yield high-$T_g$ PRP with optimized photorefractive performances.

In this connection, we have designed *hybrid organic-inorganic materials* based upon the incorporation of push-pull chromophores in a rigid amorphous inorganic matrix that can be cross-linked by smooth heating. To prepare these materials, we have used the sol-gel route allowing for mild synthesis conditions (thus preventing chromophore damaging) and for the fabrication of amorphous solids of various shapes (thin films, monoliths...). Such materials combine the excellent optical and mechanical properties of the glassy inorganic state and the "flexible" optical properties of the organic dyes. In order to elaborate stable materials with excellent nonlinearities, we have synthesized sol-gel systems wherein push-pull chromophores combining large molecular FOM and suitable thermal stability are covalently attached to the rigid backbone of a silica-based matrix also containing charge transporting carbazole units grafted as pendant side groups.

## 2. Design and evaluation of push-pull chromophores

### 2.1.   FIGURES OF MERIT OF PUSH-PULL CHROMOPHORES

Most push-pull compounds used for their quadratic nonlinear optical properties in electro-optic or photorefractive polymers have their donor-acceptor axis along the molecular long axis [21]. As a result of the 1-dimensional (1D) intramolecular charge transfer (ICT), their linear and non-linear polarizability tensors are strongly anisotropic. The diagonal component $\alpha_{//}$ of the linear polarizability along the ICT axis is larger than the one, $\alpha_{\perp}$, perpendicular to this direction. The difference between these components is noted, hereafter, $\delta\alpha$ (*i.e.* $\delta\alpha = \alpha_{//} - \alpha_{\perp}$). The quadratic polarizability tensor is approximately 1D. The main quadratic polarizability coefficient, noted $\beta$, is the diagonal component along the ICT axis. For such systems, the FOM expressions for both high- and low-$T_g$ PRP are well known.

In a pre-poled high-$T_g$ PRP, the main contribution to the PR effect arises from the PEO response, linear in the space-charge field. The FOM of push-pull molecules characterizing the PEO response is given in equation 1:
$$\text{FOM(PEO)} = \mu_g \beta(0) \tag{1}$$
In a low-$T_g$ PRP, conventionally, the PEO contribution to the FOM of a push-pull molecule is also given by Equation 1. The FOM associated with the additional contribution due to the OB mechanism is given by equation 2 [22-23]:
$$\text{FOM(OB)} = \frac{2}{9}\frac{\mu_g^2}{kT}\alpha(0) \tag{2}$$
Since the push-pull molecules are poled *in situ*, $T$ corresponds to the room temperature. Equation 1 has been derived by using the Taylor convention for defining $\beta$. The overall low-$T_g$ FOM results from the sum of the two contributions.

### 2.2.   THE TWO-LEVEL TWO-STATE MODEL : A GUIDE FOR MOLECULAR ENGINEERING OF PUSH-PULL CHROMOPHORES

The two-level two-state model (TLTS) [24-26] provides a helpful guide for the molecular engineering of the linear and nonlinear polarizabilities of push-pull compounds . This model is based upon the two-level approximation which takes into account only the ground and the first-excited state involved in the ICT phenomenon [27-28] and a simple two-valence bond-state description of the push-pull chromophores [24-26, 29-30].

The two-level approximation allows to correlate the linear and nonlinear polarizabilities with the ICT transition characteristics, namely the transition energy $E_{eg}$, the transition dipole $\mu_{eg}$ and the change of dipole between the ground and excited state ($\Delta\mu = \mu_e - \mu_g$):
$$\alpha(0) = 2\mu_{ge}^2 / E_{eg} \tag{3}$$
$$\beta(0) = 6\mu_{ge}^2 \, \Delta\mu / E_{eg}^2 \tag{4}$$
where the (hyper)polarizabilites are defined using the Taylor series expansion.

In the two valence bond state description, both the ground state and the excited state of the push-pull compound are expressed as linear combinations of the neutral (N) and the zwitterionic (Z) limiting-resonance forms :

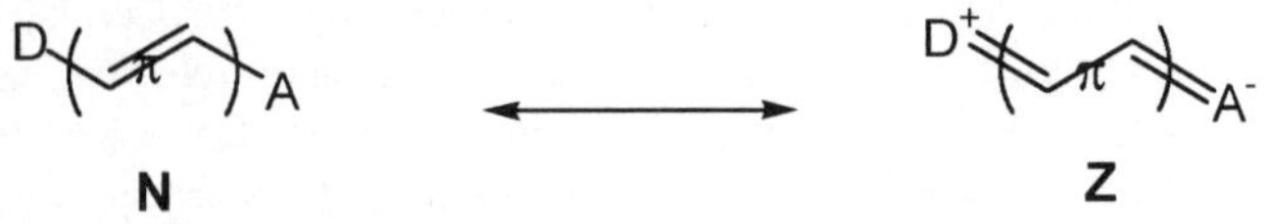

**N**                                    **Z**

The four relevant parameters of the TLTS model are the energy gap $V$ and the coupling element $t$ between the N and Z forms, and their dipole moment $\mu_N$ and $\mu_Z$. A general coordinate MIX, which characterizes the amount of mixing between the N and the Z forms, both in the ground and in the first-excited state can be defined. It corresponds to the difference between the weights of the Z and N form in the ground state and its expression is given by equation 5:

$$\text{MIX} = -\frac{V}{\sqrt{V^2 + 4i^2}} \tag{5}$$

A positive MIX corresponds to a dominant Z character in the ground state (*i.e.* $V<0$). Conversely, a negative MIX indicates a dominant N character in the ground-state (*i.e.* $V>0$). The special case MIX = 0, when both forms contribute equally (*i.e.* $V=0$), is called the cyanine limit in analogy to the symmetrical cyanines.

The TSTL model yields analytical expressions of the ground-state dipole and of the linear and the non-linear static polarizabilities, as functions of MIX [24]. In most cases, the dipole of the N form $\mu_N$ can be considered as negligible in comparison to that of the Z form $\mu_Z$ . The analytical expressions of the different figures of merits FOM(PEO) and FOM(OB) thus derived are presented in equations 6-7 .

$$\text{FOM(PEO)} = \frac{3\mu_Z^4}{16t^2}(-\text{MIX})(1+\text{MIX})(1-\text{MIX}^2)^2 \tag{6}$$

$$\text{FOM(OB)} = \frac{\mu_Z^4}{72tkT}(1+\text{MIX})^2(1-\text{MIX}^2)^{\frac{3}{2}} \tag{7}$$

The normalized variations of these two FOM are displayed in Figure 1, showing the optimum MIX values leading to either maximum FOM(PEO) or FOM(OB) [31-33]. The normalized FOM(PEO) peaks negatively for MIX=1/2 (*i.e.* halfway between the cyanine limit and the Z form) and positively for MIX=-1/3 (*i.e.* slightly closer to the cyanine limit than to the N form) where it retains about 40% of its maximum negative value. The normalized FOM(OB) peaks positively for MIX=2/5 (*i.e.* slightly closer to the cyanine limit than to the Z form).

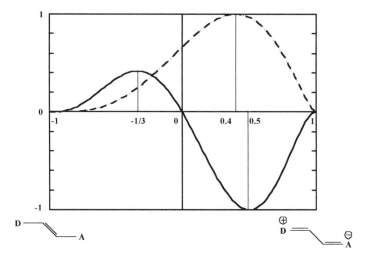

*Figure 1.* Normalized variations of FOM(PEO) (—) and FOM(OB) (- - -) versus MIX

It is interesting to note that the simple TLTS model reproduces the FOM variations as a function of a structural parameter, the bond order alternation (BOA), derived by Marder and coworkers from more refined calculations [23].

One way to design molecules with large FOM is to engineer the MIX values (by playing on the energy gap between the N and Z forms for instance) so as to approach optimum MIX values, as previously done by Marder and coworkers by tuning the push-pull structure towards optimum BOA. An additional way to engineer molecules with large figures of merit is to enhance the peak magnitudes, an interesting strategy that is overlooked in the BOA picture. In this respect, the TLTS model can be helpful since its provides insight into the parameters which control the peak magnitudes:

$$\text{FOM(PEO)}_{max} = +0.03 \frac{\mu_Z^4}{t^2}$$

$$\text{FOM(PEO)}_{min} = -0.08 \frac{\mu_Z^4}{t^2}$$

(8)

$$\text{FOM(OB)}_{max} = +0.08 \frac{\mu_Z^4}{tkT}$$

(9)

Equations 8-9 show that the peaks magnitudes of the FOM(PEO) and FOM(OB) are related, the relevant scaling factor determining the respective magnitude of FOM(PEO)$_{max}$ versus FOM(OB)$_{max}$ being $kT/t$ [31]. A simple way to enhance both the FOM(PEO) and FOM(OB) peak magnitudes is to lengthen the conjugation path since this will lead to both an increase of $\mu_Z$ and a decrease of the coupling element $t$. In addition, this is expected to lead to a steeper increase of FOM(PEO)$_{max}$ than FOM(OB)$_{max}$. A logical way for designing high-$T_g$ PRP with efficiency approaching that of prototype low-$T_g$ PRP would be to increase significantly the FOM(PEO) by increasing the conjugated chain while being close to the optimum MIX values corresponding either to positive or negative peaks. In the following, we will focus on the positive peak (i.e. predominance of the neutral form) rather than on the negative peak (i.e. predominance of the zwitterionic form), although the latter leads to roughly twice larger peaks amplitudes. This choice seems reasonable when taking into account the material stage. The large dipoles associated with the zwitterionic structures can be detrimental owing to strong dipolar interactions [33]. In addition zwitterionic structures could possibly act as ionic traps and hamper photoconduction.

## 2.3. PUSH-PULL POLYENES FOR HIGH $T_g$ PRP

Based on our previous work on push-pull polyenes, we have focused on the push-pull functionalization of the polyenic backbone in order to obtain large FOM. This strategy has already proven successful for the design of push-pull chromophores showing giant static quadratic and cubic polarizabilities [34-39]. The FOM of different series of push-pull phenylpolyenes and diphenypolyenes (Figure 2) have been calculated by using equations 1-4 and experimental data derived from absorption and electroabsorption measurements [32, 37-38, 40] conducted in dioxane.

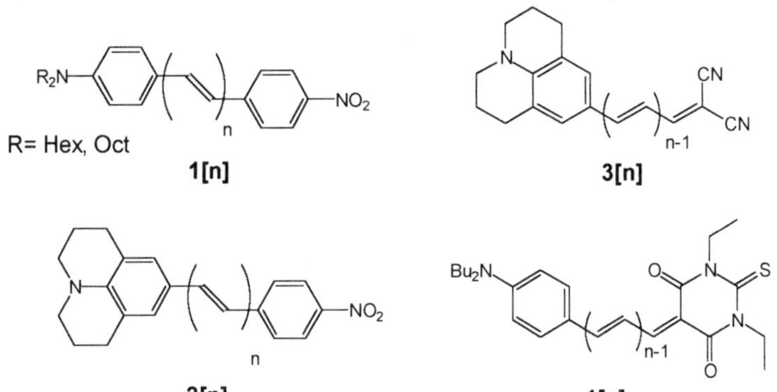

Figure 2. Structures of series of push-pull diphenylpolyenes (**1-2**) and phenylpolyenes (**3-4**)

Push-pull diphenylpolyenes of series **1-2** and phenylpolyenes of series **3-4** lie in the range of negative MIX values (i.e. predominance of the neutral form in the ground state), as shown by $^1$H NMR studies [41] and in agreement with their positive solvatochromism behaviour [37-38]. The different FOM values are collected in Table 1. The FOM derived from EOAM measurements [32, 40] performed on prototypical push-pull chromophores (Figure 3) shown to lead to low-$T_g$ polymers with very high photorefractive efficiency [10, 12-14] have also been included in Table 1, for comparison purpose. DCCEAHT is a structural analogue of the more soluble push-pull chromophore DHDC-MPN [13].

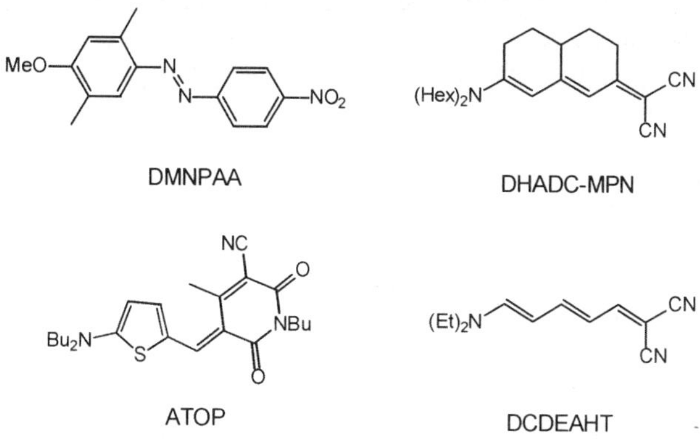

DMNPAA

DHADC-MPN

ATOP

DCDEAHT

*Figure 3.* Prototypical push-pull chromophores (DMNPAA, ATOP, DHDC-MPN) used in low-$T_g$ PRP

TABLE 1. FOM(PEO) and FOM(OB) derived from EOAM data in dioxane
for series of push-pull phenylpolyenes and diphenylpolyenes ($T$ = 300 K)

| Push-pull chromophore | FOM(PEO) =FOM (high-$T_g$) | FOM(OB) | FOM(OB)+FOM(PEO) =FOM (low-$T_g$) |
|---|---|---|---|
| | $10^{-46}$ esu | $10^{-46}$ esu | $10^{-46}$ esu |
| 1\|1\| | 18 | 54 | 73 |
| 1\|2\| | 42 | 112 | 154 |
| 1\|3\| | 60 | 146 | 206 |
| 1\|4\| | 84 | 191 | 275 |
| 1\|5\| | 111 | 232 | 343 |
| 2\|1\| | 29 | 81 | 110 |
| 2\|2\| | 51 | 126 | 178 |
| 2\|3\| | 79 | 174 | 253 |
| 2\|4\| | 97 | 206 | 303 |
| 2\|5\| | 153 | 287 | 441 |
| 3\|1\| | 14 | 128 | 142 |
| 3\|2\| | 37 | 180 | 218 |
| 3\|3\| | 81 | 273 | 353 |
| 3\|4\| | 119 | 344 | 463 |
| 3\|5\| | 238 | 518 | 756 |
| 4\|1\| | 19 | 141 | 160 |
| 4\|2\| | 61 | 258 | 318 |
| 4\|3\| | 122 | 290 | 413 |
| 4\|4\| | 206 | 379 | 585 |
| 4\|5\| | 400 | 501 | 901 |
| DMNPAA | 9 | 42 | 52 |
| ATOP | 10 | 523 | 533 |
| DCDEAHT | 20 | 319 | 339 |

It is interesting to note that both FOM(PEO) and FOM(OB) increase significantly with the length of the polyenic chain and that FOM(PEO) is found to show a steeper enhancement than FOM(OB), following the qualitative prediction of the simple TLTSM. As a result the PEO contribution amounts to about half the OB contribution for the longest derivatives of series **1-3**. This effect is even more pronounced for series **4** for which the PEO contribution of molecule **4[5]** is comparable to its OB contribution.

These experimental trends show that the molecular engineering approach towards efficient high-$T_g$ polymers based on the lengthening of the polyenic chain is a promising strategy. In this respect, we note that elongated push-pull phenylpolyenes of series **3** and **4** display FOM(PEO) (*i.e.* the relevant FOM for high-$T_g$ applications) comparable to the FOM(OB) of DHDC-MPN and approaching that of ATOP. These prototypical push-pull chromophores were designed for low-$T_g$ PRP applications [13-14] and consequently display a dominant OB contribution to their low-$T_g$ FOM. Based on the comparison between the low-$T_g$ FOM of these optimized chromophores and the high-$T_g$ FOM of elongated push-pull phenylpolyenes of series **3-4**, it should be possible to design high-$T_g$ PRP with comparable photorefractive characteristics as that of the best low-$T_g$ PRP reported recently. It should also be noted that push-pull diphenylbutadiene **2**[2] show a high-$T_g$ FOM comparable to the low-$T_g$ FOM of DMNPAA, one of the un-optimized prototypical chromophores which was used as the active NLO component in a low-$T_g$ PRP with high optical gain and diffraction efficiency [10, 12].

## 3. Design and preparation of hybrid organic-inorganic materials

We have prepared fully functionalized systems wherein both push-pull phenylpolyenes (or diphenylpolyenes) and charge-conducting carbazole units where attached to the inorganic backbone. This strategy is expected to lead to better NLO performances since it allows for higher concentration of push-pull chromophore and better temporal stability (by preventing molecular migration and thus phase segregation and aggregation processes and slowing-down orientational randomization).

We have focused on polyenes of intermediate length in order to maintain suitable thermal stability, solubility, transparency in the near I.R. and poling ability. As shown before, the lengthening of the polyenic chain leads to both a red-shift of the ICT absorption band and a decrease of thermal stability [38-39, 42]. In addition, push-pull polyenes are rod-shaped molecules whose rotational mobility is expected to decrease with increasing chain length due to the larger free volume associated with electric-field assisted dipolar orientation. This phenomenon should slow-down the poling process. This explains why the elongation strategy, which also leads to giant low-$T_g$ FOM as shown in Table 1, has serious drawbacks in the case of low-$T_g$ PRP because it would seriously delay the orientational birefringence response. On the other hand, in the case of pre-poled high $T_g$ PRP, although the elongation effect will render the poling process more arduous, it should also hinder orientational relaxation. If appropriate poling conditions could be met and combined with cross-linking reactions, materials having polar order with high temporal stability would be achieved. In this connection, hybrid-organic-inorganic materials prepared by the sol-gel route and incorporating optimized (in terms of nonlinearity-thermal stability-solubility-transparency compromise) push-pull polyenes were particularly promising.

## 3.1. SYNTHESIS OF PUSH-PULL CHROMOPHORES

In order to obtain functionalized systems wherein the push-pull chromophores are covalently linked to the polymer backbone via a flexible spacer, we have prepared push-pull chromophores bearing free hydroxyl endgroups (Figure 5).

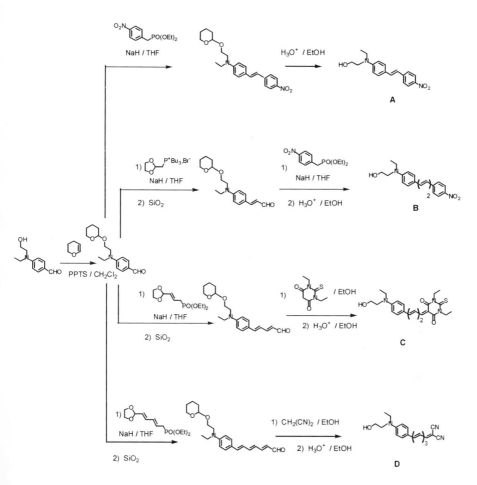

*Figure 4.* Synthesis scheme for the preparation of push-pull diphenylpoyenes **A-B** and phenylpolyenes **C-D**

Push-pull stilbene **A** (analogue to **1**[1]) was prepared via a Horner-Emmons-Wittig reaction followed by acidic cleavage of the tetrahydropyranyl protection. Push-pull diphenylbutadiene **B** (analogue to **1**[2]) and push-pull phenylpolyenes **C** and **D** (analogues to **4**[3] and **3**[4] respectively) have been synthesized following an efficient protection-vinylic extension-condensation-deprotection sequence which allows for the preparation of push-pull phenyl and diphenylpolyenes with good yields and high stereoselectivity [43].

## 3.2. SYNTHESIS OF MOLECULAR PRECURSORS

As illustrated in Figure 6, the functionalized alkoxysilane monomer bearing the charge transporting carbazole moiety attached via a flexible spacer was prepared by reacting carbazole-9-carbonyle chloride with 3-aminopropyltriethoxysilane, yielding the charge-transporter molecular precursor Si-K [44]. Alternatively, the functionalized silane monomers bearing push-pull chromophores attached via a flexible spacer were prepared by reacting the chromophores bearing free hydroxyl groups with 3-isocyanatopropyltriethoxysilane, leading to the push-pull molecular precursors Si-DA.

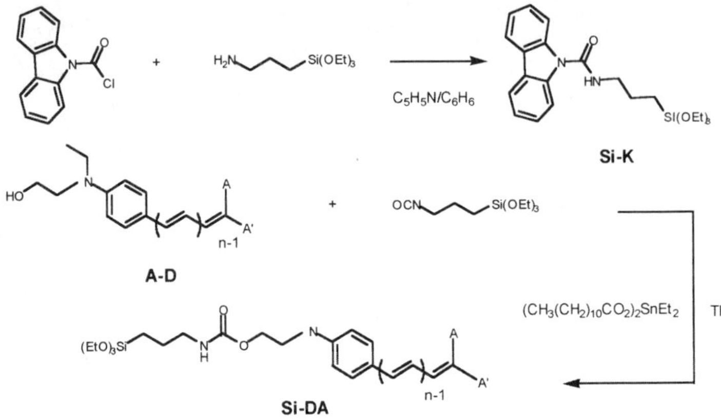

Figure 5. Synthesis of triethoxysilane functionalized monomers.

## 3.3. PREPARATION OF THE SOL-GEL THIN FILMS

The sol-gel process is based upon the sequential hydrolysis and condensation of alkoxides (such as silicon alkoxides) initiated by acidic and basic aqueous solutions in the presence of a mutual cosolvent. Given the mild synthesis conditions offered by the sol-gel route, hybrid organic-inorganic gels can easily be obtained. Organic dyes with various optical properties can be incorporated into the glassy matrix to yield, after drying, doped or functionalized xerogels with specific optical properties. In particular, there has been a growing interest for hybrid organic-inorganic for second-order NLO in the last few years [45-53].

We have used the sol-gel technique to prepare hybrid thin films incorporating both push-pull polyenic and carbazole charge-transporting units. The coating solutions were prepared according to an experimental protocole described in [44], by copolymerisation of functionalized triethoxysilane monomers with the tetraethoxysilane (TEOS) cross-linking agent. The TEOS: (Si-K + Si-DA) ratio was adjusted to 1:6, and the Si-K : Si-DA ratios were modulated in order to control the dipolar interactions in the material following the approach demonstrated by Dalton and coworkers [7].

Films of several micrometers in thickness were elaborated by spin-coating then cured for 12-24 h at 100-160 °C. Orientation of the push-pull chromophores within the material was then performed using either the single-point corona poling technique (5-6 kV/cm). Poling was carried-out for 1-2 h at 110-140 °C depending on the thermal stability of the push-pull chromophores. The electric field was maintained for 1 hour during the cooling process.

## 3.4. OPTICAL AND ELECTROOPTICAL PROPERTIES OF HYBRID THIN FILMS

### 3.4.1 Absorption properties

Push-pull chromophores **A-D** show positive solvatochromism, as indicated by the absorption data reported in Table 2 in solvents of different polarity. Such chromophores can thus serve as probes for polarity and microenvironment [54], a typical feature which is of particular interest for materials. Examination of the variation of the maximum absorption wavelength of hybrid materials functionalized with various concentrations of push-pull chromophores also provide interesting information. For push-pull diphenybutadiene **B**, the absorption wavelength remains roughly unchanged in the concentration range investigated and can be attributed to a chromophore experiencing a polar environment. In contrast, experimental studies of the sol-gel materials made from push-pull chromophore **C** show that increasing the concentration of dipolar chromophore induces changes in the absorption spectra of the chromophore as shown in Figure 7.

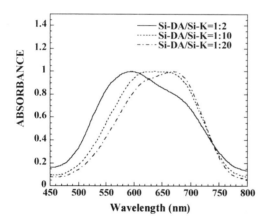

Figure 6. Absorption spectra of sol-gel films functionalized with push-pull chromophore **C** : Si-DA/Si-K =1:2 (-) , Si- Si-DA/Si-K =1:10 (- -), Si-DA/Si-K =1:20 (-.-)..

Increasing dipolar concentration leads to a broadening of the absorption band and to the progressive rise of a blue-shifted band ($\lambda_{max}$ =595 nm) overlapping with the "monomer" band ($\lambda_{max}$ = 662 nm). This blue-shifted band is predominant for a Si-**C** : Si-K ratio of 1:2 and can presumably be attributed to dipole-dipole interactions leading to H-type aggregates [55]. This effect is expected to be more pronounced for higher chromophore concentrations (since increasing concentration results in smaller average distance between dipolar chromophores) and dipole moments.

### 3.4.2. Electrooptic coefficients

Electro-optic coefficients $r_{33}$ measurements were carried-out using the reflection method [56] at $\lambda = 633$ nm or 830 nm in order to operate away from the absorption of the push-pull chromophore. The experimental results obtained with the various hybrid thin films are collected in Table 2 along with the maximum absorption wavelength ($\lambda_{max}$) of the thin-film materials.

TABLE 2. Optical linear and nonlinear characteristics of the sol-gel thin films

| Push-pull Chromophore | $\lambda_{max}$ [a] | | | DA/K [b] | $\lambda_{max}$ | $\lambda$ | $r_{33}$ |
| --- | --- | --- | --- | --- | --- | --- | --- |
| | TOL | CHCL3 | DMSO | | | | |
| | nm | nm | nm | | nm | nm | pm.V$^{-1}$ |
| A | 435 | 437 | 465 | 1/2 | 442 | 633 | 9 |
| B | 450 | 450 | 475 | 1/2 | 466 | 633 | 15 |
| B | | | | 1/5 | 466 | 633 | 6.5 |
| B | | | | 1/5 | 466 | 830 | 2 |
| C | 576 | 598 | 662 | 1/10 | 625 | 830 | 7 |
| D | 534 | 545 | 565 | 1/5 | 560 | 830 | 40 |

[a] maximum absorption wavelength of the push-pull chromophore in solution
[b] molar ratio of push-pull chromophore units versus carbazole units

Comparaison between the electro-optic coefficients obtained with the sol-gel films made from either push-pull chromophore **A** or **B** (which bear same donor and acceptor endgroups) indicates that elongation of the polyenic chain actually results in a higher electrooptic coefficient. The elongation strategy which leads to increasing static FOM(PEO), thus seems an effective way to enhance the macroscopic nonlinearity. This underlines the appeal of the approach based upon increasing the FOM. Finally, a *very large electro-optic coefficient of 40 pm.V$^{-1}$* has been obtained with sol-gel films functionalized with chromophore **D**. Comparison with the hybrid material made from chromophore **B** (which displays an electro-optic coefficient of 2 pm.V$^{-1}$ in similar conditions) shows that engineering of the molecular FOM can be highly profitable.

However, the bare molecular FOM is not sufficient to account for macroscopic nonlinearities since intermolecular interactions can play a significant role. This is clear from the examination of the materials made from chromophore **C** which has a FOM only slightly smaller than that of chromophore **D** (Table 1). Vanishing macroscopic nonlinearity is obtained for Si-**C** : Si-K ratio of 1:2 corresponding to predominant absorption of the associated species ($\lambda_{max}$ =595 nm). Accordingly, decreasing the chromophore concentration (and thus increasing the average intermolecular distance between dipolar chromophores) to a ratio of 1:10 leads to the emergence of a macroscopic nonlinearity (Table 1) At opposite, chromophore **B** leads to increasing electro-optic coefficient for increasing Si-**B**/Si-K ratio (up to 1:2) presumably due to weaker intermolecular interactions.

These observations emphasize that intermolecular interactions have to be taken into account at the material stage, as already pointed out by Dalton and coworkers [7]. In doing so, one should be aware of the role of the polar hybrid environment that is liable to increase the polarization of the push-pull chromophore via induced polarization. The magnitude of this effect depends on the polarizability of the push-pull chromophores. EOAM measurements [34-35] performed in dioxane for analogous molecules provide dipole moments and polarizabilities values of $\mu$ = 7.2, 7.5, 8.6 and 10.5 D ; $\alpha(0)$ = 20, 35, 75, 60 $10^{-24}$ esu for chromophores **A**, **B**, **C** and **D** respectively. This suggests much stronger dipolar interactions in the case of chromophores **C** and **D**, explaining why smaller optimum concentrations have to be used in the case of the latter chromophores.

### 3.4.3. Photoconductivity

The carbazole moieties are potential charge carriers. If combined with suitable photogenerators, photoconductivity can in principle be obtained. In this connection, we have prepared poled thin-film materials made of Si-**B** and Si-K (1:2 ratio) doped with 2,4,7-trinitro-9-fluorenone (TEOS : TNF = 1 : 0.05). This allows for photoinduced charge generation via irradiation in the carbazole/TNF charge-transfer complex absorption band (633 nm). As shown previously for other sol-gel films [57-58], the poled sample exhibits a non-classical behaviour : photoconduction occurs with a zero applied electric field. This result confirms the presence in the poled sample of a static internal electric field $E_i$ . This strong internal field is the sign of the polar order in the materials due to permanent orientation of push-pull chromophores.

## 4. Conclusion

By implementing a molecular engineering strategy towards large electro-optic molecular figures of merit (using a simple two-state two-level model) and suitable thermal stability, and by controlling the intermolecular dipole-dipole interactions via tuning the push-pull chromophore concentration, we have obtained *hybrid thin-films with very large electro-optic coefficients*. Photoconductivity experiments provide indication of the existence of a strong internal electric field in the poled sol-gel materials. These combined characteristics are appealing since association of photoconductivity and large electro-optic coefficients hold promise for the design of stable hybrid materials of suitable environmental stability and exhibiting a purely electro-optic *photorefractivity* without needing the application of an external electric field. We are currently investigating this route.

## 5. Acknowledgements

We thank the Ministère de l'Education Nationale, de la Recherche et de la Technologie for a PAST position to M.M. We also thank the Ministère de l'Education Nationale, de la Recherche et de la Technologie facilities and Fundacion UNAM, Conacyt, DGEP-UNAM, DGAPA-UNAM IN 10986 for a scholarship to J.R.

196

## References and Notes

1.  Heeger, A.J., Oreinstein, J. and Ulrich, D.R. eds. (1988) *Nonlinear Optical Properties of Polymers*, MRS Symposium Proceedings **109**, Materials Research Society, Pittsburg.
2.  Prasad, P.N. and Ulrich, D.R. eds. (1988) *Nonlinear Optical and Electroactive Polymers*, Plenum, New York.
3.  Messier, J., Kajzar, F., Prasad, P.N. and Ulrich, D.R. (1989) *Nonlinear Optical Effects in Organic Polymers*, Kluwer Academic Publishers, Dordrecht.
4.  Marks, T. J. and Ratner, M. A. (1995) Design, synthesis and properties of molecule-based assemblies with large second-order optical nonlinearities, *Angew. Chem. Int. Ed. Engl.*, **34**, 155-173.
5.  Marder, S.R., Kippelen, B., Jen, A.K.-Y. and Peyghambarian, N. (1997) Design and synthesis of chromophores and polymers for electro-optic and photorefractive applications", *Nature*, **388**, 845-851.
6.  see ref. 4-5 and references cited therein.
7.  Harper, A.W., Sun, S., Dalton, L., Garner, S.M., Chen, A., Kannuri, S., Steier, W.H. and Robinson, B.H. (1998) Translating microscopic nonlinearity into macroscopic optical nonlinearity : the role of chromophore-chromophore electrostatic interactions", *J. Opt. Soc. Am. B*, **15**, 329-337.
8.  Günter, P. and Huignard, J.P. eds. (1988), *Photorefractive Materials and Their Applications*, Springer-Verlag, Berlin.
9.  Moerner, W.E. and Silence, S.M. (1994) Polymeric photorefractive materials, *Chem. Rev.*, **94**, 127-155.
10. Meerholz, K., Volodin, B.L., Sandalphon, B., Kippelen, B. and Peyghambarian, N. (1994) A photorefractive polymer with a high optical gain and diffraction efficiency near 100%, *Nature*, **371**, 497-500.
11. Moerner, W.E., Silence, S.M., Hache, F. and Bjorklund, G.C. (1994) Orientationally enhanced photorefractive effects in polymers", *J. Opt. Am. Soc. B*, **11**, 320-330.
12. Grunnet-Jepsen, A., Thomson, C.L., Twieg, R.J. and Moerner, W.E. (1997) High performance photorefractive polymers with an improved stability, *Appl. Phys. Lett.*, **70**, 1515-1517.
13. Hendrickx, E., Herlocker, J., Maldonado, J.L., Marder, S.R., Kippelen, B., Persoons, A. and Peyghambarian, N. (1998) Thermally stable high-gain photorefractive polymer composites based on a tri-functional chromophore, *Appl. Phys. Lett.*, **72**, 1679-1681.
14. Meerholz, K., De Nardin, Y., Bittner, R., Wortmann, R. and Würthner, F. (1998) Improved performance of photorefractive polymers based on merocyanine dyes in a polar matrix, *Appl. Phys. Lett.*, **73**, 4-6.
15. Yu, L., Chan, W., Bao, Z. and Cao. S.X.F. (1993) Photorefractive polymers. 2. Structure design and property characterization", *Macromolecules*, **26**, 2216-2221.
16. Peng, Z., Bao, Z. and Yu, L. (1994) Large photorefractivity in an exceptionnaly thermostable multifunctionnal polyimide, *J. Am. Chem. Soc.*, **116**, 6003-6004.
17. Peng, Z., Gharavi, A.R. and Yu, L. (1996) Hybridized approach to new polymers exhibiting large photorefractivity", *Appl. Phys. Lett.*, **69**, 4002-4004.
18. Cheng, N., Swedek, B. and Prasad P.N. (1997) Thermal fixing of refractive index gratings in a photorefractive polymer, *Appl. Phys . Lett.*,**71**, 1828-1830.
19. Scholter, S., Hofmann, U., Hoechsteller, K., Baüml, G., Haarer, D., Ewert, K. and Eisenbach, C.D. (1998) Permanently poled fully functionalized photorefractive polyester, *J. Opt. Soc. Am.*, **15**, 2560-2565.
20. Nishizawa, H. and Hirao, A. (1997) Writing a grating in the photorefractive polymers with no applied field, *SPIE proc.*, **3144**, 207.
21. Blanchard-Desce, M., Marder, S.R. and Barzoukas M. (1996) Supramolecular chemistry for quadratic nonlinear optics in D.N. Reinhoudt (ed.), *Comprehensive Supramolecular Chemistry* , Vol. 10, Pergamon, 833-863.
22. Wortmann, R., Popa, C., Twieg, R.J., Geletneky, C., Moylan, C.R., Lundquist, P.M., DeVoe, R.G., Cotts, P. M., Horn, H., Rice, J.E., and Burland, D.M. (1996) Design of optimized photorefractive polymers: a novel class of chromophores, *J. Chem. Phys.* **105**, 10637-10647.
23. Kippelen, B. Meyers, F. Peyghambarian, N. and Marder, S. R. (1997) Chromophore design for photorefractive applications, *J. Am. Chem. Soc.* **119**, 4559-4560.

24. Barzoukas, M. Runser, C. Fort, A. and Blanchard-Desce M. (1996) A two-state description of (hyper)polarizabilities of push-pull molecules based on a two-form model", *Chem. Phys. Lett.* **257**, 531-537.

25. Blanchard-Desce, M. and Barzoukas, M. (1998) Two-form two-state analysis of polarizabilities of push-pull molecules, *J. Opt. Soc. Am. B* **154**, 302-307.

26. Thompson., W., Blanchard-Desce, M. and Hynes, J.T. (1998) Two valence bond state model for molecular nonlinear optical properties. Nonequilibrium solvation formulation, *J. Phys. Chem. A*, **102**, 7712-7722.

27. Oudar, J.-L. and Chemla, D. S. (1977) Hyperpolarizabilities of the nitroanilines and their relations to the excited state dipole moment, *J. Chem. Phys.* **66**, 2664-2668.

28. Oudar, J.-L. (1977) Optical nonlinearities of conjugated molecules. Stilbene derivatives and highly polar aromatic compounds", *J. Chem. Phys.* **67**, 446-457.

29. Lu, D., Chen, G., Perry, J.W. and Goddard, W. (1994) Valence-bond charge-transfer model for nonlinear optical properties of charge-transfer organic molecules, *J. Am. Chem. Soc.*, **116**, 10679-10685.

30. Castiglioni, C., Del Zoppo, M. and Zerbi, G. (1996) Molecular first hyperpolarizability of push-pull polyenes: relationship between electronic and vibrational contribution by a two state model, *Phys. Rev. B*, **53**, 13319-13325.

31. Blanchard-Desce, M. and Barzoukas, M. (1998) Molecular engineering of push-pull molecules: towards photorefractive materials, *Mol. Cryst. Liq. Cryst,.* **322**, 1-8.

32. Barzoukas, M., Blanchard-Desce, M. and Wortmann, R. (1999) Figure of merit of push-pull molecules in photorefractive polymers, *SPIE.*, **3623**, 194-204.

33. Although a large dipole is favorable for electric-field poling (this fact is accounted for by the $\mu\beta$ product, and explains why the negative peak of the FOM(PEO) is higher than its positive peak), it can become detrimental in terms of intermolecular dipole-dipole interactions which scale as $\mu^2$.

34. Barzoukas, M., Blanchard-Desce, M., Josse, D., Lehn J.-M. and Zyss, J. (1989) Very large quadratic non-linearities in solution of two push-pull polyenes series: effect of the conjugation length and of the end groups, *Chem. Phys.*, **133**, 323-329.

35. Puccetti, G., Blanchard-Desce, M., Ledoux, I., Lehn J.-M., and Zyss, J. (1993) Chain-length dependence of the third-order polarizability of disubstituted polyenes. Effects of end groups and conjugation length, *J. Phys. Chem.*, **97**, 9385-9391.

36. Blanchard-Desce, M., Lehn, J.-M., Barzoukas, M., Ledoux, I. and Zyss, J. (1994) Chain-length dependence of the quadratic hyperpolarizability of push-pull polyenes and carotenoids. Effect of End Groups and Conjugation Path, *Chem. Phys.*, **181**, 281-289.

37. Blanchard-Desce, M., Alain, V., Midrier, L., Wortmann, R., Lebus, S., Glania, C., Krämer, P., Fort, A., Muller J. and Barzoukas M. (1997) Intramolecular charge transfer and enhanced quadratic optical nonlinearities in push-pull polyenes, *J. Photochem. Photobiol. A*, **105**, 115-121.

38. Alain, V., Rédoglia, S., Blanchard-Desce, M., Lebus, S., Lukaszuk, K., Wortmann, R., Gubler, U., Bosshard, C. and Günter P. (1999) Elongated push-pull diphenylpolyenes for nonlinear optics : Molecular engineering of quadratic and cubic optical nonlinearities via tuning of intramolecular charge transfer, *Chem. Phys.*, **245**, 51-71.

39. Alain, V., Thouin, L., Blanchard-Desce, M,. Gubler, U., Bosshard, C., Günter, P., Muller, J. Fort, A. and Barzoukas M. (1999) Molecular engineering of push-pull phenylpolyenes for nonlinear optics : Towards giant quadratic and cubic optical nonlinearities and improved solubility-stability-nonlinearities characteristics, *Adv. Mater.*, in the press.

40. Würthner, F., Wortmann, R., Mastchiner, R., Lukaszuk, K., Meerholz, DeNardin, Y., Buttner, R., Bräuchle, C. and Sens, R. (1997) Merocyanine dyes in the cyanine limit: a new class of chromophores for photorefractive materials, *Angew. Int. Ed. Engl.* **36**, 2765-2768.

41. NMR studies yield indication on the bond length alternation in the polyenic chain, a structural parameter proportional to the MIX parameter for the special case of push-pull polyenes [26], via the determination of the coupling constants between vicinal vinylic protons of the polyenic chain [38 and references cited therein].

42. The thermal stability smoothly decreases with increasing polyenic chain length. However, by appropriate choice of the endgroups, suitable thermal stability can be maintained for polyenic derivatives of intermediate length [38-39].

198

43.    Mladenova, M., Ventelon, L. and Blanchard-Desce, M. (1999) A convenient synthesis of push-pull polyenes designed for the elaboration of efficient nonlinear optical materials, *Tetrahedron Lett.*, **40**, 6923-6926.

44.    Chaput, F., Darracq, B., Boilot, J.P., Riehl, D., Gacoin, T,. Canva, M., Lévy, Y. and Brun, A. (1996) Photorefractive sol-gel materials, *Mater. Res. Soc. Symp. Proc.*, **435**, 583-588.

45.    Kim, J., Plawsky, J.L., La Peruta, R. and Korenovsky, G.M. (1992) Second harmonic generation in organically modified sol-gel films, *Chem. Mater.*, **4**, 249-252.

46.    Jeng, R.J., Chen, Y.M., Jain, A.K., Kumar, J. and Tripathy, S.K. (1992) Stable second-order nonlinear optical polyimide/inorganic composite, *Chem. Mater.*, **4**, 1141-1144.

47.    Zhang, Y., Prasad, P.N and Burzynsky, R. (1992) Second-order nonlinear optical properties of N-(4-nitrophenyl)-(s)-prolinol-doped sol-gel-processes materials, *Chem. Mater.*, **4**, 851-855.

48.    Nosaka, Y., Tohriiwa, N., Kobayashi, T. and Fujii, N. (1993) Two-dimensionally poled sol-gel processing of $TiO_2$ film doped with organic compounds for nonlinear optical activity, *Chem. Mater.*, **5**, 930-932.

49.    Yang, Z., Xu, C., Wu, B., Dalton, L.R., Kalluri, S., Steier, W.H., Shi, Y. and Bechtel, J.H. (1994) Anchroring both ends of chromophores into sol-gel networks for large and stable seond-order opticla nonlinearities, *Chem. Mater.*, **6**, 1899-1901.

50.    Lebeau, B., Maquet, J., Sanchez, C., Toussaere, E., Hierle, R. and Zyss, J., (1994) Relaxation behaviour of NLO cxhromophores grafted in hybrid sol-gel matrices, *J. Mater. Chem.*, **4**, 1855-1860.

51.    Riehl, D. Chaput, F., Levy, Y., Boilot, J.-P., Kajzar, F. and Chollet, P.-A. (1995) Second-order opticla nonlinearities of azo chromophores covalently attached to a sol-gel matrix, *Chem. Phys. Lett.*, **245**, 36-40

52.    Lebeau, B., Brasselet, S., Zyss, J. and Sanchez, C. (1997) Design, characterization, and processing of hybrid organic-inorganic coatings with very high second-order nonlinearities, *Chem. Mater.*, **9**, 1012-1020.

53.    Choi, D.H., Park, J.H., Rhee, T.H., Kim, N. and Lee, S.-D. (1998) Improved temporal stability of the second-order nonlinear optical effect in a sol-gel matrix bearing an active chromophore, *Chem. Mater.*, **10**, 705-709.

54.    Schanze, K. S., Schin, D.M. and Whitten, D.G. (1985) Micelle and vesicle solubilization sites. Determination of micropolarity and microviscosity using photophysics of a dipolar olefin. *J. Am. Chem. Soc.*, **107**, 507-509.

55.    Kasha, M. (1963) Energy transfer mechanisms and the molecular exciton model for molecular aggregates, *Radiation Research*, **20**, 55-70.

56.    Teng, C.C. and Man H.T. (1990) Simple reflection technique for measuring the electro-optic coefficient of poled polymers, *Appl. Phys. Lett.*, **56**, 1734-1736.

57.    Darracq, B., Chaput, F., Lalhil, K., Boilot, J.P., Lévy, Y., Alain, V., Ventelon L. and Blanchard-Desce, M. (1998) Novel photorefractive sol-gel materials, *Optical Materials*, **9**, 265-270.

58.    Ventelon, L., Mladenova, M., Alain, V., Blanchard-Desce, M., Chaput, F., Lahlil, K., Boilot, J.-P., Darracq, B., Reyes, J., Levy, Y. (1999) Towards stable materials for electro-optic modulation and photorefractive applications: the hybrid organic-inorganic approach *Proc. SPIE*, **3623**, 184-193.

# ON THE COHERENT AND INCOHERENT IMAGE CONVERSION IN HYBRID POLYMER LIQUID CRYSTAL STRUCTURES

A. MINIEWICZ and S. BARTKIEWICZ
*Institute of Physical and Theoretical Chemistry, Wroclaw University of Technology, Wybrzeze Wyspianskiego 27, 50-370 Wroclaw, Poland*

F. KAJZAR
*LETI (CEA - Technologies Avancées) DEIN - SPE, Groupe Composants Organiques, Saclay, F91191 Gif Sur Yvette, France*

## Abstract

We report on the new hybrid photonic devices exhibiting photorefactive type behaviour. These devices known as hybrid polymer - liquid crystal structures (HPLCS) are composed of a nearly transparent thin photoconducting polymeric layer and a nematic liquid crystal layer.   In such structures an efficient coherent light amplification with a very high net exponential gain has been observed in a degenerate two-wave coupling process. The amplification results from an energy exchange between inciding beams due to diffraction on nonlocal refractive index grating formed in liquid crystal layer by molecular reorientation mechanism. HPLCS's allow also for incoherent-to-coherent image conversion. In this work we focus our attention on important issues limiting their performances in some practically important optical systems where these devices can be used.

## 1. Introduction

Among the novel organic materials attractive for photonic application photorefractive polymers [1,2], polymeric liquid crystals (PLC) [3,4], low-mass liquid crystals [5-10] and organic glasses [11,12] play a major role. Chemical tailoring and technology possibilities are matured enough to obtain materials with strong optical nonlinearities at reasonably low light intensities. In most cases, however the one- or two-photon resonance or absorption of a photon by one of the constituents is the inhibitor of the subsequent events: production of excited state, production of a pair of

199

*F. Kajzar and M.V. Agranovich (eds.), Multiphoton and Light Driven Multielectron Processes in Organics: New Phenomena, Materials and Applications, 199–212.*
© 2000 *Kluwer Academic Publishers. Printed in the Netherlands.*

charge carriers, molecular conformational change, molecular reorientation, mass transport or photochemical reaction. All these phenomena lead to a temporal or in some cases quasi-permanent light-induced spatial modification of a material linear and non-linear susceptibilities $\chi^{(1)}$, $\chi^{(2)}$ and $\chi^{(3)}$. Our research is focused on nonlinear optical characteristics of materials prepared from thin layer of photoconducting polymer adjacent to the nematic liquid crystal layer. In such a simple system the whole spectrum of nonlinear optical effects has been demonstrated like: optical phase conjugation, coherent light amplification, beam-coupling, image conversion, real-time holography, etc. [13-16]. All these effects are directly or indirectly related with the material's ability of reversible laser induced dynamic refractive index or absorption grating formation [17].

For all-optical image-processing the fundamental requirements are: high resolution and fast response. To achieve this goal many optically driven spatial light modulators (SLM's) have been successfully constructed on the basis of multiple-quantum-well (MQW) structures [18], photorefractive inorganic materials [19,20] and wide variety of liquid crystals [21-23]. Depending on construction optically addressed SLM's operate in the diffraction or transmission mode. We will focus our attention on diffraction type SLM's where the light induced holographic grating is of primary importance as it determines the time response and influences the optical resolution.

Studies of liquid crystal materials able to record holographic gratings lead us to Hybrid Polymer-Liquid Crystal Structures (HPLCS) which we have found very attractive because the operating voltage required is much lower than for polymer composites, the system allows for coherent light amplification and coherent-to-incoherent image conversion and response is relatively fast (millisecond range). Our investigations of high-performance photorefractivity in these systems based on use of high-birefringence nematic liquid crystal and thin-film organic photoconductor. When the latter, placed in the intimate contact with LC layer, is illuminated by a light intensity pattern a redistribution of photogenerated charge carriers via electric field spatial modulation reorients the LC molecules. This simple idea has been already exploited but recent progress in many laboratories working on this subject and connected with the use of new materials and configurations encouraged us to continue our efforts.

## 2. Principles of optically addressed liquid crystal spatial light modulator (OA LC SLM)

The primary mechanism underlying a functioning of liquid crystalline optically addressed spatial light modulator is a modification of the LC index of refraction according to a light illumination pattern. In the simplest possible case one can use a planarly aligned nematic liquid crystal which, when placed between glass plates equipped with polymeric orienting layers, is birefringent ($n_o$ and $n_e$ describe its ordinary and extraordinary indices of refraction). Then a dc electric field applied to the electrodes changes uniformly, over the whole modulator area, the effective extraordinary index of refraction $\langle n_e^{eff} \rangle$ given by the well known formula [24]:

$$\langle n_e^{eff} \rangle = \frac{1}{d} \int_0^d \frac{n_e n_o \, dz}{\sqrt{n_e^2 \sin^2 \varphi(z) + n_o^2 \cos^2 \varphi(z)}} \qquad (1)$$

where d is the effective thickness of the liquid crystal layer and $\varphi(z)$ is the angle between the nematic director $\hat{n} = (\cos\varphi(z), 0, \sin\varphi(z))$ and the x-axis. This formula holds when the electric field vector of incident linearly polarised light is parallel to the alignment direction of the homogeneously aligned nematic and a wave vector of light $\mathbf{k}$ is orthogonal to the electrode plane xy. In the case of nematic liquid crystal interacting with an external electric field applied along the z-axis (a classical electric Fréedericksz splay effect [24]), the molecules turn in the xz-plane according to the torque which is the result of a balance between the electric and elastic forces.

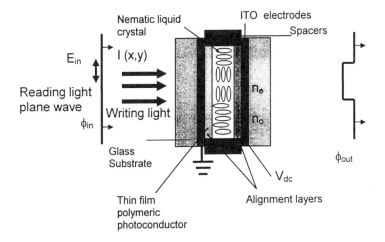

*Figure 1.* Schematic view of OA LC SLM constructed as the hybrid polymer - liquid crystal structure. The writing light induces via photoconductivity a reorientation of liquid crystal molecules. The local change of refractive index of LC layer induces for extraordinary polarised reading plane wave a respective phase change.

The maximum phase delay $\Delta\phi$ which can be imposed by action of optical field in nematic liquid crystal depends on maximum value of refractive index change $\Delta n$ which can never be higher than its birefringence $\Delta n_b = n_e - n_o$:

$$\Delta\phi = 2\pi \frac{\Delta n \cdot d}{\lambda} \qquad (2)$$

where $\lambda$ is the wavelength of the probing beam. Typically liquid crystal layers have its thicknesses in the range 1 - 50 μm and birefringence $\Delta n_b$ rarely exceeds 0.3, this gives a limits for possible phase modulation as is shown in the Table 1.

TABLE 1. Examples of limits of change of phase in LC SLM using typical nematics, thicknesses and wavelengths.

| Thickness of LC layer | LC birefringence $\Delta n_b$ | Wavelengths | Maximum phase modulation $\Delta\phi$ |
|---|---|---|---|
| 1 μm | 0.05 | 0.360 μm | 0.87 rad / 0.28 π |
| 10 μm | 0.2 | 0.532 μm | 23.6 rad / 7.5 π |
| 50 μm | 0.3 | 1.063 μm | 88.7 rad / 28 π |

However, in real optically addressed light valves the limits for $\Delta\phi$ given in the Table 1 should be treated as the most upper ones. They can be further limited as they depend also on required transverse optical resolution, switching time, operation voltage, physical properties of liquid crystal, anchoring forces of liquid crystal molecules to the substrate etc.

## 3. Diffraction grating recording in HPLCS spatial light modulators

One of the most exciting features of OA LC SLM's is their ability to serve as hologram recording media in a real-time. Holograms contain more optical information than a two-dimensional photography as one can record both intensity and phase of incoming light. This allows one to develop various all-optical devices based on light diffraction.

Let us assume that two coherent and polarised beams (the reference and the signal ones) are used for hologram recording in HPLCS. Both of the beams may contain spatial as well as phase information and they cross each other inside a recording medium. Assuming that the beams are given by the following expressions:

$$E_1 = j_1 E_1 \exp(-ik_1 r)$$
$$E_2 = j_2 E_2 \exp(-ik_2 r)$$

(3)

where $j_1$ and $j_2$ represent the polarisation states of the beams, the total intensity $I_T$ of the light incident on the photoactive medium will be spatially modulated:

$$I_T = \frac{1}{2}\left[E_1^2 + E_2^2 + j_1 j_2 E_1 E_2 \exp(-iKr) + c.c.\right]$$

(4)

Grating wavevector $K = k_1 - k_2$ determines the grating spacing $\Lambda = 2\pi/|K|$. Assuming that the energy of the writing light is sufficient for photogeneration of charge carrier pairs in a photosensitive polymer then an excess charge redistribution will occur within a thin polymeric layer. One has to take into account charge drift and diffusion effects to predict the possible spatial resolution of the electric field modulation. If the charge spreading effects are negligible, then electro-static models [25,26] can be used to determine spatial resolution of the device (i.e. spatial frequency at which the modulation transfer function MTF = $m_{out}/m_{in}$ decreases to 50%). Alternatively, the voltage distribution can be calculated by considering the conductivities of the LC layer and the dark and the illuminated regions of polymer layer [27]. In some OA LC SLM's electro-static models can properly predict the modulation transfer function (MTF) [28] in others where measured MTF's are much lower than predicted the charge spreading effects play a role [29]. As discussed in detail by Wang and Moddel [30] the resolution of the OASLM depends on the thickness of a photosensitive layer (the thinner the better), decreasing the mobility or the trapping time of charge carriers at the interface which reduces the diffusion length can significantly improve the spatial resolution R. The effect of bulk diffusion on resolution can be reduced by increasing the applied voltage to the OASLM which reduces the charge carrier transit time $\tau_{trans}$ in the polymer layer:

$$\tau_{trans} = \frac{d_p^2}{\mu_p V_p}$$

(5)

where $d_p$ is the thickness of the polymeric layer, $V_p$ is the voltage across the polymer layer and $\mu_p$ is the charge carrier mobility within the polymeric layer. Taking the typical hole mobility value for polythiophene $\mu_p = 5 \times 10^{-6} \ cm^2 V^{-1} s^{-1}$ and polymer layer thickness $d_p = 100$ nm one gets for transit time $\tau_{trans}$ values in the range 2 - 0.04 ms for electric field strengths within the polymeric layer equal to 0.1 V/$\mu$m and 5 V/$\mu$m, respectively. We should notice that the calculated transit time is large enough for being considered as a limiting factor of the dynamic response of the OA SLM.

However, the resolution of the OA SLM decreases also as a result of conversion of the interface (photoconductor – LC layer) charge distribution $\rho(x,y)$ into electric field in the light-modulating layer. This effect can be described by the charge-to-voltage MTF [31]:

$$MTF_{charge \rightarrow voltage} = \frac{\varepsilon_{lc} / d_{lc} + \varepsilon_p / d_p}{\nu\left[\varepsilon_{lc} \coth(\nu d_{lc}) + \varepsilon_p \coth(\nu d_p)\right]} \qquad (6)$$

where $\nu$ is the spatial frequency, $d_{lc}$ and $d_p$ are the thicknesses of liquid crystal and photoconducting polymer layers and $\varepsilon_{lc}$ and $\varepsilon_p$ are the respective dielectric constants. The dielectric constant of the nematic liquid crystal is dependent on the applied voltage and subjected to significant changes (e.g. from $\varepsilon_{lc,\parallel}$ to $\varepsilon_{lc,\perp}$). Finally the resolution is also limited by the elastic properties of the liquid crystal itself as was pointed out by Turalski et al. [32]. By observing the selfdiffraction efficiency in fabricated by us HPLCS structures we have found that the resolution is much better than 100 lp/mm [33-35]. In a grating with thickness d, diffraction limits the minimal feature size to $\sqrt{d \cdot \lambda}$, so the maximal expected resolution is $R \propto 1/\sqrt{d \cdot \lambda}$, this for d = 10 μm and $\lambda$ = 0.532 μm gives R = 430 lp/mm.

Other aspect of OA LC SLM's is the nature of light diffraction on generated by light periodic structures (along the x-axis) of refractive index e.g.

$$n_{mod} = n_{av} + \Delta n_G \cdot \cos\left[(2\pi / \Lambda)x\right]. \qquad (7)$$

In the above formula $n_{mod}(x)$ is the total modulated refractive index of LC, $n_{av}$ is an average index of refraction independent on x and $\Delta n_G$ represents the refractive index grating amplitude seen by the incoming light. This amplitude depends, in the case of nematic liquid crystal, on polarization of incoming light with respect to the rubbing direction as well as on incidence angle. It is obvious that for planarly aligned nematic $\Delta n_G$ will be the largest for light polarization plane parallel to the rubbing direction and zero for perpendicular polarization (cf. equation 1). Moreover, the grating amplitude $\Delta n_G$ changes along the liquid crystal layer thickness, i.e. along the z-direction. This is schematically depicted in Fig. 2 by plotting the profile of director angle $\varphi_{dark}(z)$ and $\varphi_{bright}(z)$. When the external electric field is applied to the panel the director angle profile $\varphi_{dark}(z)$ is symmetric with respect to the d/2 (curve (a) in Fig. 2). Under spatially periodic light illumination the distribution of an excess electric charge on the surface produces the excess electric field which changes the director profile. Assuming that in the bright region the field at z = 0 is smaller than the filed in dark region the director angle profile $\varphi_{bright}(z)$ will not be symmetric with respect to the d/2 and will take the form shown schematically by curve (b) in Fig. 2. The profile $\varphi(z)$ changes are currently investigated by Monte Carlo simulation methods. For normally incident light the integration over the whole LC layer thickness gives identical values of $<n_e^{eff}>$ both for dark and bright regions, so incoming light does not see the refractive index grating. However, at oblique light incidence angle the periodic structure of $\Delta n$ is visible for light and diffraction take place. For the grating spacing and the LC layer thickness ratio $\Lambda/d = 1$ the optimum light incidence angle for an observation of the maximum light diffraction is 45°. The same is true also for a process of self-diffraction of light on such a structure.

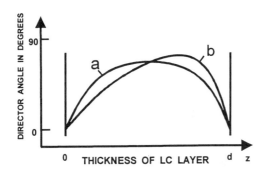

*Figure 2.* Schematic view of director angle profiles in dark (curve a) and bright (curve b) regions of the HPLCS illuminated by sinusoidal light intensity pattern. Note that at the z = 0 surface the electric field strength in the bright region is lower than in the dark region.

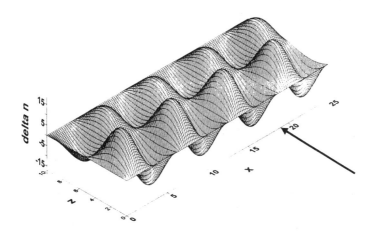

*Figure 3.* Schematic view of expected changes of effective extraordinary index of refraction $\Delta n(x,z)$ upon sinusoidal light illumination in HPLCS. Note that for the normal light incidence (cf. arrow along the z-axis) no periodic structure is seen while for the oblique incidence the periodic structure can be seen.

We assume that the light absorption in the liquid crystal layer is negligible, therefore the light induced gratings which arise due to molecular reorientation are of phase type only, i.e. are result of spatial modulation of refractive index. Depending on a dimensionless quality factor Q:

$$Q = \frac{2\pi d\lambda}{n_{av}\Lambda^2} \tag{8}$$

two types of optical diffraction can be observed. One defines the regime of Bragg diffraction for $Q \gg 1$. In this regime, multiple scattering is not permitted and only one order of diffraction is observed. For $Q < 1$ we deal with the Raman-Nath type of diffraction [17] in which the angular spread of the grating wave vector is larger than the Bragg angle, thus permitting for observation of many orders of diffraction. For typical light wavelength $\lambda = 0.532$ μm and layer thickness $d = 10$ μm with $n = 1.6$ one gets $Q = 21$ for $\Lambda = 1$ μm or 0.21 for $\Lambda = 10$ μm. This means that it is very easy to pass from one to another regime of diffraction by proper choice of parameters involved in the definition of Q factor and the intermediary diffraction is experimentally accessible. In most cases we work within a Raman-Nath diffraction regime for which a first order diffraction efficiency is described by [17]:

$$\eta = |J_1(\phi)|^2 \tag{9}$$

where $J_1$ is the 1-th order Bessel function and $\phi = 2\pi\Delta n_G \, d\lambda^{-1}$ with $\Delta n_G$ the grating amplitude. In the case of Bragg diffraction the first order diffraction efficiency $\eta$ is given by the simplified Kogelnik's formula [33] for thick transmission phase holograms:

$$\eta = \sin^2\left(\frac{\pi\Delta n_G d}{\lambda\cos\beta}\right) \tag{10}$$

where $\beta$ is the Bragg angle.

## 4. Two-beam coupling in HPLCS

The experimental setup for two-beam coupling in HPLCS is shown in Fig. 4. When the input beam polarizations are lying within an incidence plane an efficient light self-diffraction can be observed. Several orders of diffraction are visible when the Raman-Nath diffraction regime is fulfilled. Besides, depending on the sign of externally applied electric field an energy transfer from one beam to another is observed. These effects which we called coherent light amplification were the subject of our recent papers [34-36].

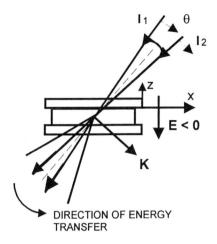

*Figure 4. Two-wave mixing experimental setup designed to study the hybrid polymer - liquid crystal structures. The self-diffraction of light and energy transfer from beam-to-beam is possible only for oblique light incidence as shown above.*

Light amplification in photorefractive materials is possible due to nonlocality of the grating recording which gives rise to a beam-coupling effect. The nonlocality is determined by a phase shift $\phi_G$ between the optical index grating and the light intensity pattern. In the case of HPLCS structures this nonlocality is connected with the specific, dependent on sign and magnitude of the applied electric field, distribution of charge carriers at the interface polymer-LC layer. Change of sign of an externally applied voltage changes the sign of carriers appearing at the polymer-LC layer interface which in consequence decrease or increase the electric potential within LC layer in the close proximity of the front (illuminated) part of the LC cell (cf. Fig. 2). Having this in mind we designed and fabricated the HPLCS using a processable photoconducting copolymer for preparation of 100 nm thick dual function photoconducting and orienting layer. In our studies we used copolymer of polyalkylthiophenes [37,38] containing NLO chromophore commercially known as Disperse Red 1. LC cells (10μm in thickness) were filled with a multicomponent nematic liquid crystal mixture E7 which is characterised at room temperature by $\Delta\varepsilon = +13.8$, birefringence $\Delta n_b = 0.2253$ at 589 nm ($n_e = 1.7464$, $n_o = 1.5211$), viscosity $\gamma = 39$ mm$^2$s$^{-1}$ and a clearing point at 331 K. The HPLCS were biased by a dc voltage and the electric field strength inside a layer was in the range (0 - 2 V/μm). Typically we used unfocused laser beams from an Ar$^+$ laser of intensities ($\approx 100$ mW/cm$^2$). The space charge electric field along the surface layer $E_x = E_{o,x} \cos(K_\beta x + \phi_G)$ is shifted in phase by an angle $\phi_G = \pi/2$ with respect to the light intensity grating when a pure diffusion process of charge carrier transport within a photoconducting layer or polymer-LC interface is considered. The sum of electric space charge field $E_x$ and externally applied electric field along the z-axis $E_a$ form a complex modulated pattern inside a dielectrically anisotropic liquid crystal. The local reorientations of nematic director are considerable only if the local total field strength exceeds the electric Fréedericksz threshold field ($E_x + E_a > E_F$). The direction of the energy flow in a two-wave mixing experiment is determined be the sign of the phase shift $\phi_G$ which can be changed by inversion of sign of the externally applied voltage.

Monitoring the intensity of the beam $I_1$ after the sample in the absence of the beam $I_2$ and its intensity $I_{12}$ in the presence of the beam $I_2$ the energy transfer between the beams was established. Setting the ratio of two input beam intensities to $m = I_{20}/I_{10} = 10$, where the subscript 0 denotes the respective beam intensity before entering the sample we were able to observe a weak beam amplification gain g = 7. Using the relation linking the gain g with an exponential gain coefficient $\Gamma$ [39]:

$$g = \frac{I_{12}}{I_1} = \frac{1+m}{1+me^{-\Gamma L}} e^{-\alpha L} \tag{11}$$

where $\alpha$ is the average absorption coefficient of a liquid crystal at the excitation wavelength, $L = d/\cos(\beta_{int})$ is the interaction length, one can establish the value of $\Gamma$ describing the strength of nonlinear material to couple coherent waves . Putting $\alpha \cong 10$ cm$^{-1}$ and L = 11.15 μm we obtained the gain coefficient in the HPLCS $\Gamma \cong 2600$ cm$^{-1}$.

This is a remarkably high coefficient close to that reported recently by Khoo et al. [16] in fullerene doped pentylcyanobiphenyl (2890 cm$^{-1}$). Despite the substantial value of an exponential gain coefficient $\Gamma$ the overall gain is much less than in the bulk photorefractive materials. Due to small thickness of the liquid crystal layer (here 10 μm) the product $\Gamma L$ is not very large and amounts to 2.9. This number cannot be easily improved as there is a limit for nematic layer thickness for which the LC alignment achieved by surface rubbing is preserved. Usually well ordered nematic layers have thicknesses up to 50 μm. Interesting is however that a real amplification ($\Gamma - \alpha$)L > 1 has been reached at the electric field strength of only 0.6 V/μm which is much lower (e.g. two orders of magnitude) than that applied to photorefractive polymers performing the same task [41]. It is interesting to estimate, on the basis of coherent light amplification experiment, the refractive index grating amplitude $\Delta n_G$ effectively created in the LC layer. For this purpose we recall that the exponential gain coefficient $\Gamma$ for a co-directional two-wave coupling is given by the following relation [17]:

$$\Gamma = \frac{2\pi\Delta n_G}{\lambda \cos(\theta_{int}/2)} \sin\phi_G, \tag{12}$$

then putting the maximum possible value of $\phi_G = \pi/2$ we arrive $\Delta n_G \approx 0.02$. It is worthwhile to notice that it is only 10% of the total E7 nematic liquid crystal birefringence $\Delta n_b$. Again basing on the model of refractive index grating formation as shown in Fig. 2 we can conclude that grating amplitude can be only a small fraction (0.1 -0.3) of a total LC birefringence. So the practically accessible value of $\Gamma \propto 2\pi\Delta n_G/\lambda$ with realistic $\Delta n_G = 0.06$ will be in the range 10000 - 5000 cm$^{-1}$ for $\lambda = 0.4$ $\mu$m and $\lambda = 0.7$ $\mu$m, respectively. This for L $\approx$ 50 $\mu$m can give an upper limit for $\Gamma$L $\approx$ 50 which would be important amplification strength of a nonlinear material.

The exponential gain coefficient $\Gamma$ depends on grating spacing $\Lambda$ showing maximum for a certain spatial frequency (here $\Lambda \cong 30$ $\mu$m). One can tentatively explain the decrease of $\Gamma$ at lower grating spacings by an elasticity of liquid crystal which resist to modulate its structure when the period of modulation becomes too small [32]. The $\Gamma(\Lambda)$ decrease at higher grating spacings we link with the properties of the photoconducting polymer itself i.e. with the effective charge transport length and amplitude of charge redistribution upon sinusoidal illumination.

## 5. Realisation of incoherent-to-coherent image conversion in HPLCS

Holographic grating recording and effective light diffraction in HPLCS structures allow for realisation of incoherent-to-coherent image conversion. The nonpixelated devices of this type are highly demanded as no side diffraction is present, which is the disadvantage of all liquid crystal matrix devices. Experimental setup used by us is shown in Fig. 5.

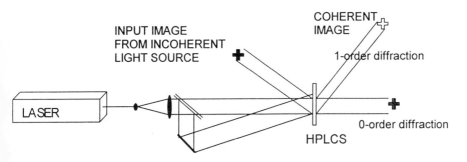

*Figure 5.* An optical system realising incoherent-to-coherent image conversion with HPLCS.

If an incoherent erasing beam is present the effective two-wave mixing coefficient $\Gamma_{eff}$ can be written as [42]:

210

$$\Gamma_{eff} = \frac{I_{coh}}{I_{coh} + I_{incoh}} \Gamma_0 \qquad (13)$$

where $I_{coh}$ is the intensity of the coherent beam, $I_{incoh}$ is the intensity of the incoherent beam, and $\Gamma_0$ is the two-wave mixing coefficient of the liquid crystal sample. This formula is valid only if one assume that both laser beams are equally effective in production of charge carriers inside the polymeric layer. From the equation 13 it is clear that the presence of the incoherent beam decreases the effective two-wave mixing coefficient. As a result one can expect the decrease of the intensity of the diffracted into 1-order beam according to the light intensity pattern of the image-bearing beam $I_{incoh}$ which is clearly shown in Figure 5. The erasing was obtained automatically when all the beams were switched off. The 1/e decay time is of the order of 100 ms in the studied system.

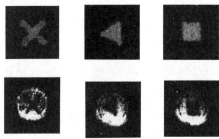

*Figure 6.* Examples of image transfer from incoherent laser beam (upper row) to coherent beam using experimental setup shown in Fig. 5.

From the experimental results shown in Figure 6 it can be seen that a transfer of an image form one to another laser beam is possible. The size of the objects shown in Fig. 6 was approximately 2 mm. This is a novel transmission type incoherent-to-coherent image converter constructed uniquely with the use of the hybrid photoconducting polymer-nematic liquid crystal panel.

### Acknowledgements
This research was supported by the Wroclaw University of Technology, Faculty of Basics Problems of Technology.

Partial financial support to S.B. and A.M. from CEA, Saclay, France and from the French-Polish programme Polonium (F.K., S.B., A.M.) is greatly acknowledged.

## REFERENCES

1. Meerholz, K., Volodin, B. L., Sandalphon, Kippelen, B. and Peyghambarian, N., (1994) *Nature* **371**, 497.
2. Liphardt, M., Goonesekera, A., Jones, B. E., Ducharme, S., Takacs, J. M. and Zhang, L., (1994) *Science* **263**, 367.
3. Shibaev, V. P. and Lam L. (1994) *Liquid Crystalline and Mesomorphic Polymers*, Springer, New York.
4. Coles, H.J., Simon, R. (1985) *Polymer* **26**, 1801
5. Kurokawa, T. and Fukushima, S. (1992) *Opt. Quant. Electr.* **24**, 1151.
6. Hawlitschek, N., Gärtner, E., Gussek, P. and Reichel, F., (1994) *Exp. Techn. Phys.* **40**, 199.
7. Rudenko, E.V. and Sukhov, A. V., (1994) *Sov. Phys. JETP (in Russian)*, **105**, 1621.
8. Khoo, I. C., (1995) *Opt. Lett.* **20**, 2137.
9. Khoo, I.C., (1995) *Liquid Crystals, Physical Properties and Nonlinear Optical Phenomena*, J. Wiley, New York, Chapter 9.
10. Miniewicz, A., Bartkiewicz, S., Turalski, W., Januszko, A. in Munn, R.W., Miniewicz, A. and Kuchta, B. (eds.) (1997) *Electrical and Related Properties of Organic Solids, NATO ASI Series*, **3/24**, Kluwer Academic Publishers, Dordrecht, pp. 323-337.
11. Zhang, Y., Prasad, P.N. and Burzynski, R. (1992) *Chem. Mater.* **4**, 851.
12. Jeng, R.J., Chen, Y.M., Jain, A.K., Kumar, J. and Tripathy, S.K. (1992) *Chem. Mater.* **4**, 972.
13. Wiederrecht, G. P., Yoon, B. A. and Wasielewski, M. R. (1995) *Science*, **270**, 1794.
14. Khoo, I. C., Li, H. and Liang, Y. (1994) *Opt. Lett.* **19**, 1723.
15. Miniewicz, A., Bartkiewicz, S., Januszko, A. and Parka, J. in Kajzar, F., Agranovich, V. M. and Lee, C. Y.-C. (eds.), (1996) *Photoactive Organic Materials Science and Application*, NATO ASI Series, Vol. **3/9**, Kluwer Academic Publishers, Dordrecht.
16. Miniewicz, A., Bartkiewicz, S. and Parka, J. (1998) *Optics Commun.* **149**, 89.
17. Eichler, H. J., Günter, P. and Pohl, D. W. (1986) *Laser-Induced Dynamic Gratings,* Springer Verlag, Berlin.
18. Bowman, S.R., Rabinovich, W.S., Beadie, G., Kirpatrick, S.M., Katzer, D.S., Ikossi-Anastasiou, K. and Adler, C.L. (1998) *J. Opt. Soc. Am.* B **15**, 640.
19. Zhang, J., Wang, H., Yoshikado, S. and Aruga, T. (1996) *Opt. Lett.* **22**, 1612.
20. Amrhein, P. and Günter, P. (1990) *J. Opt. Soc. Am.* B **7**, 2387.
21. Bahadur, B. (1992) *Liquid Crystals, Applications and Uses*, vol. 3, World Scientific, New York.
22. Fukushima, S., Kurokawa, T. and Ohno, M. (1991) *Appl. Phys. Lett.* **24**, 1151.
23. Armitage, D., Thackara, J.I. and Eades, W.D. (1989) *Appl. Opt.* **28**, 4763.

24. Khoo, I.C., (1995) *Liquid Crystals, Physical Properties and Nonlinear Optical Phenomena*, J. Wiley, New York.
25. Roach, W.R. (1974) *IEEE Trans. Electron. Dev.* **ED-21**, 453.
26. Owechko, Y. and Tanguay, Jr., A.R., (1984) *J. Opt. Soc. Am.* **A 1**, 635.
27. Flegontov, Y.A., (1993) *Sov. J. Opt. Technol.* **60**, 466.
28. Fukushima, S., Kurokawa, T. and Ohno, M. (1992) *Appl.Opt.* **31**, 6859.
29. Williams, D., Latham, S.G., Powles, C.M.J., Powell, M.A., Chittick, R.C., Sparks, A.P. and Collings, N. (1988) *J. Phys. D: Appl. Phys.* **21**, S156.
30. Wang, L. and Moddel, G. (1995) *J. Appl. Phys.* **78**, 6923.
31. Armitage, D., Anderson, W.W. and Karr, T.J. (1985) *IEEE J. Quantum Electron.* **QE-21**, 1241.
32. Turalski, W. , Mitus, A.C., Miniewicz, A. (1997) *Pure Appl. Opt.,* **6**, 589.
33. Kogelnik, H., (1969) *Bell. Syst. Tech. J.* **48**, 2909.
34. Miniewicz, A., Bartkiewicz, S. and Kajzar, F. (1998) *Nonlinear Optics,* **19**, 157.
35. Bartkiewicz, S., Miniewicz, A., Kajzar, F. and Zagorska, M. (1998) *Appl. Opt.* **37**, 6871.
36. Bartkiewicz, S., Miniewicz, A., Kajzar, F. and Zagorska, M. (1998) *Mol. Cryst. Liq. Cryst.* **322**, 9.
37. Sentein, C., Mouanda, B., Rosilio A. and Rosilio, C. (1996) *Synth. Metals* **83**, 27.
38. Binh, N. T., Gailberger, M. and Baessler, H. (1992) *Synth. Metals* **47**, 77.
39. Yeh, P., (1993) *Introduction to Photorefractive Nonlinear Optics,* J. Wiley, New York, Chapter 4.
40. Khoo, I. C., Guenther, B. D., Wood, M. V., Chen, P. and Min-Yi Shih (1997) *Opt. Lett.* **22**, 1229.
41. Grunnet-Jepsen, A., Thompson, C. L., Twieg, R. J., Moerner, W. E. (1998) *J. Opt. Soc. Am. B,* **15**, 901.
42. Kwong, S.-K., Cronin-Golomb, M., Yariv, A., (1984) *Appl. Phys. Lett.* **45**, 1016.

# OPTICALLY ADDRESSED LIQUID CRYSTAL LIGHT VALVES AND THEIR APPLICATIONS

J.-P. HUIGNARD, B. LOISEAUX, A. BRIGNON

Thomson-CSF /LCR

Domaine de Corbeville – 91404 Orsay Cedex – France

B. WATTELIER, A. MIGUS

Laboratoire LULI – Ecole Polytechnique – 91128 Palaiseau

C. DORRER

Laboratoire LOA-ENSTA – Ecole Polytechnique – 91128 Palaiseau

## 1. Introduction

Ultra intense laser chains exhibit large amounts of spatial and temporal aberrations which degrade the quality of the beam delivered by the source. These aberrations arise from the thermal lensing due to the efficient pumping of the gain media. Also, the aberrations of the optical components as well as the group delay dispersion effects due to ultra short pulse propagation in a femtosecond laser are major parameters which degrade the spatial and temporal properties of the beam. In such conditions the beam spot size after focalization and the pulse duration are far from their ultimate Fourier transform limit (1-2-3). Since now most of the applications require optimum source brightness and ultra high peak power, it is of particular importance to introduce new programmable optical components which can compensate for the spatial and temporal aberrations of the chain (4-5). The technology developed at TH-CSF/LCR consists in using an optically addressed liquid crystal spatial light modulator (OASLM) which exhibits very interesting characteristics for the control of temporal and spatial signals. After the presentation of the device operating principle, technology and performances, we highlight the following applications:

- Correction of the spatial aberrations of a 100 TW laser
- Correction of the group delay dispersion in a femtosecond laser
- Dynamic holography for image amplification.

## 2. OASLM: operating principle and technology

The early work on optical information processing at TH-CSF/LCR lead to the development of an original structure of an optically addressed spatial light modulator used as input

*F. Kajzar and M.V. Agranovich (eds.), Multiphoton and Light Driven Multielectron Processes in Organics: New Phenomena, Materials and Applications, 213–224.*
© 2000 *Kluwer Academic Publishers. Printed in the Netherlands.*

transducer for incoherent to coherent image conversion in a parallel Fourier processor (6). This is the same device structure that we now use in adaptive optics for high energy lasers and dynamic holography. The OASLM which acts as an electro-optic 2D programmable phase plate is based on the liquid crystal technology. The OASLM uses bulk monocrystalline $Bi_{12}SiO_{20}$ (BSO) both as the photoconductive material and as one of the substrate supporting a liquid-crystal layer (see Fig.1). When addressed optically with incoherent light ($\lambda_{Writing} < 500nm$), the photoconductive properties of BSO locally transfer the voltage to the liquid-crystal layer. Then, over the entire valve aperture, the liquid crystal exhibits spatial-index variation proportional to the incoherent-light spatial distribution. An IR ($\lambda_{Read} > 600nm$) beam passing through this active phase plate will have its wave front controlled according to the voltage distribution. Depending on the thickness of the liquid crystal, phase deformation of several wavelengths can be generated. Transverse resolution of the OASLM ensures the control of more than 300 x 300 pixels over the beam aperture. The optical addressing of the device is performed by means of imaging a programmable mask on BSO. In our experiments the mask is generated by a PC and displayed on an electrically addressed liquid-crystal television active matrix made of an array of 640 x 480 pixels used between crossed polarizors (see Fig.2). The addressing light originates from an incoherent low-voltage arc lamp. To avoid to create a pixelization on the valve when we are imaging the active mask, we add a slight defocusing. Following this operating principle of the OASLM, any analog light pattern generated on the mask is transferred into a spatial phase modulation of a wave front passing through the device.

Some specific characteristics of this BSO-LC SLM are given hereafter:

- Aperture size: 10 x 10 $\rightarrow$ 30 x 30 $mm^2$
- BSO thickness: 1 mm typical
- Liquid crystal: d = 8 to 10 µm - $\Delta n = 0.2$ typical
- Phase excursion: $\Delta\varphi = 4\Pi$ at $\lambda = 1,06$ µm
- Writing illumination: blue spectral range $\lambda = 450nm$
- Laser readout: $\lambda_R > 650nm$
- Operating voltage: 15 V – AC – 10 to 100 Hz
- Rise / decay time: 50 ms.

The present device exhibits a damage threshold resistance higher than $300mJ\ cm^{-2}$ with pulsed Nd-YAG laser at $\lambda = 1,06$ µm. An attractive alternative solution consists in replacing the bulk photoconductor by an organic thin film, in particular if larger aperture sizes are required.Also, a larger phase dynamic range is possible by choosing a LC material having a birefringence close to $\Delta n = 0,3$.

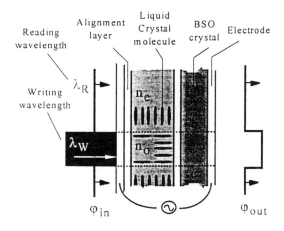

Figure 1. Operating principle of the liquid crystal OASLM.

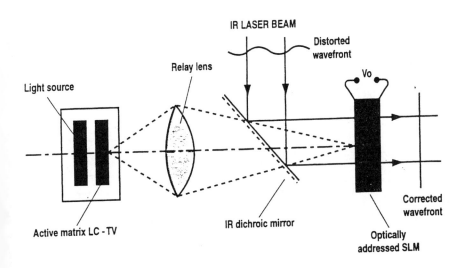

Figure 2. Optical module for adaptive control of the phase of a laser beam.

2D spatial phase modulation could also be achieved by using a bulk material with a large third order non linearity typically a photorefractive like crystal or polymer. However a determinant advantage of the light valve approach relies on the fact that each function respectively, the photoconductive and the electrooptic properties of the device can be optimized with different materials. In such conditions, the OASLM technology can exhibit better and much more flexible characteristics in comparison to a single nonlinear medium for adaptive phase control applications as presented in the following paragraphs.

### 3. Correction of the wavefront of a high energy laser

For testing the capabilities of the programmable LC OASLM to correct for phase distortions of a laser beam we tested the set-up shown in Figure 3a using a CW low power Nd-YAG laser. It consists of the following elements :

    a)   an achromatic wavefront sensor : a three beams lateral shearing interformeter.

    b)   a CCD camera for beam analysis.

    c)   the OASLM.

    d)   a processing unit.

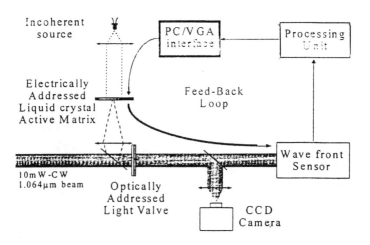

*Figure 3a. Experimental set-up showing the OASLM module in a feedback-loop.*

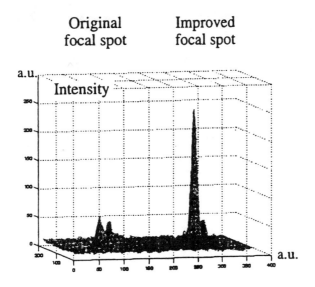

*Figure 3b. Experimental result after adaptive phase correction by the OASLM.*

The experimental set-up uses a feedback loop between the wavefront sensor and the OASLM. The procedure is thus the following.: after measuring the wavefront aberration, we display on the OASLM a 2D spatial phase law which compensates for the wavefront distortions. The role of the processor is thus to recover the phase from the interferogram delivered by the wavefront sensor, and according to this information, to generate the mask displayed on the LC-TV. Then the resulting focal spot is analyzed on a CCD camera. The experimental results firstly obtained by J.-C. Chanteloup et al in reference 7 and shown in Figure 3b clearly demonstrate the improved beam quality due to the operation with the LC light valve (Strehl ratio increases from 25% to 96%). Most recently B. Wattelier et al in reference 8 demonstrate the integration of the device in the LULI – 100 TW – 400fs. Nd-Glass laser chain whose schematic, including pulse stretchers, amplifiers, and pulse compressors, is shown in figure 4a. 3D views of wavefronts before and after correction highlight that the peak to peak values is reduced from 0,48 λ to 0,2 λ while the rms value decreases from 0,13 λ to 0,04 λ (see Figure 4b). Also an important point to notice in figure 4b is that the OASLM does not introduce spurious diffraction effects on the beam: it is a non pixelated structure which only affects the phase of the laser wavefront.

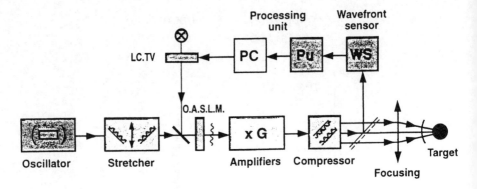

*Figure 4a. Integration of the OASLM Adaptive optical loop for spatial phase correction of a high peak power laser: 100 TW LULI laser*

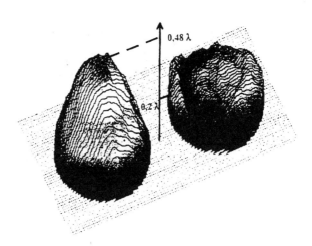

*Figure 4b. 3D views of wavefront before and after corrections.*

## 4.   Correction of spectral dispersion effects of a femtosecond laser

Remarkable breakthroughs have occured recently in the field of ultrafast laser technology in particular with Ti-Saphir crystals which have a wide spectral gain bandwidth ($\Delta\upsilon >$ 100nm). Self mode locking and chirped pulse amplification techniques (CPA) are used with success for the production of fs pulses with focused intensities exceeding $10^{20}Wcm^{-2}$. However, to attain the maximum peak power requires to deliver a gaussian pulse whose duration τ is as close as possible of the ultimate Fourier limit τ = 0.44/Δυ with high contrast ratio. In CPA systems the chirp induced by the grating stretcher and the high order dispersion effects in gain media and optical components cannot be perfectly compensated by the grating compressor. Any uncompensated dispersion prohibits pulse compression to the Fourier transform value. Therefore, as in the space domain, there is a need of a programmable 1D spatial modulator to control the phase of the each spectral component of the pulse. For that purpose, several technologies were tested such as, pixelated LCD, acousto-optic modulators or flexible piezo mirrors. In general, these devices either suffer of limited spatial resolution, or introduce parasitic diffraction due discrete pixel sampling. C. Dorrer et al in reference 9 firstly demonstrated the use of the non pixelated LC-OASLM for controlling the temporal aberrations of an ultra short pulse. The procedure described in detail in the reference 9 is the following. Before operating with the OASLM the relative phase of the spectral components of the pulse are measured by a classical spectral interferometric technique. Then, to achieve a precise control of the phase for high order dispersion compensation, we implement a feedback loop to converge toward the required phase value. In such conditions, a frequency chirped laser pulse can be recompressed by the use of a OASLM whose major characteristics have been presented in section 2.

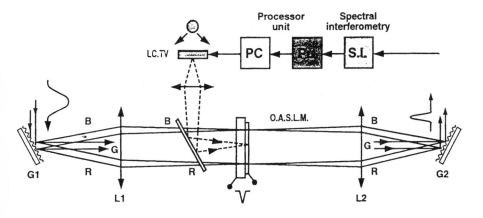

*Figure 5. OASLM in the spectral plane of a zero dispersion line: correction of group delay dispersion in a CPA for laser chain.*

The OASLM has the flexibility of classical LC modulators but it is non pixelated and it displays a spatial resolution higher than 10 $\ell$mm$^{-1}$. Its operating principle for spectral phase modulation is shown in Figure 5. The device is placed in the focal plane of a zero dispersion line including two highly dispersive gratings 1000 $\ell$mm$^{-1}$ and Fourier transform lenses $L_1 - L_2$. The spectral components are spatially dispersed over the 10 mm aperture of the light valve. Depending of the light pattern projected onto the BSO by the LC-TV, it turns into a spectral modulation of the phase of the local spectral components of the pulse delivered by the Ti-Saphir oscillator. When operating with the LC light valve, the spectral phase is rendered nearly constant over the spectral bandwidth 795-835nm and as a consequence the autocorrelation shows that the pulse is recompressed close to the Fourier limit (see Figure 6). Also it results a significant reduction of the wings of the 50fs pulse.

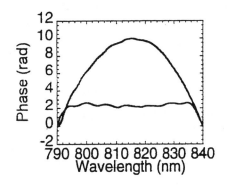 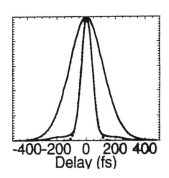

*Figure 6. Spectral phase and pulse autocorrelation without correction (a)*
*and with correction due to OASLM (b).*

The integration of the module in a CPA Ti-Saphir laser follows the schematic of Figure 4a where the module of the figure 5 is placed before the pulse stretcher. After measuring the spectral phase of the chain by spectral interferometry, the OASLM then permits to compensate the high order group delay dispersion due to propagation of the pulse through the active and passive components of the chain.

## 5. Dynamic holography for image amplification

Degenerate two-wave mixing (TWM) of two optical beams in a nonlinear medium has been extensively investigated for real time holography and wavefront amplification in particular using photorefractive materials. In a similar manner to photorefractive materials, A. Brignon et al have shown in reference 10, that an OASLM can also be used for dynamic holography with a high TWM gain as experienced in the interaction geometry of figure 7a. A low intensity signal and a pump beams interfere on the light valve and it results a periodic variation of the index of refraction of the form:

$$n(x) = n_0 + \Delta n \cos \frac{2\Pi x}{\Lambda}$$

where $\Lambda$ is the grating period. When the light valve operates at the correct bias point shown in Figure 7b, we can achieve optimum diffraction efficiency and index modulation $\Delta n$. In such conditions, the self diffracted beam will contribute to the coherent two wave interaction and the signal beam experiences a gain G given by the following relation

$$G = 1 + \left[ \frac{2\Pi}{\lambda} n_2 I_1(0) \right]^2 . L^2$$

where $n_2 = \dfrac{\partial n}{\partial I}$ is the slope of the SLM characteristics, i.e. the index versus incident pump intensity. L: interaction length in the liquid crystal. The gain equation shows that even without a phase shift between the grating and the incident interference pattern an energy redistribution is achievable, but the energy transfer is always from the strong wave to the weak one. By comparison, in conventional TWM interactions in a nonlinear medium the refractive-index grating changes in amplitude and position during the propagation of the waves in the medium. In this case it is well known that energy transfer occurs only when the induced grating is shifted with respect to the interference pattern. In the experiments, the laser used the 514nm line of a linearly polarized Ar laser. The beam was divided into two coherent beams, $A_1$ and $A_2$, and then recombined on the OASLM by a Mach-Zehnder interforemeter with the condition $A_2 \ll A_1$.

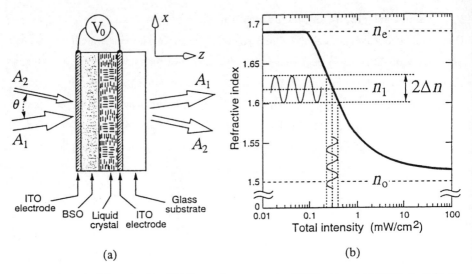

Figure 7a. Schematic of the TWM interaction geometry in a BSO liquid crystal OASLM. 7b. Refractive index as a function of the incident intensity.

For an applied voltage frequency of F = 25 Hz a maximum gain of 10 was obtained for $I_0 = 210$ µW/cm² and $V_0 = 5.5$ V. When we take into account the overall transmission of the OASLM, this result corresponds to a net gain of 6. According to the gain equation the equivalent nonlinear index coefficient $n_2$ is thus as large as 220 cm²/W. Using a beam shutter in the arm of the pump beam and with $I_0 = 210$ µW/cm², we measured a response time of 150 ms for the gain to reach 90% of its steady-state value. Signal-beam amplification depends on the grating period of the induced index grating and is limited by the resolution of the OASLM. Figure 3 shows the experimentally observed TWM gain as a function of the grating period $\Lambda$ in the weak-signal regime (pump to probe beam ratio $\beta = 225$). For $\Lambda < 80$ µm no gain was observed. The gain then increases with $\Lambda$ and reaches a constant value for a grating period larger than 400 µm. This moderate spatial resolution is attributed to the thickness of the BSO crystal. It is known that a thinner crystal or a thin film of photoconductive polymer material would certainly permit higher spatial resolution. Figure 8 illustrates the capabilities of the OASLM for image amplification when an object binary chart is inserted on the signal beam. Amplified images are due to energy transfer from the pump beam to the signal. In both cases we obtained a stationary 5X amplification of the signal wavefront in the image plane and a net gain factor of 3. These results demonstrate amplification of a laser beam by two-wave mixing in a BSO liquid crystal light valve with very low input intensities and driving voltage. Since OASLM can be designed for various ranges of wavelengths, it is also expected that this newly observed effects will be observed in spectral regimes that are not easily achieved with current nonlinear materials.

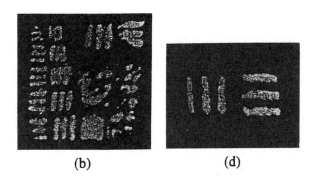

(b)                    (d)

*Figure 8. 5X-amplified image in the presence of the pump beam.*

## 6. Conclusions

The different experiments presented, based on the use of a non pixelated optically addressed liquid crystal light valve, open new attractive issues for adaptive control of laser beams. Performances demonstrated in term of optical efficiency, contrast ratio, phase excursion and damage threshold show the capabilities of the device to control and to correct the spatial and temporal aberrations araising in high energy lasers or in femtosecond chirped pulse amplification laser chains. The SLM technology permits to attain the ultimate performances of the chain while it relaxs the constraints on the quality of the optical components of the laser.

## 7. References

1. Perry, M.D. and Mourou, G. (1994) Terawatt to Petawatt subpicosecond lasers, *Science* **264**, 917–924.

2. Weiner, A.M., Heritage, J.P. and Kirschner, E.M. (1988) High resolution femtosecond pulse shaping, *JOSA-B* **19**, 1563–1572.

3.  Rockwell, D.A., Manguir, M.S. and Ottusch, J.J. (1993) Energy scaling of phase conjugate lasers and Brillouin conjugators, *Int. J. of Nonlinear Opt. Phys.* **2**, 131–155.

4.  Kudryashov, A.V., Gonglewski, J., Browne, S., Highland, R. (1997) Liquid crystal phase modulator for adaptive optics. Temporal performance characterization, *Opt. Comm.* **141**, 247–253.

5.  Wefers, M.M. and Nelson, K.A. (1995) Generation of high fidelity programmable ultrafast optical waveforms, *Opt. Lett.* **20**, 1047–1049.

6.  Aubourg P., Huignard, J.P., Hareng, M., Mullen, R.A. (1982) Liquid crystal light valve using bulk monocrystalline $Bi_{12}SiO_{20}$ as the photoconductive material, *Appl. Opt.* **21**, 3706–3712.

7.  Chanteloup, J.C., Baldis, H., Migus, A., Mourou, G., Loiseaux, B., Huignard, J.P. (1998) Nearly diffraction limited laser focal spot obtained by use of an optically addressed light valve in an adaptive optics loop, *Opt. Lett.* **23**, 475–477.

8.  Wattelier, B., Chanteloup, J.C., Zou, J.P., Sauteret, A., Migus, A. (1999) Adaptive correction of multiterrawatt Nd-Glass laser chain, *CLEO* 99 *proceedings*, Baltimore.

9.  Dorrer, C., Salin, F., Verluise, F., Huignard, J.P. (1998) Programmable phase control of femtosecond pulse by use of a non pixelated spatial light modulator, *Opt. Lett.* **29**, 709–711.

10. Brignon, A., Bongrand, I., Loiseaux, B., Huignard, J.P. (1997) Signal beam amplification by two wave mixing in a liquid crystal light valve, *Opt. Lett.* **24**, 1855–1857.

# DYE-ASSISTED OPTICAL REORIENTATION OF NEMATIC LIQUID CRYSTALS

I. JÁNOSSY
*Research Institute for Solid State Physics and Optics*
*H-1525 Budapest, P.O.B.49, HUNGARY*

## 1. Introduction

The possibility to align nematic liquid crystals by means of a laser beam was first demonstrated 30 years ago [1]. Systematic study of this topic was initiated by different groups simultaneously at the beginning of the 80's. Since then, numerous interesting optical effects were realized using nematics as nonlinear material [2].

The basic physical mechanism of optical reorientation in transparent liquid crystals is evident. In anisotropic media, the electric field of the light beam is not collinear with the induced dielectric polarization (except for special geometries). Consequently, a torque acts on the molecules, which tends to align them parallel to the polarization direction of the optical field. In liquid crystals the response of the molecules is collective: the symmetry axis of the whole system, the director, is rotated towards the electric field. Due to the realignment of the director, the effective refractive index increases within the illuminated area. The corresponding self-focusing effect can be easily observed with relatively low power c.w. lasers. In homeotropic cells (director aligned perpendicularly to the boundaries), at normal incidence, the reorientation starts at a threshold input power (optical Freedericksz transition). Quantitative data on the Freedericksz threshold are in satisfactory agreement with theoretical predictions, based on the dielectric mechanism outlined above [3,4].

The strong effect of dye-doping on optical reorientation was first observed in a commercial guest-host system, where the Freedericksz threshold turned out to be about two orders of magnitude smaller for the dye-doped material than for the pure host [5]. In subsequent studies, it was found that the enhancement of the optical torque in the presence of dyes is a general phenomenon. The magnitude of the enhancement, however, is very sensitive to the structure of the dye dopant and in most cases the sign of the dye-induced torque is opposite to the dielectric one. These findings could not be explained within the framework of the standard

225

*F. Kajzar and M.V. Agranovich (eds.), Multiphoton and Light Driven Multielectron Processes in Organics: New Phenomena, Materials and Applications, 225–236.*

continuum electromagnetic description, used for transparent nematics. Molecular models were proposed, which were based on the assumption that the dye-host interaction is modified when a dye molecule is excited by the optical field [6,7]. The models accounted for the main features of the observations and they could be applied to dyed isotropic liquids as well, where an analogous amplification of the optical Kerr effect was detected [8,9].

In the present paper, we summarize the experimental facts and the main points of the theoretical models. We consider the case of photoisomerizable dyes separately and show how the light-induced formation of isomers affects the process of optical reorientation. For a more detailed discussion of some points, we refer to a recent review paper [10].

## 2. Experimental Observations

In Table I. some representative host and dye materials are shown, used in reorientation experiments. Typical doping concentrations are in the range of 0.1-1 percent.

In Fig. 1, the optical Freedericksz threshold is plotted as a function of the concentration of the dye AQ1, in the nematic mixture E63. For the undoped liquid crystal the threshold powers are around 70mW and 120mW for the two spot sizes shown in the figure. These values are in sharp contrast with threshold powers measured for the doped material, which are in the mW range. We note that in the case of other dyes, such as D4, optical Freedericksz transition could not be observed; the homeotropic alignment remained stable up to a critical power, where laser heating caused melting into the isotropic phase.

In further investigations, self-focusing was measured in non-threshold geometrical configurations using Z-scan [11], self-diffraction [12] or pump-and-probe techniques [13]. In these experiments, the effect of any dye could be studied. Thermal contributions to the signal were separated with the help of a stabilizing external electric field that suppressed reorientation. Examples of Z-scan curves are shown in Fig.2. As one can see from the figure, for the pure host self-focusing is observed (valley before peak), as expected. The self-focusing effect is strongly amplified for the AQ2 doped material. (Note that in the later measurement the input power was reduced by a factor of 40.) For D4 self-defocusing takes place (peak before valley). This fact implies that for this dopant reorientation decreases the effective refractive index, i.e. the optical torque rotates the director away from the electric field, towards the wave vector of the beam. For a homeotropic cell, at normal incidence, a negative optical torque has a stabilizing effect on the orientation, hence in this case Freedericksz transition is absent.

TABLE 1. Chemical structures of the host and dye materials

E63 is a commercial mixture, containing cyanobiphenyls. AQ1 has the same structure as AQ2, except that the isopentyl groups are replaced by isobutyl groups.

In order to characterize the effect of dye-doping quantitatively, we introduce the enhancement factor, $\eta$, which is the ratio of the dye-assisted part of the optical torque to the dielectric torque measured in the undoped host. From the measured signals, one can deduce the coefficient $\eta$ as

$$\eta = A(S^{dye} - S^{host})/S^{host} \tag{1}$$

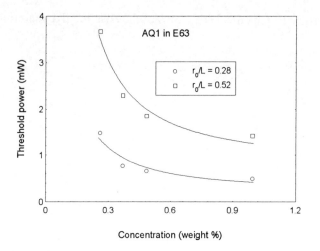

FIGURE 1. Optical Freedericksz threshold power as a function of the dye concentration in the AQ1-E63 system. $r_0$ and $L$ are the beam radius and sample thickness respectively. $\lambda = 633$ nm.

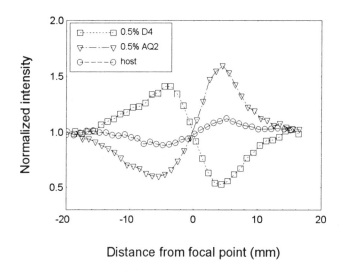

FIGURE 2. Z-scan curves for pure E63 (input power 8 mW), AQ2 doped E63 (input power 0.19 mW) and D4 doped E63 (input power 0.46 mW). Angle of incidence $42°$, $\lambda = 633$ nm.

where $S^{host}$ is the normalized peak-valley modulation for the host, divided by the input power; $S^{dye}$ is the same for the dyed sample and $A$ is a factor correcting for the attenuation of the light beam in the absorbing sample [11]. In the above relation, it is assumed that the contributions from the dye-induced and dielectric parts of the optical torque are added independently; furthermore that the experiment is carried out at low intensities, where the refractive index variation is proportional to the input power. For the examples shown in the figure, one obtains the enhancement factors 238 and -95 for the AQ2 and the D4 dye, respectively. It is clear, therefore, that the dye-assisted optical torque dominates the reorientation process even at these low concentrations.

The enhancement factor is proportional to the dye concentration. In addition, the dielectric torque is proportional to the birefringence factor $n_e^2 - n_o^2$, which is characteristic for the host, but not the dye. It is useful therefore to introduce another dimensionless quantity, the molecular coefficient $\xi$, through the relation

$$\xi = \frac{\eta}{\alpha_{av}\lambda}(n_e^2 - n_o^2) \tag{2}$$

where $\alpha_{av}$ is an orientational average of the absorption coefficient of the guest-host system, $\lambda$ is the wavelength. $\xi$ depends only on the interaction between dye and host molecules, it is independent from the concentration and the absorption cross section of the dopant. The molecular coefficients for AQ2 and D4 in E63 for the wavelength $\lambda = 633$nm are 5600 and -1160, respectively [10,13].

A number of anthraquinone dyes have been investigated quantitatively by us [11] and the Naples group [13]. Most anthraquinone derivatives show a similar effect than the D4 dye, regarding both the magnitude and the sign of the molecular coefficient. The essentially larger and positive values found for the AQ1 and AQ2 dyes are to some extent exceptional.

The molecular coefficient of a particular dye can depend on the host material and the wavelength as well. For the D4 dye, the $\xi$ coefficient is almost the same for the cyanobiphenyl hosts and MBBA; in the case of AQ2, considerably smaller $\xi$ value was found for the later host than for the former ones [13]. The wavelength dependence of the molecular coefficient for anthraquinone dyes was reported in [14]. No significant variation of $\xi$ occurred for D4 in the absorption band, while for AQ2 it increased around the absorption peak.

The temperature dependence was measured for AQ1 and AQ2 [15]. It was found that $\xi$ decreases with temperature. The enhancement ratio, however, increases near the nematic-isotropic phase transition temperature and seems to tend to a finite value at the clearing point. It is interesting to note that the dye-induced enhancement of optical reorientation exists also in the isotropic phase. It was shown that in isotropic liquids the optical Kerr effect is amplified in the presence of absorbing dyes in a similar way like the optical torque in the nematic phase [8,9]. In particular, the enhancement factor changes only about 20 percent when the nematic-isotropic transition takes place [9].

## 3. Theoretical Models

A hypothesis about the microscopic origin of the dye-induced torque was first published in 1994 [6]. The main points of the model are (i) orientationally selective excitation of the dye molecules (ii) change of the interaction between a dye molecule and the surrounding host molecules upon excitation.

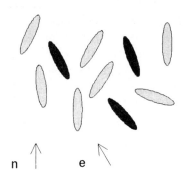

FIGURE 3. Schematic representation of the orientational distribution of the ground-state (dotted ellipses) and excited-state (dark ellipses) dye molecules. **e** is the polarization direction, **n** is the director.

In Fig. 3, the orientational distribution of the dye molecules is sketched in a pictorial way. In the absence of illumination, the distribution is axially symmetric around the director. In this case, the total torque experienced by the dye molecules on the host material vanishes by symmetry. Excitation by a polarized light beam, however, breaks the axial symmetry. The molecules, aligned with their transition dipole moment parallel to the electric vector of the optical field are excited with a high probability, while those aligned perpendicular to it do not absorb photons. Rotational diffusion restores partially the axial symmetry of the system, nevertheless deviation from the thermodynamic equilibrium distribution persists even under steady-state illumination. Due to the lack of axial symmetry, the total torque acting on the host does not vanish. This internal interaction is the source of the "dye-induced" torque, observed in the experiments.

It might appear strange at first sight that internal interactions can lead to the rotation of the whole system, as this is not allowed for rigid bodies. For systems, however, in which internal motions carrying angular momentum are possible (e.g. spin in magnetic materials), such rotations can take place. The case of nematic liquid crystals was discussed in [6,7,16].

To make the considerations quantitative, one can determine the orientational distribution of the excited and ground-state molecules, starting from a rotational diffusion equation [6,7]:

$$\frac{\partial f_e}{\partial t} + \nabla J_e = pf_g - f_e/\tau_l$$

$$\frac{\partial f_g}{\partial t} + \nabla J_g = -pf_g + f_e/\tau_l$$

(3)

Here $f_e$ and $f_g$ are the angular distribution functions of excited and ground-state dye molecules respectively; $J_e$ and $J_g$ are the corresponding rotational currents; $p$ is the excitation probability, which depends on the angle between the transition dipole moment of the dye and the polarization direction of the light beam; finally $\tau_l$ is lifetime of the excited state. The rotational currents are connected to the distribution functions through the constitutive equations:

$$J_i = -D_i \nabla f_i - D_i/kT(\nabla U_i)f_i \quad i = e, g.$$

(4)

$D_e$ and $D_g$ are the rotational diffusion constants, $U_e$ and $U_g$ are the mean-field orienting potentials for the excited and ground state, respectively. The first term on the right-hand-side of the above relation corresponds to rotational Brownian motion of the molecules with a diffusion constant of $D_i$. The second term describes the drift motion of the molecules in the nematic mean-field potential $U_i$ with a mobility $D_i/kT$. With the help of the above equations, the distribution functions can be determined and the torque acting on the host can be calculated as

$$\Gamma = -\int (f_g \nabla U_g + f_e \nabla U_e) \times s \, ds,$$

(5)

where the integration is carried out over all orientations, $s$. As pointed out earlier, this internal torque can be identified with the "dye-induced" torque, governing the director motion in the presence of absorbing dyes.

In the limit of low intensities, i.e. when the population of the excited state is small compared to the total number of dyes, the model yields for the molecular coefficient

$$\xi = \frac{Q}{\hbar D_e/kT}.$$

(6)

$Q$ is a dimensionless factor, depending on $U_i$, $D_i$, ($i = e, g$) and $\tau_l$ ; it can be calculated numerically. For typical numbers ($U_i$ few times $kT$, $D_i \approx 2 \times 10^8$ sec$^{-1}$, $\tau_l \approx 10^{-9}$ sec), one obtains that $Q$ is in the range of $0.05 - 0.2$, yielding $\xi$ values $10^3 - 10^4$. This is indeed the order of magnitude, observed in the experiments. A more precise quantitative comparison is, however, difficult, because it would require independent measurements of the molecular parameters included in the model. $U_g$ can be deduced from the ground-state dye order-parameter; $D_e$ and $\tau_l$ might be available from time-resolved polarized fluorescence studies. On the other hand, there is no direct information on the excited-state orientational energy, $U_e$, and the rotational diffusion constant of the ground-state dye molecules.

In the first theoretical model [6], it was assumed that the rotational mobility of a dye molecule is equal in the ground and excited states. If this assumption is accepted, it is possible to estimate the change of the interaction energy, with the help of the experimental value of the molecular coefficient. We note that a positive $\xi$ coefficient indicates an increase, while a negative one indicates a decrease of the orientational energy of a dye molecule upon excitation. In the case of D4 and a series of similar anthraquinone dyes, the measured negative values correspond to 10-30 percent lowering of the orienting potential in the excited state.

As mentioned earlier, the AQ2-E63 system behaves somewhat differently than other dye-doped nematics. This particular mixture was investigated in detail by Paparo et al. [8,17]. In their studies, they used nanosecond laser pulses and investigated the enhancement of the optical Kerr effect in the isotropic phase. The same theory that was used for the nematic phase could be applied to describe the observations in isotropic liquids as well; in both cases the same physical parameters are involved. With pulsed lasers, however, it is possible to go beyond the linear approximation. If the number of excited molecules is comparable to the total number of dye molecules, the optical torque is no longer proportional to the input light intensity. The investigation of the intensity dependence of the enhancement allows one to deduce from the data the change of both the interaction energy and the rotational diffusion constant. According to Paparo et al., in the case of AQ2 the enhancement is induced to a large extent by a strong decrease (by a factor of two or even more) of the rotational mobility. It is possible that such a significant reduction of the rotational constant upon excitation is due to some kind of charge transfer interaction between the excited dye molecules and the surrounding host molecules. Further investigations are needed, however, to prove this assumption.

## 4. Optical Reorientation in the Presence of Photoisomerization

In this part, we consider optical reorientation of nematic liquid crystals, doped with photoisomerizable dyes. The most frequently studied compounds of such kind are azobenzene derivatives. In the energetically more stable *trans* form, the two bonds linking the central azo group to the aromatic rings are parallel. Under illumination, after excitation the trans molecules relax with a certain probability to the metastable *cis* form, where the angle between the two bonds is 120 degrees, resulting in a V-shape of the molecule. The lifetime of the cis isomer can range from seconds to several hours, depending on the detailed chemical struture of the compound. Light-induced cis-trans transitions are also possible, and already at very weak optical fields ($<1$ $mW/mm^2$) the trans-cis equilibrium is dominated by photoinduced transitions.

It was shown by several authors that azo dyes cause a similar amplification of the optical reorientation than anthraquinone dyes [12,14,18]. Barnik et al, however, reported an interesting anomaly in connection with the angular dependence of the enhancement [19]. The effect, which was observed in a nematic doped by a di-azo derivative, is illustrated in Fig.4. As it can be seen from the figure, for this system, at small angles of incidence self-defocusing occurs, while above a critical angle self-focusing takes place. This observation seems to contradict to the models presented in the previous section, which require - from symmetry alone - that the enhancement factor should be independent from the direction of light propagation.

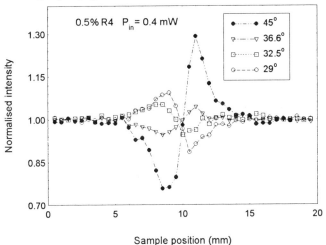

FIGURE 4. Z-scan curves for the R4-E63 system for different angles of incidence. $\lambda=514$ nm.

The reason behind this peculiar behavior is that the trans and cis isomers act as two separate dye-dopants and contribute independently to the total optical torque [20]. In a previous work [21], it was shown that in anisotropic media the steady-state concentration of the cis molecules depends on the angle between the polarization direction of the exciting light beam and the optical axis of the host system, $\Psi$. Provided that the rate of thermal cis-trans transitions can be neglected in comparison with photoinduced transitions, the fraction of cis molecules, $X$, can be given as [20]

$$X = X_{ord} \frac{1+g\cos^2\Psi}{1+h\cos^2\Psi} . \tag{7}$$

$X_{ord}$ is the fraction of cis molecules generated by an ordinarily polarized beam; $g$ and $h$ are molecular coefficients, which depend on the absorption cross sections and the order parameters of the isomers. As shown in [21], the parameters involved in the above equation can be deduced from polarized pump-probe transmission measurements.

With the help of Eq. (7), one can plot the enhancement factor as a function of the cis concentration instead of the angle of incidence. The result for the R4-E63 mixture is shown in Fig. 5. The linear fit of the experimental points corresponds to the assumption that the enhancement factor is a linear superposition of the trans and cis contributions. This assumption is appropriate for the very low concentrations of the dye used in the experiments.

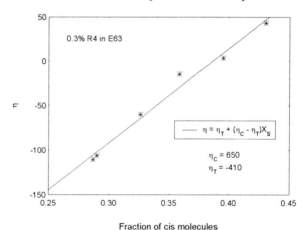

FIGURE 5. The enhancement factor as a function of the cis concentration for the R4-E63 system. $\lambda$=514 nm. From the linear fit, one obtains for the enhancement factors -410 (trans) and 650 (cis).

From the above data we conclude that the trans isomers generate a large negative optical torque, which is counteracted by the positive contribution of the cis isomers. The opposite signs might be interpreted in the following way. Trans isomers exhibit elongated shape, therefore they are relatively well ordered in the nematic host. On the other hand, the shape anisotropy in the cis form is considerably smaller than in the trans one, hence the orientational order of the cis isomers is low. The shape of the electronic excited state is intermediate between the ground-state trans and cis forms. We may suppose therefore that the orientational mean field energy decreases during a *trans ground-state → excited state → cis ground-state* process. According to the theoretical models, such a decrease implies a negative molecular coefficient. When a cis molecule is excited, the process takes place in the opposite direction; the orientational energy is increasing, hence the $\xi$ coefficient is positive. An additional factor may be the increase of rotational mobility during a trans-cis transition and a corresponding decrease during the invers transition. The relative importance of the two contributions might be separeted using pulsed lasers.

## 5. Conclusions

In this paper, we discussed a particular nonlinear optical effect taking place in dye-doped nematic liquid crystals. We pointed out that the deviation of the orientational distribution of the dye molecules from axial symmetry in the presence of an optical field, gives rise to an enhanced optical torque. The magnitude and sign of the enhancement of the optical torque depend on the change of dye-host interaction upon excitation. For many dyes it can be one or two orders of magnitude at concentrations below 1%. The mechanism is effective in dye-doped isotropic liquids as well.

A special situation occurs with isomerizable dyes, when light converts a fraction of the dye molecules into a metastable form. The two (or eventually more) isomers can be regarded as separate dopants, which contribute independently to the dye-induced torque. In this case a number of novel optical phenomena can be observed; we analysed here in detail the dependence of the nonlinearity on the angle of incidence of the light beam.

The large dye-assisted increase of the orientational nonlinearity, described here, may be interesting from the point of view of applications. In this respect it is important to note that the enhancement of the nonlinear signal does not affect the response time. The main disadvantage of the effect is that it requires the presence of absorbing dyes. The ratio of the molecular coefficient to the absorption cross section is, however, material dependent. Systematic studies of guest-host systems might help to find dyes for which the absorption losses are sufficiently low for specific purposes.

## References

1. Saupe, A. (1998), in P. Cladis and P. Palffy-Muhoray (eds) *Dynamics and Defects in Liquid Crystals - A Festschrift in Honor of Alfred Saupe*, Gordon and Breach Science Publisher p. 441.

236

2. Khoo, I.C. (1994) *Liquid Crystals: Physical Properties and Nonlinear Optical Phenomena*, Wiley Interscience, New York.
3. Csillag, L., Jánossy, I., Kitaeva, V.F., Kroó, N. and Sobolev, N.N. (1982) *Mol.Cryst.Liq.Cryst.* **84**, 125.
4. Khoo, I.C., Liu T.H.and Yan, P.Y. (1987) *J. Opt. Soc. Am.* **B4**, 115.
5. Jánossy, I., Lloyd, A.D. and Wherrett, B.S. (1990) *Mol.Cryst.Liq.Cryst.* **179**, 1.
6. Jánossy, I. (1994) *Phys.Rev.E* **49**, 2957.
7. Marrucci, L. and Paparo, D. (1997) *Phys.Rev.E* **56**, 1765.
8. Paparo, D., Marrucci, L., Abbate, G., Santamato, E., Kreuzer, M., Lehnert, P. and Vogeler, T. (1997) *Phys.Rev.Lett.* **78**, 38.
9. Muenster, R., Jarasch, M., Zhuang X.and Shen, Y.R. (1997) *Phys.Rev.Lett.* **78**, 42.
10. Jánossy, I. (1999) *J. Nonlin. Opt. Phys. Mat.* **8**, No.3.
11. Jánossy, I. and Kósa, T. (1992) *Opt.Lett.* **17**, 1183.
12. Khoo, I.C., Li, H. and Liang, Y. (1993) *IEEE J.Quantum Electron* **29**, 1444.
13. Marrucci, L., Paparo, D., Maddalena, P., Massera, E., Prudnikova, E.and Santamato, E.(1997) *J.Chem. Phys.* **107**, 9783.
14. Paparo, D., Maddalena, P., Abbate, G,. Santamato, E. and Jánossy, I. (1994) *Mol.Cryst.Liq.Cryst.* **251**, 73.
15. Jánossy, I., Csillag, L. and Lloyd, A.D. (1991) *Phys. Rev. A* **44**, 8410.
16. Palffy-Muhoray, P. and Weinan, E. (1998) *Mol.Cryst.Liq.Cryst.* **320**, 193.
17. Marrucci, L., Paparo, D., Abbate, G., Santamato, E., Kreuzer, M., Lehnert, P. and Vogeler, T. (1998) *Phys. Rev. A* **58**, 4926.
18. Kósa, T. and Jánossy, I. (1995) *Opt.Lett.* **20** 1230.
19. Barnik, M.I., Zolot`ko, A.S., Rumyantsev, V.G. and Terskov, D.B. (1995) *Kristallografiya*, **40**, 746.
20. Jánossy, I. and Szabados, L.(1998) *Phys.Rev.E.* **58**, 4598.
21. Jánossy, I. and Szabados, L.(1998) *J. Nonlin .Opt. Phys. Mat.* **7**, 539.

# PHOTOCHROMIC MATERIALS IN HOLOGRAPHY

R. A. LESSARD[a], C. LAFOND[a,b], A. TORK[a,c], F.BOUDREAULT[a],
T. GALSTIAN[a], M. BOLTE[b], A. RITCEY[c] and I.PETKOV[d]
[a] *Centre d'optique et Photonique et Laser, Département de Physique*
*Faculté des Sciences et Génie,*
*Université Laval, Québec, G1K7P4, Canada*
[b] *Laboratoire de Photochimie Moléculaire et Macromoléculaire*
*Université Blaise Pascal F-63177 Aubière Cedex France*
[c] *CERSIM, Département de Chimie, Faculté des Sciences et Génie,*
*Université Laval, Québec, G1K7P4, Canada*
[d] *Department of Organic Chemistry,*
*University of Sofia, 1 James Bouchier Avenue, 1126 Sofia, Bulgaria*

## Abstract

Fulgide Aberchrome 670-doped polymer films were studied. Two approaches were considered: photochemical characterization and holographic recording.
In PMMA matrix, the closed form presents a maximum of absorption centered at 525 nm upon irradiation at 365 nm. We have determined the photoreaction rate constants $k_{UV}$ and $k_{VIS}$ respectively for the coloring and bleaching process : $k_{UV}=1.2\times10^{-3}$ $s^{-1}$ and $k_{VIS}=11.1\times10^{-3}$ $s^{-1}$. Photochemical fatigue resistance in different polymer matrices was investigated. We found a loss of 9, 11, 13 and 35% respectively in PS, CA, PMMA and PVK. Concerning holographic recording, we obtained diffraction efficiency $\eta=0.65\%$ in PMMA films 30 μm thick.

## 1. Introduction

Properties of photochromic compounds introduced in polymer films are of great interest because of their practical use in holographic recording[1,2]. Among these systems, fulgides[3,4] are of particular concern.
Organic photochromism is usually defined as the reversible color change of a molecule upon irradiation with UV or visible light[5]. The photoreaction between the involved species proceeds via electronically excited states and causes a change on molecular structure or conformation, thereby altering the UV/visible absorption spectra of the sample.
Fulgides show completely reversible photochromism. We studied the 2-(1-(2,5-Dimethyl-3-furyl)ethylidene)-3-(2-adamantylidene) succinic anhydride[6] **1** knew in commerce as Aberchrome 670 (figure 1). The 1E-isomer (open form) presents an

*F. Kajzar and M.V. Agranovich (eds.), Multiphoton and Light Driven Multielectron Processes in Organics: New Phenomena, Materials and Applications, 237–248.*
© 2000 *Kluwer Academic Publishers. Printed in the Netherlands.*

absorption band only in UV (between 320 and 400 nm). Therefore, upon UV irradiation, at 365 nm, a coloring process due to an electrocyclic ring-closure occurs, leading to the

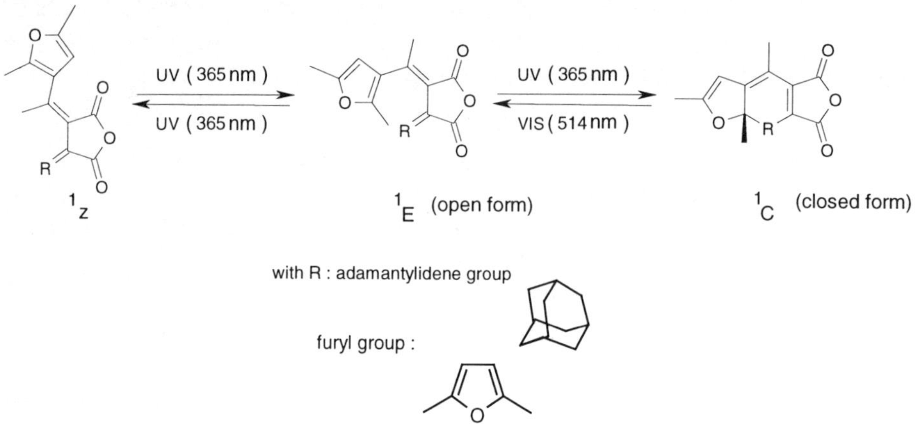

*Figure 1. Photochemical reactions of Aberchrome 670*

formation of the deeply colored 1C-isomer (closed form) which presents a strong absorption in the visible. Irradiation with visible light at 514 nm (bleaching process) induces back-conversion to the open form. However, upon UV irradiation, side reactions such as E/Z isomerisation, occurs and implies formation of non-photochromic Z-isomer but, in solid matrix, this process is less prevailing compared in solution[7].

In this study, first, we have described the photochemical characterization of Aberchrome 670 in polymethylmethacrylate, PMMA, matrix (absorption spectral changes, and kinetic study upon irradiation). Moreover, one of the indispensable properties required should be high resistance to photochemical degradation. In these conditions, we have compared the photochemical fatigue resistance of this fulgide in different polymer matrices (PMMA, polyvinylcarbazole (PVK), polystyrene (PS), and cellulose acetate CA) under UV and visible irradiation cycles.
Next, for two polymer matrices (PMMA and CA), we have compared the photochemical fatigue resistance between fulgide Aberchrome 670 and the spiropyran :1',3',3'-Trimethylspiro-8-nitro-2H-1-Ben-zopyran-2',2'-Indoline **2**. Figures 1 et 2 shows the photochemical reactions between both molecular species.

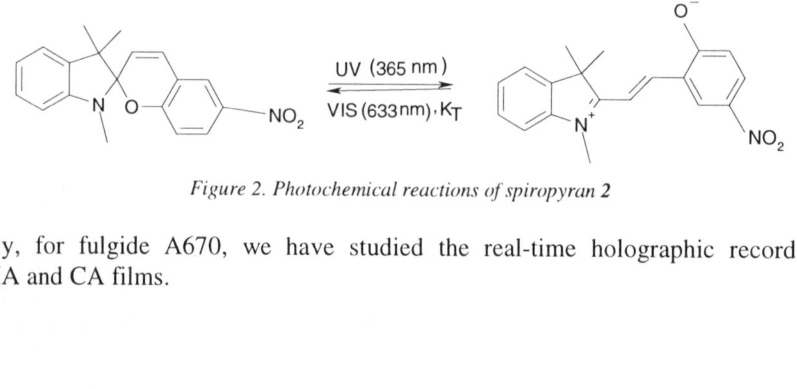

*Figure 2. Photochemical reactions of spiropyran **2***

Finally, for fulgide A670, we have studied the real-time holographic recording in PMMA and CA films.

## 2. Experimental

### 2.1. APPARATUS

For the photochemical study, we have used as irradiation set-up a Kratos 1000 W Xenon Short-Arc lamp filtered at 365 and 514 nm. Both irradiation processes were monitored by real-time UV/Visible spectrometry and the absorption spectra were recorded on a Scientech 8030 spectrometer.
Concerning the holographic study, the holograms were recorded with Argon ion laser at 514 nm, read with diode laser at 660 nm and erased by UV light.

### 2.2. PREPARATION OF PHOTOCHROMIC DYES DOPED POLYMER FILMS

A solution of polymer and fulgide (solvent: chlorobenzene for PMMA and PVK, chloroforme for PS and dioxane for CA) was stirred and heated at 40°C during 30 minutes. A volume of 1.7 ml of this coating solution was poured by gravity settling onto a glass plate (5 cm * 3.5 cm) and allowed to dry in darkness under normal laboratory conditions (T=20°C, relative humidity RH=30-40%) for 72 hours. We studied three different dye concentrations: 1%, 2% and 5% by weight of polymer and for each concentration two thicknesses of the films, measured by Sloan's Dektak II-1 profiler were found around 20 and 30 μm.

## 3. Theoretical study

### 3.1. KINETIC

Equation (1) can be used to fit the experimental data. Each irradiation process has to verify separately the following equation [8].

$$-\ln \frac{A_t - A_e}{A_0 - A_e} = kt \tag{1}$$

where, $A_t$ represents the absorbance of the closed form in the visible range at the $\lambda$max and time t. $A_e$ is the absorbance at the $\lambda_{max}$ of the closed form measured at the maximum of formation (photostationnary state : PSS) for the coloring process and at the end of the reaction for the bleaching process. $A_0$ is the absorbance at time t=0 (for the bleaching process, this value was taken as the absorbance of the UV irradiation at the photostationnary state). k represents the photoreactions rate constant : $k_{UV}$ and $k_{VIS}$ for, the coloring and the bleaching process respectively.
If the photoreactions entirely follow first-order kinetics, the left-hand side of Eq.(1) must show a linear time irradiation dependence.

## 3.2. PHOTOCHEMICAL FATIGUE RESISTANCE

In order to describe the fatigue of the polymer under cycled Write/Erase procedures, we have represented the changes in $A_n/A_0$ with UV/visible irradiation cycles numbers. $A_0$ and $A_n$ are the closed form absorbance values obtained on the first and nth cycles.

## 4. Photochemical results

### 4.1. IN PMMA MATRIX

The photochemical behavior upon UV and Visible irradiation was investigated. We have represented in figure 3a the coloring process of Aberchrome 670 in PMMA film. UV exposure causes the E-isomer at 346 nm to decrease while a visible band situated at 525 nm, which characterize the C-isomer formation, grows up simultaneously. The photostationnary state was reached after 60 minutes of exposure. The spectral changes, induced by visible irradiation at 514 nm are shown on figure 3b, the visible peak disappears and the UV band is regenerated.

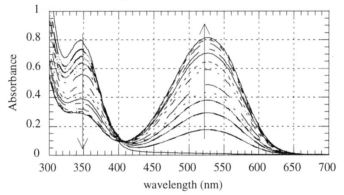

*Figure 3a. Absorption spectral changes of Aberchrome 670 in PMMA film upon irradiation at 365 nm*

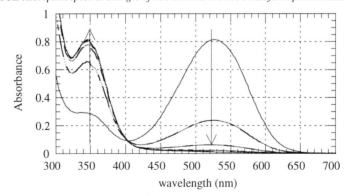

*Figure 3b. Absorption spectral changes of Aberchrome 670 in PMMA matrix upon irradiation at 514 nm*

Figure 4 shows the variations of the absorption of the closed form at 525 nm for the first cycle of UV/visible irradiation.

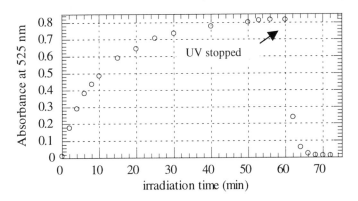

*Figure 4. Evolution of the closed form absorbance at 525 nm*
*for tne first UV/Vis irradiation cycle*
*in PMMA matrix*

For the first UV/Visible cycle irradiation time dependence of the left-hand side of Eq(1) was represented in figure 5. We can observe that both irradiation processes obey to first-order kinetic: there is a linear dependence. Moreover the bleaching process is much larger than the coloring:
$k_{UV}=1.2\times10^{-3}$ s$^{-1}$ and $k_{VIS}=11.1\times10^{-3}$ s$^{-1}$.

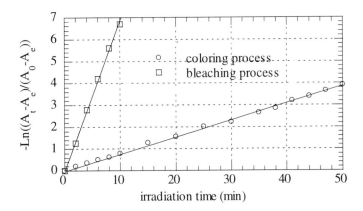

*Figure 5. Irradiation time dependence of the left-hand side of Eq(1)*

## 4.2. IN PVK MATRIX

PVK matrix induces a red-shift of the absorption maximum of the closed-form. Indeed, upon UV irradiation at 365 nm, this maximum is centered at 533 nm instead of 525 nm in PMMA matrix. Photostationnary State is reached after 40 minutes of exposure in PVK instead of 60 minutes in PMMA.

For the first UV/Vis irradiation cycle, we have represented in figure 6 the variations of the optical density of the closed-form at 533 nm.

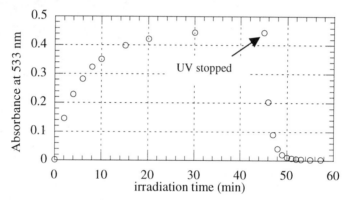

*Figure 6. Evolution of the closed-form absorbance at 533 nm*
*for the first UV/Vis irradiation cycle*
*in PVK matrix*

Figure 7 shows the irradiation time dependence of the left-hand side of Eq(1). As for PMMA films, coloring and bleaching processes follow entirely first-order reaction with rate constants: $k_{UV}=2.55\times10^{-3}$ $s^{-1}$ and $k_{VIS}=12.5\times10^{-3}$ $s^{-1}$.

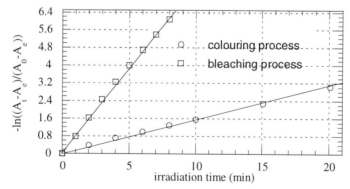

*Figure 7. Irradiation time dependence of the left-hand side of Eq(1)*

# 5. Photochemical fatigue resistance

## 5.1. EFFECTS OF THE POLYMER MATRICES ON FULGIDE A670

The photochemical fatigue resistance of fulgide Aberchrome 670 was analyzed upon repeated UV and visible exposure cycles (figure 8). We have considered four polymer matrices : AC, PS, PMMA, and PVK.

This study shows an important matrix effect between PVK and the other polymers. Indeed we have obtained, in PVK, a loss of 35% of the C-isomer initial absorption after 10 cycles against 13, 11 and 9 respectively in PMMA, CA and PS. In these conditions, PVK is not a satisfactory candidate for the recording of repeated Write/Read/Erase cycles.

Moreover, Kaneko et al[9]. found a loss of 15% for Aberchrome 540 in PMMA matrix. This value is in agreement with our result (13%). Both Aberchromes only differ by substituent R, which represents isopropylidene and adamantylidene group respectively for Aberchrome 540 and 670. Consequently, this group does not induce photochemical fatigue resistance. This phenomenon is principally caused by the aryl group, which is, on both cases, a furyl group.

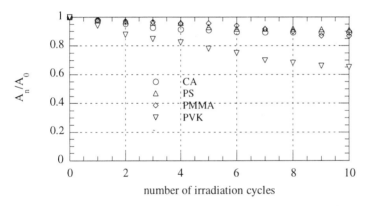

*Figure 8. Photochemical fatigue resistance of A670 in PMMA, AC, PS, and PVK films after 10 UV/Vis irradiation cycles*

## 5.2. COMPARISON BETWEEN FULGIDE A670 AND SPIROPYRAN 2

### 5.2.1. *in PMMA matrix*
As shown in figure 9, the photochemical fatigue resistance in PMMA films containing spiropyran 2 was markedly decreased in comparaison with fulgide A670 doped PMMA film. After 10 exposure cycles, we obtained a loss of 42% for spiropyran of the initial absorption instead 13% for the fulgide.

244

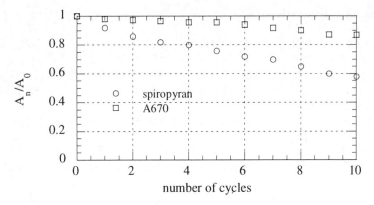

*Figure 9. Photochemical fatigue resistance in PMMA films*
*for fulgide A670 and spiropyran*

### 5.2.2. *in CA matrix*

After 10 UV/Vis exposure cycles, we have observed a strong difference between both photochromic classes. Indeed, we have obtained a disappearance percent of 72% concerning the spiropyran against 11% for the fulgide (figure 10). This deviation has been even increased compared to PMMA matrix.

In these conditions, it will be very difficult to use this spiropyran as recording material for Write/Read/Erase cycles.

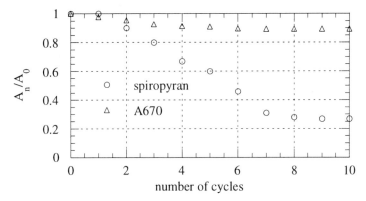

*Figure 10. Photochemical fatigue resistance in CA films*
*for fulgide A670 and spiropyran*

## 6. Holographic recording

Holographic recording using Aberchrome 670 doped PMMA films was investigated. To characterize a hologram, we have to define the diffraction efficiency[10] $\eta$:

$$\eta = \frac{I_1}{I_0}$$

$I_1$ represents the intensity of the diffracted first order beam and $I_0$ is the intensity of the incident beam.

## 6.1. EXPERIMENTAL SET-UP:

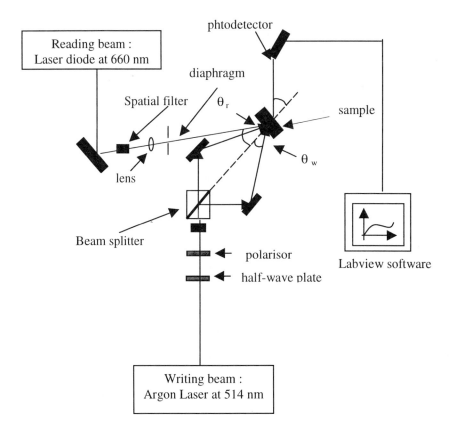

*Figure 11. Holographic recording-experimental set-up*

UV irradiation at 365nm induced the formation of a strong absorption band centered at 525 nm. In these conditions, holograms can be written with Argon ion laser at 514 nm. Diode laser at 660 nm has been used for the reading beam, indeed, at this wavelength, the photochromic compound does not absorb.

Finally, the grating can be erased upon irradiation with UV light at 365 nm.

In this study thick holograms were only considered (Quality factor[11] Q=50 for films in PMMA matrix 20 microns thick). Moreover, with a writing angle $\theta_w$=11°, holographic recording in PMMA, gives a grating with N=1100 lines/mm.

## 6.2. RESULTS IN PMMA MATRIX

The Argon and the diode lasers are polarized horizontally and vertically respectively.

First, a total of intensity of 3 mW/cm$^2$ illuminated the sample. In these conditions, we have not observed any diffraction beam for each dye concentration (1%, 2% and 5%) and for each thickness (20 and 30 μm).

The best results were found with an intensity of 10 mw/cm$^2$. The values are given in table 1 and 2.

TABLE 1. Diffraction efficiency $\eta$ for A670 dye doped PMMA films thickness 20 μm

| concentration | $\eta$ (%) |
|---|---|
| 1 | / |
| 2 | 0.4 |
| 5 | 0.5 |

TABLE 2. Diffraction efficiency $\eta$ for A670 dye doped PMMA films thickness 30 μm

| concentration | $\eta$ (%) |
|---|---|
| 1 | / |
| 2 | 0.6 |
| 5 | 0.65 |

The diffraction efficiencies measured are weak, one can observe an effect of dye concentration and thickness. For each thickness, no diffracted beam was obtained at a concentration of 1%.

If we compare the efficiencies in AC matrix, (concentration 5% and thickness 20 μm), we have obtained the same result, namely, a diffraction efficiency of 0.4%. We have even increased the thickness up to 120 μm and the diffraction efficiency has not exceed 0.6%.

## 7. Discussion

By means of two approaches (photochemistry and holography), we have characterized photochromic compound Aberchrome 6760.

Photochemically, in PMMA, the coloring and bleaching processes follow entirely first-order reaction with rate constants: $k_{UV}=1.2\times10^{-3}s^{-1}$ and $k_{VIS}=11.1\times10^{-3}s^{-1}$ : bleaching process is much faster. In PVK matrix, the reaction order is identical and for the rate constants are of the same order.

Concerning the phenomenon of photochemical fatigue resistance it becomes evident that, PVK is not suitable for the recording of W/R/E cycles. As for the other matrices studied, photochemical losses are evaluated around 10% after 10 UV/Vis irradiation cycles. For AC, we have even observed stabilization after 7 cycles. However one will have to increase the number of exposition cycles to confirm this fact.

Comparison of the photochemical fatigue resistance between spiropyran and fulgide was investigated, we can conclude that spiropyran 2 is not convenient to record Write/Read/Erase cycles : the loss of absorption is too strong after 10 UV/Vis irradiation cycles (42% and 72% respectively in PMMA and CA).

For the holographic study, diffraction efficiencies obtained in PMMA are weak ($\eta=0.6\%$ with a concentration of 5% and a thickness of 30μm). These facts show an effect of the concentration and thickness.

248

**References**

1. Ghailane, F., Manivannan, G. and Lessard, R. A. (1995) Spiropyran doped poly(vinylcarbazole): A new phtopolymer recording medium for erasable holography, *Opt. Eng*, **34**, 1683-1686.

2. Xue, S. S., Manivannan, G. and Lessard, R. A. (1994) Holographic and spectroscopic characterization of spiropyran doped poly(methylmethacrylate) films, *Thin. Sol. Film*, **253**, 228-232.

3. Whitall, J. (1990) Fulgides, in H. Dürr and H. Bouas-Laurent (eds), *Photochromism : Molecules and Systems*, Elsevier, Amsterdam, pp. 467-492.

4 Lafond, C., Lessard, R. A., Bolte, M., et Petkov, I. (1998) Characterization of dye-doped PMMA/PVK films as recording materials, *Proc. SPIE*, **3417**, 216-227.

5. Dürr, H. and Bouas-Laurent, H. (1990) *Photochromism : Molecules and Systems*, Elsevier, Amsterdam.

6. Ghailane, F. (1995) Étude du photochromisme et de la photoreactivité dans le poly(vinylcarbazole), *Ph.D. Thesis*, Université Laval, Québec.

7. Yokoyama, Y., Hayata, H., Ito, H., and Kurita, Y. (1990) Photochromism of a furylfulgide, 2-[1-(2,5-dimethyl-3-furyl)ethylidene]-3-isopropylidene succinic anhydride in solvents and polymer films, *Bull. Chem. Soc. Jpn.*, **63(6)**, 1607-1610.

8. Tsuyioka, T., Kume, M., and. Irie, M. (1997) Photochromic reactions of a diarylethene derivative in polymer matrices, *J. Photochem. Photobiol. A: Chem.* **104**, 203-206.

9. Kaneko, A., Tomoda, A., Ishizuka, M., Suzuky, H. and. Matsushima R, (1988) Photochemical fatigue resistances and thermal stabilities of heterocyclic fulgides in PMMA films, *Bull. Chem. Soc. Jpn.* **61(10)**, 3569-3573.

10. Moharam, M. G., Gaylord, T.K. and Magnusson, R. (1980) Criteria for Bragg regime diffraction by phase grating, *Opt. Commun.*, **32**, 14-18.

11. Kogelnik, H., (1969) Coupled wave theory for thick hologram grating, *Bell Syst. Tech. J.* **48**, 2909-2947.

Authors'e-mail:

ralessard@phy.ulaval.ca,          clafond@phy.ulaval.ca,
atork@chm.ulaval.ca,              galstian@phy.ulaval.ca,
mbolte@cicsun.bpclermont.fr,      aritcey@chm.ulaval.ca,
ipetkov@chem.uni-sofia.bg

# KINETICS OF PHOTOCHROMIC PROCESSES IN DIHYDROPYRIDINE DERIVATIVES

J. SWORAKOWSKI[1], S. NEŠPUREK[2], J. LIPIŃSKI[1],
E. ŚLIWIŃSKA[1], A. LEWANOWICZ[1] and A. OLSZOWSKI[1]
[1] Institute of Physical and Theoretical Chemistry, Technical University
of Wrocław, 50-370 Wrocław, Poland
[2] Institute of Macromolecular Chemistry, Academy of Sciences of the
Czech Republic, 162 06 Prague, Czech Republic

## 1. Introduction

Substitued dihydropyridines belong to a group of photoactive compounds whose properties have been relatively little known (cf. the accompanying paper [1] reviewing the structure and properties of dihydropyridines, pyrans, thiopyrans and selenopyrans). An interesting feature of the photochemical activity of many representatives of this group is that reactions following an electronic excitation are reversible (i.e., the compounds exhibit the photochromic effect), and that the effect can be observed both in liquid solutions, in solid matrices (e.g. in polymer foils), and in crystallites. Moreover, the effective rate constants deduced from the kinetics of colouring and bleaching have been shown to differ by several orders of magnitude [2-8], thus opening (at least in principle) an interesting possibility of tailoring the response kinetics of the systems to particular needs. Several molecules belonging to this group have been synthesized to date (cf. [1,9] and references therein), complex studies, however, have been performed only on few systems. Thus it comes as no surprise that the molecular mechanism of the activity of these molecules has not been unequivocally established. It has been postulated that the initial step in the photochromic cycle involves an intramolecular phenyl shift [9-11] and/or a 3,5-bridge formation [12-16]. Some further steps in the sequence of elementary processes have also been proposed, e.g., a 2,4-bridge formation [2,3,9,14] or a hydrogen shift [6,17]. It has also been believed that initial stages of the photochromic process are identical, irrespective of the nature of the heteroatom in the central ring.

The aim of this contribution is to review results of our studies of the energetics and kinetics of photochromic reactions in substituted dihydropyridines, and to put forward the underlying molecular mechanism responsible for the activity of these compounds. The results given in this contribution will mostly concern two substituted dihydropyridines: 1-methyl-2,4,4,6-tetraphenyl-1,4-dihydropyridine (hereafter referred to as DHP), and 4,4-(biphenyl-2,2'-diyl)-2,6-diphenyl-1-methyl-1,4-dihydropyridine

F. Kajzar and M.V. Agranovich (eds.), Multiphoton and Light Driven Multielectron Processes in Organics: New Phenomena, Materials and Applications, 249-260.

(BDH). Structural aspects of the reactivity being discussed in the accompanying contribution [1], we shall focus here on the results directly related to the reactions following the illumination of the samples. Molecular structures of parent molecules and some postulated products [2,3,9-17] are shown in Fig. 1.

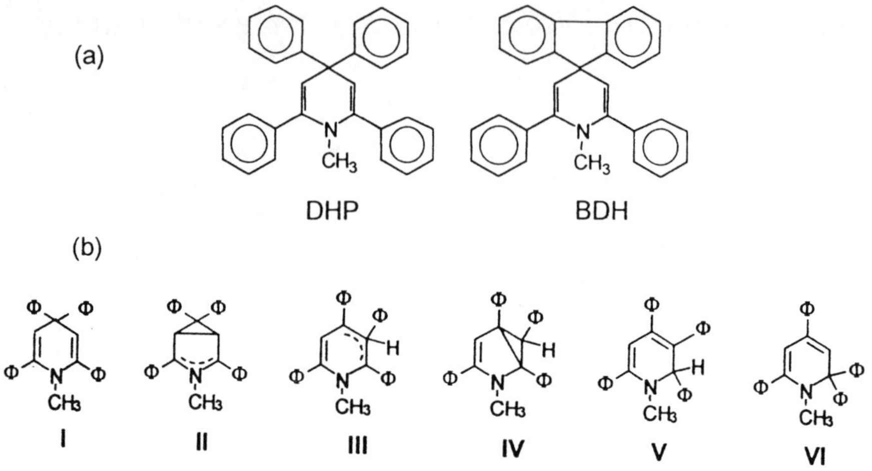

*Figure 1.* (*a*) Chemical formulae of DHP and BDH. (*b*) DHP (*I*) and possible intermediate and final products (*II-VI*) formed upon irradiation of *I*.

## 2. Review of experimental results

### 2.1. SPECTROSCOPIC MEASUREMENTS

Dihydropyridines and other compounds belonging to the class of materials under study do not absorb in the visible region unless they contain additional chromophores attached to the central ring [12,16]. Upon illumination, one may observe a buildup of bands in the visible: essentially, two broad bands appear, at 400-450 nm and around 500-600 nm (cf. Fig. 2). The process appears reversible: the bleaching is a thermally driven process, its rate depending dramatically on the environment of the reacting molecules. In spite of pronounced differences in the kinetics, the comparison of the spectra of photogenerated species in solid DHP and BDH samples and in solutions of these materials strongly supports the view that identical species are involved in photochemical reactions following the excitation. The question concerning the nature of the coloured species and the mechanism of reactions leading to their appearance and decay will be addressed in the following sections of this contribution.

### 2.2. ISOTHERMAL KINETICS

The kinetics of reactions occurring in the systems under study may be conveniently studied by following the temporal evolution of the band(s) appearing upon

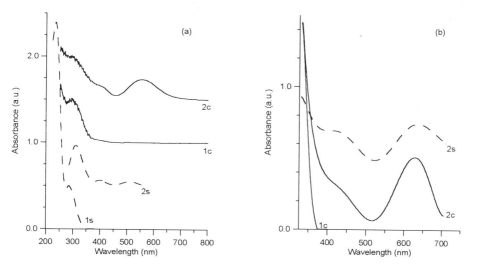

*Figure 2.* UV-Vis spectra of solutions and polycrystalline DHP (*a*) and BDH (*b*). The labels "*s*" and "*c*" refer to measurements performed on chloroform solutions and polycrystalline samples, respectively. Curves labelled "2" have been measured after UV irradiation of the samples (adapted from [3,7,18]).

illumination. Such experiments have been performed on solutions and on solid samples [2,3,7,8]. Typical results obtained for DHP are shown in Fig. 3. The decays have been measured by following the evolution of the absorbance of the band peaking at 520 nm in solutions and at 550 nm in polycrystalline samples. In solutions, the bleaching curves consist of sections of rapid and slow decay, separated by a plateau. A detailed analysis given in [8] demonstrated that such a decay can be interpreted by assuming the following sequence of reactions: reactant (absorbing at 520 nm) → intermediate (transparent) → intermediate (absorbing) → product (transparent). This scheme will be discussed in Section 4.

The solid-state bleaching curves have been qualitatively different: when plotted in the conventional semilogarithmic coordinates, they exhibit a distinct curvature which may be fitted with the 'stretched exponential' function [2]. In our previous paper [7], this feature has been attributed to a distribution of rate constants associated with a random variation of microenvironments of reacting molecules. A comparison of the solution and solid-state results clearly shows that the effective reaction rates controlling the bleaching reaction differ by several orders of magnitude: at room temperature, the rate of the bleaching reaction in solution is of the order of miliseconds, being of the order of hours in a polycrystalline sample. The activation energy of the process controlling the bleaching in polycrystalline DHP below 337 K was estimated to amount to ca. 40 kJ/mol.

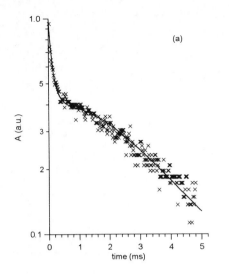
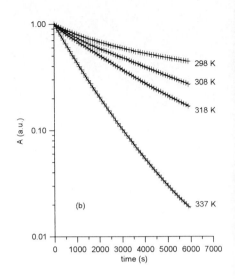

*Figure 3*. Kinetics of thermal bleaching reaction of DHP. The parameter *A* is the absorbance measured at the maximum of absorption (520 nm in solution, 550 nm in the solid sample), normalized to its initial value. (*a*) Chloroform solution. Crosses are experimental points, the full line is a fit to the points using Eq. (4). The experiment was performed at ambient temperature [8]. (*b*) A polycrystalline sample. The full lines have been calculated assuming a first-order process controlled by a Gaussian distribution of activation energies [7].

## 2.3. NON-ISOTHERMAL KINETICS

Isothermal measurements are often time-consuming; moreover, performing a reliable series of experiments may prove difficult or even impossible in some samples, due to unwanted fatigue processes. In this case, non-isothermal techniques may provide a convenient way to determine parameters characterizing the kinetics of physical or chemical processes. We have performed such measurements on polycrystalline DHP and BDH, employing the differential scanning calorimetry (DSC) [7,19]; the non-isothermal bleaching was also followed spectrophotometrically [20].

The DSC technique can be used to determine the enthalpy of a thermally driven reaction (i.e., to determine experimentally the difference of the energies of the ground states of the reactant and the product), and also to determine the activation energy (or its distribution) of the reaction. The use of the method, described in [19], may be illustrated with results shown in Fig. 4. A DSC run carried out on a polycrystalline sample, previously UV-irradiated in order to produce coloured photoproducts, exhibits the presence of a broad exothermic anomaly, observed only once after irradiation, and attributed to an exothermic reaction involving the colours species. The parameter analysed is the 'excess heat flow', i.e., a difference between the heat flow measured during the first run and that measured during any subsequent run. One may show [19] that the excess heat flow should be proportional to the reaction rate. On the other hand, the same information should be obtained from non-isothermal spectrophotometric measurements, by plotting a time derivative of the absorbance in function of temperature [20]. The comparison shown in Fig. 5 clearly demonstrates that the rate of the bleaching reaction observed by following the decay of absorbance at 550 nm is related to the heat evolved during the process observed by the DSC method. Moreover, measurements of the evolution of the shape of the signal with the irradiation time allowed us to distinguish two parallel bleaching processes [19].

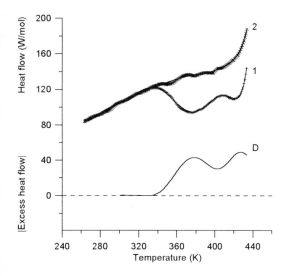

*Figure 4.* DSC curves measured in polycrystalline DHP. The numbers refer to the sequence of runs following the irradiation, the curve labelled "*D*" is the absolute value of the excess heat flow (difference between the first and second run).

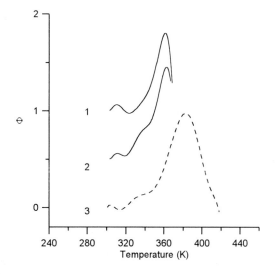

*Figure 5.* Non-isothermal bleach-ing kinetics in polycrystalline DHP measured by following the decay of the absorbance at 550 nm (curves 1 and 2), and by the DSC method (curve 3). The parameter $\Phi$ is proportional to $dA_{550}/dt$ in the curves 1 and 2, and to the absolute value of the excess heat flow in the curve 3. The heating rates amount to 0.25; 0.6; and 10 K/min, respectively. The curves 1 and 2 have been vertically shifted.

254

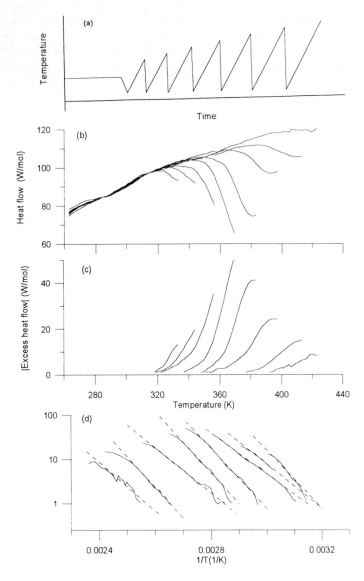

*Figure 6.* Fractional heating DSC. (*a*) A scheme of the temperature programme during a typical run; the coloured form is produced by irradiation of the sample at ambient temperature (the period corresponding to the horizontal section of the diagram). (*b*) A series of runs measured after irradiation of the sample. (*c*) Excess heat flow obtained after substraction of the baseline. (*d*) Data from (*c*) plotted in the Arrhenius coordinates.

A separation of the processes and determination of the activation energies can be achieved using so-called 'fractional heating' technique [19]. A scheme of the run and typical results obtained for polycrystalline DHP are shown in Fig. 6. It should be pointed out that the results obtained were burdened with a substantial error, due to problems with a proper setting of the baselines: our measurements revealed a buildup of an endothermic anomaly above 300 K, probably associated with a phase transition in one of the products. The average activation energy of the high-temperature process determined from measurements on several polycrystalline DHP samples, amounts to (130±30) kJ/mol. Unfortunately, we could not determine the activation energy of the low-temperature process from non-isothermal measurements.

Similar problems were encountered in case of BDH [21]: even in pristine (unirradiated) samples, we detected a small endothermic anomaly peaking at ca. 330 K (cf. Fig. 7a). Thus a reliable determination of baselines proved impossible, and the activation energies determined from runs shown in Fig. 7b should be considered as approximate values only. The average activation energy for BDH was estimated to amount to (140±50 kJ/mol).

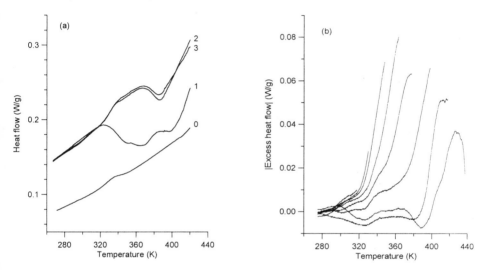

*Figure 7*. DSC curves measured on polycrystalline BDH. (*a*) A single-run DSC experiment. The numbers refer to the sequence of runs following the irradiation. The curve labelled "*0*", showing the result of measurements performed on a pristine sample, has been vertically displaced for the sake of clarity. (*b*) A fractional heating DSC experiment.

## 3. Quantum-chemical calculations

The experiments described in the preceding section were supplemented with quantum-chemical calculations. The calculations were performed employing semiempirical methods: the MNDO method [22] was used in most calculations of the ground-state geometries whereas the excitation energies were calculated using the GRINDOLmethod [23]. 200 singly excited configurations were included in the configuration interaction scheme. The results of the calculations are in a reasonable agreement with available experimental material (X-ray structure of DHP [24], positions of absorption bands [2,3,14,19]).

In case of DHP, the calculations were performed for all structures shown in Fig. 1 and for transition species between them. The energy diagrams, constructed using the results of the calculations described in [7], are shown in Fig. 8.

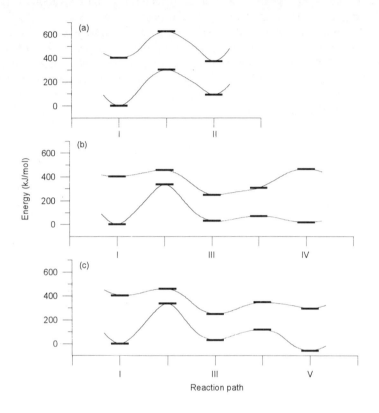

*Figure 8.* Ground and excited state energies of DHP (molecule *I*), postulated products of photochemical reactions (*II-V*), and transition species for intramolecular reactions. Only energies of two lowest singlet states ($S_0$ and $S_1$) are shown (adapted from [9]).

Moreover, basing on kinetic arguments which will be presented in the following section, similar calculations were performed for radicals, possibly formed during the photochemical reactions [8]. Of importance to the matter of the present discussion is a radical formed by breaking one of two C(4)-phenyl bonds in the parent DHP molecule (hereafter referred to as *R*).

Admitting a possibility of the radical formation opens a new path for reactions following the electronic excitation of DHP. In this case, initial stages of the reaction can be presented on a three-dimensional energy diagram, the coordinates being a distance between *R* and phenyl, and an angle between the broken bond and the bond after the radical recombination. A projection of such a diagram is shown in Fig. 9.

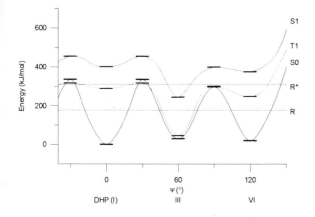

*Figure 9.* The energy diagram showing positions of the ground and excited states (S$_0$, T$_1$ and S$_1$) of DHP, some postulated products, as well as energies of the lowest states of the radical pair ($R$ + phenyl). The parameter $\psi$ is the angle between the direction of the $R$-phenyl(4) bond in the parent DHP, and the $R$-phenyl bond in molecules formed after the recombination of the radicals.

## 4. Mechanism of photochromic activity

The results of quantum-chemical calculations presented in the preceding section allow one to draw conclusions concerning the mechanism of photochemical processes following electronic excitation of DHP. First, a possibility of the 3,5-bridge closing (reaction *I* → *II*) can be ruled out because of high barriers for this reaction, both in the ground and excited state (cf. Fig. 8a). Secondly, the results shown in Fig. 9 demonstrate that the production of the coloured species (molecule *III*) seems to occur preferentially through the radical formation, and not via an intramolecular reaction as was implicitly assumed in earlier papers [2,3,6,9,14,16]. Furthermore, it comes from the results shown in Fig. 9 that the radical recombination may result in the production of molecules *I*, *III* and *VI* in their ground states (S$_0$), but also of the molecule *III* in its triplet state (T$_1$). It is important to note at this point that our results indicate that the bleaching reaction (i.e., the decay of *III*) should *not* result in formation of parent DHP as the barrier for the thermal reaction *III* → *I* is far too high (cf. Figs. 8 and 9). In other words, the photochromic effect does not result from the properties of DHP itself but rather from the properties of products of its photochemical reactions. The thermal bleaching of the coloured molecule *III* results in formation of the molecules *IV* and *V*; both products can then be excited with the UV radiation of the same energy as the parent DHP and can undergo a reaction regenerating the molecule *III*.

The model is in agreement with the results of the kinetic experiments. The results of the solution experiments described in Section 2.2 can be rationalized assuming that (*i*) the radical recombination results in production of *III* in its S$_0$ and T$_1$ states; (*ii*) the T$_1$ → T$_2$ absorption occurs in the same spectral range as the S$_0$ → S$_1$ one; (*iii*) the deactivation of T$_1$ occurs within the time scale of our experiments (milliseconds), via an intermediate; (*iv*) in solution, the radical recombination is quicker than the bleaching reactions and the triplet decay. Under these assumptions, one may derive an equation describing the temporal evolution of the absorbance [8]. The scheme of reactions, shown in Fig. 10, can be solved for the concentrations of the relevant products (the radical *R*, and the triplet- and ground state forms of *III*).

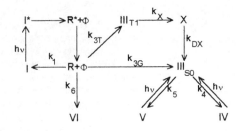

*Figure 10.* The proposed sequence of elementary processes in DHP following a UV pulse. The full lines symbolize thermally driven reactions, dashed lines indicate photochemical reactions. *X* is an unidentified intermediate product

$$\frac{dc_R}{dt} = -kc_R c_\Phi, \tag{1}$$

$$\frac{dc_T}{dt} = k_{3T} c_R c_\Phi - k_X c_T, \tag{2}$$

$$\frac{dc_G}{dt} = k_{3G} c_R c_\Phi + k_{DX} c_T - k_{DG} c_G, \tag{3}$$

Solving the above equations under appropriate boundary conditions [8], one obtains

$$A = \alpha \exp(-k_X t) + \beta \exp(-k_{DG} t) - \gamma \exp(-k_{DX} t). \tag{4}$$

In the above equations, $A$ stands for the absorbance normalized to its initial value, $\alpha$, $\beta$ and $\gamma$ are time-independent parameters containing combinations of rate constants and absorption coefficients, and $k = k_1 + k_{3T} + k_{3G} + k_6$; $k_{DG} = k_4 + k_5$. The analysis of the experimental results yields the following results: the rate constant characterizing the decay of the intermediate product, $k_{DX} = (1500 \pm 500)$ s$^{-1}$. The two remaining rate constants: $k_X$ (decay of the concentration of triplets of the molecule III) and $k_{DG}$ (the sum of rate constants of the bleaching of the ground-state *III* yielding *IV* and *V*) amount to ca. $(6000 \pm 1000)$ s$^{-1}$ and $(500 \pm 100)$ s$^{-1}$.

The bleaching process in crystals is much slower: the curves shown in Fig. 3b demonstrate that the effective rate constants are of the order of $10^{-5}$ s$^{-1}$ at ambient temperature, with the activation energy amounting to ca. 40 kJ/mol. In our previous paper [7], we tentatively attributed the latter value to the barrier for either the *III* $\rightarrow$ *IV* or the *III* $\rightarrow$ *V* reaction. However, preliminary results of EPR experiments [25] point to the presence of long-lived paramagnetic species in polycrystalline DHP. Thus it is possible that the recombination of radicals is at least one of the rate-determining process in solid DHP.

The available experimental material concerning the photochemical activity of BDH and results of quantum-chemical calculations of the parent molecule and possible products are less conclusive. The kinetics of colouring and bleaching of solid samples is qualitatively similar to that of DHP (cf. Section 2.3), the calculations show that the barriers for intramolecular reactions are also similar to those found in DHP [26]. Thus it would be reasonable to assume that the reaction paths are similar in both materials.

One must realize, however, that formation of the pair of radicals and their subsequent recombination, which we assume to be important elementary processes in DHP, are impossible in BDH. Further studies are necessary to explain the mechanism of photochemical reactions in BDH.

## 5. Acknowledgements

This work was supported by the Polish State Committee for Scientific Research (grant No 3T9A4812) and by the Grant Agency of the Academy of Sciences of the Czech Republic (grant No A1050901). In its final stages, the work was also supported by the Technical University of Wroc³aw. The authors thank Prof. J. Kuthan and Dr S. Böhm (Prague Institute of Chemical Technology) for the gift of the material used in the experiments decribed in this paper.

## 6. References

1. Nešpùrek, S., Sworakowski, J., Lipiński, J., Böhm, S. and Kuthan, J. (1999), Photo-colouration of hypervalent heterocycles. Photochromism of dihydropyridines, pyrans and thiopyrans. *These proceedings*.
2. Nešpùrek, S. and Schnabel, W. (1991), Photochromism of 1-methyl-2,4,4,6-tetraphenyl-1,4-dihydropyridine in the solid state: an example of a dispersive chemical reaction, *J. Photochem. Photobiol. A* **60**, 151-159.
3. Nešpùrek, S. and Schnabel, W. (1994) Photochromism of dihydropyridines. Formation and decay of coloured intermediates from 1-methyl-2,4,4,6-tetraphenyl-1,4-dihydropyridine on irradiation in chloroform and acetonitrile solutions and in poly(methyl methacrylate) solid solution, *J. Photochem. Photobiol. A* **81**, 37-43.
4. Šebek, P., Nešpùrek, S., Hrabal, R., Adamec, M. and Kuthan, J. (1992) Novel preparation and photochromic properties of 2,4,4,6-tetraaryl-4*H*-thiopyrans, *J. Chem Soc. Perkin Trans. II*, 1301-1308.
5. Böhm, S., Adamec, M., Nešpùrek, S. and Kuthan J. (1995) Photocolouration of 2,4,4,6-tetraaryl-4*H*-pyrans and their heteroanalogues: importance of hypervalent photoisomers, *Coll. Czech. Chem. Commun.* **60**, 1621-1633.
6. Rahmani, H. and Pirelahi, H. (1997) Kinetic study on photoisomerization of some tetrasubstituted 4-aryl-4-methyl-2,6-diphenyl-4*H*-thiopyrans, *J. Photochem. Photobiol. A* **111**, 15-21.
7. Lewanowicz, A., Lipiński, J., Nešpùrek, S., Olszowski, A., Œliwińska, E. and Sworakowski, J. (1999) Photochromic activity of dihydropyridine derivatives: energetics and kinetics of photochemically driven reactions in polycrystalline 1-methyl-2,4,4,6-tetraphenyl-1,4-dihydropyridine, *J. Photochem. Photobiol A* **121**, 125-132.
8. Sworakowski, J., Nešpùrek, S., Lipiński, J. and Lewanowicz, A. (1999) On the mechanism of bleaching reactions in a photochromic dihydropyridine derivative, *J. Photochem. Photobiol. A* **129**, 81-87.
9. Ohashi, Y. (ed.) (1994), *Reactivity in Molecular Crystals*, Kodansha and VCH, Tokyo and Weinheim.

10. Shibuya, J., Nabeshima, M., Nagano, H. and Maeda, K. (1988) Photochemical reaction of 2,4,4,6-tetrasubstituted 1,4-dihydropyridines in deaerated media: photocolouration and photorearrangement accompanying dehydrogenation, *J. Chem. Soc. Perkin Trans. II*, 1607-1612.

11. Mori, Y. and Maeda, K. (1991) Photochemical reaction of 2,4,4,6-tetraaryl-4*H*-pyrans and - 4*H*-thiopyrans with colour change by a 1,5-electrocyclic reaction. X-ray molecular structure of 4-methyl-2,3,6-triphenyl-2*H*-thiopyran, *J. Chem. Soc. Perkin Trans. II*, 2061-2066.

12. Nešpùrek, S., Schwartz, M., Böhm, S. and Kuthan, J. (1991) Solid state photochromism of 2,4,4,6-4*H*-pyran and 1,4-dihydropyridine derivatives, *J. Photochem. Photobiol. A* **60**, 345-353.

13. Vojtichovský, J., Hašek, J., Nešpùrek, S. and Adamec, M. (1992) The structure and photochromism of 2,4,4,6-tetraphenyl-4H-thiopyran, *Coll. Czech. Chem. Commun.* **57**, 1326-1334.

14. Böhm, S., Hocek, M., Nešpùrek, S. and Kuthan, J. (1994) Photocolouration of 2,4,4,6-tetraaryl-1,4-dihydropyridines: a semiempirical quantum chemical study, *Coll. Czech. Chem. Commun.* **59**, 262-272.

15. Böhm, S., Sebek, P., Nešpùrek, S. and Kuthan, J. (1994) Photocolouration of 2,4,4,6-tetraaryl-4*H*-thiopyrans: a semiempirical quantum chemical study, *Coll. Czech. Chem Commun.* **59**, 1115-1125.

16. Nešpùrek, S., Böhm, S. and Kuthan, J. (1994) Photochromism of 4*H*-thiopyrans and 1,4-dihydropyridines, *Mol. Cryst. Liq. Cryst.* **246**, 139-142.

17. Pirelahi, H., Rahmani, H., Mouradzadegun, A., Fathi, A. and Moudjoodi, A. (1996) The effect of 3,5-substitutions on the photochromism and photoisomerization of some 2,3,4,4,5,6-hexasubstituted 4*H*-thiopyrans, *J. Photochem. Photobiol. A* **101**, 33-37.

18. Nešpùrek, S. (1999), unpublished results.

19. Sworakowski, J. and Nešpùrek, S. (1998) 'Fractional heating' differential scanning calorimetry: a tool to study energetics and kinetics of solid-state reactions in photoactive systems with distributed parameters, *Chem. Phys.* **238**, 343-351.

20. Lewanowicz, A., Lipiñski, J., Olszowski, A. and Sworakowski, J. (1999) Photochemical reactions in polycrystalline 1-methyl-2,4,4,6-tetraphenyl-1,4-dihydropyridine, *Phase Trans.*, in print.

21. Œliwiñska, E. and Sworakowski, J. (1999), unpublished results.

22. Dewar, M.J.S. and Thiel, W. (1977) Ground states of molecules 38. The MNDO method. Approximations and parameters, *J. Am. Chem. Soc.* **99**, 4899-4907.

23. Lipiñski, J. (1988) Modified all-valence INDO/spd method for ground and excited state properties of isolated molecules and molecular complexes, *Int. J. Quantum Chem.* **34**, 423-435.

24. Hašek, J. and Ondráèek, J (1990) Structure of 1-methyl-2,4,4,6-tetraphenyl-1,4-dihydro-pyridine, *Acta Cryst. C* **46**, 1256-1259.

25. Lewanowicz, A., Lipiñski, J., Sworakowski, J., Komorowska, M. and Nešpùrek, S. (1999) On the mechanism of photochromic activity in substituted dihydropyridines, *Proc. Int. Workshop on Reactive Intermediates (IWRI '99)*, Szczyrk (Poland).

26. Lipiñski, J. (1999), unpublished results.

# PHOTOCOLOURATION OF HYPERVALENT HETEROCYCLES

*Photochromism of Dihydropyridines, Pyrans and Thiopyrans*

S. NEŠPŮREK[1], J. SWORAKOWSKI[2], J. LIPIŃSKI[2], S. BÖHM[3], and J. KUTHAN[3]

[1] *Institute of Macromolecular Chemistry, Academy of Sciences of the Czech Republic, 162 06 Prague 6, Czech Republic*
[2] *Institute of Physical and Theoretical Chemistry, Technical University of Wrocław, 50-370 Wrocław, Poland*
[3] *Department of Organic Chemistry, Prague Institute of Chemical Technology, 166 28 Prague 6, Czech Republic*

## 1. Introduction

The notion of hypervalent molecule is usually used to describe a chemical entity which can be characterized by two or more mesoionic valence structures and/or by a structure in which valence states of some atoms do not follow the classical rules concerning the chemical bonding [1]. Typical examples represent some isomers of heterocyclic molecules, e.g. dihydropyridines (cf. Scheme 1). While 1,2-dihydroisomer (1) and 1,4-dihydroisomer (2) do not show any hypervalence, it is possible in the case of 1,3-dihydroisomer (cf. structures (3a) - (3d)) The chemical formulas (3a) - (3d) represent formal structures; usually, information concerning a real molecular structure, i.e., the conclusion whether an N-hypervalent structure (3a) or a hybrid of mesoionic rezonant structures (3b) - (3d) is more probable, must be obtained by quantum chemical methods. It could be mentioned that "hypervalent" molecules often show properties different from non-hypervalent ones [2, 3] and they are very often unstable and coloured [4]. These materials, in agreement with theory, have relatively high internal energies and it is probable that also molecules of some photoproducts can be considered as hypervalent ones. In connexion with this fact, the phototropic behaviour of 2,4,4,6-tetraphenyl 4$H$-pyrans (4, X = 0), its thio- (X = S), and aza- (X = NR, with R specified in Table 1) analogues seems to be of interest. Some of these compounds, which become reversibly coloured under UV or solar irradiation, can be utilized in some optical devices. In this paper, the photocolouration of compounds of the type (4) is described and the mechanism of the photochromism briefly discussed. A more extensive discussion of the photochromic behaviour, including also results of model quantum chemical calculations, is given in the accompanying paper [5].

261

*F. Kajzar and M.V. Agranovich (eds.), Multiphoton and Light Driven Multielectron Processes in Organics: New Phenomena, Materials and Applications, 261–280.*

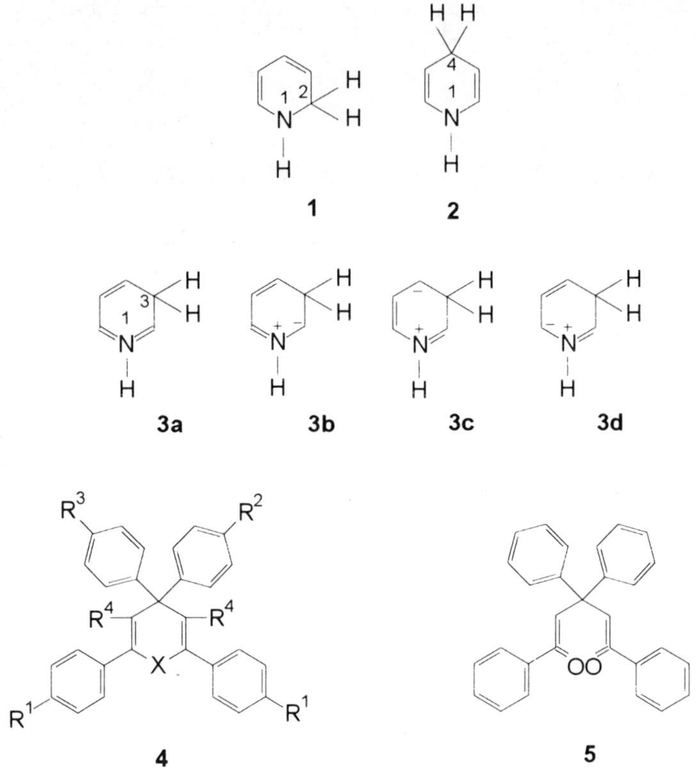

*Scheme 1*

## 2. Materials

We will discuss the behaviour of three groups of materials whose general chemical formula is represented by the structure (4) in Scheme 1:
(I) 2,4,4,6-tetraphenyl-4$H$-dihydropyridines (X = NR);
(II) 2,4,4,6-tetraphenyl-4$H$- thiopyrans (X = S);
(III) 2,4,4,6-tetraphenyl-4$H$- pyrans (X = O).

### 2.1. CHEMICAL SYNTHESIS

#### 2.1.1. *2,4,4,6-Tetraphenyl-4H-Dihydropyridines*
Compound Ia (see Table 1) was obtained by treatment of 1,3,3,5-tetraphenylpentane-1,5-dione (5) in glacial acetic acid with ammonium acetate [6]. N-substituted 1,4-dihydropyridines (Ib, Ic, Id) were prepared by cyclocondensation of diketone (6) with acetates of the respective primary amines in acetic acid. The yield of the benzyl and phenyl derivatives was lower than that of the methyl one, i.e. it followed the

increasing sterical requirements of the substituent (methyl, benzyl, and phenyl). The N-substitution had a favourable effect on the stability of the obtained heterocycles. The 1,4-dihydropyridines (Ie - Ih) were prepared in a manner similar to the compound Id [6] by the cyclocondensation of (5) with 4-substituted anilines. A mixture of diketone and an excess of the corresponding 4-substituted phenylamonium acetate in acetic acid was refluxed until the starting dioxo derivative was no longer present. Whenever necessary, the product was precipitated after dilution with water and recrystallized from an appropriate solvent [7]. Compounds Ii - It were synthesized as described in [8], from the corresponding 1,5-pentanediones (which were prepared by the reaction of the substituted acetophenone with the substituted benzophenone via heterocyclization using sodium amide) and ammonium acetate or methylammonium acetate in acetic acid [9]. All chemical characteristics, viz. analysis, NMR and IR data, are given is the above mentioned papers.

### 2.1.2. 2,4,4,6-Tetraphenyl-4H-Thiopyrans

The compounds were prepared by employing the procedure described in [8]. The synthesis was based on readily available 1,3,3,5-tetraphenylpentane-1,5-diones (5) [6,12-15], which on heating with tetraphosphodecasulfide in xylene, yield the corresponding 4H-thiopyrane [10]. This procedure allowed one to limit the amount of 2H-isomers [10-18] in the final product. An alternative heterocyclization of (5), brought about by simultaneous treatment with hydrogen sulfide and hydrogen chloride [8], resulted in lower yields of (4, $X = S$, $R^1 = R^2 = H$). From the point of view of understanding the photochromic mechanism, it seemed interesting to prepare some derivatives substituted in 3, 5 position (substituent $R^4$ in the formula (4)) [11]. These materials were prepared by substitution reactions of the parent thiopyrans or by crystalization of suitably substituted 1,3,3,5-tetraphenylpentane-1,5-diones. The details of the chemical synthesis and material characteristics are given in [11]. It is interesting to note that the substances lacking the 3,5-substituents exhibited a blue or green UV-photocolouration while the 3- and/or 5-substituted species were inactive. This feature will be discussed latter in the connection with the mechanism of the photochromism.

### 2.1.3. 2,4,4,6-Tetraphenyl-4H-Pyrans

The highest yields of 2,4,4,6-tetraphenyl-4H-pyran (IIIa) were achieved in the reaction of (5) with phosphoric oxide, while acetic anhydride with trace amount of acetyl chloride, iodine or p-toluensulfonic acid gave lower yields. The substituted 4H-pyrans IIIb - IIIg were prepared analogously by treatment of respective diketones with phosphoric oxide [6]. The paper [6] also contains results of the chemical characterization of the compounds under study.

## 2.2. MOLECULAR STRUCTURES

In this paragraph we will briefly discuss molecular conformations, crystalline properties and influence of the heteroatom X on the molecular structure, focusing on three materials: 1-methyl-2,4,4,6-tetraphenyl-1,4-dihydropyridine (DHP) [19],

TABLE 1. Solution and photochromic solid state absorption maxima of materials under study.

| Compound | Structure[a] | | | | $\lambda_{max}$ [b] | $\log \varepsilon$ [c] | $\lambda_{max}$ [d] | |
|---|---|---|---|---|---|---|---|---|
| | X | $R^1$ | $R^2$ | $R^3$ | (nm) | $\varepsilon$ (dm³/mol cm) | (nm) | |
| Ia | NH | H | H | H | 239 | 4.42 | 565 | 405 |
| Ib | NCH₃ | H | H | H | 235 | 4.45 | 550 | 405 |
| Ic | NBz [e] | H | H | H | 235 | 4.44 | 560 | 396 |
| Id | NPh [f] | H | H | H | 228 | 4.46 | 560 | 400 |
| Ie | NPh | H | H | H | 227 | 4.52 | 580 | 380 |
| If | NPh | H | H | H | 231 | 4.52 | 560 | 410 |
| Ig | NPh | H | H | H | 229 | 4.51 | 585 | 389 |
| Ih | NPh | H | H | H | 227 | 4.55 | 584 | 380 |
| Ii | NH | CH₃ | H | H | 242 | 4.48 | 627 | 432 |
| Ij | NH | F | H | H | 233 | 4.34 | 600 | 418 |
| Ik | NH | Br | H | H | 249 | 4.57 | 632 | 440 |
| Il | NCH₃ | CH₃ | H | H | 240 | 4.52 | 580 | 405 |
| Im | NCH₃ | Br | H | H | 235 | 4.42 | 575 | 410 |
| In | NCH₃ | H | H | H | 246 | 4.56 | 560 | 410 |
| Io | NH | H | CH₃ | H | 249 | 4.46 | 580 | 420 |
| Ip | NH | H | F | H | 234 | 4.48 | 572 | 408 |
| Iq | NH | H | Br | H | 236 | 4.52 | 588 | 410 |
| Ir | NCH₃ | H | H | H | 234 | 4.50 | 545 | 408 |
| Is | NCH₃ | H | H | H | 232 | 4.52 | 560 | 420 |
| It | NCH₃ | H | H | H | 233 | 4.57 | 555 | 408 |
| IIa | S | H | H | H | 250 [g] | 4.47 | 564 | 381 |
| IIb | S | OCH₃ | H | H | 260 | 4.62 | 600 | 394 |
| IIc | S | Bu [g][h] | H | H | 242 | 4.57 | 545 | 382 |
| IId | S | CH₃ | H | H | 242 | 4.57 | 572 | 399 |
| IIe | S | F | H | H | 250 [g] | 4.34 | 602 | 393 |
| IIf | S | Br | H | H | 255 [g] | 4.62 | 594 | 398 |
| IIg | S | H | CH₃ | H | 255 [g] | 4.62 | 547 | 380 |
| IIh | S | H | Cl | H | 250 [g] | 4.46 | 567 | 388 |
| IIi | S | H | Br | H | 238 | 4.53 | 564 | 387 |
| IIj | S | H | Bu | Bu | 250 [g] | 4.00 | 552 | 378 |
| IIk | S | H | CH₃ | CH₃ | 238 | 4.57 | 598 | 359 |
| IIl | S | H | F | F | 238 | 4.51 | 559 | 384 |
| IIm | S | H | Br | Br | 250 | 4.57 | 554 | 362 |
| IIn | S | CN | H | H | 260 | 4.55 | 647 | 422 |
| IIo | S | COCF₃ | H | H | 292 | 4.52 | 667 | 470 |
| IIp | S | COPh | H | H | 258 | 4.60 | 656 | 407 |
| IIq | S | CO₂H | H | H | 268 | 4.51 | 642 | 418 |
| IIr | S | [i] | H | H | 252 | 4.60 | 667 | 410 |
| IIs | S | [j] | H | H | 254 | 4.58 | 610 | 392 |
| IIt | S | [k] | H | H | 256 | 4.57 | 621 | 437 |
| IIu | S | [l] | H | H | 262 | 4.50 | 672 | 467 |

TABLE 1 (cont.)

| Compound | Structure[a] | | | | $\lambda_{max}$[b] | log $\varepsilon$[c] | $\lambda_{max}$[d] | |
|---|---|---|---|---|---|---|---|---|
| | X | R[1] | R[2] | R[3] | (nm) | $\varepsilon$ (dm$^3$/mol cm) | (nm) | |
| IIIa | O | H | H | H | 259 | 4.36 | 558 | 412 |
| IIIb | O | H | CH$_3$ | H | 249 | 4.44 | 616[m] | 410[m] |
| IIIc | O | H | Cl | H | 250 | 4.51 | 565 | 412 |
| IIId | O | H | Br | H | 253 | 4.54 | 570 | 408 |
| IIIe | O | CH$_3$ | H | H | 261 | 4.60 | 620 | 440 |
| IIIf | O | F | H | H | 249 | 4.49 | 602 | 428 |
| IIIg | O | Br | H | H | 255 | 4.59 | 630 | 445 |

[a] Structure (4) in Scheme 1. Substitutions in R$^4$ position make the molecules non-photochromic; therefore, the materials are not included in the Table. [b] Wavelengths of the low-energy maxima measured in ethanol. [c] Molar absorbance of the parent molecule. [d] Wavelengths of the photochromic band maxima measured in MgO matrix. [e] Bz ≡ benzyl group, C$_6$H$_5$CH$_2$–. [f] Ph ≡ phenyl. [g] Measured in chloroform. [h] Bu ≡ butyl group, CH$_3$–CH$_2$–CH$_2$–CH$_2$–. [i-l] Asymmetric molecules, R$^1$ ≠ R$^{1'}$. [i] R$^1$ = Br, R$^{1'}$ = CN. [j] R$^1$ = Br, R$^{1'}$ = OCH$_3$. [k] R$^1$ = CN, R$^{1'}$ = OCH$_3$. [l] R$^1$ = COCF$_3$, R$^{1'}$ = OCH$_3$. [m] Approximate values because of fast bleaching process.

2,4,4,6-tetraphenyl-4$H$-thiopyran (SPY) [20] and 2,4,4,6-tetraphenyl-4$H$-pyran (OPY) [21]. The molecular and crystal structures are shown in Fig. 1. In all materials there are no significant intermolecular contacts between non-H atoms shorter than 3.4 Å. Molecular packing and orientation of benzene rings are controlled only by van der Waals forces. No tendency for parallel stacking of the phenyl rings of symmetry-related molecules was observed.

The X-ray analysis revealed that DHP crystals are monoclinic: $P2_1/c$, $a = 1.2581(5)$ nm, $b = 1.3955(5)$ nm, $c = 1.2885(4)$ nm, $\beta = 93.63(3)°$, $V = 2.258(1)$ nm$^3$, $Z = 4$. The bonds C2–C21 and C6–C61 lie almost in the mean plane of the dihydropyridine ring (the out-of-plane angles are 2.8(2)° and 0.8(1)° respectively). The bonds C4-C421 and C4-C411 deviate from the plane by 58.0(1)°. The shortest intermolecular distance between non-hydrogen atoms is C23 - C11 (0.335(1) nm. There are two symmetrically independent molecules of DHP in the unit cell (denoted A and B in Table 2).

SPY forms orthorombic crystals, $Pna2_1$, $a = 1.7980(4)$ nm, $b = 0.6956(2)$ nm, $c = 3.4562(11)$ nm, $V = 4.323$ nm$^3$, $Z = 8$. There are also two symmetrically independent molecules $A$ and $B$ of SPY in the unit cell. The average bond lengths $<C(sp^2)-C(sp^3)> = 1.52(1)$, $<C(sp^2)=C(phenyl)> = 1.49(1)$ and $<C(sp^3)-C(phenyl)> = 1.55(1)$ Å do not differ from values observed in 1,4-dihydropyridines and 4$H$-pyrans. The average double bond length $<C(sp^2)=C(sp^2)> = 1.33(1)$ Å is shorter than the typical value. This shortening has also been observed in 1,4-dihydropyridines and 4$H$-pyrans. The average bond length $<C(sp^2)-S> = 1.75(1)$ Å. All phenyls deviate similarly from their ideal geometry, mainly due to the electron inductive effect and the rigid body thermal motion [22]. The average inner phenyl angles are 118.0(5), 120.5(9) and 119.6(6)° starting from the angle at the pivot atom. The torsion angles on the bonds connecting the phenyl to the C6 atom, i.e. C5-C6-C61-C66, are -41.1° and

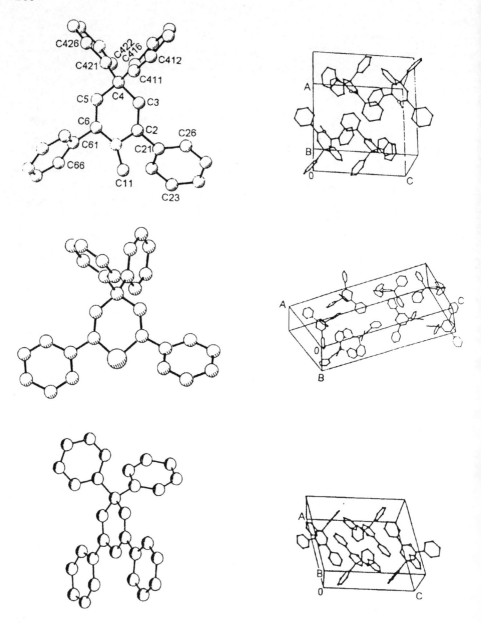

*Figure 1*. Molecular structures and unit cells of representative materials. From top to bottom: 1-methyl-2,4,4,6-tetraphenyl-1,4-dihydro-pyridine (DHP), 2,4,4,6-tetraphenyl-4*H*-thiopyran (SPY) and 2,4,4,6-tetraphenyl-4*H*-pyran (OPY).

40.3(2)° for $A$ and $B$ molecule, respectively. The corresponding values for the phenyl at the site C2, i.e. C3-C2-C21-C26, are -38.8(1)°. The orientation of the phenyls at the site C4 is described by their orientation with respect to the plane of atoms C411, C4, C421. Torsion angles C411-C4-C421-C426 and C421-C4-C11-C412, are 12.8, 62.6° and 13.0(4), 62.2(4) for $A$ and $B$, respectively.

TABLE 2. Selected conformational parameters of SPY, OPY and DHP with e.s.d.'s (in parentheses). A and B are symmetry-independent molecules in the unit cell.

| Parameter | SPY | | OPY | DHP | |
| --- | --- | --- | --- | --- | --- |
| | A | B | | A | B |
| Dihedral angles [a] | | | | | |
| $P_1 - P_2$ | 4.6(2) | 4.1(2) | 4.3(1) | 10.3(2) | 1.3(2) |
| $P_1 - P_3$ | 7.1(2) | 6.6(3) | 7.7(1) | 14.7(3) | 2.1(3) |
| $P_1 - Ph(6)$ | 43.3(2) | 41.6(1) | 3.2(1) | 49.1(1) | 52.6(2) |
| $P_1 - Ph(2)$ | 35.5(2) | 36.0(1) | 3.0(1) | 30.1(2) | 47.9(2) |
| $P_1 - Ph(41)$ | 68.3(2) | 67.8(1) | 71.8(1) | 71.1(1) | 55.9(2) |
| $P_1 - Ph(42)$ | 92.5(2) | 91.9(1) | 75.2(1) | 81.7(1) | 86.5(2) |
| Bond angles | | | | | |
| C5–C4–C411 | 110.6(2) | 111.1(3) | 111.3(2) | 113.2(2) | 112.1(2) |
| C5–C4–C421 | 106.8(2) | 106.6(2) | 107.3(2) | 105.7(2) | 108.5(2) |
| Torsion angles | | | | | |
| C5–C4–C411–C416 | 0.1(4) | 1.2(4) | 31.9(3) | 24.6(4) | 9.5(4) |
| C5–C4–C421–C422 | 69.6(4) | 67.5(4) | 63.7(2) | 73.2(3) | 66.5(3) |
| C411–C4–C421–C426 | 12.8(4) | 13.0(4) | 32.8(2) | 49.5(4) | 15.3(4) |
| C421–C4–C411–C412 | 62.6(4) | 62.6(4) | 58.3(2) | 42.2(4) | 70.2(4) |

[a] $P_1$; $P_2$ and $P_3$ denote (C2, C3, C5, C6); (C2, X, C6) and (C3, C4, C5) planes, respectively; Ph($n$) stands for a plane of phenyl at the position $n$.

OPY forms monoclinic crystals, $P2_1/c$, $a = 12.128(5)$, $b = 12.372(5)$, $c = 14.599(5)$ Å, $\beta = 105.57(3)°$, $V = 2110(1)$ Å$^3$, $Z = 4$. The 4$H$-pyran ring is approximately in plane with the phenyl rings substituted on the C($sp^2$) atoms C2 and C6. The ring shows a slightly distorted boat conformation with the boat angles 4.3(1) and 7.7(1)° on C($sp^2$) and O, respectively. The mean phenyl planes form dihedral angles of 3.2, 71.8, 75.2 and 3.0(1)° for the pivot atoms C61, C21, C411, and C421, respectively, with plane $P_1$ ($P_1$ is the mean plane of C2, C3, C5 and C6). The bonds C6-C61, C4-C411, C4-C421 and C2-C21 form angles 3.0, 62.9, 48.0 and 1.1(1)°, respectively, with the plane $P_1$. The asymmetry of phenyls at C($sp^3$) is similar to that in DHP. This follows from a comparison of the angles C5-C4-C411, 111.3(2) and 112.1(2)° with C5-C4-C421, 107.3(2) and 108.5(2)°.

It can be seen from the structures given in Fig. 1 that all molecules are non-planar. Selected conformational parameters of DHP, OPY and SPY are given in

Table 2. For the sake of simplicity, the conformation of the central ring is described here using boat angles between the mean plane of double bonds $P_1$: $C(sp^2)$, $C(sp^2)$, $C(sp^2)$, $C(sp^2)$, and the planes $P_2$: $C(sp^2)$, X, $C(sp^2)$ and $P_3$: $C(sp^2)$, $C(sp^3)$, $C(sp^2)$, respectively. All structures show boat conformations of the central ring with the boat angles listed in Table 2. The deviations of $C(sp^2)$ atoms from the double bonds mean plane $P_1$ are less than 0.012 Å for all structures. The $C(sp^3)$ atom deviates from the plane $P_1$ more than the heteroatoms S and O but less than N. With respect to the results obtained for differently substituted 1,4-dihydropyridines [23] such a high deviation of N seems to be exceptional. The orientation of the mean planes Ph(6) and Ph(2) of phenyls at the positions 6 and 2, described simply by the dihedral angles between thieir planes and the plane $P_1$, is similar in DHP and SPY but different in case of OPY (see Table 2). The rotation is rather small in OPY (about 3.2 and 3.0(1)°), and quite significant rotation in the other cases. The deviations of the bonds C6-C61 and C2-C21 from the plane $P_1$ are less than 3° for all structures. As the equivalent orientation of the mean plane of phenyls at the position 4 relative to the mean plane $P_1$ is not illustrative enough, the torsion angles on C4-C421 and C4-C411 are listed in Table 2. By comparing the bond angles C5-C4-C411 and C5-C4-C421 we observed asymmetry in the positions of the phenyl substituents.

## 2.3. ELECTRONIC STRUCTURE

UV-VIS absorption spectra of three representative materials under study, DHP [24], SPY [25] and OPY [26], together with the calculated transitions, are presented in Fig. 2, whereas the positions of UV-VIS absorption maxima and molar absorbances for the entire group of synthesized materials have already been given inTable 1. All parent materials are absorbing in UV region, being transparent in visible part of the spectrum. Calculations of the electronic spectra were performed by CNDO/S-CI method using the parameters: $\chi = 0.858$ for the integrals of $\pi$-bonds and $\chi = 1.267$ for $\gamma$-sigma types of integrals [27].

TABLE 3. Net atomic charges calculated for AM1 geometries of DHP, SPY and OPY.

| Position | DHP | | SPY | | OPY | |
|---|---|---|---|---|---|---|
| C(H$_3$) | | -0.093 | | - | | - |
| X1 | N | -0.195 | O | 0.407 | S | -0.108 |
| C6 | | +0.039 | | -0.257 | | +0.085 |
| C5 | | -0.206 | | -0.148 | | -0.234 |
| C4 | | +0.111 | | 0.094 | | +0.172 |
| C3 | | -0.206 | | -0.148 | | -0.235 |
| C2 | | +0.039 | | 0.258 | | +0.085 |

Some characteristics of the net atomic charge distributions for DHP, SPY and OPY rings are given in Table 3. High electron densities on C3 and C5 centres agree well with the known electrophilic substitutions in the positions 3 and 5 of 1,4-dihydropyridine [28]. The net charge distribution in DHP and OPY significantly differs from that of SPY, where the alternant charge distribution was not found.

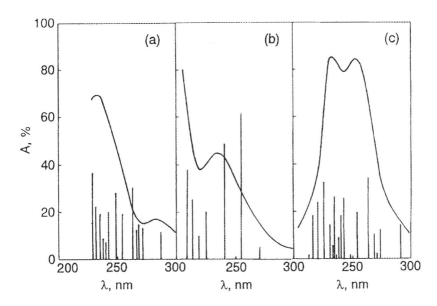

*Figure 2.* A comparison of experimental absorption spectra of 1-methyl-2,4,4,6-tetraphenyl-1,4-dihydropyridine (DHP), 2,4,4,6-tetraphenyl-4*H*-thiopyran (SPY) and 2,4,4,6-tetraphenyl-4*H*-pyran (OPY) in 1,2-dichloroethane (a) and acetonitrile solution (b,c) with the electronic transitions calculated using the CNDO/S-CI method.

## 3. Photochromic Behaviour

All three groups of materials mentioned above (X = NR, O, S, see structure (4) in Scheme 1) exhibit photochromism. The irradiation of the materials with light of sufficiently short wavelengths, exciting the molecules into their first singlet state (cf. $\lambda_{max}$ in Table 1) leads to the appearance of new transient absorption bands in the visible part of the spectrum, both in solution and in solid state. The positions of the maxima depend on the substituents R, $R_1$, $R_2$ and $R_3$ (see photochromic bands in Table 1). Because the behaviour of all three groups of the substances seems to be similar in its general features, we will discuss here in detail only the photocolouration of DHP in solution, solid state and polymer matrix.

Three types of samples were used: dilute solutions of DHP in oxygen-free chloroform and acetonitrile in a rectangular quartz cell, thin films of a solid solution of DHP in poly(methyl methacrylate) (PMMA, $M_w$ = 500 000, sample thickness about 9 μm) and polycrystalline powder mixed with MgO (1:10). The reader is referred to the original papers [29,30] for experimental details.

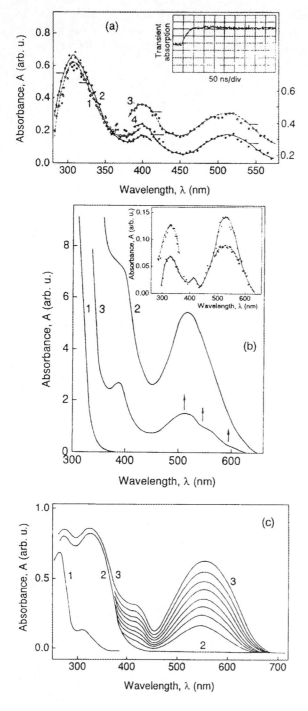

*Figure 3.(a)* Transient absorption spectrum recorded 0.8 μs (curves 1 and 3) and 7.8 μs (curves 2 and 4) after the flash for argon-saturated acetonitrile solution of DHP (*T* = 296 K, λ$_{ex}$ = 266 nm, 15 ns flash, exposure dose *P* = 210 mJ cm$^{-2}$), concentration c = 1.3×10$^{-4}$ mol dm$^{-3}$ for curves 1 and 2, and c = 3.1 × 10$^{-4}$ mol dm$^{-3}$ for curves 3 and 4.
Inset: The course of the transient absorption signal at 520 nm in acetonitrile solution after the flash.

*(b)* Absorption spectra of solid solutions of DHP in PMMA. Curve 1, before irradiation; curve 2, immediately after UV irradiation; curve 3, 30 min after UV irradiation. Inset: transient absorption spectra (λ$_{ex}$ = 266 nm; laser flash, 15 ns; *T* = 296 K): — –, 50 ns after flash; ——, 8 μs after flash.

*(c)* Absorption spectra of DHP at room temperature: curve 1, spectrum in acetonitrile solution; curve 2, spectrum in the solid state before UV irradiation; curve 3, directly after UV exposure. Curves below the curve 3 represent spectra recorded 20, 40, 60, 90, 120, 160 and 240 min after UV exposure.

## 3.1. PHOTOCHROMIC PROPERTIES OF DHP

The irradiation of DHP in acetonitrile solution with 15 ns flashes of 266 nm light leads to the appearance of a new transient absorption spectrum, different from that of parent DHP. The spectra taken at $t = 0.8$ μs and $t = 7.8$ μs after the flash are shown in Fig. 3a. The maximum of the transient absorption in the UV region is located at 310 nm and the minimum at about 385 nm. The transient absorption also shows two bands in the visible region peaking at approximately 400 and 515 nm. The threshold of the transient absorption is observed at about 620 nm. From the inset of Fig. 3a, it is evident that the formation of coloured species is completed during ca. 150 ns. After that time, the absorbance in the visible part of the spectrum (peaks at 400 and 515 nm) decreases with the time constant of the order on microseconds, whereas the UV band ($\lambda_{max} = 310$nm) slightly increases.

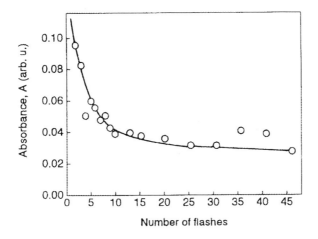

*Figure 4.* Transient absorbance at 530 nm of DHP in PMMA matrix as a function of the number of laser flashes (room temperature, $\lambda_{ex}$=266 nm; laser flash 15 ns; exposure dose $P$=50 mJ cm$^{-2}$).

Similar spectral characteristics were obtained with solid solutions of DHP in poly(methyl methacrylate). The photochromic spectral characteristics are given in Fig. 3b. The main maximum of the photochromic band in the visible region is observed at 522 nm, additional maximum at about 395 nm. A detailed analysis of the 522 nm band suggests that it consists of several peaks. The transient absorption spectra recorded 50 ns and 8 μs after the flash is given in inset to Fig. 3b. The spectra were corrected for the decay of the absorbance as a function of the number of laser flashes (Fig. 4) for each wavelength, hence the spectral dependences shown in the inset to Fig. 3b represent values averaged for several samples. The maxima are detected at about 325, 415 and 530 nm. The decrease of absorbance at $\lambda = 325$ nm (curve 1 in the inset) and a simultaneous increase at $\lambda = 530$ nm (curve 2) are clearly visible. The primary colour change is very fast, similar to that in acetonitrile and chloroform solution, as is evidenced by the transient traces in shown Fig. 5. For longer times, an additional increase in the long-wave absorption ($\lambda = 570$ nm, column c in Fig. 5) and a decrease in the short-wave absorption ($\lambda = 360$ and 420 nm, columns a and b) are observed. Thus, the polymer matrix impedes the slow process following the rapid photocolouration of DHP molecules. This feature suggests that the molecular phototransformation into the coloured species is a geometrically demanding process which needs a free volume.

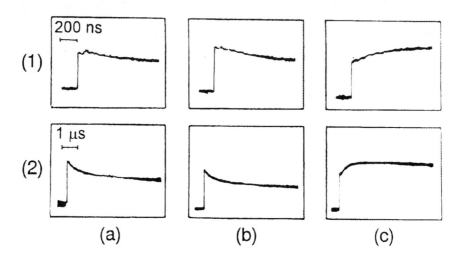

*Figure 5.* Transient absorption of a solid solution of DHP in PMMA after laser flash ($T = 216$ K, exposure dose $P = 200$ mJ cm$^{-2}$) at $\lambda = 360$ nm (a), $\lambda = 420$ nm (b), and $\lambda = 570$ nm (c). The time scale is 200 ns/div (1), and 1 $\mu$s/div (2).

The situation is more complex in polycrystalline samples. Here it was impossible to achieve an effective colouration by short-time single flashes: an irradiation over at least several seconds was necessary to reach a detection level. Absorption maxima in crystalline parent DHP appear at $\lambda = 304$ and 245 nm (Fig. 3c, curve 2). After the irradiation of DHP with light of $\lambda < 350$ nm at room temperature, the samples turn violet (cf. Fig. 3c, curve 3), the emerging absorption bands being located at 544 nm and ca. 400 nm. The colouration by UV irradiation and thermal bleaching could be repeated many times.

Interestingly, an ESR signal was observed during the coloration process under UV irradiation of solid samples on air. Figure 6 shows the ESR spectra of a sample irradiated with UV light, compared to that of a pristine sample. Two broad peaks can be identified, characterized by their $g$ factors amounting to 2.2 and 2.0 [31].

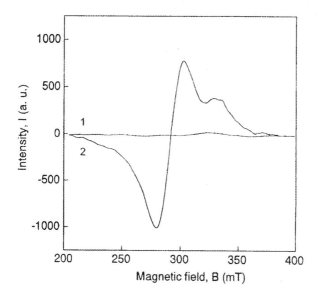

*Figure 6.* The ESR signal measured in polycrystalline DHP: 1 - pristine sample, 2 - the same sample after irradiation with UV.

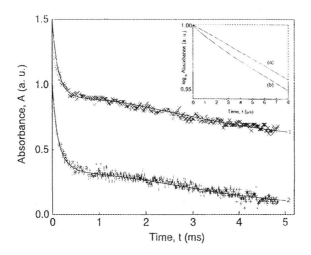

*Figure 7.* Ambient-temperature bleaching kinetics of DHP in solution measured by following the absorbance at 520 nm. 1 - pure chloroform; 2 - chloroform + PMMA (viscosity equal to $4 \times 10^{-2}$ N $m^{-2}$s). Full lines are reconstructed curves calculated using Eq. (1). The curve 1 has been vertically shifted. Inset: Time evolution of the transient absorbance at $\lambda = 520$ nm in chloroform solution of DHP (a) and in chloroform solution with PMMA added (b). Note a different time scale.

## 3.2. BLEACHING REACTION

As the photocolouration in DHP appeared to be a reversible process, the bleaching isothermal kinetics was studied in detail, both in solution and in the solid state. A typical experimental decay of the absorbance obtained for DHP in chloroform is given in Fig. 7, curve 1. The curve consists of an initial section of a rapid decay, separated from a slower decay by a clearly descernible plateau around 0.8-1.5 ms. This type of decay can be fitted with the following equation

$$A(t) = \alpha \, \exp(-k_\alpha \, t) + \beta \, \exp(k_\beta \, t) - \gamma \, \exp(-k_\gamma \, t), \tag{1}$$

where $A$ is the absorbance normalized to its initial value, and $\alpha$, $\beta$ and $\gamma$ stand for time-independent positive coefficients. Originally, the two decays were attributed to two parallel processes [32], such a scheme, however, cannot account for the presence of the plateau. In the simplest case, such a shape of the bleaching kinetics should involve a sequence of consecutive reactions, with two species absorbing at 520 nm. It should also be stressed at this point that the kinetics described by Eq. (1) requires that the absorbing molecules be disconnected, i.e., separated by at least one non-absorbing species in the sequence of elementary processes. A detailed discussion of the meaning of the coefficients $\alpha$, $\beta$, $\gamma$ and time constants in Eq. 1 is given in [33].

Taking into account results of the quantum-chemical calculations presented in [33] (see also [5]), we put forward a tentative sequence of elementary reactions as shown in Fig. 8, in which $R$ stands for a radical absorbing at 400 nm and $X$ is an unidentified intermediate product. According to the model, processes following the excitation of parent DHP consist in breaking one of the C(4)-phenyl bonds and formation of the radicals ($R$ and phenyl $\Phi$) whose recombination then may result in the production of stable DHP (molecule I), of a molecule VI and/or in formation of the coloured species (molecule III) in its $S_0$ and $T_1$ states. The calculations indicate that the optical gaps for the $S_0 \rightarrow S_1$ and $T_1 \rightarrow T_2$ absorption of molecule III are quite close (18000 cm$^{-1}$ vs. 20700 cm$^{-1}$). The bleaching reaction is mainly given by the processes III $\rightarrow$ IV and III $\rightarrow$ V. These reactions seem to be reversible: according to our calculations, the species IV and V can transform into the coloured form III upon excitation. Thus, the photochromic reversibility may be associated with the reactions III $\leftrightarrow$ IV and III $\leftrightarrow$ V. It could be pointed out that there might be a possibility for a chemical transformation IV $\rightarrow$ V involving cyclic transition states (cf. Fig. 9). Interestingly, the energy of the ground singlet state of V ($E_{S0}$ = -59 kJ/mol) is lower than that of IV ($E_{S0}$ = 17 kJ/mol).

The presented mechanism allows for explanation of the following experimental facts:
(i)   A plateau is observed in the decay curves measured in DHP solution.
(ii)  In solid solution DHP/PMMA coloured species are formed in a delayed process (several μs).
(iii) The species IV and V have been preparatively and spectroscopically detected for the following substituents: IV (X = O, S, SO$_2$; Ar$^1$ = Ar$^2$ = Ar$^3$ = Ph), [10,16,34], (X = NH; Ar$^1$ = Ar$^2$ = Ar$^3$ = Ph) [13] V (X = S, Ar$^1$ = Ar$^2$ = Ph) [16],

(X = S; Ar$^1$ = Ph) and for different Ar$^2$ = 4-CF$_3$C$_6$H$_4$ and 4-MeOC$_6$H$_4$ [35]. The photochromic activity of the species derived from DHP and other dihydropyridines should, however, be confirmed experimentally.

(iv) Because of the larger optical gap of IV (448 kJ/mol) in comparison with I (402 kJ/mol), one can expect a decrease of the quantum efficiency upon successively cycling photochromic reaction. Such a behaviour was indeed observed on solid films DHP/PMMA.

(v) Phenyl (di-π-methane) shift and the formation of 2*H*-isomers (V) needs a free volume. Thus we expected the bleaching mechanism to be sensitive to the viscosity of the solution. For this reason, the kinetic measurements were also performed in chloroform solutions of DHP additionally containing PMMA (M$_w$ ≈ 5 × 10$^5$ g/mol) at concentrations reaching 0.11 g/cm$^3$. The viscosities of the solutions obtained in such a way reached ca. 4×10$^{-2}$ Nm$^{-2}$s. The decay is presented as curve 2 in Fig. 7. We observed a systematic trend in the viscosity dependence of the parameter β as well as, possibly, the rate constant k$_{DG}$ = k$_4$ + k$_5$.

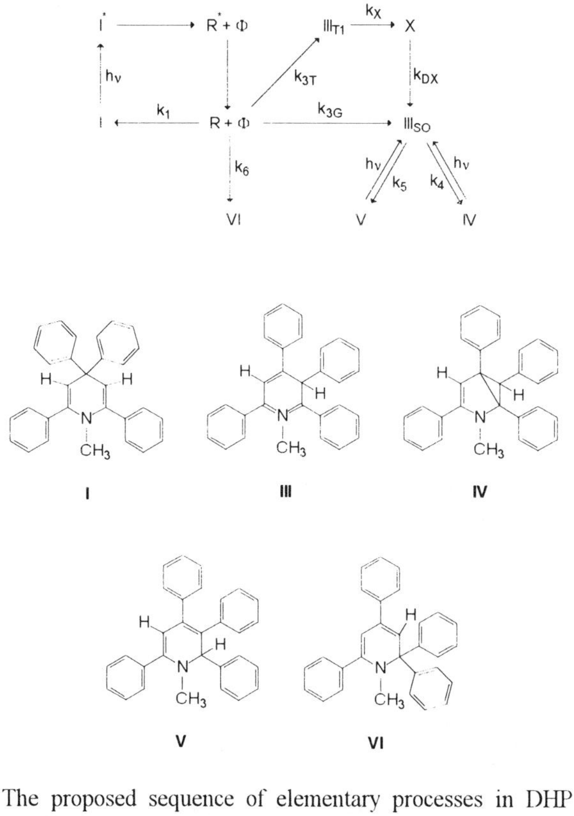

*Figure 8.* The proposed sequence of elementary processes in DHP following a UV irradiation. R is the radical, Φ is phenyl, X is an unidentified intermediate product. See text for further discussion.

According to the proposed mechanism of the di-π-methane shift, a characteristic optical absorption at about 400 nm must be detected, associated with the formation of the phenyl radicals and radicals of the heterocycle. Therefore, we performed the flash photolytic measurement with the detection of the transient absorption at 390 nm [29]. The lifetime of the transient species in acetonitrile solution was determined to be $\tau = 22 \pm 3$ μs. A similar value, 17.4 μs, was also obtained for DHP in chloroform solution.

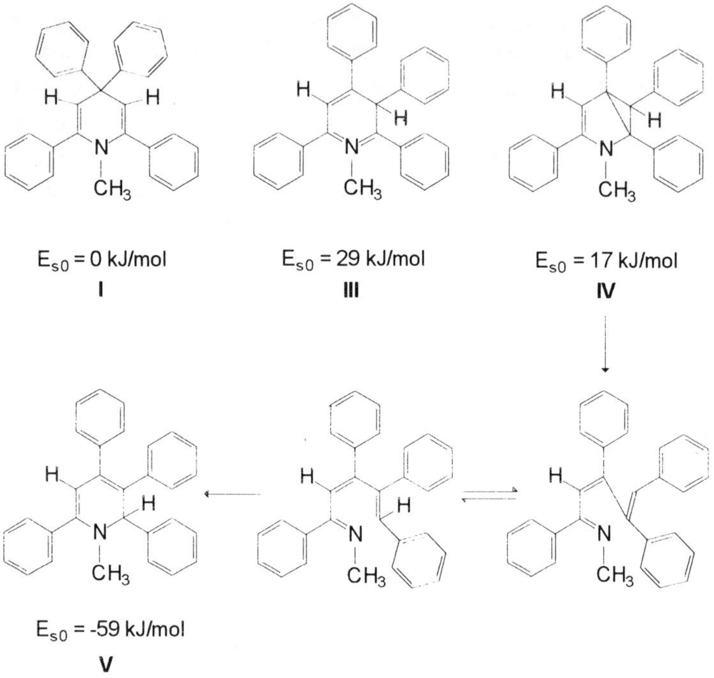

*Figure 9.* Energies of the ground states of the parent DHP and some products formed during a sequence of reactions following a UV irradiation, and a postulated sequence of processes allowing for a IV → V transformation.

The presence of PMMA in the solution changes also the character of the fast bleaching process: while the decay in DHP chloroform solution was purely exponential with the rate constant $5.7 \times 10^3$ s$^{-1}$, the addition of PMMA gives rise to a non-exponential time dependence that could be fitted with a 'stretched-exponential' function

$$[M(t)]/[M(0)] = \exp[-(\beta_1 t)^\alpha] , \qquad (2)$$

where $\alpha$ $(0 < \alpha < 1)$ is a measure of the deviation from the pure exponential behaviour (see the inset to Fig. 7).. The best fit yielded the following values: $\alpha = 0.8$ and $\beta_1 = 4.7 \times 10^3$ s$^{-1}$

The kinetics in both polymer solid solution (PMMA matrix) and in the solid state could also also be interpreted assuming dispersive first-order reactions with $\alpha$ = 0.59 and 0.79 at T = 298 K, respectively. The bleaching times were much longer in these cases in comparison with that in solution, increasing from fractions of miliseconds (in solution) to seconds (in a polymer matrix) to hundreds of minutes (in the solid state) at room temperature.

The dispersive kinetics can often be found in physical processes and chemical reactions occuring in disordered systems, even though elementary steps are monomolecular reactions. According to Richert and Baessler [36], such a kinetics may translate into a Gaussian distribution of activation energies: the time dependence of the concentration of coloured species should be a convolution of first-order decay

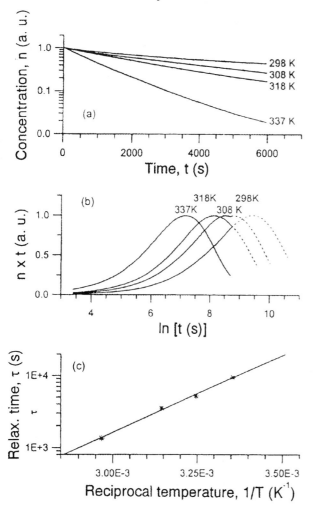

*Figure 10.* Kinetics of bleaching of a polycrystalline DHP sample calculated from the decay of absorbance at 550 nm. The decay curves plotted in the conventional semilogarithmic coordinates (a), and in the $t \times n(t)$ vs. ln $t$ coordinates (b). (c) The Arrhenius plot constructed from the results of (b).

functions and a distribution function for the reaction rates. In such a case, one can extract information concerning the position of the maximum and the width of the distribution of the activation energies using a straightforward method discussed in our earlier paper [37]. The method consists in analysis of the decays replotted in the [($Y \equiv t \times n(t)$) versus ln $t$] co-ordinates: the width of the resulting bell-shaped curves is related to the distribution width, and position of its maximum to the energy of the maximum of the distribution. We showed that the full-width at half-maximum (FWHM) of the experimental curves ($\omega_{exp}$, expressed in units of ln $t$) is related to FWHM of the distribution of activation energies ($\Omega$, assumed Gaussian and expressed in units of energy) via the equation

$$\omega^2_{exp} = \omega^2_0 + (\Omega / RT)^2 \tag{3}$$

where $\omega_0 \approx 2.45$ is the 'intrinsic' width of the $Y$ curve. Moreover, for sufficiently narrow distributions, the maxima of the $Y(\ln t)$ curves are related to the energies of the distribution maxima via a simple relation

$$E_{max} = RT \ln (A \times t_{max}), \tag{4}$$

where $A$ is the frequency factor.

The results of the kinetics of the bleaching reactions measured on polycrystalline samples given on Fig. 10a, re-plotted in the new co-ordinates and displayed in Fig. 10b, can be well fitted with three temperature-independent parameters; $E_{max} = 41.1$ kJ/mol, $\Omega = 5.3$ kJ/mol, and $A = 2 \times 10^3$ s$^{-1}$.

## 4. Acknowledgements

The research reported in this paper was supported by the Grant Agency of the Academy of Sciences of the Czech Republic (grant No A1050901 and 12/96/K), by the Polish State committee for Scientific Research (grant No 3T9A4812), and, in its final stages, by the Technical University of Wrocław.

## 5. References

1.  Ramsden, C.A. (1994), Non-bonding molecular orbitals and the chemistry non-classical organic molecules, *Chem. Soc. Rev.* **23**, 111-118.
2.  Keith, J.S., Greenberg, M.M., Goodmann, J.L., Peters, K.S., and Berson, J.A. (1986), 3,4-dimethylenefuran and 3,4-dimethylenethiophene, heterocyclic analogues of the disjoint non-Kekule hydrocarbon tetramethyleneethane, *J. Am. Chem. Soc.* **108**, 8088-8089.
3.  Cava, M.P., Pollack, M.M., and Repella, D.A. (1967), Reactive tetravalent sulphur intermediates. α-Thiaphenylene, *J. Am. Chem. Soc.* **89**, 3640-3641.
4.  Berson, J.A. (1997), A new class of non-Kekule molecules with tunable singlet-triplet energy spacings, *Acc. Chem. Res.* **30**, 238-244.

5. Sworakowski, J., Nešpůrek, S., Lipiński, J., Sliwińska, E., Lewanowicz, A., and Olszowski, A. (1999), Kinetics of photochromic processes in dihydropyridine derivatives, *This conference.*

6. Kurfürst, A., Zelený, J., Schwarz, M., and Kuthan, J. (1987), Some 2,6-di(p-x-phenyl)-4,4-diphenyl-4*H*-pyrans and analogous 1,4-dihydropyridines from 1,5-dione precursors, *Chem. Papers* **41**, 623-634

7. Nešpůrek, S., Schwarz, M., Böhm, S., and Kuthan, J. (1991), Solid state photochromism of 2,4,4,6-tetraphenyl-4*H*-pyran and 1,4-dihydropyridine derivatives, *J. Photochem. Photobiol. A: Chem.* **60**, 345-353.

8. Kuthan, J. (1983), Pyrans, thiopyrans, and selenopyrans, *Adv. Heterocycl. Chem.* **34**, 145-303.

9. Schwarz, M., Trška, P., and Kuthan, J. (1989), NMR spectroscopic investigation of p-substituted 2,4,4,6-tetraphenyl-1,4-dihydropyridines and their oxa and thia analogues, *Collect. Czech. Chem. Commun.* **54**, 1854-1869.

10. Šebek, P., Nešpůrek, S., Hrabal, R., Adamec, M., and Kuthan, J. (1992), Novel preparation and photochromic properties of 2,4,4,6-tetraaryl-4*H*-thiopyrans, *J. Chem. Soc. Perkin Trans.* **2**, 1301-1308.

11. Kroulík, J., Chadim, M., Polášek, M., Nešpůrek, S., and Kuthan, J. (1998), Synthesis of some new functionalized 2,4,4,6-tetraphenyl-4*H*-thiopyrans and study on their photo-colouration, *Collect. Czech. Chem. Commun.* **63**, 662-680.

12. Peres de Carvalho, A. (1935), δ-Diketones and γ-pyrans, *Ann. Chim. (Paris)* **4**, 449-522.

13. Shibuya, J., Nabeshima, M., Nagano, H., and Maeda, K. (1988), Photochemical reaction of 2,4,4,6-tetrasubstituted-1,4-dihydropyridines in decreated media. Photocoloration and photorearrangement accompanying dehydrogenation, *J .Chem. Soc., Perkin Trans.* **2**, 1607-1612.

14. Peres de Carvalho, A. (1934), 1,4-Pyran with simple functional groups. 2,4,4,6-tetraphenyl-1,4-pyran, *C. R. Seances Acad. Sci.* **199**, 1430-1432.

15. Peres de Carvalho, A. (1934), Phototropy: three new phototropic substances, *C. R. Seances Acad. Sci.* **200**, 60-62.

16. Mori, Y. and Maeda, K. (1991), Photochemical reaction of 2,4,4,6-tetraaryl-4*H*-pyrans and 4*H*-thiopyrans with color change by a 1,5-electrocyclic reaction. X-ray molecular structure of 4-methyl-2,3,6-triphenyl-2*H*-thiopyran, *J. Chem. Soc., Perkin Trans.* **2**, 2061-2066.

17. Price, C.C. and Pirelahi, H. (1972), Thiabenzenes. IX. The rearrangement of 1-(p-dimethylaminophenyl)-2,4,6-triphenylthiabenzene to isomeric thiopyrans, *J. Org. Chem.* **37**, 1718-1721.

18. Pirelahi, H., Abdoh, Y., and Tavassoli, M. (1977), The effect of electron withdrawing groups on the stability of thiabenzenes, *J. Heterocycl. Chem.* **14**, 199-201.

19. Hašek, J. and Ondráček, J. (1990), Structure of 1-methyl-2,4,4,6-tetraphenyl-1,4-dihydropyridine, *Acta Cryst. C* **46**, 1256-1259.

20. Vojtěchovský, J. and Hašek, J. (1990), Structure of 2,4,4,6-tetraphenyl-4*H*-pyran, *Acta Cryst. C* **46**, 1727-1730.

21. Vojtěchovský, J., Hašek, J., Nešpůrek, S., and Adamec, M. (1991), The structure and photochromism of 2,4,4,6-tetraphenyl-4*H*-thiopyran, *Collect. Czech. Chem. Commun.* **57**, 1326-1334.

22. Domenicano, A., Murray-Rust, P., and Vaciago, A. (1983), Molecular geometry of substituted benzene derivatives. IV. Analysis of variance in monosubstituted benzene rings, *Acta Crystallogr., Sect. B: Struct. Sci.* **39**, 457-468.

23. Vojtěchovský, J., Hašek, J., Huml, K., and Ječný, J. (1990), The structure of 1,2,4,4,6-pentaphenyl-1,4-dihydropyridine, *Collect. Czech. Chem. Commun.* **55**, 2059-2065.

24. Böhm, S., Hocek, M., Nešpůrek, S., and Kuthan, J. (1994), Photocolouration of 2,4,4,6-tetraaryl-1,4-dihydropyridines: A semiempirical quantum chemical study, *Collect. Czech. Chem. Commun.* **59**, 262-272.

25. Böhm, S., Šebek, P., Nešpůrek, S., and Kuthan, J. (1994), Photocolouration of 2,4,4,6-tetraaryl-4*H*-thiopyrans: A semiempirical quantum chemical study, *Collect. Czech. Chem. Commun.* **59**, 1115-1124.

26. Böhm, S., Adamec, M., Nešpůrek, S., and Kuthan, J. (1995), Photocolouration of 2,4,4,6-tetraaryl-4*H*-pyrans and their heteroanalogues: Importance of hypervalent photoisomers, *Collect. Czech. Chem. Commun.* **60**, 1621-1633.

27. Mataga, N. and Nishimoto, E. (1957), Electronic structure and spectra of nitrogen heterocycles, *Z. physik. Chem.(Frankfurt)* **13**, 140-157.

28. Schwarz, M., Šebek, P., and Kuthan, J. (1994), Electrophilic 3,5-substitution of 2,4,4,6-tetraphenyl-4*H*-pyran and some of its 1-heteroanalogues, *Collect. Czech. Chem. Commun.* **57**, 546-555.

29. Nešpůrek, S. and Schnabel, W. (1994), Photochromism of dihydropyridines. Formation and decay of coloured intermediates from 1-methyl-2,4,4,6-tetraphenyl-1,4-dihydropyridine on irradiation in chloroform and acetonitrile solutions and in poly(methyl methacrylate) solid solution, *J. Photochem. Photobiol. A: Chem.* **81**, 37-43.

30. Nešpůrek, S. and Schnabel, W. (1991) Photochromism of 1-methyl-2,4,4,6-tetraphenyl-1,4-dihydropyridine in the solid state: an example of a dispersive chemical reaction, *J. Photochem. Photobiol. A: Chem.* **62**, 151-159.

31. Lewanowicz, A., Lipiński, J., Sworakowski, J., Komorowska, M., and Nešpůrek, S. (1999), On the mechanism of photochromic activity in a substituted dihydropyridine, *Proc. Int. Workshop on Reactive Intermediates*, Szczyrk (Poland), P18.

32. Lewanowicz, A., Lipiński, J., Nešpůrek, S., Olszowski, A., Sliwińska, E., and Sworakowski, J. (1999), Photochromic activity of dihydropyridine derivatives: energetics and kinetics of photochemically driven reactions in polycrystalline 1-methyl-2,4,4,6-tetraphenyl-1,4-dihydropyridine, *J. Photochem. Photobiol. A: Chem.* **121**, 125-132.

33. Sworakowski, J., Nešpůrek, S., Lipiński, J., and Lewanowicz, A. (1999), On the mechanism of bleaching reactions in a photochromic dihydropyridine, *J. Photochem. Photobiol. A*, submitted.

34. Šebek, P., Sedmera, P., Böhm, S., and Kuthan, J. (1993), Sigma isomerization of tetraphenylcyclopentadienes: Reaction mechanism and quantum chemical treatment, *Collect. Czech. Chem. Commun.* **58**, 882-892.

35. Pirelahi, H., Rahmani, H., Mouradzadegun, A., Fathi, A., and Moudjoodi, A. (1996), The effect of 3,5-substitutions on the photochromism and photoisomerization of some 2,3,4,4,5,6-hexasubstituted 4*H*-thiopyrans, *J. Photochem. Photobiol. A: Chem.* **101**, 33-37.

36. Richert, R. and Baessler, H. (1985), Merocyanine - spiropyran transformation in a polymer matrix: an example of a dispersive chemical reaction, *Chem. Phys. Letters* **116**, 302-306.

37. Sworakowski, J. and Nešpůrek, S. (1998) A straighforward method of analysis of first-order process with distributed parameters, *Chem. Phys. Letters* **298**, 21-26.

# PHOTOCONDUCTIVITY OF ORGANIC SOLIDS AND THEIR CONFINED STRUCTURES

M. KRYSZEWSKI
*Department of Molecular Physics, Department of Chemistry,*
*Technical University of Łódź, 90-924 Łódź, Żwirki 36, Poland*
*Centre of Molecular and Macromolecular Studies, Polish Academy of*
*Sciences, 90-363 Łódź, Sienkiewicza 112, Poland*

## 1. Introduction

Photoconductivity and photoluminescence as well as photocatalysis, photosyntesis and other phenomena driven by light present a great interest for science in order to elucidate the details governing the interaction of matter with light. There is no need to enumerate here the diverse applications and devices for which the basis is light and electrical field induced phenomena [1-3].

These effects occur in inorganic and organic crystalline and amorphous materials usually being two or many component systems. Some of them are characteristic of very complicated structure e.g. photosynthetic unit. Its functions are as yet extremely difficult to mimic.

All above-mentioned effects have in common: the molecular excitations, the generation and the separation of carriers as well as their transport as characteristic feature of the system under discussion. Even the first step the excitation on a molecule by light (with some exception of gases) is not well understood as yet. This is due to the fact that chromophore molecules are surrounded with neighbours, thus not only electronic states but also vibronic states can be influenced by surrounding. We will come to this subject later on.

In spite of the fact that we will be dealing with photoconductivity in polymers and with related phenomena in confined volume, some sentences have to be devoted to photorefractive materials because their structure is in many respects similar to that of photoconductors.

The photorefractive effect is a persistent but reversible change in the refractive index of electrooptic material caused by nonuniform illumination and liberation of charges from photosensitive traps in the bright regions. In order to work with the photorefractive materials one has to have the following three properties:
(i) photoconduction, (ii) photosensitive traps and (iii) linear optical response. The liberated charges diffuse until they recombine with some ionised traps elsewhere. After some time, charges are depleted from the bright regions and added to the dark region. They create a spatialy varying space-charged fields. The space-charge formed by

*F. Kajzar and M.V. Agranovich (eds.), Multiphoton and Light Driven Multielectron Processes in Organics: New Phenomena, Materials and Applications, 281–298.*

electric field changes the refractive index (of refraction) thought the linear optical effect (Pockels effect) or the original intensity pattern-a hologram- are persistent in the dark. This is the main difference with the photoconductivity, which disappears (with some exception in living Nature in which light induced phenomena have long duration), when the light is switched off.

Polymers, which are of interest as photoconducting materials exhibit some similarities to photorefractive material: they combine the photosensitive chromophores and in general, charge transport agents. The technical requirements to apply a photoconductor are strong but considering the final aim, different.

One general remark should be stressed at the beginning. The photoconductivity is observed in single component inorganic and organic low molecular weight materials (e.g. ZnS and anthracene) and some concepts and theories elaborated for these classical photoconductores are of great importance for polymeric photoconductors too, particularly for space (and energy) confined systems. Evidently the general lack of crystalline structure makes the interpretation of photoconductivity more difficult.

## 2. General remarks on photoconducting polymers

Photoconductivity in polymers, as evidently in other materials, is defined as phenomenon of an increase of their electrical conductivity under illumination. Photoconductors are generally dielectrics in darkness. The photocurrent intensity depends in a complex way on the nature, chemical constitution and physical structure of the given material. A particular consideration should be given to all excited states and their interactions which have to be precisely defined (see e.g. [1-5] and many references given therein as well as many books and reviews dedicated to conducting polymers). In intrinsic photoconductor charges carriers and/or excitons are generated in the same material, which is also responsible for transport of charged species. In most practical photoconductors charge carriers are generated in a chromophore system consisting usually of a charge-transfer complex (CT). One of the CT components can also serve as transporting sites. Polymer in many cases is a binder but its role might also be important. Depending on the sign of majority carriers they are not lost by trapping, recombination or other immobilisation phenomena. Evidently real photoconductors are very far from such an idealised characteristics.

In general three main classes of polymers are known to exhibit semiconductivity and photoconductivity:
a) polymers with $\pi$-conjugated multiple bonds (like polyacetylene, poly(diacetylene) ($-C\equiv C-C\equiv C-$) as well as large aromatic groups e.g. anthracene and several dyes in the main chain
b) polymers with $\sigma$-bonded silicon or germanium atoms in the main chain (like polisilanes)
c) polymers with saturated backbones and pendent aromatic groups like poly (N-vinylcarbazole) (PVK) and its derivatives, which have been studied in great details (see e.g. [6] and other works quoted therein).

There are known many photopolymers consisting of various aromatic "monomers" like naphthalene, pyrene, chrysene and highly constrained structures. They will be not discussed because they show photoconductivity in UV or are intractable (one interesting exemption is poly (vinylphenylanthracene)). In this discussion we will neglect on purpose dark conducting polymers obtained by molecular doping in spite of the fact that they are the "hot" area in that field. Evidently one can distinguish n-type photoconductivity (electron) or p-type one (holes). Most polymers known are p-photoconductors. In particular case a photoconductor can be bipolar i.e. both electrons and holes are transporting charges.

Photoconductivity has been studied extensively both on organic and inorganic systems: thin films of amorphous chalcogenides ($As_2Se)_3$ [7], amorphous selenium and amorphous silicones [8]. Starting from amorphous state as an order in introduced the spacing between hopping sites and their energy levels would become more uniform and the localised states coalesce into narrow bands. These narrow bands in organics have been also postulated many years ago for molecules like anthracene [9] and this concept is not fully abandoned in spite of the fact that photoconductivity of anthracene single crystals has been studied by and elucidated by many authors. We wish to concentrate our discussion on two things:

a) on the formation of charge carriers under illumination in subpicosecond time which is a local phenomenon considering a single molecule and its closest surrounding confinement

b) on the other end we want to give some remarks on the role of order even in so called amorphous systems.

The local phenomena of excitations can be studied with new tools to follow the fate of excited molecules. The local order is seen from the confinement point of view in a little different way then before.

Mostly investigated photoconductive polymer without and with molecular dopants was PVK thus a short characteristic of this model polymer seems to be needed.

Pristine, pure PVK shows photoconductivity only in UV range. The photogeneration spectrum has a structure, which exhibit some coincidence with absorption spectrum of PVK (it means that the higher single states are excited). The primary quantum yield $\Phi$ for PVK is about 0,14 and the thermalization length $r_0$ is in the range of 22,5 Å (when excited into first single state) to 30 Å (when excited into the third single state). The absolute quantum yield for photogeneration of holes is around $10^{-3}$ at the field $10^5$ V/cm and reaches the value 0,005 at $10^6$ V/cm. Hole mobility at this field is about $10^{-6}$ cm$^2$/Vs. PVK has been doped with numerous acceptor molecules e.g. tetracyanoethylene (TCNE), chloranil and Lewis acids. The system PVK-TNF where TNF – trinitrofluorenone is photoconducting also in the visible range of light because the CT band of the complexes is extended through this part of the spectrum. This is the reason why this simple system found practical application many years ago. For PVK:TNF ratio 1:1 (monomer) the primary quantum yield $\Phi$ at $\lambda$=550nm is 0,23 and the thermalization length is $r_0$=35 Å. The absolute quantum yield at $10^6$ V/cm approaches $10^{-1}$ being the same for holes and electrons. Such system is bipolar but with the increase of TNF molecules the hole mobility decreases below $10^{-8}$ cm$^2$/Vs (for PVK:TNF ~1:1) while electron mobility increases up to $10^{-6}$ cm$^2$/Vs at the $5\times10^5$ V/cm. The charge carrier generation can be shown in the scheme presented in Fig.1.

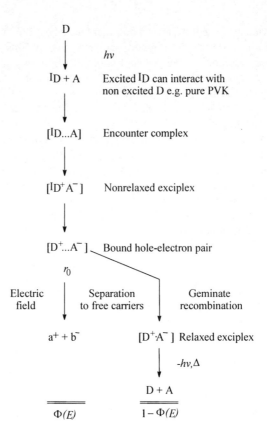

*Figure 1.* A scheme of photogeneration of charge carriers; D, electron donor;
A, electron acceptor (redrawn according to Okamoto and Itaya see [5])

In the first phase (Fig.1) the photogeneration $\Phi$ of charge carriers is field independent. The absorption of a quantum by the molecule or CT complex results in its photoionization into two opposite ion pairs or ion radicals ($D^{\circ-}$ and $A^{\circ-}$). This effect is considered to occur via unrelaxed exciplex, followed by formation of the hole-electron pair, which separates due to coulombic attraction and reaches some characteristic thermalization length $r_0$.

The photogeneration quantum yield $\Phi$ is usually determined by xerographic discharge or surface potential discharge (SPD). This technique consists in monitoring the changes of the surface potential of the investigated film (charged by corona discharge) due to illumination. The layer of ions deposited on the free surface of the sample acts as blocking electrode, so that the photoiniection can be neglected. From the initial potential decay rate dV/dt one can determine the photogeneration quantum yield $\Phi$. Such a procedure if the determined value $\Phi$ is thickness and illumination intensity independent is very simple and indicates that the recombination and trapping by deep traps does not influence the photoconductivity. These rather simple experimental conditions and the knowledge of sample surface S, dielectric constant $\varepsilon$; $\varepsilon_0$-permitlivity of free space, e-elementary charge, N-number of absorbed photon, d-thickness of the sample and $(dV/dt)_{to}$- the rate of decay of surface potential at the first moment are easy to be fulfilled thus this technique is used by many authors to obtain also mobility values as indicated above.

The second phase is electric field assisted carrier separation if they avoided geminate recombination. There is an ample literature devoted to field assisted separation and transport of carriers in a great number of systems. It seems not on purpose to analyse these results (e.g. field intensity, light intensity, temperature and structure dependence). This second step in photoconductivity is similar to ion-pair dissociation in week electrolytes (Brownian motion and superimposed field of geminate carriers) and it is well described by the Onsager theory with some exceptions which will be discussed later on.

It was a reasonable assumption that polymers are disordered solids in which the overlapping of electron orbitals is small. The created photocharge is localised (excess charge) on some molecule (host molecules). The extended states ranging over other molecules are unlikely thus band theory can not be applied (some concepts of this theory are used in description of electrical properties of organic solids but they have to be well defined).

In all polymers (intrinsic and extrinsic) without longer conjugated chains the transport occurs via thermally activated hoping among the chromophores or among the dopant molecules.

In terms of solid-state chemistry the hopping process can be described as reversible oxidation-reduction process. The localised hole forms a cation radical and can accept an electron from the next neutral molecule and so the hole (cation-radical) performs a random walk in the direction of the applied field [1]. The ion-radical state formation concerns random walk in the direction of applied electric field. In polymers with $\sigma$-bonded polymeric arrays of germanium or silicon atoms the hole transport occurs by hopping via the $\sigma$-derived states rather then via side chain groups. Thus the structure of such chains causes the transport along the chains (high, nondispersive mobility around $10^{-4}$ cm$^2$/Vs).

The validity of basic Onsager model and its many variants proposed by numerous authors (e.g. three-dimensional variant) was tested on PVK-TNF and other more complicated aromatic macromolecules as well as on silicium and germanium polymers with various substituents.

Most of the photogenerated charges undergo geminate recombination and only a small part is separated from their initial counter charge. Usually in semiconducting and photoconducting polymers the positive charges are moving thus recombination occurs when moving hole meets an immobile electron localised in the vicinity of generation site. For the materials where both holes and electrons are mobile e.g. PVK heavily doped with TNF the rate of recombination depends on the combination of carrier mobility $\mu$ and concentration n. It can be expressed by the equation:

$$\frac{dn}{dt} = -\frac{4\pi}{\varepsilon}\mu n^2$$

(1)

where $\varepsilon$ is dielectric permittivity like before.

## 3. Charge generation and transport

There are two ways of discussing charge generation and transport in amorphous solid including polymers. The first one assumes *ab initio* that we have to deal with disordered material and to look how far it is responsible for charge generation and energy transport. Are there particular conditions, which define the excitation and energy migration?

The second one consists in analysing amorphous systems in which some ordering may appear. In this case local structure is of importance for energy transfer.

Let's recall the old but classic paper of Gill [10] concerning the photoconductivity in doped amorphous polymers. The main results can be summarised as follows:

1) there is a considerable trapping and detrapping
2) there is a superposition on characteristics of log I vs. log t (where I-current density and t-time) in time of flight measurement. It means that at low signal conditions the transit time curves measured at different voltage can be overlayed on top of the other. The significance of this fit is that the trapping parameters in terms of life·time $\tau$ and trapping $\tau_T$ scale with the transit time. The superposition of characteristics is important experimentally in that it enables a consistent determination of transit time.
3) the mobility values are very low, in the range $10^{-9}$-$10^{-7}$ cm$^2$/Vs, which evidently exclude any narrow band model
4) there is a strong field dependence of mobility
5) there is a strong temperature dependence of drift mobility
6) the apparent field dependent mobilities seem to converge at some temperature for which it is difficult to give any physical significance
7) there is a correlation between concentration of acceptor molecules and electron mobility. It is very small in pure PVK but dramatically increases with addition of TNF
8) there is an exponential correlation of $\mu$ with the separation radius R (average separation of transport molecules

These observations confirm the hopping model in which localised hopping sites are associated with carbazole or with TNF molecules. This model is based on molecular picture in which all active sites are essentially localised except of for small overlap of wave functions of approximately the same energy levels between adjacent molecules.

In this context the main polymer chains consisting of -C-C- bonds hardly plays any role in the conduction process. The specific formation of charge transfer between carbazole and TNF also does not seem to seriously affect the transport properties although it plays an important role in spectral response of free carrier generation. This implies that in the ground state the number of charge carriers is small. This picture is consistent with the recent analysis of the absorption and electroabsorption. All implications above corroborated have also explained the change of majority carriers with the increase of concentration of TNF (change from p to n photoconductor).

I have mentioned before that there are similarities in photoconductivity of chalcogenides e.g. $As_2Se_3$ and amorphous Se. Amorphous Se is considered to consists of both polymeric chains and $Se_8$ ring molecules and the amorphous $As_2Se_3$ is considered to form a network structure with branching. Without going into the details one has to mention that there is a correlation between electron mobility and the density of $S_8$ rings. The idea of indentifying molecular units in these two types of materials is interesting and needs further consideration.

It is necessary to discuss the influence of disorder on charge generation and transport. There are a number of papers dedicated to this problem (see [11]); thus only important papers can be discussed in some detail.

Because of complicated structure of amorphous solids (glasses, chains, substituents and dopants) the velocity of charge carriers (drift velocity) and transit time across the sample are not of single value but are the quantities subjected to a distribution that evolves with the time. Transport of charges is thus a stochastic problem reflecting the disorder of the sample. There are various types of disorder-controlled transport phenomena e.g. time dependent (dispersive) and time independent (non-dispersive). A pertinent analysis of dispersive and nondispersive charge transport has been published by Bässler and collaborators ([12] and references therein). They have shown that the originally developed positional disorder (geometric) of hopping sites is quite general but also that various different disorder concepts can exist. They have used Monte Carlo simulation technique and analytic theory to show the influence of positional and energetic disorder on carrier motion in random solids. The multiple trapping was also considered in their discussion of transport processes.

The type of disorder, which is always present in random system, is the frozen-in positional fluctuation of the hopping sites reflecting the statistics in molecular environment (R-disorder). The energy of a site depends on the interactions with surrounding molecules e.g. polarisation energy of charge carriers or van deer Waals interaction energy of electronically excited molecule thus positional disorder results in fluctuation of inter-site transition rates ("off-diagonal disorder"). It translates into disorder of site energies of the bulk molecules (E-disorder). In such situation conduction valence or exciton bands of an organic system (e.g. crystal) are split into distribution of localised states (DOS) [12].

It was shown that under realistic conditions positional disorder would not give rise to dispersive behaviour on time scale of carrier transit. However both E-hopping and multiple trapping are reflected in dispersive transport because there is an energetic relaxation within (DOS). Since site energy depends on a large number of internal coordinates, each varying randomly by small amount the choice of Gaussian distribution of DOS seems to be justified in weakly bound molecular solids. One can find a support for such distribution from the analysis of broadening of the absorption profiles of evaporated organic layers and chromophores in glasses ([13] and references therein). Evidently one should also expect such type of distribution for trapping and detrapping times. For a Gaussian DOS of small width a dynamic equilibrium can be attained in realistic experimental conditions. It results in non-Arrhenius type of temperature dependence of charge carrier mobility $\mu$, which is often found in numerous experimental works. To conclude this remarks one should mention that the field dependence of mobility is expressed as $\ln \mu \sim E^n$ where n=0,5. This relation was verified for many systems. This very elegant interpretation of charge carrier mobility as function of disorder and external parameters (temperature and field) does not contain material specific characteristics. The further studies of Bässler's group [14] dealt with the phenomena of relaxation of a system, which as approaching equilibrium after some perturbation e.g. electronic excitation. In that case following Bässler's concepts we have several combined effects like:

  a)  instantanteous readjustment of the local electron distribution
  b)  dissipation of excess energy (electronic or vibrational)
  c)  structural response of the system on the change of electron distribution
  d)  energy loss processes related to electronic transport towards lower electronic states and finally
  e)  electronic decay followed by structural relaxation towards equilibrium state

The energy loss processes (relaxation) concern an ensemble of charged and neutral excitations, which undergo random walk among constituent elements. They are manifested as photoluminescence and transient photoconductivity. Let's discuss at first the results of studies of local energy changes and time-resolved luminescence.

In organic random solids the relaxation of energy due to weak van der Waalscoupling among the sites (the energy exchange term in energy matrix) is small, typically of the order $10^{-2}$ eV) and corresponds to disorder-induced fluctuations. The main disorder effect is therefore the modulation of the site energies. It is determined by polarisation energy of a charge carrier or Frenkel excitations. It is evidently also due to interactions with surrounding molecules. These fluctuations can be estimated to be of the order 0,1 eV for charge carriers and 0,003 eV for Frenkel excitons. As a consequence, electronic states in random organic solids can, to good approximation, be considered as completely localised. Their distribution reflects the distribution of diagonal elements in energy matrix. We will not follow the main concepts leading to temporal relaxation of the mean of an energy ensemble of random walks hopping within the Gaussian DOS and we wish only to comment as an example on the changes of phosphorescence spectrum emitted from glassy benzophenone [14]. This system offers several advantages. Benzophenone glasses are easily prepared by quenching the melt or by vapour deposition. The "hot" singlet state (excited by $N_2$-laser) results in random population of triplets $T_1$ via rapid intersystem-crossing and decays by phosphorescence emission (hopping times of $T_1$ states in crystals and their intrinsic life-time are about 10 ns and 5 ms respectively). As expected the intensity of phosphorescence is decreasing in time after excitation, which is easily seen by monitoring the temporal shift. The curves of phosphorescence intensity changes with assumption of the width of the DOS of $\sigma$=260 cm$^{-1}$ prove that the model of random walk of an exciton (within the Gaussian potential and in the limit of weak electron phonon coupling) is appropriate to treat transfer of excitation that moves via exchange excitation. The same remark concerns the host emission e.g. 2% solid solution of benzophenone in 2-methyltetrahydrofurane glass.

A very interesting example of electronic relaxation in the course of excitation hopping are energy and time resolved fluorescence studies. Particularly reach in information is Site-Selective Fluorescence (SSF) investigations. This very modern technique was recently reviewed by Bässler et al. [15] considering in particular the application of SSF to conjugated polymers and oligomers. A considerable body of information is presented in this excellent paper but we can not go into the discussion of the basis of SSF. We will only stress a possibility to use it for distinguishing the intra- and intermolecular relaxation. This question is important for us from the viewpoint of excitations leading to charge generation. The inter-molecular relaxation was detected as excimer process in crystalline pyrrene or polymers containing carbazole as chromophore [16].

Bässler group has shown that one can conclusively separate intra-from intermolecular relaxation by SSF experiments using compounds, which belong to the phenylenvinylene family e.g. 1,4 bis (p-cyano-styryl)-2,5 dioxyl benzene (CNSB). This compound can by used as dopant in methyltetrahydrofuran glass at low concentration and can be blended with polystyrene at an arbitrary mixing ratio. Fig.10 in the paper [15] shows clearly that absorption and emission spectra of CNSB in polystyrene up to 80% be weight at room temperature, are very similar. The absorption spectra are somewhat narrower as the concentration increases while fluorescence spectra show a redshift. The differentiation between the contribution of exciton migration toward low energy sites and molecular relaxation to spectral broadening SSF spectra were measured also at low concentration of CNSB in MTHF at 5 K. They have shown that in no case sharp spectra with zero phonon line were observed. The authors bear out carefully that all sites are capable of excimer formation like in a pyrene crystal or if energy transfer toward inspicient dimers is involved like in polyvinylcarbazole [17].

Until now we have discussed charge transport and later on energy migration after excitation of a molecule in a random system basing on the disorder theory of Bässler and coworkers. In the recent years much progress has been made both experimentally and theoretically on transport phenomena in molecularly doped polymers from the viewpoint of polaron effects. They have been discussed by many authors too. The very good review and comparison between disordered theory and polaron theory has been published by Schein [18]. The new interest in polaron theory is related to the observation that the change of polymer matrix may contribute to polaron binding energy. Vannikow et al. [19] have postulated that investigating electron transport in the solid solution of polystyrene (PS) and triphenylamine (TPA) one can use a polaron approach. This concept have been based on inter-molecular medium reorganisation energy, which influence the wave function decay parameter particularly at temperatures higher than the polymer matrix $T_g$ (enhanced molecular motions may reduce electrostatic energy contribution to polaron binding energy [20]. More quantitative analysis will show whether these effects are sufficiently strong to account for the observed changes in activation energy.

All before discussed models concerned in reality the transport of carriers and the dissipation (transfer) of energy in various phenomena.

Our main question concerns however the process of generation of charge carriers. At the beginning of eighties the studies of generation and localisation of carriers were devoted mainly to amorphous inorganic photoconductors (semiconductors) but the field of photoconducting organic solids was developed too e.g. poly(N-vinylcarbazole). Classic Onsager's treatment and its various variants have given a lot of information on the role of external field in charge generation assuming very fast geminate recombination. In this picture a photoexcited electron-hole pair thermalize at the distance $r_o$, which is comparable with mutual Coulomb radius $r_c$. Depending on the ratio $r_o/r_c$ some fraction of initially thermalized pairs will associate at zero field. This fraction can be increased with increasing field and asymptotes to the total dissociation of the thermalized pair. In the a-Se it was found that $r_o$ decreased from 70 Å at $\lambda$=4000 Å to $r_o$=8 Å at $\lambda$=6200 Å. More recent experiments using both xerographic discharge and delayed collection field technique show that the geminate recombination is really wavelength depended in a-Se but it is not the case of α-SiH [21]. This fact was interpreted by the difference of dielectric constant of the material under study or by specific phonon spectra. It is convincing but not fully accepted. The application of delayed field technique to study the quantum efficiency in poly (N-vinylcarbazole) revealed that for the given field it is the same if the pulse of photoexcitation occurs with the field applied or delayed up to many seconds. The suggested explanation consist in the fact that the geminate recombination rate is sufficiently slow in order to make possible the photogeneration process to be time resolved [22]. These two results are evidently not a definitive solution of the problem of charge generation particularly that they concern two different materials: simple inorganic photoconductors and much more complex material like PVK (without $O_2$) They point out however that the intrinsic properties of the material have an influence on the photogeneration of free charges.

## 4. Charge generation and spacial confinement

To summarise the discussion presented above one should stress that photogeneration process is critical for all photoconductors. The interaction of photons with a material to produce free carriers is affected by the disorder within the material and by its chemical constitution. It is accepted that geminate recombination is a limiting step where the mean free path of photocarriers is low, which is evidently the case in all amorphous materials.

It seems now necessary to collect some conclusions and to stress some critical factors. They are:

a) primary quantum yield $\Phi$ connected with production of thermalized electron-hole pair absorbed photon

b) $r_o$-thermalization distance, which is evidently correlated with free path and

c) Coulomb radius $r_c$, for which one can take $r_c \sim kT$

For simple photoconductors like a-Se the value of $\Phi$ approaches to 1 thus one can estimate using Onsager model and the appropriate value of dielectric constant the range of fields, in which $r_o$ is or is not field dependent. Typical Coulomb radii $r_c$ dependent on dielectric constant are for a-Si~49 Å, for a-Se~100 Å and for most organic materials $r_c$ ~200 Å. These values indicate also that dopping has different meaning for a-Si, a-Se and organic systems (glasses and polymers), in which dopant molecules can be at different distance from the donor or acceptor.

Let us consider at first amorphous materials, which are not photoconductors themselves but, as mentioned before, they can be doped whereby doping consists in controlled chemical oxidation of dopant molecules. The formed radical cations are equivalent to "free" charges. This is due to the fact that electrons can hop from neighbouring neutral molecules so that the free holes propagate. This picture explains why for low oxidation levels conductivity rises. However at very high concentration total oxidation occurs and there are no neutral molecules, necessary to take part in the transport, thus the conductivity must go through a maximum and then it falls down. Such a system is polycarbonate doped with tri-p-totylamine and oxidised by $SbCl_5$ [23]. It seems worthwhile to mention that there is a possibility to produce materials containing about $\sim 10^{20}$ cations per $cm^{-1}$, which Mott has designed as "degenerated polaron gas", but its conductivity depends not only on the free charge concentration but also on their mobility, which is related to "charge carrying" molecules.

Usually polycarbonate and poly(styrene) (PS) were used as matrix polymer for charge generation or charge transport [20]. These both matrices were doped with p-diethylaminobenzaldehyde diphenylhydrazone (DEH). The studies of this system are interesting because in the case of DEH doped PS it was shown that the activation energy is independent on the distance between hopping sites. It suggest that the interactions between hopping sites are minimised. The very careful experiments carried out for PS doped samples (see [20]) show that the field dependence of mobility the activation energy and the width of the hopping site manifold as well as the positional disorder parameter are all independent of the polymer host. The conclusion from this work is that only the exponential tail of the electronic wave function depends on the host polymer. This latter result seems to be significant not only for the above discussed system. The often discussed deviations from application of the Onsager theory (in its different variants) are only due the physics of excited molecule and the dielectric constant of the medium. Many examples of this kind for various photoconducting polymers have been exhaustively described by Nespurek [24]. An interesting photoconductor is poly (E, E)-[6,2] paracyclophane 1,5-diene (PDE), which is a member of a specific polymer group containing bridged ring pendent to the backbone. The polymer chain allows alignment of multiple cyclophane groups leading to extended $\pi$-interactions thought the pendent aromatic moieties. It becomes photoconductive when doped with organic acceptors such as TNF or tetracyanoethylene. The photoconductivity of PDE doped with TNF was investigated in our laboratory. It was shown that this CT complex is characteristic of high quantum yield for blue light. The electric field dependence of quantum yield can be described by Brown kinetic model of CT complex dissociation (see [25] and references within). It seems that this behaviour is due to specific strong CT complex PDE-TNF obtained by casting films in dichlorobromethane. It seems important to stress that polymers exhibit sub $T_g$-transitions, which may influence the stiftness of the matrix. There is an ample literature concerning this problem. For the sake of the compactness of the text it is necessary to give up the analysis of this problems in details. PC shows in the temperature range 150 to 230 K $\beta$-relaxation and at low temperatures (near 53 K) $\delta$ relaxation. The molecular basis for sub $T_g$-molecular relaxations consists in the case of PC of segmental motions and rotations of phenyl groups. These relaxation have not been noticed in the publications devoted to the temperature dependence of conductivity of various doped systems applying PC as a matrix. The dopants should also exhibit a cooperative motions, which make possible to consider them as being confined between polymer chain elements. Molecular motions in polystyrene have been extensively studied by Boyer ([26] and references therein). In that case the local confinement of acceptor type molecules should be stronger because of a possibility of forming CT complexes with pendant phenyl groups.

The specific confinement of the components of CT in formation of reticulate doped polymers should also be considered [27-28]. Reticulate doped polymers are the class of heterogenous conducting polymeric systems obtained by crystallisation of CT complexes during *in situ* evaporation of the solvent from common solution with polymers. The type of conducting networks or disperesed crystallites depends not only on the rate of solvent evaporation. During the increase of viscosity the interactions of the polymer chains with CT complex components result in various structures [29].

The clear influence of a confinement on the properties of various materials is seen in inclusion complexes or in layered solids (intercalates). The application and role of nanointercalates has been recently reviewed by the author [30]. This paper describes basic materials used for intercalation the course of intercalation process and interesting electrical properties of some nanointercalates. This is the reason why we will not discuss the general concepts of intercalation and of inclusion complexes formation. We wish only to present some examples showing particular properties of some confined systems.

Nature during the course of evolution organised molecular elements to form complicated - in most cases-perfectically and optimally functioning systems of unusual beauty e.g. photosynthetic unit. They are a strong stimulus for the efforts of chemists, physicists, biologists (and mathematicians) to mimic them in order to design molecular materials exhibiting all sorts of properties and serving many specific needs. As yet no system designed by scientists has the complexity and the light sensitivity and functionality of natural organisation. We have chosen the interaction with light because this is the main subject of our discussion.

Looking at the above mentioned example one can easily detect that the most important parts of this system are inside of a set of molecules thus being an inclusion or intercalate system.

The general paradigm of inclusion is our case is bringing together appropriate molecules in such a way that photons are involved in triggering and performing the expected tasks. The guest molecules are protected physically and forced both to react in chemical way and to take part in physical processes that can be controlled and predicted. The assembly of guest molecules in confined space (it seems to be adequate to use this term) has the properties which depend on the size and close surrounding cavity or layer (in "microscopic" way of looking the thin wall in the nanointercalates or nanoinclusions may be considered as a part of such system particularly when the distributions of host molecules is not homogeneous which is usually the case). There is a number of differences between a set of molecules confined in a host and the bulk material thus they can not be enumerated here. We will focus our attention on electrical and optical properties of some confined systems. Most of the experience of molecular photochemistry is connected with linear region of optical phenomena (respons is proportial to the first power of light intensity). The use of lasers with high power densities has introduced a variety of non-linear phenomena. Two photon spectroscopy, as it is discussed during this meeting, is a source of new phenomena and ideas on information storage, optical switching etc.

Second order non-linear optical (NLO) processes are directly related to hyperpolarizability (β) of individual molecules and the bulk properties of such materials are characterised with optical susceptibility (χ). The system that posses second-order NLO properties must be noncentrosymmetric. In many cases of crystalline materials polar systems are preferred. Molecules which are effective for the second order optical process (those with large β) often exhibit large degree of intermolecular charge transfer and may possess a large ground state dipole moment. They also exhibit a tendency to dimerize in a head-to-tail arrangement (in order to minimise electrostatic repulsion) thus forming crystals, which are centrosymmetric. The broadening of the class of molecules, which may be used for NLO devices can be attained by the inclusion paradigm. The first example of the application of the inclusion phenomena is their use, to design NLO materials of cyclodextrins (CD) as hosts [31]. It was shown that p-nitroaniline (PNA) forms crystalline complexes with β-cyclodextrin (β-CD) which are capable of SHG (two to four times that of area standard) when irradiated with Ne-YAG laser. To cut the story short we can say that, in spite of the fact that β-CD is only weakly hyperpolarizable the SHG of PNA arises entirely from the alignment of guest molecules in the solid. There is much other inclusion complexes known which prove the ability of inclusion complexation to align non-linear opticofores in a polar fashion.

The choice of the  guests is connected with the specific structural features (channels rather than alternative cage type). Orientation within the inclusion structure is believed to be favoured because of electrostatic interactions among the guests. Usually the organisation is such that the positive pole of a neighbour molecules is translated half of molecular length down to the channel, then the cross-channel nearest molecule interactions are attractive, rather than repulsive.

Several polymeric hosts have been shown to form inclusion complexes with guests. The structure of these complexes is often not known precisely. Observation of SHG is important not only from materials perspective but also represents a good source of information about the nature of interactions between the host and the guest. It is well known that there is a possibility to grow oriented crystals in a polymeric matrix by doping an organic material into polymer matrix and then orienting the film uniaxially e.g. by stretching. In the literature of the subject under discussion the older works on crystallisation of organic molecules and CT-complexes *in situ* of the polymer matrix (see e.g.[27]) are usually neglected. The crystallites have been discussed mainly from the viewpoint of electrical properties but still being a stimuls for studies on crystallisation of NLO molecules. Before mentioned composites have been investigated by many authors, but it should be stressed, that in the case of electrical field pooling the orientation is achieved in the crystallite nucleation and growth process. It is initiated by the softening of the matrix and crystallite orientation is forced by external electric field.

There are several examples known in which the polymer does play a genuine role is inclusion process. It was shown that poly(ε-caprolactone) (PCL) films containing PNA exhibit substantial SHG (115 x urea for 500μ film containing 20-25% PNA by weight).

The above discussed examples concerned mainly the NLO properties of various compound confined in polymer matrix.

Adsorption of molecules on a surface can also be considered as a confinement effect. It is well evidenced in the studies of fluorescent molecules adsorbed on layered material surfaces e.g. fluorescence of pyrene on layered silicate sheets. In this type of studies mica, montmorillonite or laponit (commercial material, which is a synthetic hectorite) are usually used. It seems not necessary to give here the complicated chemical formule of these aluminosilicates. Usually these layers are charged so only charged molecules are strongly bound with mineral thin sheet surface. The studies of optical properties of adsorbed molecules are carried out on aluminosilicates with particle size within the micron range. The investigated substance is added to the quasi colloidal dispersion of mineral particles in water. Pyrene (Py) itself is poorly soluble in water and poorly adsorbed on conventional clay (without special pretreatment). In such a case one uses polar pyrene-substituted compounds of general composition: $Py-(CH_2)_n-NH_3^+X^-$. Depending on the number of $(CH_2)$ groups the fluorescence of adsorbed molecules changes. In the presence of only one $(CH_2)$ group (short flexible chain) between pyrene and ammonium group no excimer emission is observed. In the case of long flexible chain $(CH_2)_8$ the folding of these group allows for interaction between pyrene and $NH_3^+$ with some quenching effect but the excimeric emission appears clearly. The heterogenous adsorption may also result in clusters. The clustering can be related to the inhomogeneity of the mineral sheet. In the case of pyrene derivatives the clustering may also be related to the minimalisation of the contact between the hydrophobic part of pyrene derivative and surrounding water phase [32]. The detailed discussion of these clustering effect is out of scope of this paper but the above presented results show the importance of confinement effect at the surface.

One of the very clear effects of inclusion complexation show thin film photoconductors consisting of thiapyrilium dye (TPD) and aromatic amine 4,4'-diethylamino-2'2-dimethyltriphenylen methane (LG) molecules dispersed in polycarbonate PC [33]. The ratio of component js 3:40:60 (TPD-LG-PC) on a weight basis. The glassy film with good optical properties absorbs in the range 565 nm being a weak photoconductor. A strong intramolecular CT transition from the donor (dialkilamino) function to acceptor pyrilium ring is responsible for the colour of this system. Light absorption by the dye in an excess of good electron donor LG results in electron transfer quenching of the CT excited state. The ion-radicals that are formed (as well as their ionic decomposition products) are responsible for charge conduction (in high fields e.g. electrophotographic technique). In that respect this film is not different from (PVK-TNF) photoconductor. When the TPD-LG-PC film is exposed to vapours of methylene chloride it crystallises [33] (see the analogy to the systems described in [27-28]. The formation of the crystalline phase is accompanied by important changes of absorption spectrum and of photoconductivity. More detailed studies of the phase structure of the system under discussion revealed the existence of the amorphous phases (TPD-PC with $T_g$ equal to 125-135°C) and of the crystalline complex. From the viewpoint of confinement effects the crystalline phase 1:1 TPD:PC seems to be important. It was shown that these aggregates consists of TPD molecules arranged in chains, the donor ($Me_2N-$) end of one molecule atop the acceptor pyrilium ring. In order to elucidate this structure studies have been carried out on the model crystalline system

TPD$_2$-DPBC (where DPBC is diphenylcarbonate of bisphenol A). TPD is present in this crystal as perchlorate. It was shown that DPBC molecules pack in congruent layers of dicarbonate chains, alternating with TPD. The complex is centrosymmetric (P2$_{1/a}$). Because DPBC is a model of polycarbonate one can conclude that in this photoconducting system the dye is included partially in the crystalline phase within the amorphous matrix. Taking into consideration that the model crystals of DPBC-TPD$_2$ show photoconductivities three orders of magnitude higher than their dark conductivities one see clearly that the polymer backbone has organised the system to promote effective charge separation and transport.

It seems very interesting to note that Wang [34] has observed a SHG activity in the system mentioned above, based on thiapirilium tetraborate in polycarbonate. The signal from a 5µ film containing 1,5 wf % of the dye was equivalent to that from a 250 µ thick urea crystals. It means about 1000-fold enhencement over urea powder. It was also indicated that the model complex (P2$_{1/a}$) is centrosymmetric thus one should try to find an explanation to this observation. One can speculate that the origin of acentricity is related to internal structure of the complex. One can consider that the conformation of DPCB molecules is twisted with carbonate functions out of the conjugation with phenylene rings. Weak charge transfer interactions with the dye molecules may be responsible for this effect.

In the case of polymeric system one could expect that polycarbonate chain units can be hellicaly twisted within small domains. In such a case the domains would lack a center of symmetry because of the helical pitch of the polymeric host structure. Assuming this organisation (similar to aggregation and crystallisation of iodine-poly(vinylalkohol)-boric acid complex) helical regions of TPD-polycarbonate may become associated and the overall crystalline structure could be acentric. Thus the observed SHG would be similar to traditional polycrystalline materials. The appearance of SHG is in this case a clear confinement effect.

## 5. Conclusions

Concluding remarks will be connected particularly with the last part of this work it is with confinement effects in respect to the optical and conducting properties of polymers. The general the beheviour of doped amorphous polymeric photoconductors seems to be rather clear particularly in view of the transport of charges. There are evidently some examples, which favour various types of dispersive or non-dispersive transport with DOS or polaron mechanism. The intrinsic problem of charge generation is always infered from models (Onsager model with its all variants or Braun model in particular cases). The site-selective fluorescence spectroscopy of conjugated polymers and oligomers removes inhomeogeneities effects on emission spectra. It makes possible to distinguish spectral shifts due to conformational changes of a chromophore after excitation from those arrising from energy transfer.

In general the confined space influence on photogeneration of carriers is not very easy to be confirmed. Our studies of this effect on the organised surfaces (clays) does not give a clear answer, however one sees the influence of the one dimensional constrains on one component of CT complex (pyrene).

The results of the investigations of the influence of polymer matrix on SHG generation are briefly discussed. They show that the polymeric matrix can be an important factor influencing the NLO properties. One clear example of such influence on optical and electrical properties is the systeme polycarbonate-thiapyrilium dyes. It shows that the polymer matrix, under specific conditions can lead to formation of crystalline inclusions with high conductivity and NLO properties, evidently being an example of spatial confinement. The confinement effects are connected with particularly spacialy organised systems e.g. with respect to the adsorbing surface or volume.

In the case of non-porous inorganic oxides such as silica, alumina or aluminosilicates only a small surface is interacting with adsorbed atoms or molecules. These surfaces provide one dimensional organised medium for reaction and interactions. Clays offer both internal and external surface as a medium for interactions. Most clays posseses layerd structure and therefore the microcavities, wherein the guests are accomodated, can be consider as two dimensionally organised or confined. A different situation is in zeolites. Internal micropore surface is accessible for the adsorbent. The pore dimension is of molecular size and this unique arrangement provides three dimensional constrains for interacting or reacting molecules.

Generally, the attempts to modify photophysical behaviour of the guest molecule involved the restriction of rotational and/or translational motions utilising an appropriate constrained medium. Layerd solids present significant opportunity for the development of new types of materials. Through the process of intercalation it is possible to engineer solids with desired physical properties to prepare chemicaly modified materials. The transparency of layerd systems and the ability to adsorb and or intercalate wide range of organic, inorganic and organometallic species suggests that a great number of materials with flexible properties may be obtained. Particular interest is related to luminescent and photoconducting systems, in which complex energy transfer and charge transfer processes are possible.

It is clear, however, that for the majority of layerd materials structural characterisation is not straightfoward. Frequently the detailed structure of the host matrix itself (including polymers) is unknown or unclear. Standard techniques such as X-ray and neutron diffraction will continue to be important tools. Nuclear magnetic resonance will be also used as an important technique for structure elucidation as well as monitoring the motions and interactions of guest molecules. In many cases of photoconducting materials including polymers the formation of radicals or ion radicals is of primary importance. The EPR techniques are often difficult with regard to short living species. Recently for these purposes muon chemistry methods are proposed [35]. The discussion of these methods should be however a matter for another lecture.

## 6. Acknowledgement

It is a pleasure to thank dr J.K. Jeszka for his help in the preparation of this article.
I have been fortunate to be able to interact periodically with Prof. G. Wegner whose global vision has helped me to shape my thoughts in this area.
This work was supported by KBN under the project 7 TO8E 066 14p02.

# 7. References

1. Mort, J. and Pfister, G. (eds.) (1982) *Electronic Properties of Polymers*, Wiley & Sons, New York.
2. Kryszewski, M. (1980) *Semiconducting Polymers*, PWN-Polish Scientific Publishers, Warszawa.
3. Mort, J. and Pai, D.M. (1976) *Photoconductivity and Related Phenomena*, Elsevier Science Publishing, New York.
4. Seanor, D.A. (ed.) (1982) *Electrical Properties of Polymers*, Academic Press, New York.
5. Ulański, J. and Kryszewski, M. (1994) Polymers, electrical and electronic properties, in *Encyclopedia of Applied Physics*, VCH Publishers, Weinheim, Vol.14., p. 497.
6. Pearson, J.M. and Stolka, M. (1981) *Poly (N-vinylcarbazole)*, Gordon and Breach, New York.
7. Enek, R.G. and Phister, G. (1976) Amorphous chalcogenides, in J. Mort and D.M. Pai (eds.), *Photoconductivity and Related Phenomena*, Elsevier, Amsterdam, p. 215.
8. Spear, E.W. and Le Comber, P.G. (1976) Amorphous tetrahedrally bonded solids, in J. Mort and D.M. Pai (eds.), *Photoconductivity and Related Phenomena*, Elsevier, Amsterdam, p. 185.
9. Inokuchi, H. and Murayama, Y. (1976) Molecular crystals, in J. Mort and D.M. Pai (eds.), *Photoconductivity and Related Phenomena*, Elsevier, Amsterdam, p. 155.
10. Gill, W.D. (1976) Polymeric photoconductors, in J. Mort and D.M. Pai (eds.), *Photoconductivity and Related Phenomena*, Elsevier, Amsterdam, p. 303.
11. Mott, N.F. and Davies, E.A. (1971) *Electronic Processes in Non-Crystalline Solids*, Clarendon Press, Oxford.
12. Bässler, H. (1989) Dispersive and non-dispersive processes in Polymers, *Progress Colloid & Polymer Science* **80**, 35.
13. Eichner, A. and Bässler, H. (1987) *J. Chem. Phys.*, **112**, 285.
14. Bässler, H. (1994) Exciton and charge carrier transport in random organic solids, in Richert, R. and Blumen, A. (eds.) *Disordered Effects in Relaxational Processes*, Springer-Verlag, Berlin-Heidelberg.
15. Bässler, H. and Schweitzer, B. (1994) *Acc. Chem. Res.*, **32**, 173.
16. Mort, J. and Pfister , G. (eds.) *Electronic Properties of Polymers*, Wiley, New York, p. 169.
17. Rauchez, U. and Bässler, H. (1990) Site-selective fluorescence spectroscopy of polyvinylcarbazole, *Macromolecules*, **23**, p. 398.
18. Schein, L.B. (1992) Comparison of charge transport models in molecularly doped polymers, *Phil. Mag. B.*, **69**, p. 795.
19. Vannikov, A.V., Krybukov, A.Yu., Tyuring, A.G. and Zhuraleva, T.S. (1989) Influence of the medium on electron transport in polymer systems, *Phys. Stat. Sol. (a)*, **115**, K47.
20. Schein, L.B. and Borsenberger, P.M. (1993) Hole mobilities in a hydrazone-doped polycarbonate and poly (styrene), *J. Chem. Phys.*, **177**, p. 773.
21. Mort, J., Chen, S., Grammatica, S. and Morgan, M. (1982) Geminate recombination-controlled photogeneration in amorphous solids, *Appl. Phys. Lett.*, **40**, 980.
22. Mort, J., Morgan, M., Grammatica, S., Noolandi, J. and Hong, K.M. (1998) Time resolution of carrier photogeneration controlled by geminate recombination, *Phys. Rev. Lett.*, **48**, 1411.
23. Mort, J. and Knight, J. (1981) Localisation and electronic properties in amorphous semiconductors, *Nature*, **290**, 659.
24. Nespurek, S. (1996) Photoconductive polymers and their use in molecular devices, in F. Kajzar, V.M. Agranovich and C.Y.-C. Lee (eds.), *NATO ASI Series*, Kluwer Academic Publishers, Dordrecht, p.411.
25. Jung, J., Głowacki, J. and Ulański, J. (1999) Analysis of electric-field assisted photogeneration in poly paracyclophane doped with 2,4,7 trinitrofluorenone, *J.Chem. Phys.*, **110**, 7000.
26. Boyer, R.F. and Turley S.G. (1978) Molecular motions in polystyrene, in D.J. Meier (ed.), *Molecular Basis of Transitions*, Gordon and Breach Publ., London.
27. Jeszka, J.K., Ulański, J. and Kryszewski, M (1981) Conductive polymer: reticulate doping with charge transfer complexes, *Nature*, **289**, p. 188.
28. Ulański, J. and Kryszewski, M. (1994) Reticulate composites, in D. Bloor, R.J. Brook, M. Flemings, S. Mahajan, R.W. Cahn (eds.) *The Encyclopedia of Advanced Materials*, Elsevier, London, p. 2301.
29. Tracz, A. Ulański, J. and Kryszewski, M. (1983) Conditions for reticulated crystalline doping polymer with charge transfer complexes, *Polymer Journal*, **15**, p. 635.
30. Kryszewski, M. (1999) Nanointercalates-novel class of materials, in B. Kuchta and A. Miniewicz (eds.) *Electrical and Related Properties of Organic Solids -ERPOS-8, Synthetic Metals*, in press.

31. Zyss, J. and Chemula, D.S. (1987) *Nonlinear Optical Properties of Organic Molecules and Crystals*, Academic Press, New York.
32. Kryszewski et al., unpublished results
33. Perlstein, J. (1982) Structure and charge generation in low-dimensional organic self-assemblies, in D.A. Seanor (ed.), *Electrical Properties of Polymers*, Academic Press, New York, p. 59.
34. Wang,Y. (1987) Pyrilium Dye Nonlinear Optical Elements, *U.S. Patent 4*, 692,636 Sept 8.
35. Roduner, E. (1988) The positive muons as a probe for radical chemistry. Potential and limitations of the μ SR techniques, *Lecture notes in chemistry*, Vol. 49, Springer, Heidelberg.

# POLARIZATION GRATINGS IN DISPERSE RED 1 DOPED POLYSTYRENE

A. MINIEWICZ, S. BARTKIEWICZ and A. APOSTOLUK
*Institute of Physical and Theoretical Chemistry, Wroclaw University of Technology, Wybrzeze Wyspianskiego 27, 50-370 Wroclaw, Poland*

F. KAJZAR
*LETI (CEA - Technologies Avancées) DEIN - SPE, Groupe Composants Organiques, Saclay, F91191 Gif Sur Yvette, France*

## Abstract

Polystyrene films containing commercially available azobenzene dye Disperse Red 1 were illuminated with interference patterns of polarized laser beams with wavelengths near the maximum of chromophore absorption. Laser induced mixed amplitude and phase gratings in this system were observed by the degenerate two-wave mixing (DTWM) and the four-wave mixing (DFWM) techniques. These experiments allowed for measurement of the gratings build-up time constants and the phase conjugate (PC) signal. We determined which of the gratings is responsible for the main contribution to the PC signal in DFWM process. The volume gratings were selectively destroyed by changing the polarization of respective incoming light beams from being parallel (full intensity modulation) to orthogonal (no intensity modulation) configuration while monitoring the PC signal. Initially apparent disagreement between the measurements and the simple model proposed by us for the whole process lead us to the hypothesis of the presence, in the studied system, of the efficient polarization gratings. This finding was confirmed by a separate two-wave mixing experiment and when taken into account could reasonably fair explain the obtained in DFWM results.

## 1. Introduction

The interaction of polarized light with azo-dyes embedded in various matrices has been extensively studied in order to develop novel holographic recording materials required

*F. Kajzar and M.V. Agranovich (eds.), Multiphoton and Light Driven Multielectron Processes in Organics: New Phenomena, Materials and Applications, 299–310.*

300

for processing of optical information in the real-time regime [1-6]. Materials designed for construction of all-optical devices should reversibly change their optical properties like refractive index, birefringence or absorption coefficient under irradiation. A polymeric film containing a reversibly photoisomerisable dye molecule is a material of choice, however the proper matching of properties of polymer matrix and the photosensitive chromophore is a difficult task to achieve with respect of the expected performances. The photoresponse should be fast and a very low light scattering and low absorption are highly required for such applications as an optical pattern recognition, image intensification, construction of associative optical memories, front wave restoration, etc. [7-10].

Numerous papers has been published so far reporting on various azo dyes showing photoinduced dichroism which is a consequence of the *cis-trans* photoisomerisation with respect to the N=N double bond [11]. Disperse Red 1 chromophore (DR1, 4-N-(2-hydroxyethyl)-N-ethyl)-amino-4'-nitroazobenzene) has been widely used as a photoactive moiety in various polymeric matrices like poly(methyl metacrylate) [12,13], poly(vinyl alcohol) [1,14,15], polyimide [16] and polystyrene (PS) [17] - either as a dopant or as a side group chemically attached to the main chain. Its chemical formulae and electronic spectrum measured in polystyrene matrix at room temperature are given in Figure 1.

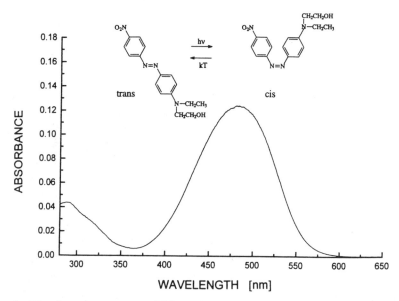

*Figure 1.* The absorption spectrum of 35 μm thick DR1 doped (1.5 wt. %) polystyrene film and isomerisation of DR1.

The presence of the ground-state *trans* form of DR1 is characterized by the absorption band centered at 490 nm. Excitation of a molecule to the first excited state by a photon of suitable energy (~2.4 eV) leads to its subsequent decay to one of the two possible ground states: stable *trans* form or metastable *cis* form. The latter, at room temperature, relaxes spontaneously to the *trans* form with the environment-dependent rate. In the *trans* form, molecules are strongly anisotropic (rod-like) while in the *cis* form they are bent and their anisotropy is less pronounced (cf. Fig. 1).

In this paper which follows the previously published one [6], we first describe the optical phase conjugation effect in the polystyrene films containing admixed DR1 with the emphasis put on the evaluation of the main grating contribution to the observed phase conjugate signal. We try to achieve this goal by selective erasure of the respective grating by changing the recording conditions while observing the PC signal. Then we develop the simple model predicting the PC response equivalent with our assumption of the major role played by the grating having the highest period. Finally, seeing the disagreement between the model and experimental results we performed two-wave mixing experiment with orthogonally polarized input beams revealing the presence of polarization gratings in DR1 doped polystyrene. Then extension of our model allowed us to properly describe the obtained results under low excitation regime.

## 2. Experimental

### 2.1. SAMPLE PREPARATION

Atactic amorphous polystyrene PS ($M_w \cong 250\ 000$) was obtained by purification of a commercial resin. Free-standing films, typically 28 to 36 μm thick, were prepared by casting from a mixture of polystyrene and DR1 dissolved in chloroform. The solution was spread over a glass plate, dried in air and then the film was stored for 24 h at 100 °C under vacuum. Then the film was removed by immersing the plate into water. The content of the dye in samples ranging from 0.5 to 2.5 wt. %. A few samples of neat polystyrene have been also prepared in a similar way. The polymer films were entirely amorphous, with the glass transition temperature $T_g \cong 363$ K as was established by X-ray diffraction patterns and differential scanning calorimetry. Refractive index n = 1.596 at $\lambda = 589.3$ nm and dn/dT = -1.42 x $10^{-4}$ $K^{-1}$ [18]. Absorption spectra of polystyrene foils containing DR1 dye were measured with a UV-VIS Shimadzu spectrometer in the spectral range 300 - 700 nm and a typical spectrum is shown in Fig. 1. The absorption maxima at $\lambda_{max} = 490$ nm scaled linearly with the dye concentration. At 532 nm the absorbance of a 35 μm thick polystyrene sample containing 2.5 % DR1 amounted to 2.20 ($\alpha_o = 629$ $cm^{-1}$).

### 2.2. OPTICAL MEASUREMENTS

In this work we have used degenerate two- as well as four-wave mixing experimental set-ups, the latter is shown schematically in Fig. 2.

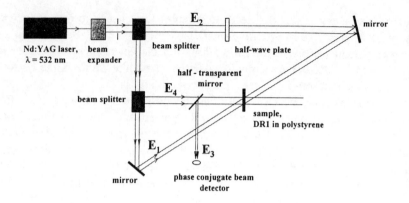

*Figure 2.* Scheme of the experimental setup for measuring degenerate four wave mixing processes. $E_i$ - light beam labeling: $E_1$ and $E_2$ - the pump beams, $E_4$ - the probe beam and $E_3$ - the phase conjugate signal beam.

A linearly polarized frequency doubled Nd:YAG continuous wave laser (Coherent Inc., model DPSS 532, 50 mW, $\lambda = 532$ nm, coherence length ~ 150 m) served as a light source. In DFWM experiment two counter-propagating pump beams with electric field amplitudes $E_1$, $E_2$ and the probe beam of amplitude $E_4$ of similar apertures (4 mm in diameter) were all supplied by the same laser source. The probe and pump beams intersected in the volume of the polymer film making in air angles $\theta_{air} = 4.5°$. All the beams could be cut off simultaneously or separately and their polarization plane could be changed by introduction of the half-wave plate into optical path. Intensities of all the beams and especially the phase conjugated $E_3$ beam were carefully measured with the Labmaster Ultima (Coherent Inc.) two-head laser power meter and were averaged in steady state conditions over 20 seconds. PC signal build-up and decay times were evaluated using a fast photodiode coupled with an A/D converting card and monitored using a simple computer program. The normal reflections from the polymer film were carefully removed and the phase conjugate signal intensities were always calculated by subtracting background intensity level. Neither self-diffraction nor generation of phase conjugated signal were observed for pure polystyrene films. The experiments were usually performed at ambient temperature on freestanding foils.

Typical configurations and coordinate system used throughout this work for a degenerate two-wave mixing are given in Fig. 3a.

As is shown in Fig. 2, depicting DFWM experiment, the interacting beams 1, 2,

3 and 4 are of the same angular frequency ω and basically of the same polarization (s-polarization was employed for the basic studies). The electric field amplitudes $E_j(r)$ associated with a j-th beam in a DFWM experiment are assumed to be

$$E_j(r,t) = E_j(r) \exp[i(k_j r - \omega t)] + \text{c.c.} \tag{1}$$

Here $k_j$ is the j-th light wavevector and the propagation directions satisfy the relations $k_1 + k_2 = 0$ and $k_3 + k_4 = 0$. In general (cf. Fig. 3b) the interference between the beams gives rise to set of six spatial light intensity modulations in the polymer bulk. Two sets, between $E_1$ and $E_4$ and between $E_2$ and $E_3$, form transmission gratings, while the remaining four form reflection gratings.

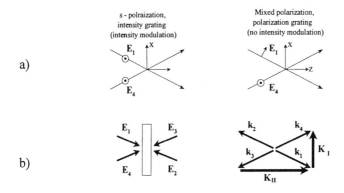

a)

b)

*Figure 3.* Degenerate two-wave mixing schemes showing two configurations for s- and mixed polarization grating recording (a); Labeling scheme for DFWM experiment and directions of the wave vectors of light ($k_i$) and the principal spatial grating wave vector ($K_i$) (b).

The wave vectors of these spatial gratings $K_\Gamma = 2\pi/\Lambda_\Gamma$ are known from geometrical considerations of interacting light wave vectors $k_i$ ($|k_i| = n\, 2\pi/\lambda$ where $\lambda$ is the wavelength in free space) and can be written in the form:

$$K_I \equiv k_4 - k_1 = k_2 - k_3 \quad , \qquad K_{II} \equiv k_1 - k_3 = k_4 - k_2 \tag{2}$$
$$K_{III} \equiv k_1 - k_2 \quad , \qquad K_{IV} \equiv k_4 - k_3$$

Light interference patterns produce in the material mixed gratings in which the complex index of refraction $\underline{n}$ is modified. The latter parameter is defined here as $\underline{n} = n - i\kappa$, where $n$ is its real part and its imaginary part $\kappa$ is related to the absorption coefficient $\alpha = 4\pi\kappa/\lambda$.

## 3. Results and discussion

### 3.1. IN RESONANCE S-POLARIZATION DEGENERATE FOUR-WAVE MIXING PROCESS IN POLYSTYRENE FILMS CONTAINING DR1

For a single grating one can assume that the complex refraction index of a polymer in the steady state illumination is spatially modulated according to local intensity changes

$$\underline{n}(\mathbf{r})_\Gamma = \underline{n}_o + \underline{n}_\Gamma \cos(\mathbf{K}_\Gamma \cdot \mathbf{r}), \tag{3}$$

where $\underline{n}_\Gamma$ is the amplitude of the spatial modulation of the complex index of refraction in a grating with a wavevector $\mathbf{K}_\Gamma$, and $\underline{n}_o$ is the average value of the complex index. When a four-wave interaction is considered, the resulting light-induced complex index of refraction of the medium can be written as consisting of four Fourier components [7]

$$
\begin{aligned}
\underline{n} = \underline{n}_o &+ \frac{n_1 e^{i\Phi_I}}{2} \frac{(E_1^* E_4 + E_2 E_3^*)}{I_o} e^{iK_I \cdot r} + \frac{n_{II} e^{i\Phi_{II}}}{2} \frac{(E_1 E_3^* + E_2^* E_4)}{I_o} e^{iK_{II} \cdot r} \\
&+ \frac{n_{III} e^{i\Phi_{III}}}{2} \frac{E_1 E_2^*}{I_o} e^{iK_{III} \cdot r} + \frac{n_{IV} e^{i\Phi_{IV}}}{2} \frac{E_3^* E_4}{I_o} e^{iK_{IV} \cdot r} + c.c
\end{aligned}
\tag{4}
$$

where $I_o \equiv \sum_{j=1}^{4} I_j$, $I_j = |E_j E_j^*|$, $\Phi_\Gamma$ is the phase shift between the index or absorption modulation and the intensity modulation. Within the model of a real-part refractive index modulation under low-excitation regime one may assume that the response in polystyrene films containing DR1 is local [13] and in the first approximation one may put $\Phi_\Gamma = 0$ for any $\Gamma$. Assuming weak phase conjugate signal i.e. $I_3 \ll I_1, I_2, I_4$ (this was proved in the experiment) all terms with $E_3^*$ can be neglected in Eq. 4. Further we assume that one can treat the four-wave mixing in this system as being analogous to real-time holography in the sense that the hologram written in the sample by the input reference beam $I_1$ and the signal beam $I_4$ ($\mathbf{K}_I$ grating) is reconstructed by the reference beam $I_2$ propagating in the direction opposite to that of $I_1$. So we neglect other possible (e.g. reflective type) contributions to the PC signal. As result the PC signal $E_3$ which is proportional to the complex conjugate of the incident field $E_4(0)$ originates from the $E_2$ wave diffraction on the $\mathbf{K}_I$ grating. Under these assumptions the PC reflectivity (defined as $R_c = I_3 / I_4$) is proportional to the square of the complex $\mathbf{K}_I$ grating amplitude $\underline{n}_1^2$ established in the sample. The maximum PC reflectivity of 0.1 % has been measured for a polystyrene film containing 2.5 % of DR1. The PC signal was exactly counter-propagating to the input laser beam $I_4$.

In Fig. 4 we show typical PC signal build-ups and decays observed in polystyrene doped with DR1 [6] under conditions when a steady-state PC signal was disturbed by sequential shutting and opening of the three beams: $I_1 = 7$ mW, $I_2 = 14$ mW and $I_4 = 6$ mW, respectively.

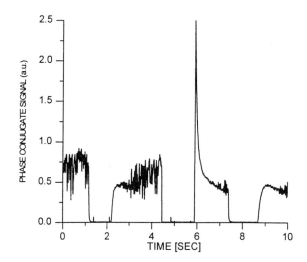

*Figure 4.* Time evolution of PC signal ($I_3$) in DR1 doped polystyrene film as measured in DFWM experiment with sequentially blocked beams: $I_1$, $I_2$ and $I_4$, respectively.

An detailed analysis of resulting changes has been given by us in the preceding paper [6]. Here we would like to note that the rise and decay time constants ($\tau_R$ and $\tau_D$, respectively) are within millisecond regime: $\tau_R \cong 100$ ms and $\tau_D \cong 20$ ms. It can be noticed that the build-up times of PC signal are usually much longer than the respective decay times. The latter ones are connected with natural grating destruction mechanisms and are dependent on a light assisted *cis-trans* conversion rate and thermal relaxation time. The decay time constant of thermal gratings is governed by thermal diffusion decay time [10] given by the formula:

$$\tau_{thD} = \frac{\gamma C_p \Lambda^2}{4\pi^2 \kappa} \tag{5}$$

where $\gamma$ is the density, $C_p$ is the specific heat and $\kappa$ is the thermal conductivity of the polymer matrix. Taking, $\Lambda = 4.5 \times 10^{-6}$ m and typical values for atactic polystyrene [19] $\gamma = 1050$ kgm$^{-3}$, $C_p = 1.21$ kJ kg$^{-1}$K$^{-1}$ at 20 °C , $\kappa = 1.1$ Wm$^{-1}$K$^{-1}$ at 20 °C, one obtains $\tau_{thD} \cong 6$ μs. This means that the thermal gratings, if present, should decay within a few useconds.

Of interest is the sharp spike of a PC signal (cf. Fig. 4) observed after opening the $I_2$ beam. Its appearance points out to the fact that the $K_I$ grating amplitude in steady state conditions with all the beams present is lower than that when $I_2$ is off. Then just after opening of the $I_2$ beam we observe diffraction on the existing strong $K_I$ grating followed by fast reduction of its amplitude in the presence of $I_2$. From the experiment shown in Fig. 4 we concluded that the main contribution to the PC signal comes from the $I_2$ diffraction on the $K_I$ grating. This can be rationalized if one takes into account fairly strong light absorption in the sample. Moreover, it is obvious that the $K_{II}$ grating formed by the interference of $I_2$ and $I_4$ beams has the smaller amplitude than the $K_I$ grating which is dominant near the surface. Also smaller grating period of $\Lambda_{II} = 0.2$ μm ($\Lambda_I = 4.5$ μm) make that this grating is more efficiently erased by vibrations and thermal relaxation (thermal diffusion decay time is about 500 times faster for $K_{II}$ grating than for $K_I$ grating).

## 3.2. IN RESONANCE MIXED-POLARIZATION DEGENERATE FOUR-WAVE MIXING PROCESS IN POLYSTYRENE FILMS CONTAINING DR1

In order to prove experimentally the hypothesis of the dominant role of the $K_I$ grating in the generation of PC wave we used a half-wave plate and inserted it into path of the $I_1$ and $I_4$. By turning this plate in the range of 0 - 45° we were able to change the polarization of one of the beams from being parallel to orthogonal to each other. As it is well known for the s-polarization of the $K_I$ grating forming beams (polarization of the beams $I_1$ and $I_4$ is horizontal with respect to the incidence plane xz) there is an intensity modulation and the interference tensor $\Delta M$ degenerates into the one-element tensor [10]:

$$\Delta M = \begin{bmatrix} 0 & 0 & 0 \\ 0 & \Delta I & 0 \\ 0 & 0 & 0 \end{bmatrix} \tag{6}$$

where $\Delta I = \sqrt{I_1 \cdot I_4}$ .

Assuming that (i) DR1 doped polystyrene forms an intensity dependent grating e.g. via mechanism of selective light absorption along the long molecular axis of the chromophore and (ii) that the dominant contribution to the PC signal comes from the diffraction of the $I_2$ beam on $K_I$ grating one can calculate the expected PC response according to the formula:

$$I_3 \propto I_2 \cdot \underline{n}_I{}^2 \propto I_2 \cdot m_I^2 \tag{7}$$

where $\underline{n}_I$ is the saturated value of $K_I$ grating amplitude recorded in the sample due to interference of two beams $I_1$ and $I_4$ having parallel polarizations. However, assuming a linear material response this amplitude will be a function of light modulation factor m, given by [10]:

$$m_I = \frac{\sqrt{I_1 \cos^2\alpha \cdot I_4}}{I_1 \cos^2\alpha + I_4} \tag{8}$$

where $\alpha$ is the angle between polarization planes of $\mathbf{E}_1$ and $\mathbf{E}_4$ and half-wave plate is placed on the beam $I_1$. The expected dependence $I_3(\alpha)$ calculated according to our assumptions is shown qualitatively in the Figure 5a. The quantitative calculations are far more complicated as

they require a knowledge about the diffraction process and mechanism of grating recording in the medium. In the Figure 5b we show the experimentally obtained curves $_3(\alpha)$ for half-wave plate placed on beam $I_1$ and turned from 0° to 45°. Qualitatively similar curve was observed when a half-wave plate was placed on $I_4$ beam.

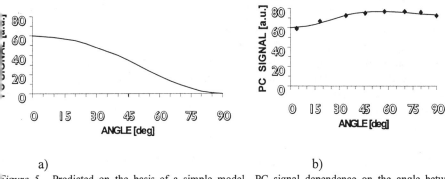

a)                                                                        b)

*Figure 5.* Predicted on the basis of a simple model PC signal dependence on the angle between polarization planes of two incoming light beams $I_1$ and $I_4$ (a). Experimentally observed dependence in DR1 doped polystyrene, a half-wave plate is placed on the $I_1$ beam.

An obvious disagreement between the predicted and observed results pressed us to perform a two-wave mixing experiment under conditions of orthogonal beam polarizations. In the case of mixed polarization for example when $E_1 \| y$ and $E_4 \perp y$ no intensity modulation occurs for any value of the angle between the incident beams. The interference tensor is [10]:

$$\Delta M = \tfrac{1}{2} c\varepsilon_0 \begin{bmatrix} 0 & 0 & 0 \\ E_1 E_{4x} & 0 & E_1 E_{4z} \\ 0 & 0 & 0 \end{bmatrix} \tag{9}$$

and only the polarization sensitive optical medium is able to record such a type of grating. In DR1 doped polystyrene we observed quite efficient self-diffraction in mixed configuration i.e. when ($E_1 \| y$ and $E_4 \perp y$) and when ($E_4 \| y$ and $E_1 \perp y$). The first order

diffraction efficiency in that case was even about 10 - 20 % higher than in the case o
$(E_1 \| y$ and $E_4 \| y)$. One can suspect that because in the former case there is no therma
grating present then there is also no internal stress or strain trying to diminish th
amplitude of light induced grating.

Now, with the knowledge taken from the described above two-wave mixing
experiment we could complete our model and take into account also the contribution to
the PC signal coming form the $I_2$ diffraction on the polarization grating. Assuming, in
the first approach, that these processes are independent of each other (then a simple
addition of diffracted intensities can be used) we arrive to the following relation:

$$I_3 \propto I_2 \cdot \left( \underline{n}_I^{\,2} + \sigma \cdot \underline{n}_{I,mixed}^{\,2} \right)$$ (10)

where the $\underline{n}_{I,mixed}$ is the saturated value of $K_I$ grating amplitude recorded in the sample
due to interference of two beams $I_1$ and $I_4$ having orthogonal polarizations. The
parameter $\sigma$ corresponds to weighting factor and depends on efficiency of ligh
scattering on these two types of gratings recorded in different configurations o
incoming recording beams having either parallel or orthogonal light polarizations.

In Figure 6 we present results of fitting, according to the formula (10), of the
experimentally obtained curves $I_3(\alpha)$. As one can see the fitting is reasonable with the
parameter $\sigma = 1.10$.

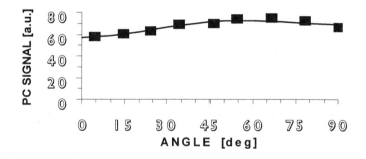

*Figure 6.* Results of measurements of phase conjugate signal $I_3$ in function of angle $\alpha$ between
polarization planes of incoming $K_I$ writing beams ( points) and fitting of the experimental data according
to the equation 10, a half-wave plate is placed on the $I_4$ beam.

There are two main mechanisms which should be considered in the discussion of
the origin of the optical phase conjugation in polystyrene doped with DR1: scalar
(intensity -dependent) and vector (polarization-dependent) formation of gratings
Angularly selective photoexcitation and molecular reorientation may both be
responsible for polarization grating recording in the studied system. The photobleaching

or chromophore diffusion may be responsible for intensity gratings. We believe that both processes are of importance in DR1 doped polystyrene.

## 4. Conclusions

We observed the optical phase conjugation effect in DR1-containing polystyrene films. The underlying optical nonlinearity is basically resonant being connected with the process of light absorption by the dye molecules. The light induced intensity-dependent gratings are due to DR1 *trans-cis* photoisomerisation and the electric field-dependent gratings are due to molecular reorientation and angularly selective photoexcitation. The dynamics of the mixed (refractive index and absorption) grating formation is relatively fast and characteristic exponential build-up time constant of the order of 100 ms and erasing times of the order of 20 ms were measured at low intensity illumination conditions ( $I < 200$ mW/cm$^2$). The phase conjugate (PC) reflectivities R$_c$ reached 0.1 % in 35 μm thick polystyrene films containing 2.5 wt. % of the dye. Polarization sensitive hologram recording in DR1 polystyrene make this material a promising candidate for applications in integrated optical devices.

ACKNOWLEDGMENTS

We thank José A. Giacometti and Mauro M. Costa of Departamento de Física e Ciencia dos Materiais, Instituto de Física de São Carlos, Universidade de São Paulo, São Carlos, Brazil for supplying us with the polymer samples.

This work was sponsored by the Wroclaw University of Technology and partially by the French-Polish program Polonium (A.M, S.B, F.K.).

## REFERENCES
1. Todorov, T., Nikolova, L. and Tomova, N. (1984) *Appl. Optics* **23**, 4309.
2. Sekkat, Z. and Dumont, M. (1992) *Appl. Phys.* B **54**, 486.
3. Charra, F., Kajzar, F., Nunzi, J.M., Raimond, P. and Idiart, E. (1993) *Opt. Lett.* **18**, 941.
4. Kim, D.Y., Li, L., Kumar, J. and Tripathy, S.K. (1995) *Appl. Phys. Lett.* **66**, 1166.
5. Pham, V.P., Galstyan, T., Granger, A. and Lessard, R.A. (1997) *Jpn. J. Appl. Phys.* **36**, 429.
6. Miniewicz, A., Bartkiewicz, S., Sworakowski, J., Giacometti, J. A. and Costa, M.M., (1998) *Pure Appl. Opt.*, **7**, 709.
7. Günter, P. and Huignard, J-P, (eds) (1989) *Photorefractive Materials and Their Applications II*, Topics in Applied Physics, vol.62, Springer Verlag, Berlin.
8. Yeh, P. (1993) *Introduction to Photorefractive Nonlinear Optics,* J. Wiley, New York.

9. Feinberg, J. (1982) *Opt. Lett.,* **7**, 486.

10. Eichler, H. J., Günter, P. and Pohl, D. W. (1986) *Laser-Induced Dynamic Gratings,* Springer Verlag, Berlin.

11. Rau, H. in *Photochemistry and Photophysics,* (1990), vol. II, Chap. 4, Ed. J. F. Rabek, CRC Press Inc. (and references therein).

12. Sekkat, Z., Morichere, D., Dumont, M., Loucif-Saibi, R. and Delaire, J. A. (1991) *J. Appl. Phys.* **71**, 1543.

13. Loucif-Saibi, R., Nakatani, K., Delaire, J. A., Dumont, M. and Sekkat, Z. (1993) *Chem. Mater.* **5**, 229.

14. Solano, C., Lessard, R. A. and Roberge, P. (1987) *Appl. Optics* **26**, 1989.

15. Lessard, R. and Couture, J. J. (1990) *Mol. Cryst. Liq. Cryst.* **183**, 451.

16. Jeng, R. J., Chen, Y. M., Jain, A. K., Kumar, J. and Tripathy, S. K. (1992) *Chem. Mater.* **4**, 1141.

17. Hampsch, H. L., Yang, J., Wong, G. K. and Torkelson, J. M. (1990) *Macromolecules* **23**, 3648.

18. Piorkowska, E. and Galeski. A., (1986) *J. Appl. Phys.* **60**, 493.

19. Brandrup, J. and Immergut, E. H. (Eds) (1989) *Polymer Handbook* J. Wiley & Sons, New York.

# MODELING PTCDA SPECTRA AND POLYMER EXCITATIONS

Z.G. Soos and M.H. Hennessy
Department of Chemistry
Princeton University, Princeton N.J. 08544

**Abstract:** The spectra of PTCDA crystals and multiple quantum wells are modeled using Frenkel and charge-transfer excitons in molecular stacks with vibronic coupling to a local mode. They correspond to intra and interchain excitations, respectively, of conjugated polymers with precisely defined chromophores and contacts. Mixed Frenkel-CT vibronics with vanishing dispersion at k = 0 lead to a dimer that accounts for absorption and electroabsorption, while fluorescence indicates a delocalized Frenkel exciton with a band width of 0.28 eV. In contrast to polymer films, the model parameters for PTCDA are largely fixed by structural, molecular and solution data.

## 1. Introduction

Conjugated polymers share important aspects of molecules, inorganic semiconductors and flexible chains. Their electronic structure, including nonlinear optical (NLO) spectra and photophysics, are due to delocalized π-electrons and offer promising applications for optoelectronics and light-emitting diodes (LEDs). Recent reviews [1,2] summarize contrasting approaches to electronic processes in polymers. Their conformational degrees of freedom require extensions of molecular or crystal states, and interchain processes depend sensitively on the local arrangement of chains.

We develop in this paper two connections between electronic excitations of conjugated polymers and other materials. First, since large molecules are almost as delocalized as polymers, their π-π* spectra are similar. The 2.2 eV optical gap of perylenetetracarbo-xylic dianhydride (PTCDA, Fig. 1) is comparable to poly-*para*-phenylene-vinylene and substituted PPVs. Ordered PTCDA films make possible the extensive spectroscopic studies reviewed by Forrest [3]. Second, charge-transfer (CT) states of molecular crystals correspond to interchain polaron pairs [4] in polymers and resolved PTCDA spectra are ideal for demonstrating CT contributions [5]. Molecular crystals have van der Waals contacts in three dimensions rather than in two for extended strands. An effective conjugation length and backbone structure are the basis for invoking either bands and semiconductors or molecules and disorder. Organic crystals such as PTCDA combine molecular features and exciton bands, while multiple quantum wells [3] contain stacks as short as three PTCDA molecules.

*F. Kajzar and M.V. Agranovich (eds.), Multiphoton and Light Driven Multielectron Processes in Organics: New Phenomena, Materials and Applications, 311–323.*
© 2000 *Kluwer Academic Publishers. Printed in the Netherlands.*

With the notable exception of polydiacetylene (PDA, Fig. 1) single crystals, conjugated polymers are amorphous and interchain interactions depend on chain conformation and preparation conditions. The optimization of polymers used in LEDs will clarify such contributions. Extended planar chains may resemble semiconductors, as emphasized by Heeger for stretch-oriented polymers [6], but structural perfection has not been achieved.

The broad absorption and narrow emission of polymer films indicate a distribution of chromophores, or segment lengths, and rapid (Förster) energy transfer to the longest segments [7]. Such qualitative ideas predate current NLO or LED studies. The relevant conjugated fragments contain roughly 20 double bonds or 5 phenyls. The spectra show vibronic structure, notably a 0.18 eV mode associated with the effective conjugation coordinate [8,9]. This well-characterized mode is an out-of-phase C=C and C–C stretch that modulates the Hückel gap [10]. In contrast to wide band semiconductors, the lowest triplet, singlet and charge-carrying excitations of conjugated polymers are different, as in molecules. Centrosymmetric backbones have separate thresholds [11] for $A_g$ and $B_u$ singlets, with $S_1 = 1B_u$ in fluorescent polymers and $S_1 = 2A_g$ otherwise. The $1B_u/2A_g$ crossover is closely linked [9] to alternating transfer integrals $t(1\pm\delta)$ along the backbone or to effective $\delta$ due to phenyls or thiophenes.

## 2. Molecular and Extended π-Electronic States

### 2.1. EVEN AND ODD π-ORBITALS FOR POLYCYCLIC SYSTEMS

Hückel theory is the starting point for π-systems. We introduce $t(R)$ between adjacent carbons in Fig. 1, with t for intermediate benzene bonds and $t(1\pm\delta)$ for partial double

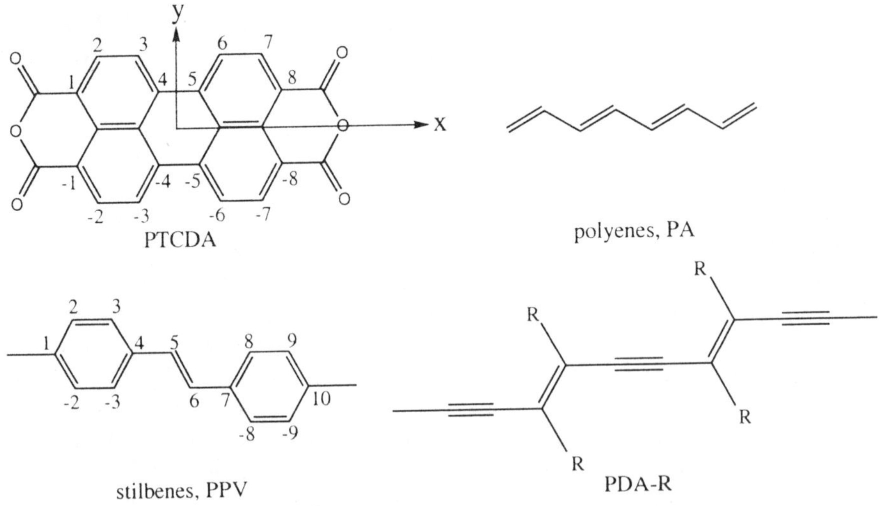

Fig. 1. Schematic representation of conjugated systems with side-groups R.

and single bonds. The Su-Schrieffer-Heeger (SSH) model [12] for polyacetylene (PA) emphasizes linear electron-phonon (e-ph) coupling $\alpha = (dt/dR)_0$. Raman and infrared spectroscopy are excellent probes of $\pi$-electron fluctuations [9] and e-ph contributions to NLO coefficients is a current topic [13]. For noninteracting $\pi$-electrons, the simple transformation at rings in Fig. 1

$$e_{n\sigma}^+, o_{n\sigma}^+ = (a_{n\sigma}^+ \pm a_{-n\sigma}^+)/\sqrt{2} \tag{1}$$

generates even and odd orbitals, respectively, with respect to reflection. We obtain [11] localized odd MOs with energy $\pm t$ at each phenyl and extended even MOs with $t\sqrt{2}$ at all bridgeheads in PPV. Extended sites (1) explicitly relate $\pi$-MOs of polycyclic polymers to linear chains with alternating t's.

Perylene, the PTCDA $\pi$-system, has two *peri*-linked naphthalenes. Its odd $\pi$-MOs have nodes in the xz plane and map exactly [14] into the Hückel model for octatetraene with extended sites $\pm n$. Odd MOs of longer rylenes map into longer polyenes and their intense $\pi$-$\pi^*$ excitation to $1B_u$ is in the odd manifold. This explains coupling to the 0.18 eV vibration featured in polyenes. The smaller PTCDA displacement discussed below is consistent with delocalization on two strands.

The Pariser-Parr-Pople (PPP) model describes $\pi$-electronic excitations and is necessary for even-parity states [15,9]. It extends Hückel theory by adding on-site Coulomb interactions U and intersite interactions $V_{nm}$. Conjugated hydrocarbons are sufficiently similar to have transferable parameters that allow PPP predictions of excitation energies and transition moments in conjugated polymers, including $\sigma$-conjugated polysilanes [7]. By contrast, the diversity of solid-state systems treated by Hubbard and extended Hubbard models precludes transferable parameters.

Octatetraene, longer polyenes and PA have $2^1A_g$ below $1^1B_u$, in accord with exact PPP results. This correlation effect is striking because it reverses the order given by one-electron theory, even at the Hartree-Fock level. Stronger effective alternation due to $t\sqrt{2}$ at bridgeheads places $1^1B_u$ below $2^1A_g$ in *trans*-stilbene and PPV. The same PPP ordering is found for extended sites [14] with reduced U in rylenes, whose intense $1^1B_u$ is ~1 eV lower than the corresponding polyene. Accordingly, perylene and PTCDA have high quantum yield for fluorescence in solution.

## 2.2. FRENKEL AND CT EXCITONS IN PTCDA STACKS

The connection between PTCDA crystals and polymer films can now be stated precisely. Instead of poorly characterized conjugation lengths, each PTCDA is a chromophore that corresponds to an *identical* segment. $1^1B_u$ is the M* excitation in Fig. 2. We model [5] it as a harmonic oscillator with $\hbar\omega = 0.18$ eV and displacement g from the ground state M. As in polymers, the $\pi$-$\pi^*$ transition dominates the absorption and is polarized along the chain, the long (x) molecular axis.

PTCDA crystals contain [3] face-to-face stacks that are nearly along **z** in Fig. 1. The interplanar spacing R = 3.38Å is uniform and below van der Waals contact, while all interstack contacts are normal. All "interchain" interactions are equal by symmetry for adjacent PTCDAs. An ion pair ..MM⁺M⁻M.. is a CT state that, like polaron pairs, has spin-1/2 partners. Its energy in Fig. 2 is 2Δ from M*; its displacements are $g_+$ and $g_-$ along the same coupled mode. We will use PTCDA spectra to evaluate 2Δ directly. The relative energy of excitons and P⁺P⁻ pairs in PPV and related polymers is much less accessible: a segment distribution implies a range of exciton energies and additional assumptions are needed for the size and separation of P⁺ and P⁻.

PTCDA is an excellent and highly anisotropic hole conductor, consistent with a one-dimensional structure. As sketched in the middle column of Fig. 2, the hopping integral $J = \langle M^*M|H|MM^*\rangle$ generates a Frenkel-exciton band along the stack (z) axis. Since π-π* is polarized normal to the stack, the dipole-allowed k = 0 transition is at the top and k = π is at the bottom. An exciton band immediately rationalizes the weak, red-shifted fluorescence of PTCDA crystals and films [16].

The mixing of Frenkel and CT excitations is a general excited-state process for finite intermolecular or interchain overlap. The relevant matrix elements are [4]

$$t_e\sqrt{2} \;=\; \left\langle M_p M_{p+1}^* \left| H \right| M_p^+ M_{p+1}^- \right\rangle \quad and \quad t_h\sqrt{2} \;=\; \left\langle M_p^* M_{p+1} \left| H \right| M_p^+ M_{p+1}^- \right\rangle , \qquad (2)$$

where $|M^+M^-\rangle$ is the singlet linear combination of radical ions. Such integrals cannot be evaluated accurately. They are approximated in semiempirical theory as proportional to LUMO-LUMO and HOMO-HOMO overlaps, respectively. Translational symmetry for the infinite stack leads to states $|Fr,k\rangle$ and $|CT,k,\pm\rangle$ with wavevector $-\pi < k \leq \pi$ in

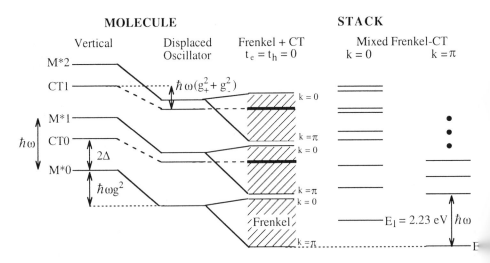

Fig 2. Evolution of molecular states in a PTCDA stack with Frenkel and CT excitons.

the first Brillouin zone. Frenkel excitons are linear combination of ..MM*M.. with one excited molecule, while the CT excitons have adjacent $M^+M^-$ or $M^-M^+$ along the stack. The matrix element for Frenkel-CT mixing is [5]

$$\langle Fr, k' | H | CT, k, \pm \rangle = \delta_{kk'} (t_e \pm t_h)[\exp(ik/2) \pm \exp(-ik/2)] \qquad (3)$$

The k' = k restriction is standard; k-dependent mixing is more interesting. Only the even combination mixes at k = 0 and does so as $t_e + t_h$. At $k = \pi$ the mixing is $t_e - t_h$ for the minus state. Semiempirical (AM1) calculation of overlaps at the crystal structure gives comparable $t_e$, $t_h$ with the same sign. Mixing is strong for absorption at k = 0 and weak for emission from $k = \pi$. As sketched in Fig. 2, we model emission entirely in terms of Frenkel excitons, without CT. Mixed vibronics at k = 0 are treated numerically in terms of the exciton-phonon-CT dimer defined in the next Section.

## 3. PTCDA Spectra: Frenkel-phonon-CT dimer and Holstein model

PTCDA absorption and fluorescence in $CH_2Cl_2$ solution [17] are shown in Fig. 3a. At this resolution, the spectra are consistent with a displaced harmonic potential for M* with $\hbar\omega = 0.18$ eV. The relative intensities of 0-n vibronics give the displacement g = 0.84. As in polyenes [18], the emission has the same g and somewhat smaller vibrational spacing. The 70 meV Stokes shift between the 0-0 lines represents the collective contribution of other molecular vibrations.

The perturbation of *molecular* spectra in Fig. 3a by the formation of stacks with R = 3.38 Å is a textbook application of molecular exciton theory. Single-crystal absorption [5] shows the entire 2-3 eV band in Fig. 3b is polarized in the PTCDA plane and thus associated with the molecular $\pi$-$\pi^*$ transition. We summarize below the relevant models and collect parameters in Table 1. All simulations are relative to $E_1 = 2.23$ eV, the $1^1B_u$ energy in solution. The PPP excitation for perylene is 3.07 eV, slightly above the observed 0-0 of 2.86 eV. Molecular excitations are adjustable in exciton theory, which deals instead with solid-states shifts and splittings.

Table 1. Parameters for the spectra of PTCDA stacks.

| Parameter | Role | Source or related quantity |
|---|---|---|
| $\hbar\omega = 0.18$ eV | vibrational quantum | solution and polyenes |
| g = 0.84 | excited-state displacement | 0.84 in $CH_2Cl_2$ |
| J = 204 meV | exciton hopping | transition dipoles, 220 meV |
| $\Delta$ = 121 meV | CT position | adjustable, model |
| $|t_e + t_h|/2 = 62$ meV | Frenkel-CT mixing | adjustable, model |
| $(g_+ + g_-)/2 = 0.76$ | ionic displacement | adjustable, relaxation energies |
| $E_1 = 2.23$ eV | absorption origin | free |
| $\delta_S = 70$ meV | Stokes shift | solution, empirical |

$J = \langle M^*M|H|MM^* \rangle$ generates a band of Frenkel excitons and uniform spacing ensures a single J. The integral is related to the transition dipole, $\mu_{1B} = \langle M^*|\mu|M \rangle$, that is accurately given in PPP theory; J is also related to Förster transfer at large separation R and does not require intermolecular overlap. PPP gives [19] J = 220 meV for extended sites in perylene at R = 3.38 Å.

## 3.1. FLUORESCENCE IN FILMS AND MULTIPLE QUANTUM WELLS

The three parameters J, g, and $\hbar\omega$ define the Holstein model [20]

$$H/\hbar\omega = \sum_p J(a_p^+ a_{p+1} + a_{p+1}^+ a_p) + g a_p^+ a_p (b_p^+ + b_p) + \sum_n b_n^+ b_n , \qquad (4)$$

where $a_p^+$ generates M* at site p and the bosons $b_n^+$ create a vibrational quantum at site n. Accurate *independent* estimates of $\hbar\omega$, g and J set PTCDA apart from many other Holstein systems. Our best current values in Table 1 are the expected ones. We

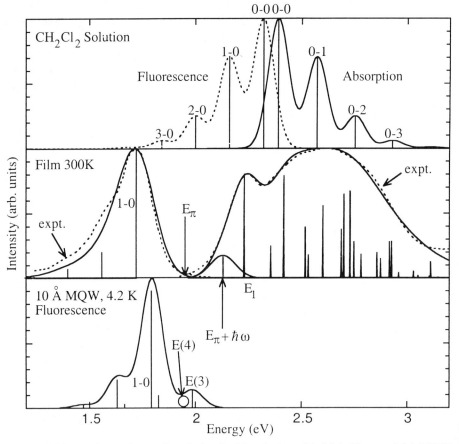

Fig. 3. PTCDA absorption and emission in (a) solution, (b) thick films and (c) MQWs.

have a localized M* for J = 0 and a delocalized band for g = 0. Approximate solutions of (4) for a single electronic excitation and arbitrary e-ph coupling g have been extensively discussed in connection with exciton or polaron transport. We use a variational solution [19] for the infinite chain and both direct and variational solutions for finite stacks in multiple quantum wells.

The film emission in Fig. 3b is based on Table 1 and reproduces the large red shift observed in PTCDA films [16]. The bottom of the band at $E_\pi = 1.95$ eV is dipole forbidden; 1-0 emission at 1.71 eV becomes allowed [19] on creating a $q = \pi$ phonon and 2-0 requires two phonons with total $q = \pi$. The stick spectra in Figs. 3a and b show the n-0 intensity changes on going from a PTCDA molecule to a stack with J = 204 meV. Figure 3c simulates the emission of 10 Å layers (stacks of three PTCDA) in a multiple quantum well (MQW) at 4.2 K. The even-odd symmetry of finite stacks makes 0-0 weakly allowed for an odd number of PTCDA and forbidden for an even number. The simulation is a 1:5 combination of N = 3 and 4 stacks. This incorporates roughly the expected energy transfer to long segments seen in conjugated polymers. The blue shift of short stacks agrees with experiment, but the observed [3] 2-0 intensity is 2-3 times higher than calculated.

Absorption at $E_\pi + \hbar\omega = 2.13$ eV is dipole allowed through a $q = \pi$ phonon and appears in the red edge of Fig. 3b. Strong support for this interpretation is provided by the 20-fold increase of the fluorescence quantum yield for 2.10-2.15 eV photons [16]. Direct absorption of a vibronic in the $\pi$ manifold bypasses any relaxation or exciton scattering following k = 0 absorption at the top of the band.

## 3.2. ABSORPTION AND ELECTROABSORPTION

The asymmetric dispersion of Frenkel bands in Fig. 2 follows from direct solutions of (4) with finite N and cyclic boundary conditions [21]. The physical interpretation is that, since J > 0 is antibonding, k = 0 is lowered by out-of-phase displacements of molecules adjacent to M*. Smaller Franck-Condon overlap reduces electronic energy at the expense of greater strain. The actual balance depends on the wavevector k as well as J, g, and $\hbar\omega$. Reduced dispersion at k = 0 implies a large effective mass for the Frenkel exciton. Frenkel-CT mixing with a narrow CT band further increases the $m_{eff}$ and vanishing dispersion at k = 0 amounts to a dimer approximation [5].

With M* in Fig. 2 as the reference energy, we consider linear coupling to both CT and molecular excitations at two sites 1,2. There are four electronic states, all singlets, $|r\rangle =$ $|M^*M\rangle$, $|MM^*\rangle$, $|M^+M^-\rangle$, and $|M^-M^+\rangle$, and two harmonic oscillators with displacements 0, g, $g_+$ or $g_-$. The exciton-phonon-CT dimer is [5]

$$H_d / \hbar\omega = \sum_{rs} h_{rs} a_r^+ a_s + \sum_r a_r^+ a_r \Lambda_r + b_1^+ b_1 + b_2^+ b_2$$

$$\Lambda_r = g_{1r}(b_1^+ + b_1) + g_{2r}(b_2^+ + b_2)$$

(5)

with $h_{11} = h_{22} = 0$ for M*, $h_{33} = h_{44} = 2\Delta$ for CT, $h_{12} = J$, $h_{34} = 0$, mixing terms $h_{13} = h_{24} = t_h\sqrt{2}$ and $t_{23} = t_{14} = t_e\sqrt{2}$, and e-ph coupling $g_{1r}, g_{2r} = (g,0), (0,g)$ for $r = 1,2$ and $(g_+,g_-), (g_-,g_+)$ for $r = 3,4$. The complete basis increases as $4\Sigma_p(p+1)$ on including states up to $p_{max}$ vibrational quanta; $p_{max} = 10$ gives converged energies and transition moments for the parameters in Table 1.

The dimer absorption in Fig. 3b closely follows experiment. The stick spectrum gives the Frenkel component, since all intensity is due to the $\pi$-$\pi^*$ transition. Frenkel-CT mixing completely alters the regular vibrational spacing found in solution. The absorption is considerably broader in films, as expected for increased displacements. It is sharper at 4.2 K, where the close correspondence [3] between 500 Å films and 10 Å MQWs supports our neglect of dispersion at $k = 0$. Additional support for the dimer approximation is the <10 meV shift at $k = 0$ found in direct solution [22] of a 4-site version of (5) with cyclic boundary conditions and Table 1.

In the electroabsorption (EA) spectrum, comparable Stark shifts are found for applied E field in or normal to the PTCDA plane [23]. The normal spectrum is due to CT, since large $\pi$-electron polarizability is confined to the molecular plane. The sensitivity of EA and the large dipoles of ion pairs allow the detection of weak CT transitions in organic crystals against the strong background of Frenkel excitons and their vibronics. To model EA, we add $\pm eRE$ to the CT energies $h_{33}$ and $h_{44}$ in (5), solve as before, and plot the difference spectrum. The calculated [5] Stark shift of $E_1$ of an isolated dimer indicates a CT polarizability increase of 150 Å$^3$ over the ground state. Quantitative comparisons to the actual $\Delta\alpha \sim 300$ Å$^3$ require improved analysis of local fields and dielectric contributions. Such general questions apply to all microscopic models.

A dielectric continuum is used for Wannier excitons in inorganic semiconductors, where hydrogenic 1s radii are large compared to lattice spacing. The isotropic Stark shift is proportional to $\varepsilon$. In the discrete limit, adjacent ions are not shielded and the Stark shift scales as $\varepsilon^2$, the square of the effective field $\varepsilon E$. The dielectric constant of PTCDA is 1.9 along the stack and ~4.5 in the molecular plane [24]. Thus $\varepsilon$ increases the dimer's Stark shift by an amount that remains to be found. An in-plane $\Delta\alpha$ ~50 Å$^3$ is likely for the $\pi$-$\pi^*$ excitation of isolated dimers and is quantum chemistry problem.

## 4. Excited states of conjugated polymers

The spectroscopy of PTCDA crystals and MQWs has several implications for excited states of conjugated polymers. First, crystal states of stacks are quite different from excitations of disordered segments. Second, the combined modeling of PTCDA spectra yields direct evidence of CT contributions ($\Delta$, $t_e$, $t_h$ in Table 1) that supports and makes more precise current discussions of polaron pairs. Third, PTCDA spectra indicate phonon-assisted processes consistent with strong e-ph coupling in conjugated polymers; they are difficult to include explicitly.

The fits in Fig. 3 rely on the Frenkel-CT band width, $E_1 - E_\pi = 0.28$ eV, and on the redistribution of vibronic intensities at $k = 0$ due to Frenkel-CT mixing. Such detailed comparisons exploit the structural, electronic and vibrational simplicity of PCTDA stacks. They underscore the qualitative nature of conjugated-polymer discussions in terms of excitons and polarons with unspecified mixing or vibronics. PTCDA and other organic molecular crystals are model compounds for demonstrating and quantifying these electronic processes.

Figure 4a sketches the elementary excitations of conjugated polymers whose backbones have inversion symmetry. In contrast to wide band semiconductors, the energy $E_g$ for direct generation of charge carriers is rarely known in polymer films. The $1^1B_u$ exciton is polarized along the chain, or nearly so. Thus $k = 0$ is at the bottom of an intrachain band. Linear e-ph coupling leads to displaced harmonic oscillators, with equal ground and excited-state frequencies and relaxation energy $\hbar\omega g^2$ in Fig. 2. The lowest two-photon state $2^1A_g$, triplet excitation $1^3B_u$ and interchain excitation $P^+P^-$ also couple to the effective conjugation coordinate with different g's. *Ab initio* results for PTCDA relaxation energies [19] are consistent with the spectroscopic g's in Table 1. The *intrachain* relaxation of $1^1B_u$ in extended chains is conceptually the same, but is more difficult to evaluate because the region of $1^1B_u$ displacements must be found. Polymers also have *interchain* relaxation processes that, while smaller, may be crucial for nearly degenerate states such as $1^1B_u$ and $P^+P^-$. The exciton binding energy of PPV and related polymers, $E_b$ in Fig. 4a, has been repeatedly studied without reaching consensus [1].

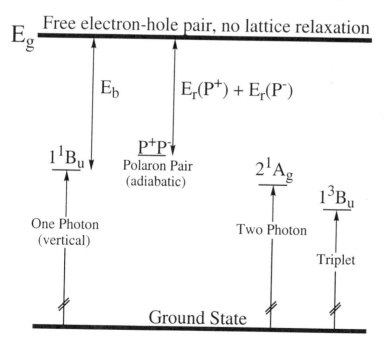

Fig. 4. Schematic excitation of conjugated polymers.

## 4.1. SEGMENTS, BANDS AND CHARGE CARRIERS

A host of pump-probe experiments have been performed on PPVs, PDAs, polythiophenes and other polymers to identify excitations up to ~ 6 eV [1,2]. The limitations of simple excitons, polarons or bipolarons become evident as the number of states and relaxation pathways increases. Interchain or intersegment interactions are widely recognized to be important. Perhaps by default, recent interpretations emphasize molecular or oligomeric states, including pairwise interactions between chromophores. The tacit assumption is that strong disorder in polymer films, whether due to conformational defects or other sources, precludes the formation of excited-state bands. In effect, the uncorrelated SSH $\pi$-bands of extended chains have been replaced with oligomer or segment excitations. This neglects exciton bands of correlated excited states that appear in organic crystals.

Intrastack interactions in PTCDA and in ion-radical or CT crystals containing stacks of $\pi$-donors and acceptors lead to band widths of ~0.5 eV. Smaller $1^1B_u$ splittings are likely for interchain interactions in conjugated polymers, while $\pi$-electron delocalization indicates wider intrachain bands. As emphasized by Weiser [25], the band gap $E_g$ of PDA single crystals is accurately known, varies with the nonconjugated side group R in Fig. 1, and gives $E_b$ ~ 0.5 eV. The enormous EA signal [25] starting at $E_g$ clearly implies band states with small effective mass ~$0.2m_e$ and is very sensitive to sample quality. Large side groups in PDA crystals limit interchain interactions or any transverse band width.

$E_g$ involves both molecular and lattice relaxation in conjugated polymers or organic molecular crystals. The dielectric constant $\varepsilon$ ~ 3 also has a central role. Charge separation is accurately represented [24] by $-e^2/\varepsilon R$ down to lattice spacings in a self-consistent analysis of polarizable lattice points. This places $E_g$ more than an eV above $1^1B_u$, which is unreasonably large. More realistic representations of molecules and accurate relaxation energies are needed. The local $C_2O_3$ dipoles of PTCDA require major extensions of the self-consistent analysis. The photoconduction [26] closely follows the linear absorption in Fig. 3b, but such charge generation in organic crystals involves exciton diffusion to the surface [27]. PTCDA offers an opportunity for an accurate $E_g$ determination and better integration of excited states in Fig. 4.

## 4.2. VIBRONIC STRUCTURE AND MIXING

Vibronic structure is characteristic in conjugated polymers and sets limits on the extent of delocalization. Linear e-ph coupling in the Holstein model (4) conserves the total displacement g, which is confined to one molecule at J = 0 and spreads out with increasing J. The top of Frenkel bands has *increased* g(J) at M*, out-of-phase displacements at neighboring sites M, and increased vibronic width in PTCDA stacks. In polymers, transition moments along the chain place k = 0 at the bottom of the band and produce in-phase displacements that decrease with increasing J and go as g/N in the limit of large J. Only the 0-0 transition persists in infinite chains or in wide band semiconductors. Proponents [28] of polymer bands assign vibronic structure to localized states in band edges. Such states may correspond to conjugated segments separated by conformational defects, but neither defects nor localized edge states have been modeled.

The polymer problem is difficult even for PDA single crystals. Their well-resolved absorption and EA show characteristic C≡C and C=C sidebands with displacements g ~ 1 that are comparable to oligomers and that only increase slightly in films [9]. The $1^1B_u$ exciton in PDA crystals is a k = 0 state with modest electron-hole separation of ~10 Å, just as the absorption of PTCDA stacks is a k = 0 process with displacements limited to adjacent molecules. Vibronic coupling in excited states of extended system may produce localized displacements and vibronics, although we are not aware of explicit models. Chromophores associated with segments in polymer films are easy to rationalize. At least in PDAs, however, the comparable $1^1B_u$ vibronics of crystals and films implies comparable excited-state displacements.

Frenkel-CT mixing in PTCDA or $1^1B_u$ mixing with $P^+P^-$ in conjugated polymers is allowed by symmetry and can be substantial, even for small interchain overlap, when the states are nearly degenerate. Intra and interchain excitations may fail in zeroth-order, as shown by PTCDA absorption in Fig. 3b. Such limitations are worth recognizing. Applied perturbations can mix states of different symmetry. Overlapping $2^1A_g$ and $1^1B_u$ vibronics of the 0.18 eV mode follow directly from the linear and two-photon absorption of PDA-PTS single crystals [29]. Small energy denominators of overlapping vibronics more than triple the three-photon resonance in third-harmonic-generation and contributes to four-wave mixing [9]. Simulated EA spectra [30] are very sensitive to $2^1A_g/1^1B_u$ mixing, again due to near degeneracies, and reproduce experiment in the 2.0-2.4 eV interval associated with $1^1B_u$ in PDA crystals. Such mixing is an alternative interpretation to a Stark shift due to a higher-energy $A_g$ state at $E_g$ ~ 2.5 eV, which implies a huge polarizability change.

## 4.3. CONCLUDING REMARKS

Electronic spectra of PTCDA crystals and multiple quantum wells show mixed π-π* and CT excitations of molecular stacks and vibronics of the characteristic 0.18 eV mode of polyenes. The structural, electronic and vibrational simplicity of PTCDA allows detailed modeling. We find an exciton band with strong dispersion at the bottom, as seen in emission, and negligible dispersion at the top, as seen in absorption. Molecular and CT excitations of organic crystals parallel excitons and polaron pairs in conjugated polymers, with well-defined chromophores and intermolecular interactions rather than a distribution of segments and interchain interactions. Excited-state mixing and vibronics raise challenging questions even in simple structures with localized and extended states.

**Acknowledgements:** We thank V. Bulovic for digitized film spectra and S.R Forrest, R.A. Pascal, Jr., and D.M. McClure for stimulating discussions. We gratefully acknowledge support from the National Science Foundation through DMR-9530116 and the MRSEC program under DMR-9499362.

322

## 5. References

1. Sariciftci, N.S. (ed. 1997) *Primary Photoexcitations in Conjugated Polymers: Molecular Exciton versus Semiconductor Band Model* , World Scientific, Singapore.
2. Skotheim, T.A., Elsenbaumer, R.L., and Reynolds, J.R. (eds. 1998) *Handbook of Conducting Polymers*, Second Ed., Marcel Dekker, New York.
3. Forrest, S.R. (1997) Ultrathin organic films grown by organic molecular beam deposition and related techniques, *Chem. Rev.* **97**, 1793-1896.
4. Conwell, E. (1997) Intramolecular Excitons and Intermolecular Polaron Pairs as Primary Photoexcitations in Conjugated Polymers, in Ref. 1, pp. 99-114.
5. Hennessy, M.H., Soos, Z.G., Pascal, R.A., Jr. and Girlando, A. (1999) Vibronic structure of PTCDA stacks: The exciton-phonon-charge-transfer dimer, *Chem. Phys.* **245**, 199-212.
6. Halverson, C. and Heeger, A.J. (1993) Two-photon absorption spectrum of oriented trans-polyacetylene, *Chem. Phys. Lett.* **216**, 488-92.
7. Kepler, R.G. and Soos, Z.G. (1993) Electronic Properties of Polysilanes: Excitations of σ-Conjugated Chains, in T. Kobayshi, (ed.) *Relaxation in Polymers*, World Scientific, Singapore, pp. 100-133.
8. Gussoni, M., Castiglione, C., and Zerbi, G. (1991) Vibrational spectroscopy of polyconjugated materials: Polyacetylene and polyenes, in R.J.H. Clark and R.E. Hester, (eds.) *Advances in Spectroscopy: Spectroscopy of Advanced Materials*, Wiley, New York, pp. 251-353.
9. Soos, Z.G., Mukhopadhyay, D., Painelli, A. and Girlando, A. (1998) π-Electron Models of Conjugated Polymers: Vibrational and Nonlinear Optical Spectra, in Ref. 2, pp. 165-196.
10. Ehrenfreund, E., Vardeny, Z., Brafman, O. and Horovitz, B. (1987) Amplitude and phase modes in *trans*-polyacetylene: Resonant Raman scattering and induced infrared activity, *Phys. Rev.* **B36**, 1533-53.
11. Soos, Z.G., Etemad, S., Galvão, D.S. and Ramasesha, S. (1992) Fluorescence and topological gap of conjugated phenylene polymers, *Chem. Phys. Lett.* **194**, 341-46; Soos, Z.G., Galvão, D.S. and Etemad, S. (1994) Fluorescence and excited-state structure of conjugated polymers, *Adv. Mater.* **6**, 280-87.
12. Heeger, A.J., Kivelson, S., Schrieffer, J.R. and Su, W.P. (1988), Solitons in Conducting Polymers, *Rev. Mod. Phys.* **60**, 781-850.
13. Painelli, A. (1998) Vibronic contribution to static NLO properties: Exact results for the DA dimer, *Chem. Phys. Lett.* **285**, 352-58.
14. Soos, Z.G., Hennessy, M.H. and Wen, G. (1997) Perylenes and Polyenes: A second π-electron approximation, *Chem. Phys. Lett.* **274**, 189-95.
15. McWilliams, P.C.M., Hayden, G.W. and Soos, Z.G. (1991) Theory of even-parity states and two-photon spectra of conjugated polymers, *Phys. Rev.* **B43**, 9777-91.
16. Bulovic, V., Burrows, P.E., Forrest, S.R., Cronin, A.J. and Thompson, M.E. (1996) Study of localized and extended excitons in 3,4,9,10-perylenetetracarboxylic

dianhydride (PTCDA). 1. Spectrocopic properties of thin films and solutions, *Chem. Phys.* **210**, 1-12.

17. Gomez, U., Leonhardt, M., Port, H. and Wolf, H.C. (1997) Optical properties of amorphous ultrathin films of perylene derivatives, *Chem. Phys. Lett.* **268**, 1-6.

18. Granville, M.F., Kohler, B.E. and Snow, J.B. (1981) Franck-Condon analysis of the $1^1A_g$ and $1^1B_u$ absorption in linear polyenes with two through six double bonds, *J. Chem. Phys.* **75**, 3765-69.

19. Hennessy, M.H., Pascal, R.A., Jr. and Soos, Z.G. (1999) Vibronic model of PTCDA stacks: Fluorescence and relaxation energies, SPIE meeting, Denver.

20. Holstein, T. (1959) Studies of polaron motion. Part 1. The molecular-crystal model, *Ann. Phys.* **8**, 325-42; Emin, D. (1975) Phonon-assisted transition rates. I. Optical-phonon-assisted hopping in solids, *Adv. Phys.* **24**, 307-48.

21. Wellein, G. and H. Fehske, H. (1997) Polaron band formation in the Holstein model, *Phys. Rev.* **B56**, 4513-17; Zhao, Y., Brown, D.W. and Lindenberg, K. (1997) Variational band theory for polarons: mapping polaron structure with the Merrified method, *J. Chem. Phys.* **106**, 5622-30.

22. Hennessy, M.H. (1999) PhD thesis, Princeton University (unpublished).

23. Haskal, E.I., Shen, Z., Burrows, P.E. and Forrest, S.R. (1995) Excitons and exciton confinement in crystalline organic thin films grown by organic molecular-beam deposition, *Phys. Rev.* **B51**, 4449-62.

24. Shen, Z. and Forrest, S.R. (1997) Quantum size effects of charge-transfer excitons in nonpolar molecular organic thin films, *Phys. Rev.* **B55**, 10578-92.

25. Weiser, G. and Horváth, A. (1997) Electroabsorption spectroscopy or $\pi$-conjugated polymers, in Ref. 1, pp. 318-62.

26. Bulovic, V. and Forrest, S.R. (1996) Study of localized and extended excitons in 3,4,9,10-perylenetetracarboxylic dianhydride (PTCDA) II. Photocurrent response at low electric fields, *Chem. Phys.* **210**, 13-25.

27. Kepler, R.G. and Soos, Z.G. (1997) The role of excitons in charge carrier production in polysilanes, in Ref. 1, pp. 363-383.

28. Hagler, T.W., Pakbac, K. and Heeger, A.J. (1995) Polarized electroabsorption spectrum of highly ordered poly (2-methoxy, 5-(2'ethyl-hexoxy)-$p$-phenylene vinylene) *Phys. Rev.* **B51**, 14199-206.

29. Lawrence, B., Torruellas, W.E., Cha, M., Sundheimer, M.L., Stegeman, G.I., Meth, J., Etemad, S. and Baker, G.L. (1994) Identification and role of two-photon excited states in a $\pi$-conjugated polymer, *Phys. Rev. Lett.* **73**, 597-604.

30. Soos, Z.G., Hennessy, M.H. and Mukhopadhyay, D. (1997) Correlations in conjugated polymers, in Ref. 1, pp. 1-19.

# PHYSICS OF ORGANIC ELECTROLUMINESCENCE

J. KALINOWSKI

*Department of Molecular Physics*
*Technical University of Gdańsk*
*Ul. G. Narutowicza 11/12*
*80-952 Gdańsk, Poland*

## 1.  Introduction

A huge number of papers which appears every year on organic electroluminescence (EL) is focused on improvement of various performance parameters by synthesis of new materials for organic light-emitting-devices (LEDs) and variation of the LEDs structure. Figure 1 shows examples of emission from organic EL devices.

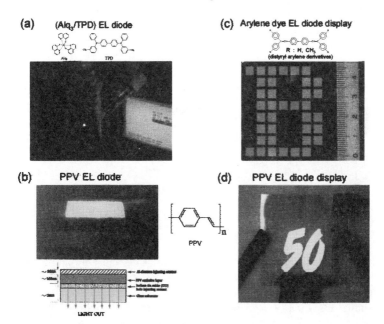

*Figure 1.* Examples of EL emission and LEDs displays obtained from: (a) Istituto FRAE, C.N.R. Bologna (Italy), (b) Cavendish Laboratory, Cambridge (England), (c) Central Research Laboratories, Idemitsu-Kosan Co., Ciba (Japan), (d) Department of Experimental Physics II, University of Bayreuth (Germany).

*F. Kajzar and M.V. Agranovich (eds.), Multiphoton and Light Driven Multielectron Processes in Organics: New Phenomena, Materials and Applications, 325–344.*
© 2000 *Kluwer Academic Publishers. Printed in the Netherlands.*

In part (a) an Alq$_3$/TPD junction-based LED reveals brightness that exceeds brightness of the screen of an ordinary computer monitor; yellow EL emission from PPV formed into 50 cm$^2$ display is shown in part (b) and (d); a display of a letter "B" in part (c) shows a set of blue emitting LEDs based on a distyryl arylene derivative. Behind these impressive EL emission patterns complicated physical phenomena are hidden. These are they that decide about LED performance and they must be understood in detail in order to tailor the properties of organic light emitting devices.

Since in electroluminescence electrical energy is converted directly into light three fundamental processes are of essential interest. These are: (i) electrical energy supply, (ii) excitation mode of emitting states, and (iii) light generation mechanism itself (see Figure 2).

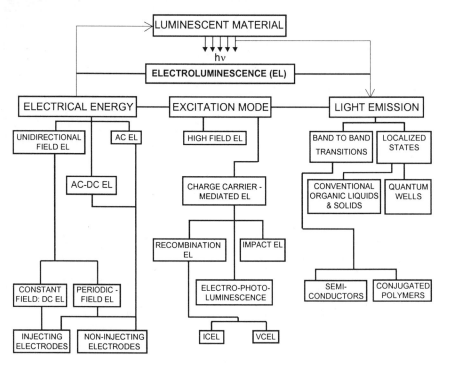

*Figure 2.* Fundamental processes and classification of EL phenomena.

Dependent on electrical energy supply steady-state (DC) and alternating current (AC) EL can be distinguished which often determine the excitation mode of emitting states being excited directly by external electric fields producing high-field EL or charge carriers-mediated EL. Finally, the spectral composition of EL light is determined by the nature of excited states being either free electron states producing EL in band-to-band transition modes or localized states which generate EL resembling light emission from isolated molecules in conventional organic liquids or solids. As can be seen from the scheme of Figure 2, even very brief description of various mechanisms of EL would need a series of lectures and certainly exceeds greatly the time allotted for this talk. Many of them have been discussed in the introducing chapter of a recent book on organic EL [1]. Here, I would like to select one type, perhaps most common EL excited by a steady-state electric field applied to electron and hole injecting electrodes, which leads to electron-hole recombination producing localized excited states. Their radiative decay emit photons creating light called commonly organic electroluminescence.

## 2. EL Output

The EL flux per unit area ($\Phi_{EL}$) being an integral of EL yield over the total thickness of the sample (d) [1] is determined by decay mechanisms of excited states through the ratio $k_F/k_S$ of radiative ($k_F$) to overall ($k_S$) decay rate constants for singlet excitons (that is photoluminescence quantum yield, $\varphi_{PL}$), recombination processes through the product $P_S\,\gamma$ with $P_S$ standing for the probability to create a singlet exciton in the bimolecular recombination process ($\gamma$ - second order recombination coefficient), and injection properties of the electrode contacts through the concentration of holes ($n_h$) and electrons ($n_e$) within the sample:

$$\Phi_{EL} = \varphi_{PL} P_S \gamma \, n_h \, n_e \, d \tag{1}$$

To describe the fate of charge carriers injected from the electrodes into insulating samples, the rate equations for electron and hole concentrations under steady-state conditions may be analyzed. When diffusion effects can be neglected, they are expressed by simple equations composed of carrier generation terms and carrier decay terms:

$$\frac{j_i^e}{ed} - \left( \frac{1}{\tau_t^e} + \frac{1}{\tau_{rec}^e} \right) n_e = 0$$

$$\tag{2}$$

$$\frac{j_i^h}{ed} - \left( \frac{1}{\tau_t^h} - \frac{1}{\tau_{rec}^h} \right) n_h = 0$$

While the generation terms are associated with injection currents of holes ($j^h$) and electrons ($j^e$), there are two decay channels for the carriers : discharge at the electrodes characterized by their transit time $\tau_t = d/\mu F$ ($\mu$-carrier mobility, F-electric field within the sample) and mutual recombination in the sample volume, characterized by the recombination time $\tau_{rec} = (\gamma n)^{-1}$.

Two important limiting cases for EL can be distinguished basing on the relation between transit and recombination time.

(i) Injection-controlled EL (ICEL) for $\tau_t < \tau_{rec}$. The free carriers decay efficiently on electrode contacts forming leakage currents – currents that avoid recombination. EL output is then a function of two injection currents and mobilities which like the currents depend, in general, on electric field:

$$\Phi_{EL} = \varphi_{PL} P_S \gamma \frac{j_i^h j_i^e}{e^2 \mu_e \mu_h F^2} d \tag{3}$$

(ii) Volume-controlled EL (VCEL) for $\tau_t > \tau_{rec}$. In contrast to the former case, in the present case overwhelming majority of the carriers decay by the recombination process in the sample bulk, EL flux is a linear function of the current (which is virtually the recombination current).

$$\Phi_{EL} = \varphi_{PL} P_S \frac{j}{e}; \qquad j_i^e = j_i^h = j \tag{4}$$

It is clear that the VCEL mode, obviously desired in high-brightness light-emitting diodes, requires efficient injection contacts and high recombination coefficients to make $\tau_{rec}$ as low as possible.

## 3.  EL Quantum Yield

One of the most important performance parameters of LEDs is the EL quantum yield defined as the number of photons (hv) emitted per one carrier (e) passing through the EL cell, expressed usually in percentage representation, reads

$$\varphi_{EL} = \frac{EL / h\nu}{j / e} 100 \quad [\%] \tag{5}$$

By definition the EL quantum yield is described by the product of photoluminescence (PL) quantum yield ($\varphi_{PL}$), $P_S$, and recombination probability ($P_R$).

$$\varphi_{EL} = \varphi_{PL} P_S P_R \tag{6}$$

The recombination probability defined as the ratio of the recombination rate constant to the sum of the rate constants for recombination and sucking up of the carriers from the sample volume, being expressed by appropriate time constants, makes it a simple function of the recombination-to-transit time ratio:

$$P_R = \frac{k_{rec}}{k_{rec} + k_t} = \frac{\dfrac{1}{\tau_{rec}}}{\dfrac{1}{\tau_{rec}} + \dfrac{1}{\tau_t}} = \frac{1}{1 + \dfrac{\tau_{rec}}{\tau_t}} \qquad (7)$$

Let us note that our two limiting cases can now be distinguished with the value of $P_R=1/2$. The ICEL mode reveals $P_R<1/2$ whereas the VCEL mode appears for $P_R>1/2$. Obviously, for $\tau_{rec}/\tau_t \ll 1$ $P_R \to 1$. In the latter case $\varphi_{EL}=\varphi_{PL}P_S=const$ and depends on PL quantum yield of the material and $P_S$. If due to spin statistics $P_S=1/4$, $\varphi_{EL}=0.25\varphi_{PL}$, that is $\varphi_{EL}$ can reach 25% of the $\varphi_{PL}$, which for $\varphi_{PL}=1$ gives the theoretical limit of $\varphi_{EL}=25\%$ photons/carrier. Every 100 carriers passing through the sample produces 25 photons. The quantum yield can be strongly affected by reabsorption, scattering and waveguiding losses of emitted light in the sample and glass substrate, making the external (available for measurements) efficiency much lower,

$$\frac{\varphi_{EL}^{ext}}{\varphi_{EL}} = \underbrace{(1-R)(1-\cos\theta_c)}_{\text{total internal reflectance}} \overbrace{\exp(-\alpha x)}^{\text{extinction losses}} < 1 \qquad (8)$$

where $\theta_c = arc\sin(n_c^{-1})$ - critical angle ($n_c$ - refractive index). Typical values of $n_c$ and $\alpha$ lead to the average limit of $\varphi_{EL}^{ext} \cong 5\%$.

All these seem to represent a simple physical picture of the recombination EL. The question thus arises what is the reason for a huge number of papers appearing every year on organic EL. What do researchers look for in the phenomenon ? Obviously, to manufacture stable EL devices reaching this limit has been one of the important reasons. However, the value 7.1% obtained recently with an organic LED based on small molecules of a highly photoluminescent derivative of quinoline Al complex exceeds the limit 5% [2]. This, on one hand, seems to solve one of the technological problems for organic LEDs but, on the other hand, suggests existence of additional physical processes enhancing the calculated limit. These might be excitonic interactions.

## 4.  Excitonic Interactions

As already mentioned in Section 3, due to spin statistics, three times more triplets than singlets are produced in the free electron-hole recombination process, that is the creation probability of singlets equals ¼ and triplets ¾ , whenever energy gap (that is electron-hole energy) is larger than singlet exciton energy. Till now, triplets have been considered as lost for optical emission. In essence, they can contribute to the emission directly, producing phosphorescence, or indirectly, producing delayed fluorescence by triplet-triplet annihilation of which one third leads to emitting singlets. Consequently, the overall singlet production is not 25% but 35% (see Figure 3), that is the upper limit of the quantum yield increases.

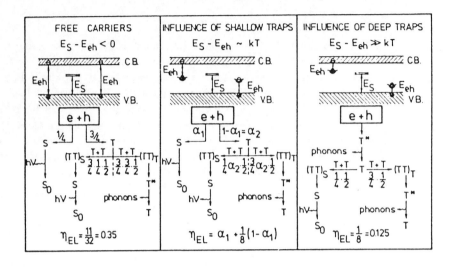

*Figure 3.* Triplet-triplet annihilation contribution to the emitting states in the absence and presence of carrier traps.

On the other hand, the presence of deep traps can change the triplet-to-singlet branching ratio completely [1]. On the extreme, when the electron-hole energy is much lower than the singlet exciton energy due for example to deep trapping of electrons, only triplets are energetically feasible and possible emission is purely phosphorescence or delayed fluorescence. In the latter case the recombination yield of singlet exciton production drops down to 12.5% (Figure 3). Phosphorescence, neglecting triplet-triplet annihilation, would give the emission yield limited only by phosphorescence quantum yield; for all emitting triplets it would be as much as 100%. Therefore, unexpectedly, to further improve the EL quantum yield, a highly phosphorescent material with deep one carrier traps must be applied as an emitter in organic LEDs.

The triplet-triplet interaction postulated to explain increased quantum yield of thin film organic LEDs has been well known in EL of organic single crystals [1,3]. One of the most spectacular manifestation of this type excitonic interactions is spatial distribution of EL emission (Figure 4). Above certain voltage, the EL emission zone in tetracene single crystals splits up into three regions seen in part (c) of Figure 4. The second anodic region has been shown to be dominated by the delayed fluorescence due to triplet-triplet annihilation [3]. A strong drop in EL intensity around 20μm is interpreted as due to quenching of triplets by non-homogeneously distributed space charge at the anode. The near-electrode regions of emission indicate the recombination

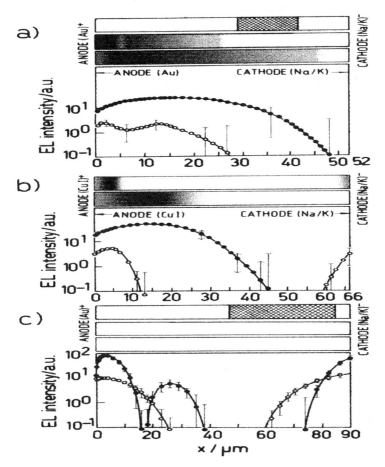

*Figure 4.* Spatial distribution of the EL output in anthracene (a,b) and tetracene (c) single crystals. The black patterns over each of the graphs simulate the light intensity distribution at different voltages. The upper parts in (a) and (c) illustrate the position and width of the recombination zone as predicted for trap-free crystals and ohmic contacts [1].

as being due to free-trapped carrier synthesis, electrons trapped at the cathode and holes trapped at the anode. The intermediate third emission region disappears at lower voltages. Moreover, it does not appear at all in case of anthracene crystals (Figures 4a and 4b). This observation turns us back to the charge carrier supply aspects of organic EL. Since the gold anode and sodium/potassium alloy cathode form quasi-ohmic electrical contacts to tetracene [3], the voltage lowering causes lowering of the near-electrode charge density which, in turn, reduces the quenching efficiency of triplets, the minimum at 20μm disappears. The same metals do not form ohmic contacts to anthracene [3] and, therefore, the charge density at both anode and cathode is not high enough to quench triplets with a rate sufficient to produce the emission minimum and, consequently, the third emission zone seen in tetracene.

## 5. Injection Conditions for EL

The charge injection at electrical contacts is of essential importance for organic EL, but also attracts researchers by its pure theoretical aspects. Highly efficient EL devices require high-injection efficiency, since then recombination-to-transit time ratio becomes very small. Let us recall that the recombination probability is a function of the recombination-to-transit time ratio dependent on injection conditions and transport properties of the material (see Equation (7)),

$$\frac{\tau_{rec}}{\tau_t} = \frac{\mu_{e,h} F}{\gamma n_{h,e} d} \tag{9}$$

If , for example, electrons are majority carriers it can be expressed by the ratio of the electron space-charge-limited (SCL) current ($j_{SCL} = (9/8)\varepsilon_o\varepsilon\mu_e F^2/d$) to the actual current ($j \cong j_e = e\mu_e n_e F$) in the cell

$$\frac{\tau_{rec}}{\tau_t} = \frac{8e\mu_h}{9\gamma\varepsilon_0\varepsilon} \frac{j_{SCL}}{j} \tag{10}$$

As a rule $j < j_{SCL}$, so that the minimum value of the ratio occurs for $j = j_{SCL}$ :

$$\left(\frac{\tau_{rec}}{\tau_t}\right)_{min} = \frac{8e\mu_h}{9\gamma\varepsilon_0\varepsilon} \tag{11}$$

Then, the ratio is independent of the current (voltage) whenever $\mu_e(F)$=const and $\gamma(F)$=const. Consequently, recombination probability and EL quantum yield are field (and current) independent. Otherwise, when $j < j_{SCL}$ the ratio

$$\frac{j_{SCL}}{j} \cong \frac{\dfrac{9}{8}\varepsilon_0\varepsilon\mu_e F^2}{j(F)d} \tag{12}$$

decreases as F increases because all the injection-describing functions increase with field much stronger than squared F [1]:

$$j_1(F)=AF^{3/4}exp(aF^{1/2}) \qquad \text{- Schottky-type function,} \quad (13)$$

$$j_2(F)=BF^2exp(-b/F) \qquad \text{- Fowler-Nordheim-type function,} \quad (14)$$

$$j_3(F)=j_0exp(-c/F^{1/2}) \qquad \text{- hot-carrier injection,} \quad (15)$$

This would mean that $\tau_{rec}/\tau_t$ decreases with field and $P_R$ and $\varphi_{EL}$ increase with field. What does experiment tell us ? In Figure 5 the EL quantum yield versus electric field for a single (SL) and double (DL) layer LEDs based on TPD as hole-transporting material and PBD as electron-transporting material are presented.

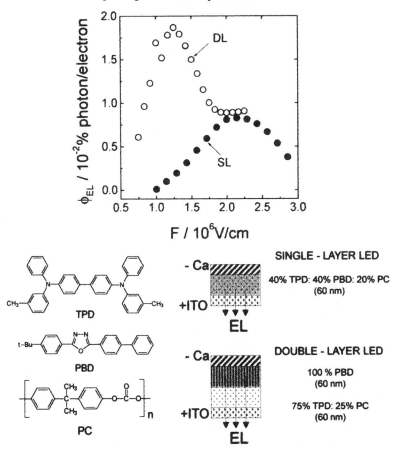

*Figure 5.* EL quantum yield as a function of electric field for a single and double layer organic LEDs [4].

In both cases the quantum yield increases at low fields, then reaches maximum and decreases at high fields. Of three types injection mechanisms (13), (14), (15), current-field characteristics follow well a Schottky-type function as it is apparent from Figure 6.

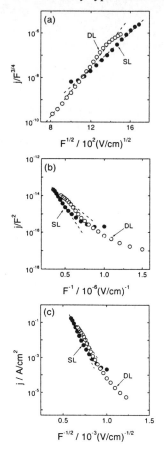

*Figure 6.* Current-field characteristics for the LEDs from Fig. 5 in different representations corresponding to the three injection mechanisms described by equations (13) (a), (14) (b) and (15) (c).

## 6. Origin of the Field Dependence of $\varphi_{EL}$

Introducing function (13) in the expression for the recombination-to-transit time ratio (10) yields a function which decreases with electric field according to

$$\frac{\tau_{rec}}{\tau_t} = \frac{e\mu_h\mu_e F^{1/4}}{\gamma Ad\,\exp(aF^{1/2})} \tag{16}$$

The decreasing ratio (16) leads to increasing EL quantum yield (cf. Equations (6) and (7)) as observed in experiment at low fields (Figure 5).

In order to make $\varphi_{EL}$ decreasing the recombination-to-transit time ratio must increase. This happens only if the mobility product divided by the recombination coefficient, $\mu_e\mu_h/\gamma$, increases with electric field. The Langevin recombination mechanism assumes $\gamma$ to be determined by the carrier motion ($\gamma=e\mu_e/\varepsilon_0\varepsilon$) [5], thus the factor $\mu_e\mu_h/\gamma=(\varepsilon_0\varepsilon/e)\mu_e$ can be field dependent only through a field-dependent carrier mobility $\mu_e(F)$. In fact, the experiment shows that in the $10^5$V/cm field regime the mobility increases with field, following a Poole-Frenkel type function $\mu=\mu_0\exp(\beta_\mu F^{1/2})$ [6]. Then

$$\frac{\tau_{rec}}{\tau_t} = \frac{\varepsilon_0\varepsilon F^{1/4}}{Ad}\exp\left[\left(\beta_\mu - a\right)F^{1/2}\right] \tag{17}$$

and the $\tau_{rec}/\tau_t$ ratio increases with field solely when $\beta_\mu \geq a = 9.4\times10^{-3}$ (cm/V)$^{1/1}$, the value obtained from the slope of the $\lg(j/F^{3/4})$-$F^{1/2}$ plot of Figure 6. For materials with diagonal energetic (E) disorder, $g(E)=(2\pi\sigma)^{-1/2}\exp(-E^2/2\sigma^2)$, $\beta_\mu = C(\sigma/kT)^2$ which with C = $4.5\times10^{-4}$(cm/V)$^{1/4}$ requires $\sigma \geq 0.114$ eV, the value typical for organic films, found from mobility measurements [6]. However, if $\mu_e$, $\mu_h$ are field independent or decrease with field as would come out from high-field saturated carrier velocity [7], or $\beta_\mu<a$, one would expect $\gamma$ to decrease with electric field. This would mean that the Einstein relation between diffusion coefficient (D) and carrier mobility (D=$\mu$kT/e) does not hold any more and Thomson-type approach [5] is more appropriate to describe the recombination process rather than Langevin formalism. This is another challenge for both theoreticians and experimentalists to provide more experimental data and to apply the Thomson model to weak-bonded organic solids. Some our new, still unpublished data on EL quantum yield of DL LEDs, are shown in Figure 7. The LEDs consist of TPD-doped PC hole-transporting-layer (HTL) and vacuum-evaporated Alq$_3$ electron-transporting layer (ETL) (also acting as the emitter). The maximum quantum yield appears for 100% as well as 25% of TPD in polycarbonate (PC). Its electric field position moves towards high fields as the concentration of TPD decreases. Once having the measured quantum yield versus electric field, the field dependence of the recombination-to-transit time ratio can be obtained from

$$\varphi_{EL} = \frac{\varphi_{PL}P_S}{1+\dfrac{\tau_{rec}}{\tau_t}} \tag{18}$$

336

Since in the LEDs of Figure 7 emission comes out from Alq$_3$ layer the PL efficiency of Alq$_3$ films, $\varphi_{PL}$=12%, has been used together with P$_S$=1/4 to calculate $\tau_{rec}/\tau_t$ presented in Figure 8. Apparent minima correspond to observed maxima of $\varphi_{EL}$(F) (Figure 7).

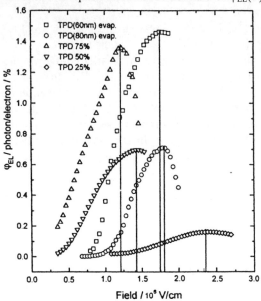

*Figure 7.* EL quantum yield vs applied field for a series of DL LEDs based on (PC+%TPD)/Alq$_3$ junction [8].

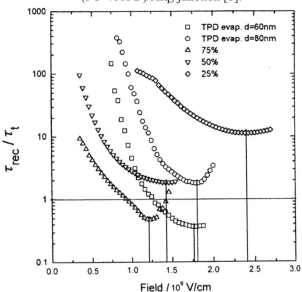

*Figure 8.* Recombination-to-transit time ratio as a function of electric field for the LEDs from Figure 7.

Apart of evaporated 100% TPD hole-transporting layer, the minimum shifts towards higher fields as concentration of TPD in PC decreases. At the same time the recombination-to-transit time ratio increases by more than one order of magnitude. Mostly, this ratio is greater than unity, only for the highest 75% and 100% concentrations of TPD it drops down below approaching 0.5 – the limit for VCEL operation mode (see Section 3). For low $\varphi_{EL} < 0.1\%$ it can be as high as 100.

## 7. Electric Field Effect on EL Spectra

There is a series of experiments on EL spectra of mixed systems supporting the idea of Thomson-like recombination mechanism. Since it is controlled by the ultimate recombination step rather than the carrier motion, one would expect such a process to occur in the locally defected environment, where obstacles for the carrier capture can be created as, for example, at dopant molecular recombination centers or molecular solid interfaces. In fact, the voltage evolution of the emission spectra from such sites seems to confirm this supposition. The EL spectrum of lightly $Alq_3$-doped TPD single-layer cell shows two maxima corresponding to the blue emission from TPD and green emission from $Alq_3$ (Figure 9). The latter diminishes with increasing voltage.

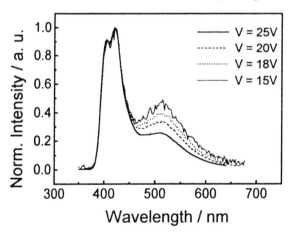

*Figure 9.* Voltage evolution of the EL spectra in a lightly $Alq_3$-doped TPD single layer LED: ITO/TPD+$10^{-6}$ $Alq_3$/Mg [9].

This can be explained by decreasing recombination probability at the $Alq_3$ dopant.
In a triple-layer system TPD/$Alq_3$/PBP there are two emission bands belonging to $Alq_3$ (green) and perylene derivative dye (red) (Figure 10). The red emission is practically eliminated at higher voltages, which could mean that the hole recombination at negatively charged centers at the $Alq_3$/PBP interface is effectively suppressed, the emission color of the LED becomes green.

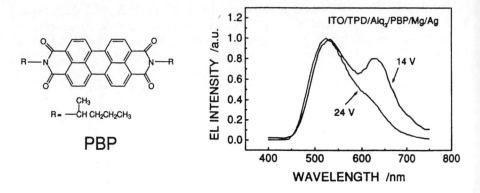

*Figure 10.* Voltage-induced change in the EL spectrum of a triple-layer LED ITO/Alq$_3$/PBP/Mg [10].

## 8. Nature of the Emitting States

Recently, a number of papers has appeared showing that the EL spectrum is completely different from that of molecular components of the cells. The most typical is EL from double-layer LEDs when emission comes out from the organic solid interface. The EL spectra are often much broader and red shifted with respect to those for separated organic solids. An example is shown in Figure 11.

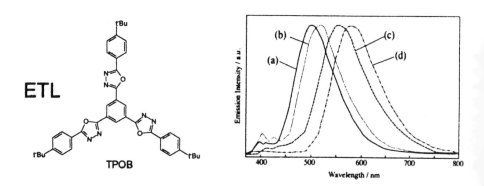

*Figure 11.* EL spectra of the bilayer device ITO/HTL/TPOB/MgAg. HTL: (a) TCTA, (b) TPD, (c) *p*-DPA-TDAB, and (d) *m*-MTDATA [11]. The molecular structures of HTLs are shown in Figure 12.

The broad band EL spectra are due to DL LEDs based on different HTL and TPOB ETL also serving as the emitter. A weak structures around 400nm correspond to PL spectra of separated material components as shown in Figure 12. They are relatively narrow and

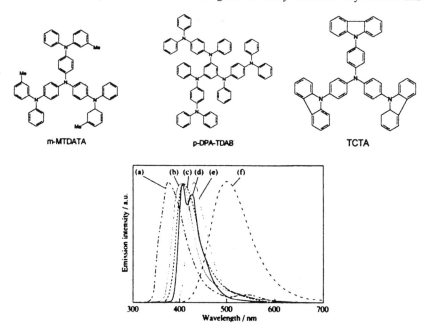

m-MTDATA     p-DPA-TDAB     TCTA

*Figure 12.* PL spectra of the material components forming DL LEDs from Figure 11: (a) TPOB, (b) TCTA,, (c) p-DPA-TDAP, (d) TPD, (e) m-MTDATA, and (f) TPOB:TCTA equimolar mixture in their solid films prepared by spin coating [11]. Molecular structures of the HTLs are shown in the upper part of the figure.

emit in blue region. However, a mixture of the ETL material with a HTL one yields a broad red shifted PL spectrum identical with the EL spectrum for the solid state interface of these materials in the DL LED (Figure 11a). This strongly suggests the EL spectrum to be underlaid by exciplexes formed between HTL and ETL materials. But, on the other hand, in some cases the intermediate EL bands from DL LEDs do not have equivalent PL bands in the two-layer material mixtures. Such a situation has been observed with DL LEDs, where a conjugated polymer (PTOPT) was used as HTL (Figure 13). The long-wavelength maximum corresponds to the red PL emission of this polymer, the short-wavelength maximum corresponds to the emission of the electron-transporting layer (PBD). The green emission at about 550 nm could not be observed in the PL spectrum of combined HTL and ETL materials composing the LED [12]. This rules out the exciplex as possible emission source and a sort of "cross reaction" between LUMO of PBD and HOMO of PTOPT, called "electroplex" has been introduced to explain the emission band (Figure 14).

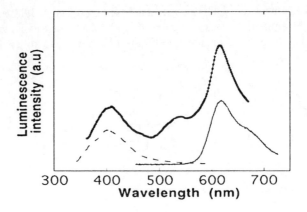

*Figure 13.* The EL spectrum of a PTOPT/PBD junction-based DL LED (upper curve) and PL spectra of PTOPT and PBD forming HTL and ETL of the LED, respectively, (lower displayed curves) [12].

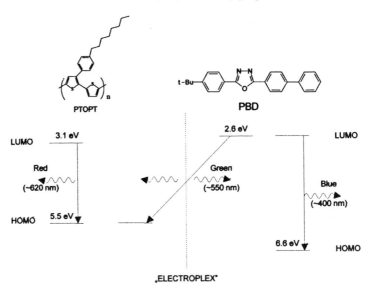

*Figure 14.* Cross reaction underlying "electroplex" emission from the DL LED based on the PTOPT/PBD junction (cf. Figure 13).

In contrast to PTOPT, molecules of TPD are able to form with PBD both exciplexe and electroplexes. The EL SL (40%TPD+40%PBD+20%PC)(60nm) and Dl (75%TPD+25%PC)/PBD EL devices reveal broad complex spectra which can b decomposed into 3 and 4 components, respectively, which are assigned to th monomolecular emission of TPD (band 1), well defined exciplex (band 2), electroplex (band 3) and trapped electroplex (band 4) (Figure 15).

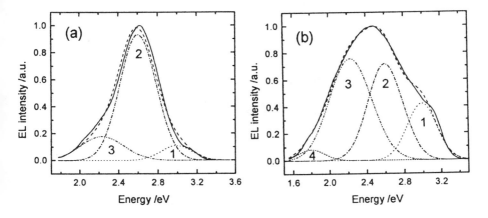

*Figure 15.* Combined molecular - exciplex – electroplex EL emission from SL (a) and DL (b) organic LEDs [13].

Electric field variation in their contribution makes the EL spectra to be dependent on the applied field. The open question is to understand the mechanism underlying the branching ratio between exciplex and electroplex emission components. An even more extreme examples of the specificity of excited states formed under electron-hole recombination conditions provide EL from single-component, single layer devices [14,15]. Long-wavelength narrow bands of their EL spectra rule out the concept of exciplex and electroplex emission as a possible explanation for the spectra and suggest unidentified molecular species to emit EL light.

These examples of spectral properties of organic LEDs show the importance of generation mechanisms of excited states which cannot be simply divided into two groups of extended and localized single component emitting species. The local interactions must be taken into account, their influence on formation rate constants of various type of complex excited states being of crucial importance for EL yield.

## 9. Structural Modifications of Organic LEDs

In closing , efforts to improve organic LEDs by various modifications of their structure must be mentioned.. The light-guiding effects which in the face vertically emitting LEDs have led to emission losses (cf. Section 3) can be turned out to enhance the emission when observed from the edge of the devices. There are numerous solutions to construct edge emitting organic LEDs with enhanced light output and reduced spectral width [16]. Fabrication of an electrically-pumped all organic LASER would be a breakthrough in comparing applicability of organic materials in microelectronics. Recently, a device emitting LASER-like light has been reported [17]. It is a DL structure composed of $Alq_3$-Nile Blue (NB) mixture acting as hole-transporting layer and NB forming an electron-transporting layer. Indium and Al electrodes are considered as Schottky contacts supplying symmetrically holes and electrons (Figure 16).

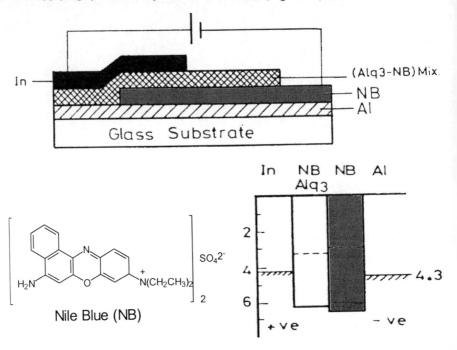

*Figure 16.* A double-layer structure with EL blue edge emission showing laser light features as demonstrated in Figure 17 [17].

The edge emitting device shows a sharp increase of the output power above the threshold current at about 90 µA (Figure 17). The strong narrow band emission of blue light characteristic of Nile-Blue is considered by the authors as the laser beam. The unexpectedly low threshold current density (<1A/cm$^2$) and a dramatic decrease of the optical power at higher voltages make the lasing process in this edge emitting LED questionable, and further efforts are necessary to establish reproducible conditions for electrically-pumped organic lasers. The nature of the emitting species in the device is not yet clear as well.

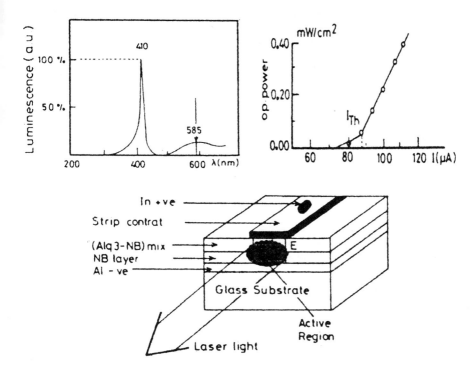

*Figure 17.* The emission spectrum (a) of the light beam (b) at the current of 0.11 mA above the threshold current (c) from the edge emitting diode as described in Figure 16 .

## 10. Final Remarks

From the above overview of basic phenomena underlying organic electroluminescence the following remarks can be made:

i)   A simple picture of physical phenomena including charge carrier injection, recombination of free carriers and emission from fully extended or well localized molecular states do not explain many characteristics of organic LEDs as e.g. EL quantum yield, space and spectral distribution of EL intensity or low-current threshold of lasing.

ii) Mechanisms of charge injection and recombination processes are not yet established (Langevin versus Thomson recombination is of particular interest).

iii)   The nature of emitting species formed in electron-hole recombination differ, in general, from those excited by light and need theoretical description.

iv)   Extended quantitative description of physical processes underlying organic EL would lead to improvement of organic EL devices.

344

## 11. References

1.  Kalinowski, J. (1997) Electronic processes in organic electroluminescence, in S. Miyata and H.S. Nalwa (eds), *Oganic Electroluminescent Materials and Devices,* Gordon & Breach Science Publishers, Amsterdam, pp.1-72.
2.  Kido, J. (1998) Fabrication of highly efficient organic electroluminescent devices, *Applied Physics Letters* **73**, 2721-2723.
3.  Kalinowski, J. (1994) Space-resolved recombination electroluminescence in organic crystals, *Synthetic Metals* **64**, 123-132.
4.  Kalinowski, J., Cocchi, M., Giro, G., Fattori, V. and Di Marco, P., The nature of emitting species in organic light emitting diodes based on polymer dispersion of an aromatic diamine and an oxadiazole derivative, *to be published.*
5.  Pope, M. and Swenberg, C.E. (1982) *Electronic Processes in Organic Crystals,* Clarendon Press, Oxford, p.501.
6.  Borsenberger, P., Magin, E.H., Van der Auweraer, M. and De Schryver, F.C. (1993) The role of disorder on charge transport in molecularly doped polymers and related materials, *Physica Status Solidi (a)* **140**, 9-47.
7.  Albrecht, U. and Bässler, H. (1995) Langevin-type charge carrier recombination in a disordered hopping system, *Physica Status Solidi (b)* **191**, 455-459.
8.  Cocchi, M., Giro, G., Kalinowski, J., Di Marco, P. and Fattori, V., *unpublished results.*
9.  Kalinowski, J., Camaioni, N., Di Marco, P., Fattori, V. and Giro, G. (1996) Operation mechanisms of thin film organic electroluminescent diodes, *International Journal of Electronics* **81**, 377-400.
10. Kalinowski, J., Di Marco, P., Fattori, V., Giulietti, L. and Cocchi, M. (1998) Voltage-induced evolution of emission spectra in organic light-emitting diodes, *Journal of Applied Physics* **83**, 4242-4248.
11. Ogawa, H., Okuda, R. and Shirota, Y. (1998) Tuning of the emission color of organic electroluminescent devices by exciplex formation at the organic solid interface, *Applied Physics A* **67**, 599-602.
12. Inganäs, O. (1997) Making polymer light emitting diodes with polythiophenes, in S. Miyata and H.S. Nalwa (eds.), *Organic Electroluminescent Materials and Devices,* Gordon & Breach Science Publishers, Amsterdam, pp. 147-175.
13. Giro, G., Cocchi, M., Kalinowski, J., Di Marco, P. and Fattori, V. (1999) Multicomponent emission from organic light-emitting diodes based on polymer dispersion of an aromatic diamine and an oxadiazole derivative, *Applied Physics Letters*, submitted.
14. Rommens, J., Vaes, A., Van der Auweraer, M. and De Schryver F.C. (1998) Dual electroluminescence of an amino substituted 1,3,5-triphenylbenzene, *Journal of Applied Physics* **84**, 4487-4494.
15. Kalinowski, J., Cocchi, M., Giro, G., Di-Nicoló, E., Fattori, V. and Di Marco, P. Disparity in electroluminescence and photoluminescence spectra from vacuum evaporated films of 1,1-bis((di-4-tolylamino)phenyl) cyclohexane (TAPC), *to be published.*
16. Kalinowski, J. (1999) Electroluminescence in organics, *Journal of Physics D: Applied Physics*, accepted for publication.
17. El-Nadi, L., Al-Houty, L., Omar, M.M. and Ragab, M. (1998) Organic thin film materials producing novel blue laser, *Chemical Physics Letters* **286**, 9-14.

# POLYDIACETYLENE MICROCRYSTALS AND THEIR THIRD-ORDER OPTICAL NONLINEARITY

H. Kasai, S. Okada, and H. Nakanishi
*Institute for Chemical Reaction Science, Tohoku University;*
*Sendai, 980-8577, Japan*

Aiming at future photonic device, we have studied on the improvement of third-order nonlinear optical properties of polydiacetylene crystals. For molecular design, we could prepared a variety of ladder-type PDAs to have larger $\chi^{(3)}$ values. For morphological engineering, we have established the effective and simple reprecipitation method to prepare organic microcrystals with controlled size and shape. The highly concentrated poly(DCHD)-microcrystal thin film with good optical quality and large area could be obtained, and the films were found to show three order of magnitude larger $\chi^{(3)}$ compared with the original water dispersion. In the Fabry-Perot experiment, materials based on PDA microcrystals were found to show the possibility for achieving the optical switching devices.

## 1. Introduction

Fine particles or microcrystals have been extensively studied to understand the intermediate state between corresponding bulk crystals and the isolated atom or molecule, and are expected to exhibit some interesting properties [1,2]. For example, it is well-known that, in semiconductor nanoparticles, the enhancement of nonlinear optical (NLO) properties was confirmed in less than 10 nm diameters [3,4]. Organic microcrystals, however, have attracted even little attention so far due to the difficulties of preparing them.

We have demonstrated that organic microcrystals could be easily prepared by a reprecipitation method [5], and some unique properties of π-conjugated organic and polymeric microcrystals were revealed when the crystal size was reduced to less than 200 nm [6,7]. Among them, the study on microcrystals of polydiacetylene (PDA) [8], which is obtained by solid-state polymerization of diacetylene, is attractive [9]. Because polydiacetylene crystals are expected to show superior third-order NLO susceptibilities and ultrafast optical response, leading to future photonic technologies [10]. In this paper, molecular design for novel PDAs to enhance NLO properties will be introduced, and recent progress on the third-order NLO of PDA microcrystals will be also described in detail.

*F. Kajzar and M.V. Agranovich (eds.), Multiphoton and Light Driven Multielectron Processes in Organics: New Phenomena, Materials and Applications, 345–356.*
© 2000 *Kluwer Academic Publishers. Printed in the Netherlands.*

## 2. Molecular Design

We have been studying novel PDA derivatives as third-order NLO materials mainly from molecular design and morphological engineering. The former is important to increase molecular performance, and the latter is an inevitable subject to apply the materials for devices.

For PDA molecular design to enhance third-order nonlinear optical susceptibilities ($\chi^{(3)}$s), modification in their substituents is interesting. For example, when the $\pi$-conjugated substituents are directly bound to the PDA backbone, the electronic state of the backbone is naturally perturbed resulting in variation of linear and nonlinear optical properties. In this regard, we have prepared several PDAs possessing $\pi$-conjugated interaction between polymer backbone and side chains, e.g. PDAs substituted by aromatic groups [11-14] or by acetylenic groups [13-18]. In the case of aromatic-ring-substituted butadiyne monomers, obtained crystals are often not polymerizable, because the monomers in suitable arrengement in crystals, i.e. about 0.5-nm translation distance between butadiyne moieties of adjacent molecules and around 45° angle between the translation axis and the butadiyne moiety, only show solid-state polymerizability [19] and many of the aromatic-ring-substituted butadiyne monomers do not satisfy above criterion. Thus, empirical rules to realize the polymerizable stack of monomers in crystalline lattice have been proposed [12].

Recently, we are focusing on ladder-type PDAs, in which there are two PDA backbones per monomer unit [17,18,20,21]. One of the advantages of the ladder-type PDAs is considered to be increase of $\pi$-conjugated backbone density. If two butadiyne moieties are incorporated in a fixed length of a monomer molecule, doubled backbone density compared with the single chain PDA is expected. In addition to this virtue, the $\pi$-conjugation between polymer backbones is also possible, when $\pi$-conjugated moieties are used to link two PDA backbones. These strategy to prepare novel PDAs for third-order NLO materials are schematically displayed in Figure 1.

Single Chain PDAs

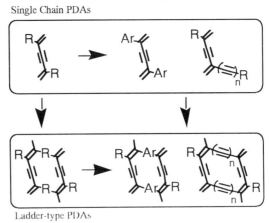

Ladder-type PDAs

Figure 1. Strategy to prepare novel PDAs for third-order NLO materials.
R and Ar represent aliphatic and aromatic groups, respectively.

# 3. Morphological Engineering

For morphological engineering of PDAs, growth of monomer single crystals is one of the most important tasks, because PDA single crystals succeed the size, shape and perfection of the corresponding monomer single crystals. However, few reports have been published on monomer single crystal growth of PDA related compounds with large enough size and good optical quality [22]. Since crystal growth ability fairly depends on monomer molecules, molecular engineering for butadiyne monomers to satisfy high polymerizability together with good crystallization properties is requested. However, achievement of both requirements is no so easy. Thus, we have been studying microcrystals as an alternative way to obtain good optical quality materials. This is because light scattering from crystals is known to be considerably diminished, if the size becomes far less than input light wavelength.

## 3.1. MICROCRYSTAL LIQUID DISPERSION

Here, we focus on microcrystallization of 1,6-di(N-carbazolyl)-2,4-hexadiyne (DCHD) [9,23-25]. Figure 2 shows morphological change of DCHD microparticles with time elapsed after the reprecipitation by SEM observation. Just after reprecipitation from acetone solution in water, the shape of microparticle was roughly sphere as shown in Fig. 2(a). At this stage, solid-state polymerization did not occur at all by UV irradiation, and the DCHD dispersion showed only pail yellow color. However, over 10 minutes, the monomer dispersion gradually turned into blue dispersion by UV irradiation, indicating that the solid-state 1,4-addition polymerization has progressed. After 20 minute, the shape of reprecipitated DCHD was perfectly changed into a rectangle, similar to DCHD bulk crystals (Fig. 2(b)) [24,25].

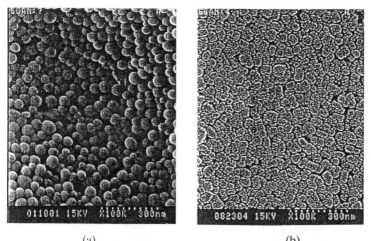

(a)                                    (b)

Figure 2. SEM photographs of DCHD microparticles under the different retention times: (a) 0 min and (b) 20 min, before UV irradiation to occur solid-state polymerization is carried out.

348

These facts suggest that, just after reprecipitation, DCHD monomer microparticles were amorphous and that they turned into microcrystals. When we measured absorption spectra of DCHD water dispersion after UV irradiation with differnt retention time from reprecipitation, the absorbance measured at 650 nm increased with the increasing retention time. This arises from exciton absorption in the $\pi$–conjugated chain of poly(DCHD) microcrystals. The absorption became constant with a retention time of *ca.* 10 minutes, suggesting that DCHD microcrystallization is completed within that period. Powder XRD measurements also support these considerations. That is, the diffraction peaks of microparticles were not detected just after reprecipitation, but the diffraction patterns very similar to those of DCHD bulk crystals appeared after retention times of more than 10 minutes.

Figure 2 also indicates that the average size of particles was almost the same before and after crystallization. This fact means that size of amorphous particles should be controlled in order to control final DCHD microcrystal size. The concentration of the injected DCHD solutions should be lowered to reduce the crystal size. Figure 3 shows differences of normalized intensity of scattered light ($I_s/I_o$) by changing solution concentration and the injection amount. According to Rayleigh scattering relation, $I_s/I_o$ is proportional to $NV^2$; where N and V are the number and the volume of setting objects, respectively. Experimentally, $I_s/I_o$ has been proportional to the injected amount at whole range of concentration, which implies increase of scattering objects *i.e.*, increase of number of DCHD microparticles. On the other hand, $I_s/I_o$ decreased with reducing concentration, which means the reduction of the volume of DCHD microparticles under a constant injection amount condition. In fact, the crystal size was controlled to be *ca.* 50 nm when the solution concentration was 2.5 mM [24,25]. DCHD amorphous microparticles with much smaller size are stably obtained when an anionic surfactant of SDS (sodium dodecylsulfate) was added in the acetone solution. In this system, the crystal size of poly(DCHD) microcrystals was much reduced to be *ca.* 15 nm, as shown in Figure 4(a) [25,26].

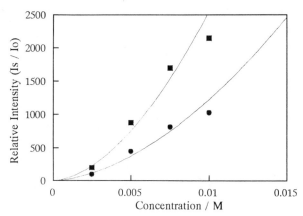

Figure 3. Relationship between concentration of injected solution and normalized intensity of scattered light ($I_s/I_o$). The solution amount was (●) 100 μl and (■) 200 μl, respectively.

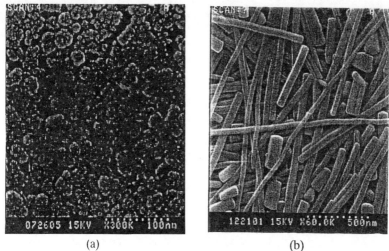

(a) (b)

Figure 4. SEM photographs of (a): 15 nm-DCHD microcrystals and
(b): DCHD crystalline microfiber.

In addition, when crystallization of DCHD microparticles dispersed in water were kept at 60 °C, the rate of crystallization became very slow, and fibrous DCHD microcrystals (60 nm × 60 nm × 5-10 μm in size) could be finally obtained, as shown in Figure 4(b). The microcrystals formed at the earlier stage are considered to play a role of nucleus for growing crystalline microfiber, which may be accelerated by the collision between DCHD particles and microcrystals [27]. These fibrous DCHD microcrystals were also solid-state polymerizable by UV-irradiation.

Figure 5 shows the excitonic absorption peak position ($\lambda_{max}$) of poly(DCHD) microcrystals dispersed in water depending on crystal size. The band gap energy of excitonic absorption was shifted to higher energy side with reducing crystal size [23,25,26]. The similar absorption shift was also found in microcrystals of perylene or cyanine dyes [6,28,29]. These phenomena looks like the well-known quantum size effect of semiconductor microparticles. However, the quantum size effect is generally observed only for semiconductor microparticles smaller than 10 nm. On the other hand, the size effect of poly(DCHD) microcrystals was exhibited at one order of magnitude larger crystal size than that of semiconductors, which is unknown and novel experimental results and is impossible to explain by use of conventional quantum size effect theory. This phenomenon could be attributed to strong exciton-phonon coupling. In the case of microcrystals with 30-nm size, excitonic band intensity is small compared with the corresponding phonon sideband intensity (See the spectrum in Figure 8(a)) [30]. This indicates that lattice vibration of small microcrystals comparatively larger than that of large microcrystals, and π-conjugated system along polymer backbone is less ordered in small microcrystals.

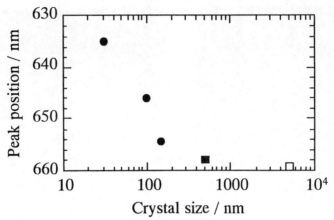

Figure 5. Relationship between crystal size and the excitonic absorption peak position of DCHD microcrystals.

## 3.2 MICROCRYSTALS IN SOLID MATRIX

Since the concentration of the compound in microcrystals in liquid dispersion was limited to be relatively low around $10^{-2}$ mol/l, we have tried to increase density of microcrystals to enhance the physical properties of bulk materials. First, we prepared microcrystal-dispersed polymer film. Namely, poly(DCHD) microcrystal water dispersion was concentrated and gelatin was added to it. Spin coating of the mixture on a quartz substrate gave a thin film containing microcrystals. However, the content of microcrystals in the film could not be more than several weight percents, if one obtains uniform films. Therefore, further density increase of microcrystals was performed using a different method.

Electrostatic adsorption of polyanions and polycations on oppositely charged substrates has been reported as a convenient technique to construct layered polymer assemblies [31]. A merit of this technique is to build well defined mulitilayer structures by piling uniform monolayer. The dispersion of organic microcrystals are so stable that the surface potential on the microcrystals is high. Actually, the poly(DCHD) microcrystals in water were found to have ζ-potential of about -40 mV. By utilizing this properties, the alternating layer-by-layer deposition of polycation and negatively-charged microcrystals successfully gave high-density microcrystal thin films [32]. First, polycation monolayer was deposited onto a clean glass slide by simply putting the glass into polydiallyl ammonium chloride (PDAC) aqueous solution. After rinsing, this glass was immersed into poly(DCHD) microcrystal-dispersed water, then rinsed in pure water again. This cyclic process could be repeated. After the deposition has been completed, the samples were dried and characterized. A SEM photograph of poly(DCHD) microcrystals electrostatically adsorbed on PDAC thin film after 12 cycles are shown in Figure 6. The poly(DCHD) microcrystals trapped were highly condensed and covered completely everywhere in the area of the SEM photograph. Figure 7 shows UV-VIS-absorption spectra of the poly(DCHD) microcrystal thin film in different deposition times. Absorbance at $\lambda_{max}$ was increased almost linearly without changing $\lambda_{max}$. Because the highly concentrated poly(DCHD) microcrystal thin film can be fabricated in large area with good optical quality, this film is much useful for NLO measurements. In extension, any desired pattern and super-lattice structures as well might be attainable for sophisticated optics.

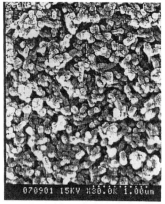

Figure 6. A micrograph of poly(DCHD) microcrystals deposited on PDAC thin film at 12 cycles

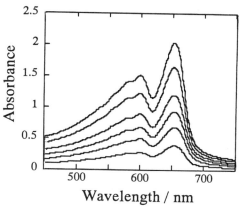

Figure 7. UV-VIS-absorption spectra of poly(DCHD) microcrystal thin film in different deposition times

## 4. NLO Properties

By using PDA microcrystal dispersion, Optical Kerr shutter experiment [33] was first performed for evaluation of third-order NLO properties [34]. In this experiment, light wavelength used for probe and pump beams were 813 nm and 1064 nm, respectively. For PDA microcrystal water dispersions in concentration of about $10^{-3}$ mol/l regarding to a monomer unit, obtained $\chi^{(3)}$ values were around $10^{-12}$ esu. When the $\chi^{(3)}$ values are normalized by the concentration of the dispersion system, they become about two order of magnitude larger than those for solution system of low-molecular-weight organic compounds. Namely, if the PDA concentration increases without increasing scattering loss, PDA microcrystals seems to be applied for NLO devices.

Third-order NLO property of PDA microcrystal dispersion was also investigated by z-scan method [35] for two different size of crystals. The input pulses for z-scan measurement were obtained by an optical parametric generator pumped by the third-harmonic wave from a mode-locked Nd:YAG laser with 20 ps pulse width and 10 Hz pulse repetition. Figure 8 shows $\chi^{(3)}$ values of two samples of poly(DCHD) water dispersion together with their absorption spectra. Averaged microcrystal sizes were 30 nm and 100 nm. For $Re\chi^{(3)}$ values, the sign is negative from the absorption edge to the absorption maximum, and it becomes positive in the wavelength region shorter than the absorption maximum for 100-nm size microcrystals. The sign change could be interpreted by three-level model [30]. Though the sign of $Re\chi^{(3)}$ is negative around the

352

excitonic band for 30-nm size microcrystals, it seems due to smaller excitonic peak at 635 nm than their phonon sideband peaks appearing below 600 nm. Regarding to $Im\chi^{(3)}$ values, the sign is negative in the excitonic band for both samples, indicating saturable absorption.

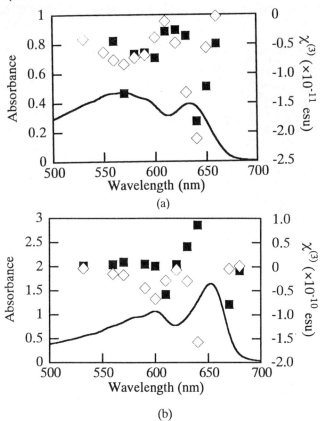

(a)

(b)

Figure 8. Absorption spectra and $\chi^{(3)}$ values evaluated by z-scan method for water dispersion of poly(DCHD) microcrystals with size of about 30 nm (a) and about 100 nm (b). Closed and open squares indicate $Re\chi^{(3)}$ and $Im\chi^{(3)}$, respectively.

TABLE 1. The $Re\chi^{(3)}$ values of poly(DCHD) microcrystals evaluated by z-scan method

| Crystal size (nm) | Sample form | Wavelength (nm) | $Re\chi^{(3)}$ (esu) |
|---|---|---|---|
| *ca.* 30 | Water dispersion ($4.4\times10^{-4}$ mol/l) | 640 | $-1.8\times10^{-11}$ |
| *ca.* 100 | Water dispersion ($7.4\times10^{-4}$ mol/l) | 670 | $-7.8\times10^{-11}$ |
| | | 640 | $8.7\times10^{-11}$ |
| *ca.* 100 | Dispersion in gelatin film (2.4wt%) | 670 | $-8.8\times10^{-10}$ |
| *ca.* 100 | Alternately deposited film with polycation | 670 | $-1.8\times10^{-7}$ |

TABLE 1 summarizes the $Re\chi^{(3)}$ values of poly(DCHD) microcrystals evaluated by z-scan method in various sample forms. Large $Re\chi^{(3)}$ values are required for ultrafast all-optical switching. For the 100-nm size poly(DCHD) dispersion, the largest $Re\chi^{(3)}$ values around the excitonic absorption band were $-7.8 \times 10^{-11}$ esu at 670 nm and $+8.7 \times 10^{-11}$ esu at 640 nm. The $Re\chi^{(3)}$ values of 30-nm size microcrystals were less than those of 100-nm size microcrystals, and is seems to be consistent with the excitonic absorption intensity as mentioned above. For a gelatin film containing 2.4wt% of poly(DCHD) microcrystals with 100-nm size, the $Re\chi^{(3)}$ value increased about one order of magnitude compared with water dispersion systems, i.e. $-8.8 \times 10^{-10}$ esu at 670 nm [36]. For a layer-be-layer deposited film containing the 100-nm size microcrystals in high density showed much larger $Re\chi^{(3)}$ value of $-1.8 \times 10^{-7}$ esu at 670 nm. The enhancement in $\chi^{(3)}$ value for the films could be simply explained by microcrystal density increase in the films. In these samples, microcrystals were randomly oriented, and a little more enhanced $Re\chi^{(3)}$ value may be anticipated, if the orientational control of microcrystals becomes possible on a substrate.

Using a poly(DCHD) microcrystal dispersed in gelatin film as a nonlinear spacer layer in a Fabry-Perot cavity, a large Fabry-Perot fringe shift was observed recently [37]. In this experiment, a poly(DCHD) microcrystal film was sandwiched between reflecting dielectric-coated substrates, which had more than 95% of reflectivity between 700 and 800 nm and nearly transparent at 645 nm. The pump-probe technique using resonant pump pulses at 645 nm and white-light probe pulses was employed to confirm its optical switching behavior. When a pump pulse with a duration of 800 fs, a peak irradiance of 40 MW/mm$^2$ and energy per pulse of 4 $\mu$J was irradiated, the peak at 686 nm, which is one of the transmission maximum peaks of the Fabry-Perot fringe, was shifted to shorter wavelength by 4.2 nm. This fringe shift was observed to recover within 3 ps. This blue shift was calculated to correspond to refractive index change $\Delta n$ of 0.009 for a 850-nJ absorbed energy. The magnitude of the refractive cross-section, i.e. change in refractive index per generated excitation carrier per unit volume, was calculated to be $1.6 \times 10^{-21}$ cm$^3$ at 670 nm. These values were found to satisfy criterion for the materials to be used as optical switching devices.

## 5. Concluding Remarks

We could produce novel PDA derivatives and microcrystals as third-order NLO materials. based on molecular design and morphological engineering. For molecular design, we have succeeded to prepare a variety of PDA derivatives with $\pi$-conjugation between polymer backbone and substituents and ladder-type PDAs. Though the $\chi^{(3)}$ values of these new PDAs was not mentioned in this article, some of them have already been found to have larger $\chi^{(3)}$ values than the conventional PDAs [13,16]. Combination of these designed PDAs and microcrystallization is the remaining interesting subject. For morphological engineering, we have established the effective and simple reprecipitation method to prepare organic microcrystals with controlled size and shape. The highly concentrated poly(DCHD)-microcrystal thin film with good optical quality and large area could be obtained, and the films were found to show three order of magnitude larger $\chi^{(3)}$

compared with the original water dispersion. In the Fabry-Perot experiment, materials based on PDA microcrystals were found to satisfy criterion for the optical switching devices[37].

## 6. Acknowledgments

We express our sincere gratitude to Drs. H. Matsuda, T. Fukuda, R. Rangel-Rojo, E. Van Keuren and S. Yamada of National Institute of Materials and Chemical Research; Drs. T. Kaino and H. Kanbara of Opto-Electronics Laboratories, NTT; Drs. S. K. Tripathy, J. Kumar and S. Balasubramanian of Department of Chemistry and Physics, University of Massachusetts Lowell; Drs. S. Wherrett, A. K. Kar and S. C. Smith of Department of Physics, Heriot-Watt University; Drs. H. Oikawa and H. Katagi of Institute for Chemical Reaction Science, Tohoku University, for collaborative research cited in this paper.

## 7. References

1. Kubo, R. (1962) Electric properties of metallic fine particles. 1., *J Phys Soc Jpn* **17**, 975-86.
2. Hanamura, E. (1988) Very large optical nonlinearity of semiconductor microcrystallites, *Phys. Rev.* **B37**, 1273-1279.
3. Tokizaki, T., Akiyama, H., Tanaka, M., Nakamura, A. (1992) Linear and nonlinear optical properties of CdSe microcrystallites in glasses, *J. Cryst. Growth* **117**, 603-607.
4. Tang, Z.K., Nozue, Y., Goto, T. (1991) Quantum size effect on the exciton energy and the oscillator strength of $PbI_2$ clulasters in Zeolites, *J. Phys. Soc. Jpn* **60**, 2090-2094.
5. Kasai, H., Nalwa, H.S., Oikawa, H., Okada, S., Matsuda, H., Minami, N., Kakuta, A., Ono, K., Mukoh, A. and Nakanishi, H. (1992) A novel preparation method of organic microcrystals, *Jpn. J. Appl. Phys.* **31**, L1132-L1134.
6. Kasai, H., Yoshikawa, Y., Seko, T., Okada, S., Oikawa, H., Matsuda, H., Watanabe, A., Ito, O., Toyotama, H., and Nakanishi, H. (1997) Optical properties of perylene microcrystals, *Mol. Cryst. Liq. Cryst.* **294**, 173-176.
7. Kasai, H., Kamatani, H., Yoshikawa, Y., Okada, S., Oikawa, H., Watanabe, A., Ito, O., and Nakanishi, H. (1997) Crystal size dependence of emission from perylene microcrystals, *Chem. Lett.*, 1181-1182.
8. Wegner, G. (1969) Topochemische Reaktionen von Monomeren mit konjugierten Dreifachbindungen I. Mitt.: Polymerisation von Derivaten des 2.4-Hexadiin-1.6-diols im kristallinen Zustand, *Z. Naturforsch.* **24b**, 824-832. [in German]
9. Iida, R., Kamatani, H., Kasai, H., Okada, S., Oikawa, H., Matsuda, H., Kakuta, A., and Nakanishi, H. (1995) Solid-state polymerization of diacetylene microcrystals, *Mol. Cryst. Liq. Cryst.* **267**, 95-100.
10. Sauteret, C., Hermann, J.P., Frey, R.,. Pradere, F., Ducuing, J., Baughmann, R.H., and Chance, R.R. (1976) Optical nonlinearities in one-dimensional-conjugated polymer crystals, *Phys. Rev. Lett.* **36**, 956-959.

11. Matsuda, H., Nakanishi, H., Hosomi, T., and Kato, M. (1988) Synthesis and solid-state polymerization of a new diacetylene: 1-(N-carbazolyl)penta-1,3-diyn-5-ol, *Macromolecules* **21**, 1238-1240.

12. Nakanishi, H., Matsuda, H., Okada, S., and Kato, M. (1989) Preparation and nonlinear optical properties of novel polydiacetylenes, in T. Saegusa, T. Higashimura, and A. Abe (eds.), *Frontiers of Macromolecular Science*, Blackwell Scientific Publications, Oxford, pp.469-474.

13. Okada, S., Ohsugi, M., Masaki, A., Matsuda, H., Takaragi, S., and Nakanishi, H. (1990) Preparation and nonlinear optical property of polydiacetylenes from unsymmetrical diphenylbutadiynes with trifluoromethyl substituents, *Mol. Cryst. Liq. Cryst.* **183**, 81-90.

14. Sarkar, A., Okada, S., Nakanishi, H., and Matsuda, H. (1998) Polydiacetylenes from asymmetrically substituted diacetylenes containing heteroaryl side groups for third-order nonlinear optical properties, *Macromolecules* **31**, 9174-9180.

15. Okada, S., Hayamizu, K., Matsuda, H., Masaki, A., and Nakanishi, H. (1991) Structures of the polymers obtained by the solid-state polymerization of diyne, triyne, and tetrayne with long alkyl substituents, *Bull. Chem. Soc. Jpn.* **64**,. 857-863.

16. Okada, S., Doi, T., Mito, A., Hayamizu, K., Ticktin, A., Matsuda, H., Kikuchi, N., Masaki, A., Minami, N., Haas, K.-H., and Nakanishi, H. (1994) Synthesis and third-order nonlinear optical properties of a polydiacetylene from an octatetrayne derivative with urethane groups, *Nonlinear Opt.* **8**, 121-132.

17. Okada, S., Hayamizu, K., Matsuda, H., Masaki, A., Minami, N., and Nakanishi, H. (1994) Solid-state polymerization of 15,17,19,21,23,25-tetracontahexayne," *Macromolecules* **27**, 6259-6266.

18. Matsuzawa, H., Okada, S., Matsuda, H., and Nakanishi, H. (1996) Synthesis and optical properties of polydiacetylenes from dodecahexayne derivatives, in R.A. Lessard (ed.), *Photopolymer Device Physics, Chemistry, and Applications III, Proc. SPIE* **2851**, SPIE, Bellingham, pp.14-25.

19. Enckelmann, V. (1984) Structural aspects of the topochemical polymerization of diacetylenes, in H.-J. Cantow (ed.), *Polydiacetylenes, Adv. Polym. Sci.* **63**, Springer-Verlag, Berlin, pp. 91-136.

20. Matsuzawa, H., Okada, S., Sarkar, A., Matsuda, H. and Nakanishi, H. (1999) Synthesis of ladder polymers containing polydiacetylene backbones connected with methylene chains and their properties, *J. Polym. Sci. Pt. A: Polym. Chem.* **37**, 3537-3548.

21. Matsuzawa, H., Okada, S., Sarkar, A., Nakanishi, H., and Matsuda, H., Synthesis of polydiacetylenes from novel derivatives having two diacetylene units linked by an aromatic group, in preparation.

22. Thakur, M. and Meyler, S. (1985) Growth of large-area thin-film single crystals of poly(diacetylenes), *Macromolecules* **18**, 2341-2344.

23. Nakanishi, H. and Kasai, H. (1997) Polydiacetylene microcrystals for third-order nonlinear optics, *ACS. Symp. Ser.* **672**, 183-198.

24. Katagi, H., Kasai, H., Okada, S., Oikawa, H., Komatsu, K., Matsuda, H., Liu, Z., and Nakanishi, H. (1996) Size control of polydiacetylene microcrystals, *Jpn. J. Appl. Phys.* **35**, L1364-L1366.

25. Katagi, H., Kasai, H., Okada, S., Oikawa, H., Matsuda, H., and Nakanishi, H. (1997) Preparation and characterization of poly-diacetylene microcrystals, *J. Macromol. Sci.-Pure Appl. Chem.* **A34**, 2013-2024.

26. Kasai, H., Katagi, H., Iida, R., Okada, S., Oikawa, H., Matsuda, H., and Nakanishi, H.(1997) Preparation of polydiacetylene microcrystals and their properties, *Nippon Kagaku Kaishi* , 309-317. [in Japanese]

27. Oshikiri, T., Kasai, Katagi, H., H., Okada, S., Oikawa, H., Nakanishi, H. (in press) Fabrication of polydiacetylene fibrous microcrystals by the reprecipitation method, *Mol. Cryst. Liq. Cryst.*

28. Kasai, H., Kamatani, H., Okada, S., Oikawa, H., Matsuda, H., and Nakanishi, H. (1996) Size-dependent colors and luminescences of organic microcrystals, *Jpn. J. Appl. Phys.* **35**, L221-223.

29. Kamatani, H., Kasai, H., Okada, S., Matsuda, H., Oikawa, H., Minami, N., Kakuta, A., Ono, K., Mukoh, A., and Nakanishi, H., (1994) Preparation of J-aggregated microcrystals of pseudoisocyanine, *Mol. Cryst. Liq. Cryst.* **252**, 233-241.

30. Rangel-Rojo, R., Yamada, S., Matsuda, H., Kasai, H., Nakanishi, H., Kar, A.K., and Wherrett, B.S. (1998) Spectrally resolved third-order nonlinearities in polydiacetylene microcrystals: Influence of particle size, *J. Opt. Soc. Am. B* **15**, 2937-2945.

31. Decher, G. (1997) Fuzzy nanoassemblies: Toward layered polymeric multicomposites, *Science* **277**, 1232-1237.

32. Tripathy, S.K., Katagi, H., Kasai , H., Balasubramanian, S., Oshikiri, H., Kumar, J., Oikawa, H., Okada, S., and Nakanishi, H. (1998) Self assembly of organic microcrystals 1: Electrostatic attachment of polydiacetylene microcrystals on a polyelectrolyte surface, *Jpn. J. Appl. Phys.* **37**, L343-L345.

33. Duguay, M.A. and Hansen, J.W. (1969) An ultrafast light gate, *Appl. Phys. Lett.* **15**, 192-196.

34. Kasai, H., Iida, R., Kanbara, H., Okada, S., Matsuda, H., Oikawa, H., Kaino, T., and Nakanishi, H. (1996) Organic microcrystals as optical Kerr shutter materials, *Nonlinear Opt.* **15**, 263-266.

35. Sheik-Bahae, M., Said, A.A., Wei, T.-H., Hagan, D.J., and Van Stryland, E.W. (1990) Sensitive measurement of optical nonlinearities using a single beam, *IEEE J. Quantum Electron.* **26**, 760-769.

36. Matsuda, H., Yamada, S., Van Keuren, E., Katagi, H., Kasai, H., Okada, S., Oikawa, H., Nakanishi, H., Smith, E.C., Kar, A.K., and Wherrett, B.S. (1997) Nonlinear refractive indices of polydiacetylene microcrystals, in M.P. Andrews (ed.), *Photosensitive Optical Materials and Devices*, *Proc. SPIE* **2998**, SPIE, Bellingham, pp.241-248. Polydiacetylene microcrystal content in a gelatin film in this paper should be corrected to 2.4wt%.

37. Bakarezos, M., Camacho, M.A., Blewett, I.J., Kar, A.K., Wherrett, B.S., Matsuda, H., Fukuda, T., Yamada, S., Rangel-Rojo, R., Katagi, H., Kasai, H., Okada, S., and Nakanishi, H. (1999) Ultrafast nonlinear refraction in integrated Fabry-Perot etalon containing polydiacetylene, *Electron. Lett.* **35**, 1078-1079.

# STRONG COUPLING IN ORGANIC SEMICONDUCTOR MICROCAVITIES BASED ON J-AGGREGATES

D. G. LIDZEY[*a], D. D. C. BRADLEY[a], A. ARMITAGE[a], T. VIRGILI[a], M. S. SKOLNICK[a] AND S. WALKER[b]

[a]*Department of Physics and Astronomy, University of Sheffield, Hicks Building Hounsfield Road, Sheffield S3 7RH, U.K*
[b]*Department of Electronic and Electrical Engineering, University of Sheffield, Mappin Street, Hounsfield Road, Sheffield S1 3JD, U.K*

[*] author for correspondance : d.g.lidzey@sheffield.ac.uk

**Keywords / Abstract:** microcavity / cavity polariton / organic materials / J-aggregates / electroluminescence / LED

We report a room temperature study of the strong coupling regime in a planar microcavity, using 'J-aggregates' of cyanine dyes. The characteristic features of energetic anticrossing between photon and exciton clearly observed, indicating the formation of cavity polaritons. Rabi-splittings as large as large 80 meV are measured. We generate polariton emission via non-resonant excitation and find that emission from the lower polariton branch dominates. We present light emitting diodes (LEDs) based on J-aggregates / polymer composite films, indicating that electrically generated polariton emission may also be possible.

## 1. Introduction

Planar semiconductor microcavities (structures that consist of wavelength dimension semiconductor layers positioned between two mirrors) allow systematic control over solid-state exciton-photon interactions. The cavity quantises the local electromagnetic field into a discrete set of resonant photon modes that interact with the

*F. Kajzar and M.V. Agranovich (eds.), Multiphoton and Light Driven Multielectron Processes in Organics: New Phenomena, Materials and Applications, 357–370.*

optical transitions of the semiconductor. The result is a modification of the semiconductor absorption and emission characteristics that is of both fundamental [1] and practical interest [2,3]. Specific examples of microcavity device structures include surface emitting lasers [4] and resonant light emitting diodes for displays [5].

Within a microcavity there are two regimes into which interactions between the confined electromagnetic field (cavity photon mode) and the optical transitions of a material can be classified, namely weak- and strong-coupling [6]. The strong coupling regime is realised when the interaction strength (Rabi-frequency $\Omega$) between photon and exciton is larger than both the inverse cavity photon lifetime and inverse exciton dephasing time. The dephasing time is a measure of how quickly an ensemble of excitons loose mutual coherence in their wavefunctions (e.g. by phonon scattering or other relaxation processes). Under such conditions, a coherent superposition of the exciton and photon states can occur. The emission of a photon by an exciton is then no longer an irreversible event; the exciton emits the photon into a single mode of the cavity and is then free to subsequently reabsorb it. The strong coupling regime can be identified by the observation of energetic anti-crossing between the photon and exciton modes [7]. At the minimum separation between photon and exciton, a doublet of cavity polariton states separated by the vacuum Rabi-splitting energy is observed. The linewidth of the polariton states is the average of that of the uncoupled photon and exciton modes [8]. Recently, the quasi-particle nature cavity polaritons has been verified [9] by the observation of stimulated scattering of excitons into polariton (bosonic) states.

In the weak coupling regime, the interaction strength ($\Omega$) between the photon and exciton is smaller than the inverse cavity lifetime and inverse exciton dephasing time. In this limit, the photon and exciton become mutually dephased before they can form a coherent superposition of states. In such circumstances, the emission of a photon by an exciton is an irreversible event.

In this paper, we discuss our experiments to access the strong coupling regime using organic materials. This is illustrated by data showing energetic anticrossing linewidth averaging and also optically-excited room-temperature polariton emission. We further show that it may be possible to electrically generate polariton emission.

## 2. Background

The strong coupling regime in a microcavity was first observed in microcavities containing multiple inorganic quantum wells (QW) [7]. As stated above, the interaction strength (expressed as a frequency) between the photon and exciton must be greater than the inverse exciton dephasing time and inverse photon damping time. In typical microcavity QW structures (utilising GaAs/InGaAs), Rabi splittings of around $\hbar\Omega = 5$ meV have been observed (at 20K). Expressing the Rabi-splitting as an inverse frequency, this energy corresponds to oscillations with a period of 800 fs. Inorganic QW exciton states can have dephasing times of ~ 1 ps [6] i.e. on a longer time-scale. Using sophisticated semiconductor growth techniques it is possible to define microcavity structures with very narrow cavity-photon linewidths (down to ~ 0.1 meV) [10], corresponding to cavity-photon damping rates of ~ 6 ps, again satisfying the conditions discussed above. In addition to limitations of exciton dephasing and photon damping rates, it is necessary that the inhomogeneous linewidth of the exciton transition be narrower than the Rabi-splitting. This is because a broad linewidth effectively encompasses a large distribution of narrower (homogeneous) states, which in a cavity results in a distribution of coupling strengths, effectively masking the anticrossing behaviour.

It is because of restrictions regarding linewidths, that it was generally assumed that organic materials could not be used in a microcavity to access the strong coupling regime. Organic materials generally have inhomogeneous transition linewidths between 0.5 and 1 eV, the result of both heterogeneity in the molecular environment and the presence of a vibronic progression. There are however exceptions to this; in an ordered crystalline form, some organic materials [12] have linewidths as small as 1 meV (at 2K). However such crystalline solids frequently have many closely spaced optical transitions which for reasons discussed above make the observation of strong coupling more problematic. Organic materials also have fast dephasing times, which are often around 100 fs [11], therefore this also would appear problematic for strong coupling.

In order to investigate whether strong-coupling coupled be achieved using organic materials, we have identified and utilised [13,14] a number of different materials with linewidths between ~ 60 and 90 meV, and very strongly suppressed vibronic replicas. These linewidths are still much larger than inorganic QW exciton linewidths and the Rabi-splittings observed in inorganic QW microcavities. However as will be evidenced below, the Rabi-splittings that we observe using organic semiconductors are often over two orders of magnitude larger than in inorganic microcavities due to the large oscillator strengths of organic materials. As a consequence we have been able to demonstrate strong coupling in organic semiconductor microcavities for the first time. Giant Rabi-splittings ($\leq$ 180 meV) that persist to room temperature (due to the large exciton binding energy) are found. In this paper, we describe our work on strong coupling J-aggregates of a cyanine dye, and present new data on their potential application in electrically addressed devices.

## 3. Experimental Methods

The cyanine dye that we have used is (2,2'dimethyl-8-phenyl-5,6,5',6'-dibenzothiacarbocyanine chloride) [manufacturer's code name NK2567]. The chemical structure of this material is shown in the inset to figure 1. In common with other cyanine dyes it carries a net charge which in suitable solvents drives a self-association of the molecules to form a J-aggregate structure [15]. It is this structure that possesses the desired spectral features that we wished to utilise for our experiments. To form stable thin films containing the aggregates, the cyanine dye was dissolved with polyvinyl alcohol [PVA] in an 50:50 mixture of water and methanol. All solutions were heated to ~80°C to dissolve the dye, and then filtered using a 5μm pore-size filter. The solution was then spin coated onto either quartz or dielectric mirror substrates to form thin films of cyanine dye J-aggregates dispersed in a PVA matrix. Excellent optical-quality films of controlled thickness could be prepared by adjusting the spin speed.

Figure 1 shows the room temperature optical absorption and photoluminescence emission from a PVA/NK2567 film prepared on a quartz substrate. The spectra are typical of J-aggregates having an absorption (emission) linewidth of 58 meV (42 meV). As expected for the J-aggregate structure the Stokes shift between absorption and emission maxima is small compared to the absorption linewidth (13 meV) and the vibronic replica in both absorption and emission is weak.

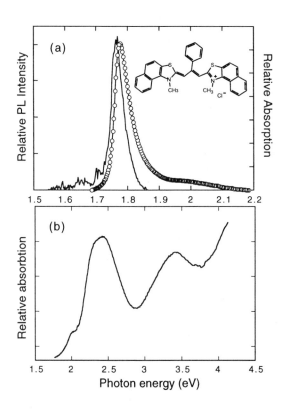

*Figure 1: (a) Photoluminescence (line) and optical absorption (open-circles) of J-aggregates of the dye NK2567.*

*(b) Optical absorption of amorphous cyanine dye film.*

The structure of our cavity devices is shown in figure 2, along with a theoretical prediction of the optical structure of the device at the cavity mode wavelength. Microcavities were fabricated using 1cm$^2$ dielectric mirror substrates consisting of 9 alternating $\lambda/4$ pairs of $SiO_2$ (n=1.45) and $Si_xN_y$ (n=1.95) deposited onto quartz substrates via plasma enhanced chemical vapour deposition. The cyanine dye dispersed in PVA matrix was deposited by spin coating on top of the dielectric mirror and then a silver top mirror was evaporated onto the PVA/Cyanine film. The organic films had a thickness of ~ 180 nm (n ~ 1.6) which formed a $\lambda/2$ cavity (with $\lambda$ ~ 710 nm at normal incidence). This resulted in the antinode of the confined photon field positioned in the middle of the organic layer, hence ensuring good coupling with the cyanine dye aggregates. Cavities had a photon linewidth of ~ 18 meV equivalent to a Q factor of ~ 90.

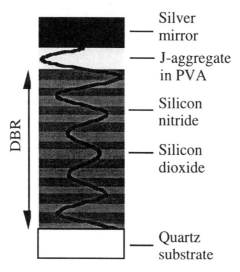

*Figure 2: Generic structure of the microcavities used in this work.*

We have also investigated using J-aggregates as the active emissive materials in organic light emitting diodes (LEDs). In our devices, we have used a cyanine dye closely related to NK2567, having a peak absorption / emission at around 670 nm (1.85 eV). We use the same combination of PVA / cyanine dye that was used in the microcavity experiments in the LED as a single active layer. This is not anticipated to give optimal device performance as the PVA host matrix is essentially insulating. However this combination does undergo strong-coupling in an appropriate microcavity format, in principle demonstrating the possibility of generating electrical emission in a strongly coupled LED.

Devices were fabricated by spin-coating a 100 nm (PVA / cyanine dye) layer onto an indium tin oxide (ITO) coated glass substrate and then coating the organic film with a final layer of calcium. Electrical contact was made to both the ITO (anode) and calcium (cathode). Measurements of the current-voltage characteristics were made, and electroluminescent emission spectra from the device were collected. All fabrication and testing were done in a nitrogen filled glove box to reduce device degradation.

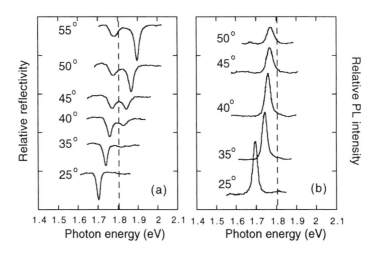

*Figure 3: (a) Reflectivity and (b) emission of the cavity as a function of angle.*

## 4. Results

To confirm strong coupling within a microcavity, it is necessary to observe energetic anticrossing between the photon and exciton modes. This requires the relative energy separation between the cavity-photon and exciton to be varied. One convenient method is by 'angle-tuning' [16]; this makes use of the variation of the photon mode energy as $1/\cos\theta_{int}$ where $\cos\theta_{int}$ is the internal angle between the cavity axis and the measurement direction. By adjusting the measurement angle one is thus able to map out the dispersion of the photon mode and observe its behaviour as it approaches resonance with the non-dispersing exciton mode.

The cavity was designed such that at normal incidence the photon mode was approximately 150 meV lower in energy than the exciton mode. Figure 3(a) shows the variation in room temperature reflectivity spectra as the photon mode is angle-tuned through the exciton mode energy. The approximate peak of the exciton absorbance is marked by a vertical dashed line.

At 25° the photon mode is ~ 80 meV from resonance with the exciton mode and can be identified as a sharp reflectivity dip with 20 meV full width half maximum. At such large detunings, the interaction between the photon and exciton is weak and thus the exciton cannot be detected outside the cavity. As the measurement angle increases the photon mode approaches resonance with the exciton and the signatures of strong coupling are observed. The reflectivity spectrum evolves into a pair of reflectivity dips as the exciton mode takes on a photon character. The lower energy feature broadens and loses intensity. At 45° two equal intensity and equal linewidth features separated by 84 meV are observed. These features can be formally described as an equal mixture of a photon and an exciton. For larger angles beyond resonance, the higher energy mode reverts to being 'photon-like', gains in intensity and narrows in linewidth.

The separation between the polariton modes as a function of angle is plotted in the form of a dispersion curve in figure 4(a). The horizontal dashed line corresponds to an energy very close to the peak of the exciton absorbance. A clear anti-crossing of the photon and exciton is observed. The closest approach between the polariton modes occurs between 40° and 42.5° and is approximately 80 meV. This defines zero detuning between the photon and exciton and hence resonance.

The observation of anti-crossing provides conclusive evidence for strong-coupling [8]. The correct description of the coupled modes is then as cavity polaritons, and we find a room temperature Rabi-splitting of 80 meV. The magnitude of the Rabi-splitting is approximately 20 times larger than that seen in microcavities based on (InGaAs/GaAs ) QWs [7].

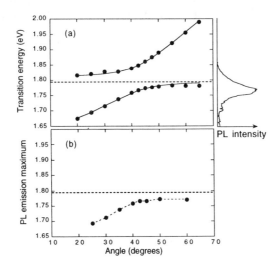

*Figure 4: Microcavity dispersion of (a) reflectivity and (b) emission.*

The large increase in Rabi-splitting is principally due to the very much enhanced oscillator strength of organic semiconductors. We have calculated (for similar cavities instead containing a Zn porphyrin dye), that the exciton oscillator strength of the organic films can be up to 100 times larger than that of multiple InGaAs QWs. Thus in our device whilst the inhomogeneous linewidth of the J-aggregates is perhaps 60 times larger than typical inorganic QW excitons, their enhanced splitting allows the strong-coupling regime to be resolved. The large Rabi-splittings imply much smaller Rabi oscillation-periods ($2\pi/\Omega$). For example, splittings with $\hbar\Omega = 80$ meV observed here, would correspond to an oscillation period of 50 fs. Thus whilst the dephasing time of organic excitons is short ~ 100 fs, the enhanced oscillator strength results in a *smaller* Rabi-period. Ultra-fast measurements to determine the temporal properties of these cavities are planned.

Additional experiments were made to investigate the c.w. emission properties of the strongly coupled system. Excitation was with 5 mW cm$^{-2}$ radiation from a HeNe laser (1.96 eV, 632.8 nm). Experiments were performed at room temperature with the excitation light directed onto the microcavity at normal incidence. At 1.96 eV the reflectivity of the dielectric mirror is substantially reduced from its peak value. This allows the light to efficiently enter the cavity and be absorbed by the cyanine dye aggregates. The absorption spectrum in figure 1 shows that at 1.96 eV the absorption coefficient is only 7% of its peak value but this is sufficient to generate PL emission.

The PL emission spectra collected for a range of measurement angles are shown in figure 3(b). This is, to our knowledge, the first report of room temperature cavity-polariton emission from any semiconductor. The emission peak disperses as a function of angle as shown in figure 4(b). It is clear that the emission follows the dispersion of the lower polariton branch with essentially no emission seen from the upper branch. Similar behaviour has been previously reported for PL from InGaAs quantum well microcavities in the strong-coupling regime, where the emission closely followed the polariton absorption multiplied by a Boltzman occupancy factor [17]. In this case, the large reservoir of uncoupled excitons can scatter into both the upper and lower polariton branches. The linewidth of an inorganic exciton (which is determined by a Boltzman distribution) determines the relative proportion of excitons scattered into each polariton branch and hence the distribution of PL.

In our case, the emission comes from a slightly relaxed form of the lattice. This small but finite Stokes shift between absorbance and emission can be seen in figure 1(a). To the right of figure 4(a), the PL emission intensity from uncoupled excitons is re-plotted. It can be clearly seen that around the resonance angles (40 to 45°) the uncoupled excitons can only energetically scatter into the lower polariton branch, and hence the emission from the upper branch is essentially zero. The emissive species extends down in energy to 1.55 eV, and therefore excitons can scatter into the lower polariton branch even at large negative detunings (corresponding to angles of 25°). At such large negative detunings, the lower branch has a large photon-like component to its character, and thus it can easily couple out of the cavity by photon emission. In contrast, at these large detunings, the upper-branch is principally exciton like and cannot be easily detected from outside the cavity in reflectivity, and also cannot easily couple out of the cavity. Thus even though there is a small overlap between the emissive species and the upper branch at 25°, the 'exciton-like' polaritons cannot couple out of the cavity.

Figure 5 shows the absorption of the J-aggregate film used in the LED, along with its electroluminescence (EL) spectrum. The inset to the figure shows the current-voltage characteristic of the device. The absorption and emission look qualitatively similar to the absorption and PL emission plotted in figure 1(a), however we note that the EL emission linewidth is slightly broader than its normal fluorescence (not shown). The broadening and redshift of emission can be explained by strong self-absorption within the device, as this has the effect of increasing the relative proportion of the lower-energy tail in the spectrum. Using a calibrated integrating sphere, we have measured the EL quantum efficiency of the LED and the PL quantum efficiency of the J-aggregate film fluorescence. We find maximum values of $\Phi_{EL}$ of 0.008% and $\Phi_{PL}$ = 15%.

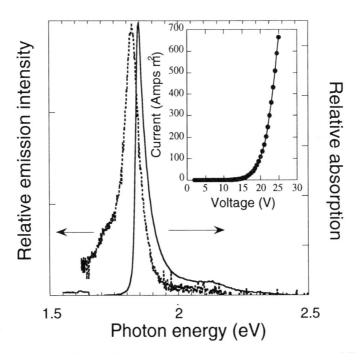

*Figure 5: Absorption and electroluminescence emission from a J-aggregate LED. Inset shows the current-voltage device characteristics.*

We have incorporated this J-aggregate in the microcavity shown in figure 2, and have successfully observed strong coupling phenomena similar to those described above. Therefore by modifying the microcavity structure to include an ITO (anode) layer and utilising the top mirror to inject electrons, a strongly coupled LED could be created. This would allow the possibility of electrically generated room temperature polariton emission.

## 5. Discussion

It is apparent that the J-aggregate structure is very significant in allowing the observation of strong-coupling. Figure 1(b) shows the absorption spectrum of the cyanine-dye used in these experiments, deposited onto a room-temperature quartz substrate by vapour deposition. The vapour deposition technique produces an essentially amorphous film, with little opportunity for the molecules to self-organise and produce J-aggregates. The absorption spectrum is fairly typical of molecular materials with a series of broad absorptive features. The linewidths of the transitions in this material are typically around 700 meV derived both from inhomogeneous broadening and from the presence of vibronic replicas (note the spectra shown in fig 1(a) and (b) are displayed on different energy scales). This contrasts with the optical and electronic structure of the aggregate, where exciton coupling leads to a band of delocalised states of which only the lowest lying have dipole allowed transitions to the ground state. Such exciton exchange phenomena result in a red-shifted and much narrower absorption peak and leads to super-radiant emission at low temperature [18]. Thus whilst the isolated molecules would be poorly suited for observing strong-coupling, the J-aggregate structure makes this regime readily accessible. We have also shown that J-aggregates of other organic semiconductors are also good candidates for the observation of strong-coupling.

We have demonstrated that materials which undergo strong coupling can also emit electroluminescence. It is interesting to speculate about the possible applications of such systems. It has been shown that the lifetime of strongly coupled excitons is very different from uncoupled excitons. For example [6], a strongly coupled multiple

QW system was excited both on resonance and above resonance. On resonance, up conversion measurements demonstrated the lifetime of the polaritons to be 260 fs (which is essentially governed by the lifetime of the cavity-photon component of the polariton). When excited above resonance, the fluorescence lifetime increased to 600 ps. It was concluded that the increase in decay rate was dominated by the time taken for the uncoupled excitons to scatter into polariton states.

In principle, by coupling an exciton to a photon, the decay rate of the polariton (into a photon) is very much faster than the normal radiative rate (~1 ns), and can in fact be shorter than non-radiative decay channels that can compete with fluorescence. If the density of polariton states is not prohibitively low, then this may offer a route to increase the emission quantum yield. This could aid the construction of improved efficiency LEDs and also reduce the lasing thresholds of vertical cavity surface emitting lasers (VCSELS). However, to achieve such an enhanced decay rate by non-resonant excitation (such as achieved by electrical injection) the scattering rate of excitons (by interaction with phonons) into polariton states has to be very much faster than the normal fluorescence decay processes. This was not the case in the experiments described by Norris [6], as the relaxation of inorganic (Mott-Wannier) excitons by emission of acoustic phonons (to conserve momentum) is often slow resulting in the formation of bottlenecks. Such a situation does not occur with organic (Frenkel) excitons, as momentum is no longer a good quantum number to describe the exciton state, and thus energetic relaxation processes are much faster. Decay of excitons through vibronic levels can often occur over time-scales of ~100 fs, in inorganic materials this can take between 1 and 10 ns.

In conclusion we have reported the observation of strong-coupling in a microcavity containing cyanine dye J-aggregates. A giant room-temperature Rabi-splitting of 80 meV was obtained and room-temperature cavity polariton emission was observed for the first time. The data demonstrate the generality of strong-coupling in appropriately chosen organic semiconductors and establish J-aggregates as a desirable molecular structure with which to study the strong-coupling regime.

## Acknowledgments

DGL would like to thank the Lloyd's of London Tercentenary fund for a Research Fellowship. We thank the U.K. Engineering and Physical Sciences Research Council (GR/L 01916) "Conjugated Polymer Microcavities: Light Emitting Diodes and New Physics" for support of this work.

## References

1. M.S. Skolnick, T.A. Fisher, D.M. Whittaker (1998) *Semiconductor Science and Technology* **13** (7) 645-669

2. E.F. Schubert, A.M. Vredenberg, N.E.J. Hunt, Y.H. Wong, P.C. Becker, J.M. Poate, D.C. Jacobson, L.C. Feldman, G.J. Zydzik (1992) *Applied Physics Letters* **61** (12) 1381-1383

3. M.S. Ünlu, S. Strite (1995) *Journal of Applied Physics* **78** (2) 607-639

4. e.g. see "Surface-emitting lasers: a new breed" J. Jewell (July 1990) *Physics World* 28-30

5. D.G. Lidzey, D.D.C. Bradley, S.J. Martin, M.A. Pate (1998) *IEEE Journal of Selected Topics in Quantum Electronics* **4** (1) 113-118

6. T.B. Norris (1995) *Confined Electrons and Photons* Edited by E. Burstein and C. Weisbuch, Plenum Press, New York

7. C. Weisbuch, M. Nishioka, A. Ishikawa, Y. Arakawa (1992) *Physical Review Letters* **69** 3314-3317

8. Y. Zhu, D.J. Gauthier, S.E. Morin, Q. Wu, H.J. Carmichael, T.W. Mossberg (1990) *Physical Review Letters* **64** (21) 2499-2502

9. P. Senellart, J. Bloch (1999) *Physical Review Letters* **82** (6) 1233-1236

10. R.P. Stanley, R. Houdré, U. Oesterle, M. Gailhanou, M. Ilegems (1994) *Applied Physics Letters* **65** (15) 1883-1885

11. T. Hattori, T. Kobayashi (1987) *Chemical Physics Letters* **133** (3) 230-234

12. A.I. Attia, B.H. Loo, A.H. Francis (1973) *Chemical Physics Letters* **22** (3) 537-542

13. D.G. Lidzey, D.D.C. Bradley, M.S. Skolnick, T. Virgili, S. Walker, D.M. Whittaker, (1998) *Nature* **395** 53-55

14. D.G. Lidzey, D.D.C. Bradley, T. Virgili, A. Armitage, M.S. Skolnick (1999) *Physical Review Letters* **82** (16) 3316-3319

15. T. Kobayashi (Ed.) (1996) *J-aggregates* World Scientific Publishing Co. Singapore

16. D. Baxter, M.S. Skolnick, A. Armitage, V.N. Astratov, D.M. Whittaker, T.A. Fisher, J.S. Roberts, D.J. Mowbray, M.A. Kaliteevski (1997) *Physical Review* **B56** R10032

17. R.P. Stanley R. Houdré, C. Weisbuch, U. Oesterle, M. Ilegems (1996) *Physical Review* **B53** 10995-11007

18. E.O. Potma, D.A. Wiersma (1998) *Journal of Chemical Physics* **108** 4894

# ELECTRON ACCEPTORS OF THE FLUORENE SERIES IN PHOTONICS AND ELECTRONICS: RECENT ACHIEVEMENTS AND PERSPECTIVES

I.F. PEREPICHKA

*L.M.Litvinenko Institute of Physical Organic and Coal Chemistry,
National Academy of Sciences of Ukraine, Donetsk 340114, Ukraine
e-mail: i_perepichka@yahoo.com*

**Abstract:**      The recent progress in the chemistry of electron acceptors of the fluorene series and their applications for electronics and photonics is described, together with results from our laboratory. It is demonstrated that acceptors incorporating electron donating groups that results in intramolecular charge transfer are promising candidates for advanced materials with specific electrical and optical properties.

## 1. Introduction

Substantial progress in the chemistry and physics of organic $\pi$-electron acceptors in the last 3–4 decades is partly encouraged by their applications in molecular electronics. Most organic acceptors used in organic conducting materials belong to a class of polycyano compounds. This is due to strong electron withdrawing ability of the cyano group ($\sigma_p = 0.66$, $\sigma_m = 0.56$, $\sigma_I = 0.59$ [1]) as well as its planarity and small size that minimizes steric hindrance for charge transfer salt formation with $\pi$-electron donors. Such structures are represented below and several recent reviews describe current progress in the field [2,3,4]:

TCNE (2.75 eV)     HCBD (3.2 eV)     TCB (2.55 eV)     TCBQ (3.45 eV)

TCNQ (2.8 eV)     DCNQI (2.8 eV)

*F. Kajzar and M.V. Agranovich (eds.), Multiphoton and Light Driven Multielectron Processes in Organics: New Phenomena, Materials and Applications, 371–386.*

Although the nitro group ($\sigma_p = 0.78$, $\sigma_m = 0.71$, $\sigma_I = 0.63$ [1]) is stronger than cyano, it is not employed in olefinic or quinoid-type acceptors due to high reactivity and low stability of the compounds. In contrast, polynitro-aromatics are much more stable compounds and rather strong acceptors. The most known of this class of acceptors are electron acceptors of the fluorene series. This is a unique class of acceptors for organic photoconductors and electron transport materials used in electrophotography and related applications. Thus, a charge-transfer complex (CTC) of poly-N-vinylcarbazole (PVK) and 2,4,7-trinitro-9-fluorenone (TNF, **1**) was used in IBM Copier 1 series introduced in 1970 [5].

PVK

PEPC

This work delights the recent progress in the chemistry and materials applications of fluorene acceptors, and some of recent results of our laboratory in this field together with literature data.

## 2. Synthesis, Functionalization and Structure-Property Relationships

Nitrosubstituted 9-fluorenones and 9-dicyanomethylenefluorenes (**1–6**) are the most known fluorene acceptors during the 40's–60's [6]. Later, nitrosubstituted 9-X-fluorene-4(2)-carboxylic (**7**, **8**) [7,8], 9-X-fluorene-2,7-dicarboxylic (**9**) [7], and 9-X-fluorene-2,7-disulfonic acids (**10**) [9] derivatives were synthesized. Although both $CO_2R$ or $SO_2R$ substituents are poorer electron withdrawing groups as compared to $NO_2$ [1], they allow easy variations in functionality of the acceptors bringing new important properties for materials applications (solubility, compatibility with a polymer matrix, adhesion to a base, supramolecular architecture etc.).

### 2.1. ELECTRON AFFINITY

Fluorene acceptors can be characterized as medium to strong electron acceptors (EA ~ 2.0–2.75 eV). The vast majority of data, however, on electron affinities (EA) of different various of acceptors have been obtained by indirect methods (*e.g.* by cyclic voltammetry/polarography or CTC methods [10]) which do not provide absolute EA values. Nevertheless, satisfactory correlations are observed between indirect EA in solutions and EA in gas phase, or any theoretical parameters ($\Sigma\sigma$ of substituents in an acceptor or its $\varepsilon_{LUMO}$). Indirect EA values obtained by different methods are also in good agreement with each other [11,12]; thus, EA for a number of polynitrofluorene

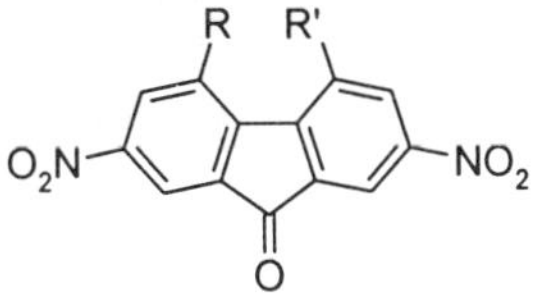

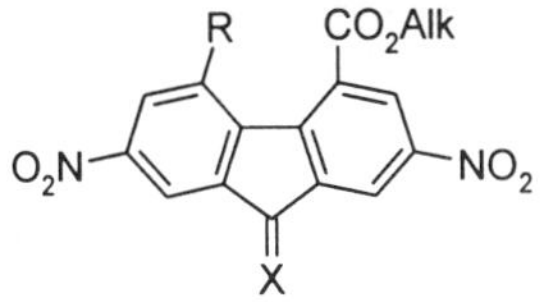

1, DNF:   R = R' = H
2, TNF:   R = NO$_2$, R' = H
3, TENF: R = R' = NO$_2$

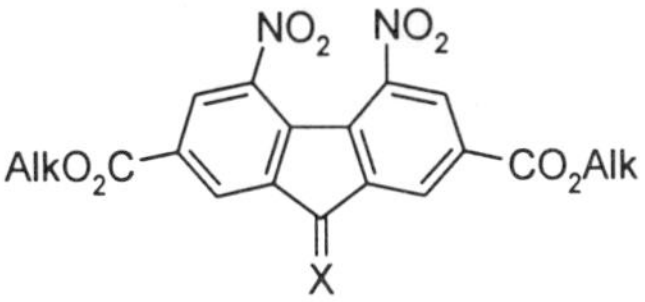

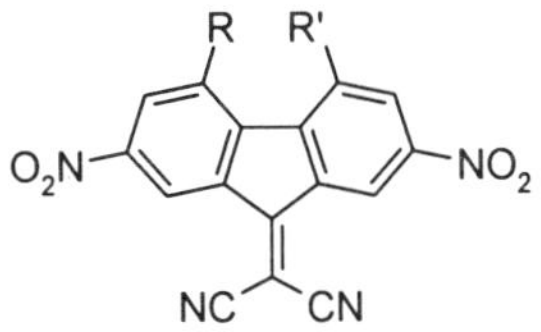

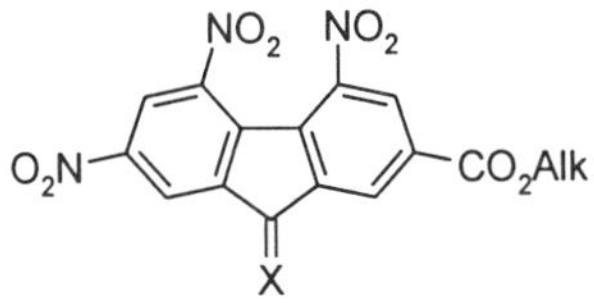

7

9

4, DDF:   R = R' = H
5, DTF:   R = NO$_2$, R' = H
6, DTEF: R = R' = NO$_2$

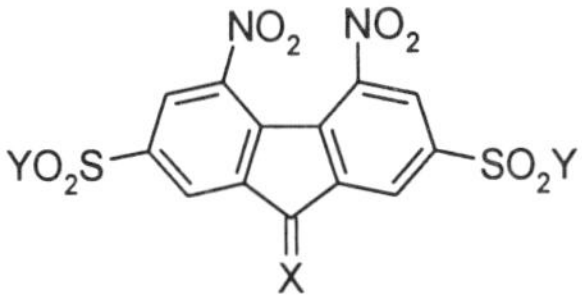

R = H, NO$_2$; X = O, C(CN)$_2$; Y = OAlk, NAlk$_2$

8

10

derivatives determined by CTC method are in good linear relationship with cyclic voltammetry reduction potentials (E$_{1/2}^{1red}$) of the acceptors[12]:

$$EA = (2.59 \pm 0.01) + (1.06 + 0.04)\, E_{1/2}^{1red} \qquad (1)$$

Each additional NO$_2$ group in 9-fluorenones increases EA by ≈ 0.18–0.24 eV and substitution of an oxygen of the carbonyl group by dicyanomethylene group increases EA by ≈ 0.40–0.45 eV [6b,9,11,12,13].

## 2.2.  CHARGE TRANSFER COMPLEXATION

Fluorene acceptors readily form CTC with electron donors that can be manifested by an appearance of additional bands (CTC bands) in the visible region of their electron absorption spectra (Fig. 1).

The energies of the long wavelength CT adsorption are used for evaluation of electron affinities of acceptors by the so called "CTC method", which is based on an assumption that solvation energies of related acceptors are close and the difference in the electron affinities of the acceptors is proportional to the differences in their CT energies with the same donor [10]. Quantitative estimation of CTC formation of fluorene acceptors ($h\nu_{ICT}$, stability constant and extinction coefficients for CTC) was a subject of many investigations [6a,8c,9,13,14,15]. Just such a CTC formation with PVK [16,17,18] is responsible for the photoconductivity of the polymers in the visible region of the spectrum (discussed in sections 3.3). Solid state studies of CTC of fluorene acceptors are discussed in section 4.

374

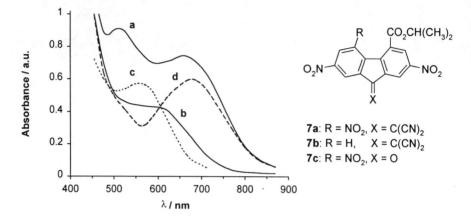

*Figure 1.* Electron absorption spectra of 1:1 CTC of (a) **7a** ($\lambda_{max}$ = 513, 659 nm) and (b) **7b** ($\lambda_{max} \approx$ 570 nm) with N-propylcarbazole and (c) **7c** ($\lambda_{max}$ = 559 nm) and (d) **7a** ($\lambda_{max}$ = 678 nm) with anthracene in 1,2-dichloroethane, 25 °C.

## 2.3. INTRAMOLECULAR CHARGE TRANSFER

Introduction of electron donating group into the acceptor molecule results in intramolecular charge transfer (ICT) which is manifested by an appearance of long-wavelength absorption bands in the visible (or even near-IR) region. Several such structures (**11–16**) are presented below (see also structures **17–19, 20**, and **21**) together with electron absorption spectra for some of them demonstrating ICT absorption (Fig. 2).

Due to a high efficiency of this class of acceptors in photothermoplastic and NLO materials (see sections 3.3.2 and 5) as well as theoretical interest in push- pull

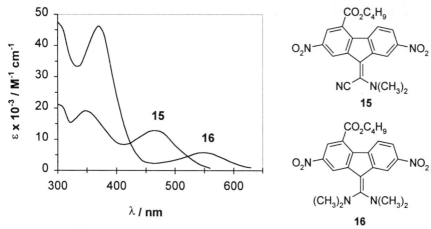

*Figure 2.* Electron absorption spectra of compounds **15** and **16** [19b] in acetonitrile, 25 °C.

compounds with specific electronic properties we studied structural and positional regularities of the effects of various electron donating groups on energies and intensities of ICT bands in fluorene acceptors [19,20,21,22,23,24].

## 3. Photoconductive Materials

### 3.1 ELECTRON TRANSPORT COMPOUNDS

The most widely investigated fluorene acceptor with respect to its electron transport properties is TNF (**2**) [25]. Electron drift mobility, $\mu_-$, of crystalline (at 25°C) and melted (at 173°C) TNF were measured as $6\times10^{-2}$ and $4\times10^{-4}$ cm$^2$ V$^{-1}$ s$^{-1}$, respectively [26]. Both values were found to be independent of the applied field. Substantially lower electron mobility is observed in amorphous TNF, that can be explained by a disorder argument. Electron mobility was found electric field dependent varying in the range of $\mu_- \approx 10^{-6} - 10^{-5}$ cm$^2$ V$^{-1}$ s$^{-1}$ at applied fields $E \approx (1-3)\times10^5$ V cm$^{-1}$ [27]. Similar values of electron mobility was reported later: $\mu_- \approx 6\times10^{-5}$ cm$^2$ V$^{-1}$ s$^{-1}$ at $E = 4\times10^5$ V cm$^{-1}$ [28]. Bipolar (electron and hole) transport was first described by Gill for mixtures of TNF and PVK [17]. Both electron and hole mobilities are field dependent, varying as $\log \mu \propto \beta E^{1/2}$. The temperature dependencies are described by Arrhenius relationships with zero-field activation energies of $\approx 0.7$ eV for both carriers. At $E = 5\times10^5$ V cm$^{-1}$ hole mobilities in pure PVK are $\mu_+ \approx 10^{-5}-10^{-6}$ cm$^2$ V$^{-1}$ s$^{-1}$ ($\mu_-$ is too low to be measured) which decrease with increasing TNF contents in the mixture. At molar ratio TNF:PVK $\approx$ 1:2 mobilities of both carriers are comparable: $\mu_+ \approx \mu_- \approx 10^{-7}$ cm$^2$ V$^{-1}$ s$^{-1}$.

Much lower electron mobilities $\mu_- \approx 3\times10^{-8}$ cm$^2$ V$^{-1}$ s$^{-1}$ ($E = 10^5$ V cm$^{-1}$; 25°C) were reported for 2,5,7-trinitro-4-butoxycarbonyl-9-dicyanomethylenefluorene **7d**, R = NO$_2$, X = C(CN)$_2$, Alk = C$_4$H$_9$) [29]. It is surprising as this value is of *ca.* 3 order of magnitude lower than observed in TNF, although **7d** is a stronger electron acceptor than TNF (based on the correlation described in section 2.1, we estimate the difference EA(**7d**) – EA(TNF) ≈ 0.5 eV). Alkyl-substituted polynitro-9-fluorenones synthesized with the expectation of improved solubility showed even somewhat improved charge mobilities compared to TNF [30]. Recently, Matsui *et al* [31] reported on the synthesis and electron transport of 2,4,7-trinitrofluorenone-9-arylimines (**17**), as well as their application in positive charge electrophotography.

## 3.2 POLYMERIC ACCEPTORS

Electron transport materials based on fluorene acceptors dispersed in an inert polymer matrix or in a polymeric photoconductor have several disadvantages in practical applications. One of them is limited solubility of low-molecular compounds in polymeric binders, that limits the level of electron mobilities of the material, and crystallization of an acceptor from the films with time. To overcome these problems long chain substituents in the acceptors were introduced, increasing the solubility and decreasing the melting points of the acceptors [7,8,14]. Another promising approach is the design of polymeric acceptors; such acceptors would be attractive due to no necessity for an inert polymeric matrix which decreases the concentration of the acceptor units in the film. Some of such acceptors containing polynitro-9-fluorenonecarboxylic acid units were synthesized [32,33]. They, however, showed low electron mobilities [34] as compared to TNF. Recently, fluorene-functionalized trithiophene **18** was synthesized which was electrochemically polymerized to electroactive polymer with a bandgap of 1.7 eV (spectroscopy) [35].

## 3.3 SENSITIZATION OF PHOTOCONDUCTIVITY OF HOLE TRANSPORT POLYMERS.

### 3.3.1 ELECTROPHOTOGRAPHY
An intrinsic photoresponse of carbazole-containing polymers like PVK, poly-N-(2,3-epoxypropyl)carbazole (PEPC), etc. lies in the UV region of the spectrum. It can be substantially increased and, through the CTC mechanism, its spectral response can be shifted to the visible region by sensitizing a polymer with fluorene acceptors. In sensitization the photoconductivity of PVK and related polymers by electron acceptors,

two steps can be singled out, *i.e.* charge carrier photogeneration and their transport through the layer in the applied electric field. At low acceptor concentrations in the mixture, holes are the main carriers and their mobilities do not change drastically when variations in acceptor concentrations are ranged from 0 to 10% [17,25]. Therefore, an influence of acceptors on electrophotographic response of such a materials are mainly dictated by changes in photogeneration efficiency. A large number of electrophotographic compositions containing fluorene acceptors as sensitizers are described in the literature, including patents; since 70's, several tens of citations appear each year, that makes it impossible to delight this in the present paper. Among them, TNF is the most widely investigated and used sensitizer of PVK and related polymers [5a,25,36].

### 3.3.2. PHOTOTHERMOPLASTIC STORAGE MEDIA

Although the photothermoplastic process of recording optical information is related to electrophotography, it has some important advantages such as real-time scaling and cycling the process of recording/erasure on the same material. On the other hand this requires the specific properties of the components and the whole PTSM (its structure is represented on Fig. 4).

Recording optical information on PTSM (Fig. 4) includes the following main steps:

a)  charging the surface of photothermoplastic recording film (PTRF) by positive corona (depending on the dark conductivity of PTRF, surface potential varies in the range of $E \approx 80–200$ V $\mu m^{-1}$);

b)  exposure of the PTSM resulting in electrostatic image of the original picture (as a measure of the efficiency of this step, electrophotographic response, $S_{\Delta V}$, is used; below presented values of $S_{\Delta V}$ correspond to 20% decay in the surface potential under the illumination);

c)  visualization of the latent image by heating to $T_g$ in the dark, where electrostatic forces form a relief (as a measure of the efficiency of this step, we used holographic response (He-Ne laser, $\lambda = 632.9$ nm), $S_\eta$, at 1% level of diffraction efficiency of plane wave holograms and maximal diffraction efficiency, $\eta_{max}$, of PTSM);

d)  erasure of the relief image by heating higher than $T_g$ after which the PTSM is ready to be used for recording another image.

The first two steps, physically based on photoconductive properties of PTRF (i.e. on its great difference between the dark and under the illumination conductivities), are the same as in electrophotographic process. Fixation of the image by formation of a relief (step "c") is based on specific rheological properties of PTRF. Although generally processes "b" and "c" can be realized in different layers when PTSM consists of photoconductive and thermoplastic materials as separate films, such an approach has serious disadvantages due to its technological difficulties and increases the cost of the PTSM. Recent progress in more cost-effective single-layer PTSMs (showed on Fig. 4) allowed to reach characteristics of two-layers PTSM at appropriate choice of components [14,37].

Requirements of low $T_g$ of a polymer to decrease the thermal relaxation of charges during the development of the image (step "c") dictates to change the mostly known polymer PVK to another one, namely PEPC which have $T_g$ around 85 °C when obtained by anionic polymerization of the monomer. Fig. 4 demonstrates electrophotographic response of PEPC sensitized by the electron acceptors of 9-fluorenone (1–3) and 9- dicyanomethylenefluorene (4–6) series. For both series, increasing the number of nitro group results in a bathochromic shift of the red limit of a response, in accordance to EA values of acceptors.

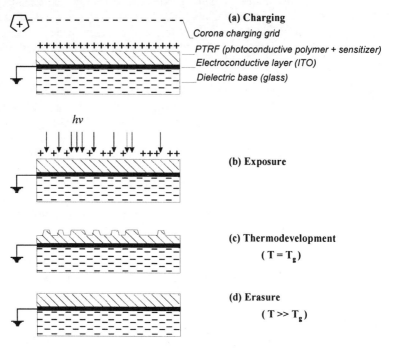

*Figure 3.* Main steps of photothermoplastic process of recording optical information.

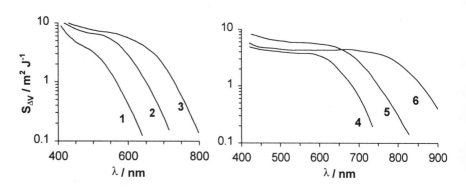

*Figure 4.* Spectral distribution of electrophotographic response of PTSM based on PEPC sensitized by 2 mol. % of fluorene acceptors 1–6.

However, holographic response of PTSM with polynitrofluorenes 1–6 is moderate due to observable charge thermorelaxation of latent electrostatic image during its thermodevelopment (Fig. 4, step "c"). $S_\eta$ and $\eta_{max}$ are in the range of 15–30 m$^2$ J$^{-1}$ and 10–15%, respectively, at low concentration of acceptors (1–3wt%) but sharply decrease with increasing the concentration of acceptors becoming 5–10 m$^2$ J$^{-1}$ and 4–8%, respectively, at 5–7wt% of acceptors. This problem can be overcome by improvement of rheological properties of the film by adding a placticizer, which, however, dilute the layer resulting in a decrease of $S_{\Delta V}$. Therefore, we realized another approach, i.e. modification of the acceptors with long-chain substituents (general structures 7–10) which drastically decreases their melting points and improved the solubility. By this way a number of efficient sensitizers were prepared [38] allowing elaborate high-sensitive PTSMs [39] with $S_\eta$ and $\eta_{max}$ of 80–120 m$^2$ J$^{-1}$ and 20–25%, respectively.

Push-pull fluorene acceptors were found to show an increased photoconductivity in the spectral region of their ICT. Electrophotographic response of PEPC films for some of such acceptors are represented in Fig. 5. At certain choice of electron donating moiety, PTSM containing with fluorene acceptors with ICT gave $S_\eta$ and $\eta_{max}$ as high as 200–300 m$^2$ J$^{-1}$ and 25–27%, respectively (p-alkoxybenzylidene and 1,3-dithiol-2-ylidene moieties was found to be mostly effective [21,40,41]) .

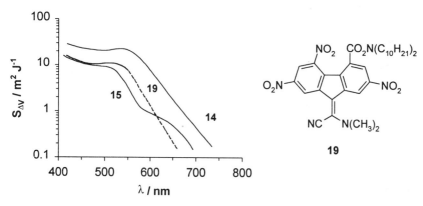

*Figure 5.* Spectral distribution of electrophotographic response of PTSM based on PEPC sensitized by 4–5mol% of fluorene acceptors with ICT (**14, 15**, and **19**).

## 3.4 PHOTOREFRACTIVE MATERIALS

Due to its excellent charge transport properties PVK is a very popular polymer for photorefractive materials, especially in guest-host composites with NLO chromophores [42]. To increase charge generation efficiency PVK was sensitized by TNF (**1**) which was usually used in small amounts of 0.05–1.3% [43,44,45]. The photorefractive material [43a] based on PVK, which was doped with the optically nonlinear chromophore F-DEANST (33%) and sensitized with TNF (1.3%), exhibited a diffraction efficiency as high as 1% (125μm sample, λ = 647 nm, writing power 1 W

$cm^{-2}$ and applied field 40 V $\mu m^{-1}$: grating growth times $\approx$ 100 ms, beam-coupling gain coefficient > 10 $cm^{-1}$) which is much higher than previously reported values for any organic polymers and comparable with some inorganic photorefractive materials such as $BaTiO_3$ or $Bi_{12}SiO_{20}$ [43c,46]. Further increasing in the diffraction efficiency (up to 5%) was achieved for the system PVK:TNF:DMNPAA:ECZ [45a] when another chromophore (DMNPAA) and plasticizer (N-ethylcarbazole, ECZ) were used.

F-DEANST                    DMNPAA

We believe that, taking into account NLO activity in some fluorene push-pull compounds (see section 5) [47] and plasticizing effects in fluorene acceptors with long-chain substituents we recently demonstrated in PTRF for holography [14], one can expect further progress in photorefractive materials based on these or related fluorene derivatives which can act in composition combining charge generation/NLO active/ plasticizing properties in one component.

## 3.5  PHOTOELECTRIC DEVICES

TNF:PVK systems were also studied in photovoltaic cells [48] demonstrating that solar energy conversion efficiencies of the order of $10^{-2}$ to 1 % might be realized in thin (< 1 $\mu m$) films. Other fluorene acceptors were almost not studied in photovoltaics. In the meantime, their improved characteristics (as compared to TNF) in related applications in electronics demonstrate their high potential for solar cells application.

## 4.  Molecular Conductors Based on Charge Transfer Salts with Tetrathiafulvalenes

Extensive work has been carried out in the design of organic metals and superconductors based on charge transfer complexes and salts of tetrathiafulvalene (TTF) derivatives as donors [49]. TTF-TCNQ was the first CTC which showed metallic type conductivity [50] and since then wide variation in TTF structures were performed whereas acceptors were mainly TCNQ or DCNQI derivatives as well as their analogs of the series of polycyano compounds [2]. Recently, formation of CTC between fluorene acceptors 2, 3, 5, and 6 and BEDO-TTF as donor was reported: TNF (2) and TENF (3) formed semiconductive complexes of A:D = 1:1 composition with $\sigma_{rt}$ $\approx 10^{-10}$ S $cm^{-1}$ whereas DTF (5) and DTEF (6) gave metallic-type A:D = 1:2 complexes with electrical conductivity in compressed pellets $\sigma_{rt}$ = 65 and 18 S $cm^{-1}$ [$\sigma_{max}$ = 390 (at 8 K) and 32 (at 94 K) S $cm^{-1}$], respectively [51].

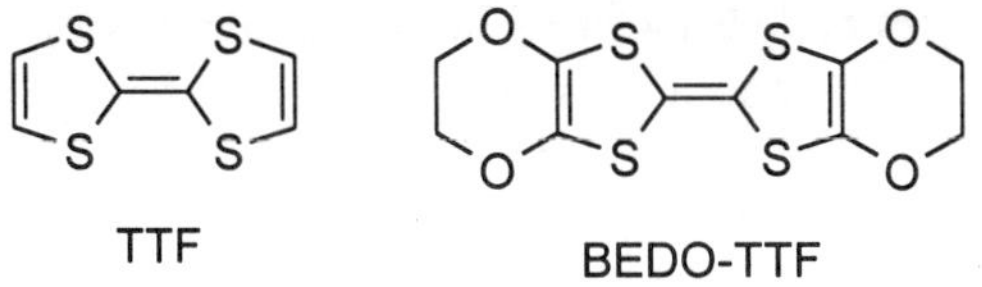

TTF             BEDO-TTF

Unfortunately, no X-ray crystallographic data were reported for the complexes although substantial anisotropy of the conductivity (characteristic for $\pi$–$\pi$ CTC) is expected, with highest values along the axis with $\pi$–$\pi$ overlapping between the donor and the acceptor molecules. Because the information on the crystal structures of CTC is important for design of conducting and semiconducting complexes of such type, we attempted to grow single crystals of fluorene acceptors with TTF donors and to determine their structures by single crystal x-ray diffraction [13,52,53]. Strongest acceptors of the fluorene series, i.e. DTEF (6), formed 1:1 CTC with TTF characterized by mixed ADAD stacking motif with weak interplanar charge transfer [13]. Variation in structures of either donor or acceptor resulted in a change in stoichometry of CTC enabling to form A:D = 1:2 (ADDADD mixed stacks) [52,] or 2:3 (ADADD mixed stacks) [53]. For all single crystals of fluorene acceptors with TTF donors obtained to the date only mixed stacks which are characteristic for organic semiconductors[54] are observed. In the meantime, the possibility of easy variation of stoichometry and $\pi$-orbital overlapping in CTC by structural variations in donor or acceptor seems a promising indication on the possibility to design segregated-stack CTCs of fluorene acceptors with TTFs which possess metallic-type electrical conductivity.

## 5. NLO Materials

Push-pull molecular systems of the A–$\pi$–D type are promising class of compounds for second order non-linear optical materials [55]. Intramolecular charge transfer in such type of compounds which can be described in terms of resonance between neutral and zwitter-ionic $A^+$–$\pi$–$D^-$ structures can be particularly controlled by variations in an electron acceptor ability of the " A " moiety and fluorene acceptors are good candidates for this.

Synthesis of a fluorene-containing NLO active polymer 20 with high $T_g$ = 205 °C was reported in the literature [56]. It exhibits second order hyperpolarizability

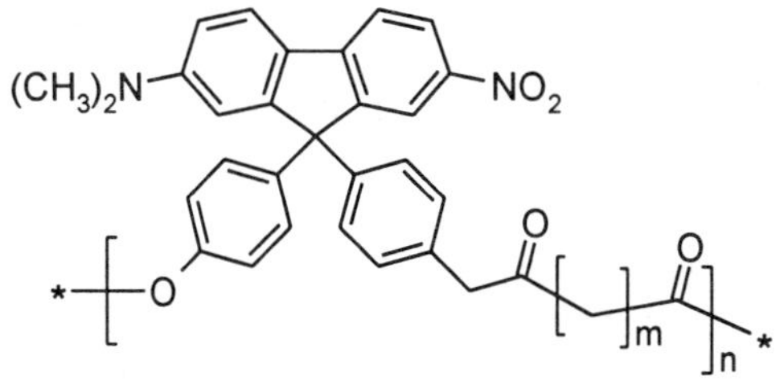

20

originating from the push-pull character of the fluorene moiety in the main chain of the polymer: $r_{33}$ = 2.6 pm $V^{-1}$ and 11 pm $V^{-1}$ for poling fields 50 and 250 V $\mu m^{-1}$, respectively (at 810 nm).

Recently we showed that fluorene acceptors which incorporate ferrocene units as the donor moiety show substantially high levels of second harmonic generation efficiency (which depends on the length and the nature of the π-bridge between donor and acceptor moieties) [47]. Two examples of fluorene compounds with ICT (**15** and **16**) which show high second order hyperpolarizability are shown below:

**15**, $\lambda_{ICT}$ = 600 nm,
$\mu\beta(0)$ = 800 × $10^{-48}$ esu

**16**, $\lambda_{ICT}$ = 660 nm,
$\mu\beta(0)$ = 2400 × $10^{-48}$ esu

Lengthening the chain of the ester substituent in the fluorene moiety substantially increased the solubility of the compound.

These results indicate that the fluorene moiety is a promising acceptor terminal substituent for second order NLO materials. One of its advantage is that the fluorenyl anion is an aromatic system (14π-electron system) that additionally stabilizes bipolar structure $A^+$–π–$D^-$ of the molecule. Moreover, the fluorene moiety can act as electron acceptor unit for specific donor-acceptor interactions possessing unusual physical, electric and optical properties to the material.

Thus, combination of NLO and photoconductive properties in the same molecule can be utilized for design of photoconductors (e.g. holographic PTSMs; see section 3.3.1) with photoresponse in the near-IR region. If the A–π–D molecule displays ICT in the visible region (500–800 nm) with high level of photoresponse at $\lambda_{ICT}$ and due to extended π-bridge shows high second order hyperpolarizability, it can udergo two-photon absorption in the near-IR region (IR laser with an irradiation at 1000–1600 nm, e.g. Nd:YAG). Doubling the frequency due to NLO effect results to emission of photons at visible region which can excite the molecule at its $\lambda_{ICT}$ (one photon excitation) resulting to the photoconductivity. In spite of much lower efficiency of such a process than normal photoresponse of the materials in the visible region, it looks attractive for many advanced applications (e.g., now PTSMs for holographic interferometry in the near-IR region are unavailable).

# 6. Conclusions

Recent advances on synthetic, theoretical, physical-organic investigations as well as applications of electron acceptors of the fluorene series demonstrate their high potential for electronics and optics. Wide synthetic possibilities of variations in structures of fluorene acceptors to control their electronic properties make them very attractive for practice. Growing interest in fluorene acceptors with intramolecular charge transfer in the past decade resulted in excellent materials for photothermoplastics and non-linear optics and new applications in advanced technologies are expected.

# 7. Acknowledgements

We thank to the Royal Society for a Exchange Fellowship for visiting Durham University, to C.I.E.S. for funding the visit to CEA Saclay/France, and to Alexander von Humboldt Foundation for the Fellowship to conduct some of a research in Würzurg University (1995-97) and for an electrochemical equipment donation.

# 8. References

1   Hansch, C., Leo, A., and Taft, R.W. (1991), *Chem. Rev.* **91**, 165–195.
2   (a) Ogura, F., Otsubo, T., and Aso, Y. (1992), *Sulfur Reports* **11**, 439–464. (b) Khodorkovsky, V., and Becker, J.Y. (1994), in J.-P. Farges (ed.), *Organic Conductors. Fundamentals and Applications*, Marcel Dekker, New York, pp. 75–114.
3   (a) Hünig, S. (1995), *J. Mater. Chem.* **5**, 1469–1479. (b) Martín, N., and Seoane, C. (1997), in H.S. Nalwa (ed.), *Handbook of Organic Conductive Molecules and Polymers*, vol. 1, Wiley, Chichester, pp. 1–86. (c) Martín, N., Segura, J. L., and Seoane, C. (1997), *J. Mater. Chem.* **7**, 1661–1676.
4   (a) Zhao, H., Heintz, R.A., Ouyang, X., Grandinetti, G., Cowen, J., Dunbar, K.R. (1999), in J. Veciana, C. Rovira and D.B. Amabilino (eds.), *Supramolecular Engineering of Synthetic Metallic Materials: Conductors and Magnets. NATO ASI Series, Vol. 518*, Kluwer Academic Publishers, Dodrecht, pp. 353–376. (b) Wudl, F., Yu, H., Fourmigue, M., Hicks, R. (1999), in J. Veciana, C. Rovira and D.B. Amabilino (eds.), *Supramolecular Engineering of Synthetic Metallic Materials: Conductors and Magnets. NATO ASI Series, Vol. 518*, Kluwer Academic Publishers, Dordrecht, pp. 393-407.
5   (a) Strohriegl, P., and Grazulevicius, J.V. (1997), in H.S. Nalwa (ed.), *Handbook of Organic Conductive Molecules and Polymers*, vol. 1, Wiley, Chichester, pp. 553–620. (b) Shattuck, M. D., and Vahtra, U. (1969), U.S. Patent 3,484,327.
6   (a) Mukherjee, T.K., and Levasseur, L.A. (1965), *J. Org. Chem.* **30**, 644–646. (b) Mukherje, T.K. (1968), *Tetrahedron* **24**, 721.–728. (b) Mukherje, T.K. (1966), *J. Phys. Chem.* **70**, 3648–3652. (c) Orchin, M., and Woolfolk, E.O., (1946), *J. Am. Chem. Soc.* **68**, 1727–1729. (d) Woolfolk, E.O., and Orchin, M. (1955), *Org. Syntheses*, Coll. Vol. III, 837–838.
7   Sulzberg, T., and Cotter, R.J. (1970), *J. Org. Chem.* **35**, 2762–2769.
8   (a) Turner, S.R. (1977), U.S. Patent 4,062,886. (b) Bloom, M.S., and Groner, C.F. (1977), *Research Disclosure*, 32–36. (c) Mysyk, D.D., Sivchenkova, N.M., Kampars, V.E., and Neilands, O.Ya. (1987), *Izv. Akad. Nauk Latv. SSR, Ser. Khim.*, 612–626 (in Russian).
9   Mysyk, D.D., Perepichka, I.F., and Sokolov, N.I. (1997), *J. Chem. Soc., Perkin Trans. 2*, 537–545.

384

10   Kampars, V.E., and Neilands, O.Ya. (1977), *Usp. Khim.(Russ. Chem. Rev.)* **46**, 945–966 (in Russian).

11   Kuder, J.E., Pochan, J.M., Turner, S.R., and Hinman, D.-L.F. (1978), *J. Electrochem. Soc.: Electrochem. Sci. And Technology* **125**, 1750–1758.

12   Mysyk, D.D., Perepichka, I.F., Edgina, A.S., and Neilands, O.Ya. (1991), *Latvian J. Chem.*, 727–735 (in Russian).

13   Perepichka, I.F., Kuz'mina, L.G., Perepichka, D.F., Bryce, M.R., Goldenberg, L.M., Popov, A.F., and Howard, J.A.K. (1998), *J. Org. Chem.* **63**, 6484–6493.

14   Perepichka, I.F., Mysyk, D.D., and Sokolov, N.I. (1995), in N.S. Allen, M. Edge, I.R. Bellobono and E. Selli (eds.), *Current Trends in Polymer Photochemistry*, Ellis Horwood, New York, pp. 318–327.

15   (a) Mysyk, D.D., Perepichka, I.F., Sivchenkova, N.M., Kampars, V.E., Neilands, O.Ya., and Kampare, R.B. (1984), *Izv. AN Latv. SSR, Ser. Khim.*, 328–331 (in Russian). (b) Mysyk, D.D., Sivchenkova, N.M., Kampars, V.E., Neilands, O.Ya., and Kampare, R.B. (1984), *Izv. AN Latv. SSR, Ser. Khim.*, 332–335 (in Russian).

16   Weiser, G. (1972), *J. Appl. Phys.* **43**, 5028–5033.

17   Gill, W.D. (1972), *J. Appl. Phys.* **43**, 5033–5040.

18   Enomoto, T., and Hatano, M. (1974), *Makromol. Chem.* **175**, 57–65.

19   (a) Perepichka, I.F., Popov, A.F., Artyomova, T.V., Vdovichenko, A.N., Bryce, M.R., Batsanov, A.S., Howard, J.A.K., and Megson, J.L. (1995), *J. Chem. Soc., Perkin Trans. 2*, 3–5. (b) Perepichka, I.F., Popov, A.F., Orekhova, T.V., Bryce, M.R., Vdovichenko, A.N., Batsanov, A.S., Goldenberg, L.M., Howard, J.A.K., Sokolov, N.I., and Megson, J.L. (1996), *J. Chem. Soc., Perkin Trans. 2*, 2453–2469.

20   Perepichka, I.F., Mysyk, D.D., and Sokolov, N.I. (1999), *Synth. Metals* **101**, 9–10.

21   Perepichka, I.F., Perepichka, D.F., Bryce, M.R., Goldenberg, L.M., Kuz'mina, L.G., Popov, A.F., Chesney, A., Moore, A.J., Howard, J.A.K., and Sokolov, N.I. (1998), *Chem. Commun.*, 819–820.

22   Perepichka, D.F., Perepichka, I.F., Bryce, M.R., Popov, A.F., Chesney, A., and Moore, A.J. (1998), in *Structures of Organic Compounds and Reaction Mechanisms*, Inst. Physical Organic & Coal Chemistry, Donetsk, p. 94–100.

23   Skabara, P.J., Serebryakov, I.M., and Perepichka, I.F. (1999), *J. Chem. Soc., Perkin Trans. 2*, 505–513.

24   (a) Mysyk, D.D., and Perepichka, I.F. (1994), *Phosphorus, Sulfur, and Silicon* **95-96**, 527–529. (b) Mysyk, D.D., Perepichka, I.F., Perepichka, D.F., Bryce, M.R., Popov, A.F., Goldenberg, L.M., and Moore, A.J. (1999), *J. Org. Chem.* **64**, 6937–6950.

25   (a) Borsenberger, P.M., and Weiss, D.S. (1998) *Organic Photoreceptors for Xerography*, Marcel Dekker, New York. (b) Borsenberger, P.M., and Weiss, D.S. (1993) *Organic Photoreceptors for Imaging Systems*, Marcel Dekker, New York.

26   Gill, W.D. (1975), in J. Stuke and W. Brenig (eds.), *Amorphous and Liquid Semiconductors*, Taylor and Francis, London, p. 135.

27   Gill, W.D. (1974), in J. Stuke and W. Brenig (eds.), *Proceedings of the Fifth International Conference on Amorphous and Liquid Sermiconductors*, Taylor and Francis, London, p. 901.

28   Emeraldt, R.L., and Mort, J. (1974), *J. Appl. Phys.* **45**, 3943–3945.

29   Turner, S.R. (1980), *Macromolecules* **13**, 782–785.

30   (a) Loutfy, R.O., Hsiao, C.-K., Ong, B.S., and Keoshkerian, B. (1984), *Can. J. Chem.* **62**, 1877–1885. (b) Loutfy, R.O. and Ong, B.S. (1984), *Can. J. Chem.* **62**, 2546–2551. (c) Ong, B.S., Keoshkerian, B., Martin, T.I., and Hamer, G.K. (1984), *Can. J. Chem.* **63**, 147–152.

31   (a) Matsui, M., Fukuyasu, K., Shibata, K., and Muramatsu, H. (1993), *J. Chem. Soc., Perkin Trans. 2*, 1107–1110. (b) Matsui, M., Shibata, K., Muramatsu, H., and Nakazumi, H. (1996), *J. Mater. Chem.* **6**, 1113–1118.

32   (a) Turner, S.R., and Pochan, J.M. (1977), U.S. Patent 4,056,391. (b) Pochan, J.M., and Turner, S.R. (1977), U.S. Patent 4,063,947.

33 Bulyshev, Yu.S., Kashirskii, I.M., Pashkin, I.I., Andrievskii, A.M., and Tverskoi, V.A. (1990), *Khim. Vys. Energ.* **24**, 232–236.

34 Pearson, J.M. (1977), *Pure and Appl. Chem.* **49**, 463–477.

35 Skabara, P.J., Serebryakov, I.M., and Perepichka, I.F. (1999), *Synth. Metals* **101**, 1336–1337.

36 (a) Pravednikov, A.N., Kotov, B.V., and Tverskoi, V.A. (1978), *Zh. Vses. Khim. Obshch. im. D.I. Mendeleeva* **23**, 524–536 (in Russian). (b) Hoegl, H., Barchietto, G., and Tar, D. (1972), *Photochem. and Photobiol.* **16**, 335–352.

37 (a) Getmanchuk, Yu.P.; Sokolov, N.I.(1983), in *Fundamentals of Optical Memory and Medium*, issue 14, Vyshcha Shkola, Kiev, pp. 11–20 (in Russian). (b) Sokolov, N.I., Barabash, Y.M., Poperenko, L.V., Perepichka, I.F., Mysyk, D.D., and Kostenko, L.I. (1997), *Proc. SPIE* **3055**, 171–180. (c) Sokolov, N.I., Barabash, Yu.M., Poperenko, L.V., Perepichka, I.F., Mysyk, D.D., and Komarov, V.A. (1998), *Functional Materials* **5**, 441–446.

38 (a) Mysyk, D.D., Perepichka, I.F., Romashev, V.E., Andrievskii, A.M., and Kostenko, L. I. (1986), USSR Patent 1,241,673. (b) Mysyk, D.D., Perepichka, I.F., Kostenko, L.I., Sokolov, N.I., Perelman, L.A., and Grebenyuk, S.A. (1988), USSR Patent 1,441,718. (c) Perepichka, I.F., Mysyk, D.D., Sokolov, N.I., and Barabash, Yu.M. (1990), USSR Patent 1,637,244. (d) Perepichka, I.F., Mysyk, D.D., Sokolov, N.I., and Bazhenov, M.Yu. (1991), USSR Patent 1,658,599. (e) Perepichka, I.F., Mysyk, D.D., Sokolov, N.I., and Bazhenov, M.Yu. (1992), USSR Patent 1,777,487.

39 (a) Kostenko, L.I., Perelman, L.A., Perepichka, I.F., Mysyk, D.D., Grebenyuk, S.A., Kotelenets, M.I., Popov, A.F., Sokolov, N.I., Kuvshinsky, N.G., and Bazhenov, M.Yu. (1986), USSR Patent 1,228,672. (b) Kostenko, L.I., Perepichka, I.F., Perelman, L.A., Mysyk, D.D., Grebenyuk, S.A., Popov, A.F., Sokolov, N.I., Kuvshinsky, N.G., and Bazhenov, M.Yu. (1988), USSR Patent 1,441,964. (c) Manushevich, G.N., Buzurnyuk, S.A., Mysyk, D.D., Kostenko, L.I., Perelman, L.A., Perepichka, I.F., and Sivchenkova, N.M. (1988), USSR Patent 1,471,874. (d) Bazhenov, M.Yu., Barabash, Yu.M., Perepichka, I.F., Kostenko, L.I., and Sokolov, N. I. (1991), USSR Patent 1,702,850. (e) Perelman, L.A., Sokolov, N.I., Mysyk, D.D., Kostenko, L.I., Miroshnichenko, A.A., Kharaneko, O.I., Perepichka, I.F., Bazhenov, M.Yu., and Barabash, Yu.M. (1991), USSR Patent 1,729,277. (f) Perepichka, I. F., Mysyk, D.D., Sokolov, N.I., Kostenko, L.I., Perelman, L.A., Grebenyuk, S.A., Popov, A.F., Bazhenov, M.Yu., and Barabash, Yu.M. (1992), USSR Patent 1,743,300. (g) Malakhova, I.A., Bodrova, N.A., Pavlov, A.V., Koshelev, K.K., Orlova, L.I., Petrova, M.I., and Perepichka, I.F. (1992), USSR Patent 1,802,969. (g) Perepichka, I.F., Mysyk, D.D., Sokolov, N.I., Kostenko, L.I., Perelman, L.A., Grebenyuk, S.A., Popov, A.F., Sivchenkova, N.M., Bazhenov, M.Yu., and Barabash, Yu.M. (1992), USSR Patent 1,814,409.

40 (a) Semenenko, N.M., Abramov, V.N., Kravchenko, N.V., Trushina, V.S., Buyanovskaya, P.G., Kashina, V.L., and Mashkevich, I.V. (1985), *Zh. Obsch. Khim.* **55**, 324–330. (b) Abramov, V.N., Andrievskii, A.M., Bodrova, N.A., Borodkina, M.S., Kravchenko, I.I., Kostenko, L.I., Malakhova, I.A., Nikitina, E.G., Orlov, I.G., Perepichka, I.F., Pototskii, I.S., Semenenko, N.M., and Trushina, V.S. (1987), USSR Patent 1,343,760.

41 (a) Mysyk, D.D., Neilands, O.Ya., Kuvshinsky, N.G., Sokolov, N.I., and Kostenko, L.I. (1987), USSR Patent 1,443,366. (b) Belonozhko, A.M., Davidenko, N.A., Kuvshinsky, N. G.; Neilands, O.Ya., Mysyk, D.D., and Prizva, G.I. (1989), USSR Patent 1,499,553. (c) Mysyk, D.D., Neilands, O.Ya., Khodorkovsky, V.Yu., Kuvshinsky, N.G., Belonozhko, A.M., and Davidenko, N.A. (1991), USSR Patent 1,665,678.

42 Zhang, Y., Wada, T., and Sasabe, H. (1998), *J. Mater. Chem.* **8**, 809–828.

43 (a) Donckers, M.C.J.M., Silence, S.M., Walsch, C.A., Hache, F., Burland, D.M., Moerner, W.E., and Twieg, R.J. (1993), *Opt. Lett.* **18**, 1044–1046. (b) Silence, S.M., Donckers, M.C.J.M., Walsch, Burland, D.M., Twieg, R.J., and Moerner, W.E. (1994), *Appl. Opt.* **33**, 2218–2222. (c) Burland, D.M., Bjorklund, W.E., Moerner, W.E., Silence, S.M., and Stankus, J.J. (1995), *Pure Appl. Chem.* **67**, 33–38. (d) Burland, D.M., Devoe, R.J., Geletneky, C., Jia, Y., Lee, V.Y., Lundquist, P.M., Moylan, C.R., Poga, C., Twieg, R.J., and Wortmann, R. (1996), *Pure Appl. Opt.* **5**, 513–520.

386

44    (a) Malliaras, G.G., Krasnikov, V.V., Bolink, H.J., and Hadziioannou, G. (1994), *Appl. Phys. Lett.* **65**, 262–265. (b) Malliaras, G.G., Krasnikov, V.V., Bolink, H.J., and Hadziioannou, G. (1995), *Proc. SPIE* **2526**, 94. (c) Malliaras, G.G., Krasnikov, V.V., Bolink, H.J., and Hadziioannou, G. (1995), *Appl. Phys. Lett.* **66**, 1038–1040. (d) Malliaras, G.G., Krasnikov, V.V., Bolink, H.J., and Hadziioannou, G. (1995), *Proc. SPIE* **2527**, 250. (e) Malliaras, G.G., Krasnikov, V.V., Bolink, H.J., and Hadziioannou, G. (1995), *Appl. Phys; Lett.* **67**, 455–457. (f) Malliaras, G.G., Krasnikov, V.V., Bolink, H.J., and Hadziioannou, G. (1995), *Phys. Rev. B* **52**, 14324–14327. (g) Malliaras, G.G., Angerman, H., Krasnikov, V.V., ten Brinke, G., and Hadziioannou, G. (1996), *J. Phys. D: Appl. Phys.* **29**, 2045–2048. (h) Malliaras, G.G., Krasnikov, V.V., Bolink, H.J., and Hadziioannou, G. (1996), *Pure Appl. Opt.* **5**, 631–643.

45    (a) Kippelen, K., Sundalphon, K., Peyghambarian, N., Lyon, S.R., Padias, A.B., and Hall, H.K., Jr. (1993), *Elect. Lett.* **29**, 1873–1874. (b) Meerholz, K., Volodin, B.L., Sundalphon, K., Kippelen, K., and Peyghambarian, N. (1994), *Nature* **371**, 497–500. (c) Sundalphon, K., Kippelen, K., Peyghambarian, N., Lyon, S.R., Padias, A.B., and Hall, H.K., Jr. (1994), *Opt. Lett.* **19**, 68–70. (d) Volodin, B.L., Meerholz, K., Sundalphon, K., Kippelen, K., and Peyghambarian, N. (1994), *Proc. SPIE* **2144**, 72. (e) Sundalphon, K., Kippelen, K., Meerholz, K., and Peyghambarian, N. (1996), *Appl. Opt.* **35**, 2346–2354. (f) Kippelen, K., Sundalphon, K., Meerholz, K., and Peyghambarian, N. (1996), *Appl. Phys. Lett.* **68**, 1748–1750.

46    Yeh, P. (1987), *Appl. Opt.* **26**, 602–604.

47    Perepichka, I.F., Perepichka, D.F., Bryce, M.R., Chesney, A., Popov, A.F., Khodorkovsky, V., Meshulam, G., and Kotler, Z. (1999), *Synth. Metals* **102**, 1558–1559.

48    (a) Reucroft, P.J., Takahashi, K., and Ullal, H. (1974), *Appl. Phys. Lett.* **25**, 664–666. (b) Reucroft, P.J., Takahashi, K., and Ullal, H. (1975), *J. Appl. Phys.* **46**, 5218–5223.

49    (a) Coronado, E., and Gómez-García, C.G. (1998), *Chem. Rev.* **98**, 273–296. (b) Day, P., and Kurmoo, M. (1997), *J. Mater. Chem.* **8**, 1291–1295. (c) Otsubo, T., Aso, Y., and Takimiya, K. (1996), *Adv. Mater.* **8**, 203–211. (d) Bryce, M.R. (1995), *J. Mater. Chem.* **5**, 1481–1496.

50    Ferraris, J.P., Cowan, D.O., Walatka, V.V., and Perlstein, J.H. (1973), *J. Am. Chem. Soc.* **95**, 948–949.

51    Horiuchi, S., Yamochi, H., Saito, G., Sakaguchi, K., and Kusunoki, M. (1996), *J. Am. Chem. Soc.* **119**, 8604–8622.

52    (a) Moore, A.J., Bryce, M.R., Batsanov, A.S., Heaton, J.N., Lehmann, C.W., Howard, J.A.K., Robertson, N., Underhil, A.E., and Perepichka, I.F. (1998), *J. Mater. Chem.* **8**, 1541–1550. (b) Bryce, M.R., Moore, A.J., Batsanov, A.S., Howard, J.A.K., Robertson, N., and Perepichka, I.F. (1999), in J. Veciana, C. Rovira and D.B. Amabilino (eds.), *Supramolecular Engineering of Synthetic Metallic Materials: Conductors and Magnets. NATO ASI Series, Vol. 518*, Kluwer Academic Publishers, Dordrecht, pp. 437-449.

53    Perepichka, I.F., Popov, A.F., Orekhova, T.V., Bryce, M.R., Andrievskii, A.M., Batsanov, A.S., Howard, J.A.K., and Sokolov, N.I., *J. Org. Chem.*, submitted.

54    (a) Gutman, F.E., and Lyons, L.E. (1967) *Organic Semiconductors*, Wiley, New York. (b) Mayoh, B., and Prout, C.K. (1972) *J. Chem. Soc., Faraday Trans. 2* **68**, 1072–1082. (c) Torrance, J.B. (1979) *Acc. Chem. Res.* **12**, 79–86.

55    (a) Nalwa, H.S., and Miyata, S. (eds.) (1997) *Nonlinear Optics of Organic Molecules and Polymers*, CRC Press, Boca Raton. (b) Bosshard, Ch., Sutter, K., Prêtre, Ph., Hulliger, J., Flörscheimer, M., Kaatz, P., and Günter, P. (1995) *Organic Nonlinear Optical Materials (Advances in Nonlinear Optics Series. Vol. 1)*, Gordon and Breach Publishers, Basel. (c) Marder, S.R., Sohn, J.E., and Stucky, G.D. (eds.) (1991) *Materials for Nonlinear Optics. Chemical Perspectives. ACS Symposium Series 455*, ACS, Washington.

56    Nahata, A., Wu, C., Knapp, C., Lu, V., Shan, J., and Yardley, J.T. (1994), *Appl. Phys. Lett.* **64**, 3371–3373.

# DESIGN AND SYNTHESIS OF PHOTOCHROMIC SYSTEMS

IVAN PETKOV
*University of Sofia, Faculty of Chemistry, Department of Organic Chemistry, 1, James Bourchier Av., 1126, Sofia, Bulgaria*

The development of the synthesis of new organic compounds with unusual properties has long been a desirable goal for organic chemists. In particular, the chemistry of the photosensitive compounds has gained considerable prominence in recent years because of their usefulness in different scientific directions. High sensitivity and multifunctional characteristics of such organic compounds becomes a primary requirement in the design and development of materials for microelectronics and photonics, for analytical chemistry, biochemical and environmental sensors and indicators. Additionally, the ability of specific properties has provided increased impetus in the investigations of their photophysical and photochemical properties and development of new systems on their basis [1-3].

Of the various classes organic compounds available to answer of the specific requirements of the different areas of the applications the bridged styrylheterocyclyc salts are the most attractive class of compounds because of their multifunctional nature. Some typical features of the compounds are: *the ease of fabrication, the possibility to shape the compounds into the desired structures by molecular engineering, the introduction and fine tuning of a large variety of physical properties by small changes in the molecular structure and the construction and characterization of simple isolated systems, which can provide solutions for fundamental problems.* That can be illustrated with **Scheme 1**.

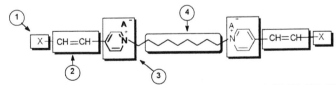

① Substituted aromatic, heterocyclic and nonaromatic radiacals. Substituents: OH, CN, $NO_2$, $NH_2$, $N(R)_2$, $C(F)_3$, alkyl and alkenyl, halogens, others. Position: orto, meta, para.

② $-(CH=CH)_n-$ ; $-CH=N-$ ; $-N=N-$

③ Heterocyclic bases: pyridine, quinoline, acridinium, others

④ Alkyl and alkenyl chains.

## Scheme 1

*F. Kajzar and M.V. Agranovich (eds.), Multiphoton and Light Driven Multielectron Processes in Organics: New Phenomena, Materials and Applications, 387–402.*

Since, multifunctional nature of the compounds is associated with the structure, the changes in the fourth positions strongly will influence on the properties. The multifunctional properties might be based on various properties of molecules like photoinduced electron transfer, *cis-trans* isomerization, differences in anions behaviour, photochromic, electrochromic properties, whereas light, heat, pressure, magnetic or electric fields, chemical reactions etc. can be used to achieve the change in the molecules [4,5]. The actual compounds must have the capability to react against an outside influence at the molecular level, giving unusual molecular structures with a desirable architecture, and allow a photochemical control of their properties. Then only the necessary conditions will be fulfilled for a successful development of photosensitive molecular devices. The changes of the structure can be directed and depending on the properties, which must have the end products. Such intelligent functions are based on the conformational and the electronic changes of the structure and the compounds can be named **Intelligent compounds.**

### SYNTHESIS

The synthesis of the bridged styrylpyridinium, styrylquinolinium and styrylacrydinium salts can be separated in two stages [6-9]:
*-Interaction of the alkylene dibromides with two moles of the tertiary base to form the bis–quaternary ammonium salts (Scheme 2).*

$$Br(CH_2)nBr + 2\,N \longrightarrow {}^-Br\,N-(CH_2)n-N\,Br^-$$

**Scheme 2**

*-Interaction of the bis-quaternary ammonium salts with substituted benzaldehydes (Scheme 3).*

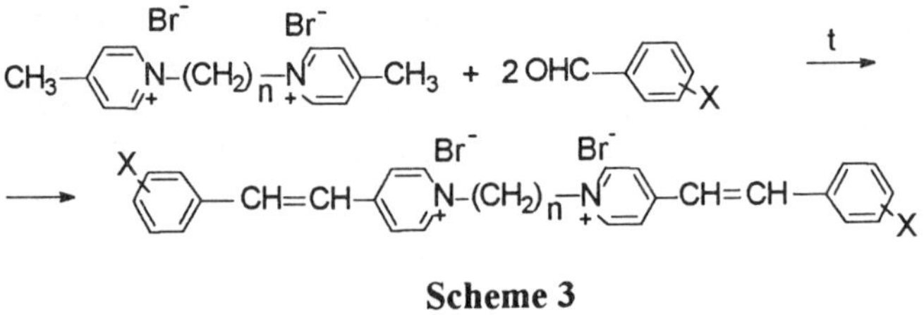

$$CH_3-\langle\;\rangle N-(CH_2)_n-N\langle\;\rangle-CH_3 + 2\,OHC-\langle\;\rangle_X \xrightarrow{t}$$

$$\longrightarrow X\langle\;\rangle-CH=CH-\langle\;\rangle N-(CH_2)_n-N\langle\;\rangle-CH=CH-\langle\;\rangle_X$$

**Scheme 3**

There are several important notes about these chemical interactions. First, the reaction of the dibromides with pyridines yields hygroscopic solids. Such dibromides cannot be characterized as anhydrous salts. This fact requires the dibromides to be converted into corresponding salts as perchlorates or tetraphenylborates. The interaction in **Scheme 2** will depend on the kind of the dibromides. For example, it is anticipated that the yields of bis-salts with the secondary bromides would be less than with the previously used primary bromides. In these cases, the steric hindrance will play important role in the course of the reaction. It is very important during the course of the first reaction to be ensured conditions for the receiving of the bis-product.

## General synthetic scheme

1. $CH_3$—N + $Br(CH_2)_nBr$ + N—$CH_3$ $\xrightarrow[\text{without solvent}]{t}$ $CH_3$—N–$(CH_2)_n$–N—$CH_3$  (Br⁻, Br⁻)

2. $CH_3$—N–$(CH_2)_n$–N—$CH_3$ (Br⁻, Br⁻) + $(Ph)_4B$ Na⁺ $\xrightarrow{MeOH}$ $CH_3$—N–$(CH_2)_n$–N—$CH_3$ (Ph)₄B, B(Ph)₄

hygroscopic

3. $CH_3$—N–$(CH_2)_n$–N—$CH_3$ (Ph)₄B⁻, B(Ph)₄⁻ + 2 OHC—X $\xrightarrow[\substack{\text{piperidine}\\\text{DMSO}}]{t}$

$\xrightarrow[\substack{\text{piperidine}\\\text{DMSO}}]{t}$ X—CH=CH—N–$(CH_2)_n$–N—CH=CH—X  (Ph)₄B⁻, B(Ph)₄⁻

## DESIGN OF THE STRUCTURE

The general scheme for the construction and the design of the photochromic systems is forwarded containing three subunits:
- *Initiated structural changes which at least in one way must be driven by light*
- *Functional groups which convert structural changes into the appropriate physical or chemical properties or behaviour*
- *Medium (solvent, polymers, surfaces) to mediate the effects and to optimize and amplify the molecular processes*

In this connection our efforts were directed to be synthesized bridged styrylheterocyclic salts as the central functional unit and their potential to stimulate chemical and physical processes on the basis of the changes in the position 1,3 and 4 (**Scheme 1**).

*Position 1. Modification of the structure with substituents as: OH, NO₂, N(CH₃)₂, H and the change of the phenyl with pyridil ring.*

X—CH=CH—N–$(CH_2)_n$–N—CH=CH—X  (Ph)₄B⁻, B(Ph)₄⁻

**Position 3.** *Modification of the structure with heterocyclic bases as: pyridinium, quinolinium and acrydinium.*

**Position 4.** *Modification of the structure with alkyl bridges as:*
*1,3-propyl-, 2,4-dimethylpentyl-, 2,5-dimethylhexyl-, 1,4-butyl-, 1,5-pentyl-, 1,10-decyl.*

## UV SPECTRAL DATA

The main absorption peaks of the substituted compounds in DME, methanol and aceto-nitrile showed changes in absorption spectra in wide spectral range from 350 to 600 nm depending on the substituents.

Table 1. UV spectral data of 2,5-bis[4-(4-X-styryl)pyridinium] hexane tetra-phenylborate in acetonitrile

| X | $\lambda_{max}$, nm |
|---|---|
| $C_6H_5$ | 338 |
| $O_2N\text{-}C_6H_4$ | 345 |
| $HO\text{-}C_6H_4$ | 390 |
| $(CH_3)_2N\text{-}C_6H_4$ | 490 |
| $C_5H_4N$ | 490 |

There is not influence of the chain length of the methylene bridge on the absorption of the compounds.

Table 2. UV spectral data of styrylpyridinium alkanes tetraphenylborates in acetonitrile.

| Alkyl Bridge | $NO_2$ | H | $\lambda_{max}$ nm OH | $N(CH_3)_2$ | pyridil |
|---|---|---|---|---|---|
| 1,3- | 345 | 355 | 400 | 490 | 325 |
| 1,4- | 345 | 350 | 400 | 485 | 323 |
| 1,5- | 345 | 350 | 399 | 482 | 320 |
| 1,10- | 345 | 350 | 400 | 490 | 315 |
| 2,4- | 345 | 350- | - | - | - |
| 2,5- | 345 | 350 | 398 | 485 | - |

There are no serious influences of the solvent and the structure on the $\lambda_{max}$ (**Table 3**). In this connection it is interesting to note that with 2,4- compounds the position of the absorption band depends on the time. In freshly prepared $CH_3CN$ solution absorption band is at 346 nm. After 24 hours the band changes to 337 nm.

These results suggested small electronic interaction of the chromophores in bridged compounds. The change of the spectral picture for **2,4-bis[4-(4-nitrostyryl)pyridinium]pentane tetraphenylborate** can be connected with the influence of the structure – presence of **meso-** and **d,l** forms and the solvent polarity.

**Table 3. UV spectral data for 1,3-, 2,4- and 2,5-bis[4-(4-nitrostyryl)pyridinium]alkane tetraphenylborate salts.**

| Solvent | 1,3- | 2,4- | 2,5- |
|---------|------|------|------|
| DME | 346 | 346 | 346 |
| $CH_3CN$ | 347 | 346* | 347 |
| PMMA | 350 | 350 | 350 |

### PHOTOCHEMICAL INVESTIGATIONS

Upon irradiation (150 W Hg-Xe lamp trough an L-39 cut filter, $\lambda_{ex} > 365nm$ and an IR cut filter) the spectral behaviour of the compounds is different depending on the solvent and the presence of oxygen. Sample solutions were stirred with a magnetic stirrer during irradiation unless otherwise indicated. The irradiation of the compounds under air in solution leads only to its **trans-cis** isomerization (**Figures 1-3**).

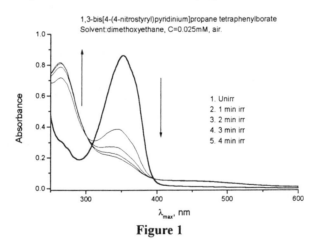

1,3-bis[4-(4-nitrostyryl)pyridinium]propane tetraphenylborate
Solvent:dimethoxyethane, C=0.025mM, air.

1. Unirr
2. 1 min irr
3. 2 min irr
4. 3 min irr
5. 4 min irr

**Figure 1**

Upon exposure to visible light, absorption spectrum of the compounds gradually changed, showing a weak blue shift of the maximum at 350-365nm. The clear isosbestic points suggested the presence of two-component system. The same behaviour was observed for all compounds and the effect of the alkyl chain was negligible. Since only this isomerization process was detected during the irradiation, the formation of other absorption maxima in the spectral region above 400 nm was neglected under the experimental conditions used here. Weak changes were registered in the area after 400 nm at higher concentration. Although a detailed discussion of the process of isomerization will be reported elsewhere, it is important to point out here that the photoinduced isomerization is the first process registered here. When there are structural reasons, as with the compound **2,4-bis [4-(4-nitrostyryl)pyridinium]pentane tetraphenylborate** this process can to be passed thermodynamically. The process is reversible and several times can be repeated.

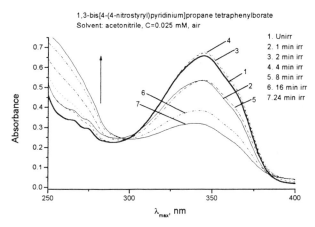

1,3-bis[4-(4-nitrostyryl)pyridinium]propane tetraphenylborate
Solvent: acetonitrile, C=0.025 mM, air

1. Unirr
2. 1 min irr
3. 2 min irr
4. 4 min irr
5. 8 min irr
6. 16 min irr
7. 24 min irr

**Figure 2**

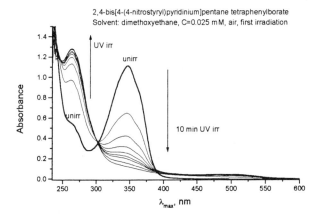

2,4-bis[4-(4-nitrostyryl)pyridinium]pentane tetraphenylborate
Solvent: dimethoxyethane, C=0.025 mM, air, first irradiation

**Figure 3**

Upon irradiation of these salts in inert atmosphere new absorption peaks at about 450, 580 and 650nm were observed in the visible region. The maxima have been assigned to nitrostyrylpyridinyl radicals formed by photoinduced electron transfer reaction from tetraphenylborate anion to styrylpyridinium cation. These photogenerated styrylpyridinyl radicals are fairly stable mainly because of oxidative decomposition of tetraphenylborate anion after photoinduced electron transfer. A new peak at about 532 nm gradually increased in the dark with the decay of radicals. It strongly suggested the formation of a new species via radicals, but it is not assigned yet. The species at 650 nm can be assigned to J-like aggregates of nitrostyrylpyridinyl radicals. During further storage in the dark at room temperature, the absorption at 650 nm decreased accompanying an increase of absorption in a shorter wavelength region with an isosbestic point. This suggests further changes in the structure of aggregates or other reactions, the details of which have not been clarified yet (**Figures 4-12**).

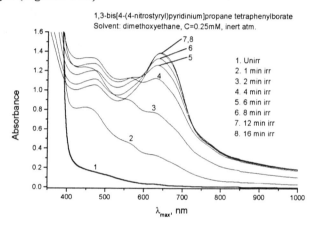

**Figure 4**

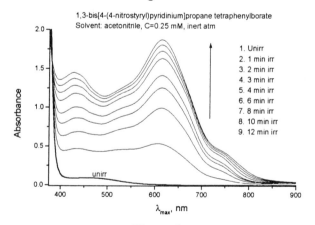

**Figure 5**

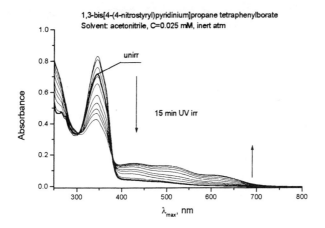

**Figure 6**

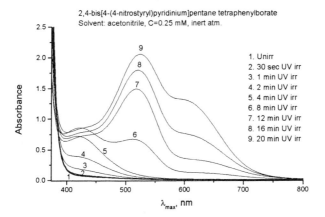

**Figure 7**

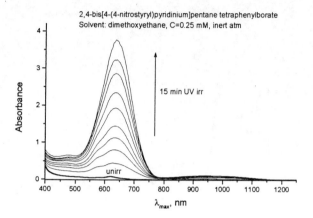

**Figure 8**

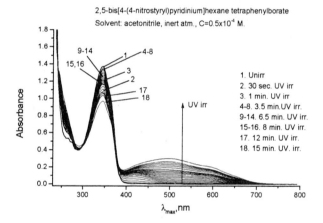

**Figure 9**

As can be seen from the figures the dynamics of the receiving of the radicals and the dimmers is different and depend on the structures.

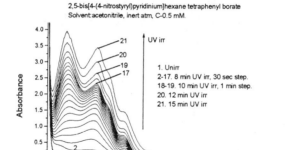

**Figure 10**

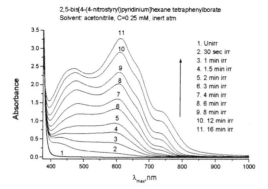

**Figure 11**

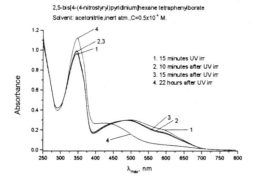

**Figure 12. Dark reaction.**

The present investigations suggest that it is possible to control the product of the photolysis towards either unimolecular (photoisomerization) or bimolecular (photodimerization) reactions.

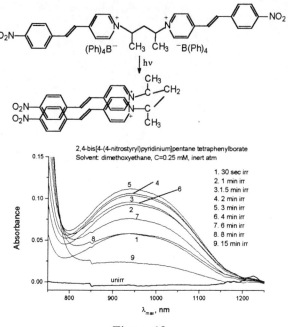

**Figure 13**

The observed new band at about 1000nm was attributed to the CR band due to the electronic interaction between photogenerated radicals and its parent cations according with the results of T. Nagamura et al.[10](**Figure 13**).

The nitro group is much stronger electron-withdrawing group, which most probably contributed to the stabilization of photogenerated styrylpyridinil radicals. Higher stabilization of the system can be expected from such long and planar structures with extensive charge delocalization, which favour the electronic interaction [11]. So-called π-π interaction are dominated by electrostatics, which means the most favorable interaction of the systems is found when the substituents are both strongly electron withdrawing. It is interesting to be note that the formation of the radical cations is during the first 3 minutes irradiation. After 15 minutes irradiation the maximal decreases of the absorbance is close to the initial unirradiated solution. These results indicate that the intermolecular interaction of photogenerated styrylpyridinyl radicals depend on their concentration.

Our investigations in polymer (PMMA) films were directed to the possibilities of the hard matrix for the stabilization of the species of the irradiation compounds. As that can be seen on **Figure 14** the spectral changes are similar to these in solution on air and they are connected with the changes of the colour.

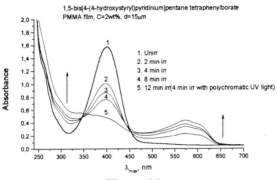

**Figure 14**

The photochromic changes are different when as polymer matrix is used some PVC material. Under the influence of the UV irradiation, temperature, gamma radiation in PVC take place intensive process of dehydrochalogenation. HCl, product of this process can interact with the compounds incorporated into the PVC matrix and on the base of the chemical reaction the colour changes to be others. These possibilities show that with the change of the polymer matrix can be controlled the photochemical and the photophysical properties of the compounds. And indication of these changes can be the change of the colour. The modification of the structure, for example on the base of the change of the substituents is other possibility for such control. The presence of the hydroxyl group in the structure of the compound is reason the photochromic changes to be in the colour direction: yellow-green-blue-violet.

Other very important possibilities of the compounds are the presence of the fluorescence, which in some cases is too intensive (**Figure 15**). That experimental fact shows that there are other ways for the dissipation of the energy in the molecules.

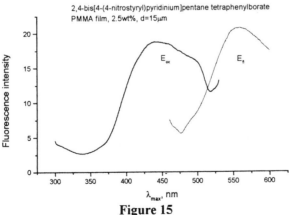

**Figure 15**

The photochromic characteristics of the changes are similar to these of tetrakis[3,5-bis(trifluoromethyl)phenyl]borate as a counter ion of 4,4'-bipyridinium due only to the photoinduced electron transfer and thermal reverse reactions in solutions, in polymer and Langmuir-Blodgett films[12-15]. During our investigations with the bridged styrylheterocyclic salts we receive stable photochromic picture. In inert atmosphere are necessary 20 hours for the reversible reaction and 1 or 2 hours for the reversible reaction in air. Very stable picture we found in polymer film (PMMA). Several months after irradiation the more compounds incorporated in polymer films, keep the changes. These results show that the structure of the salts is very important factor for the stability of the changes. The variation of the substituents or heterocyclic bases is reason for the photochemical and photophysical properties of the compounds - in solution or polymer films.

## CONCLUSION AND PROSPECTS

Multifunctional systems on the base of organic compounds and ingredients will beyond doubt play a key role in future (nano)-technology development.

The photochromism of the compounds is based on a single electron transfer, via the excited state of an ion-pair charge transfer complex between pyridinium ion and tetraphenylborate. The irradiation of this salt at the CT absorption band ($\lambda$>365nm) generated new absorptions in the visible region with main peaks at 430, 580 and 650 nm, observed by a colour change from pale yellow to blue. These peaks are characteristic of a pyridinium radical cation and its associates. The generation and reverse reaction of these radical cations could be repeated for more than ten cycles by irradiation at room temperature to induce the reverse thermal process. These compounds can be also electrochromic, which is the property to change colour reversibly in response to an applied external electric field.

Important drawbacks of this system are:
- *Sensitivity towards oxygen, which destroys the radical formed*
- *Limited life-time of the radical cation.*

But the influence of the oxygen can be limited with the using of the quenchers, as for example tiolate complexes as quainter ions. The lifetime could be drastically improved by the use of the polymer matrix - incorporation of the salts in the polymer matrix.

It is very interesting that the different polymer matrix can be reason for the different photochromic changes.

One of the most attractive properties of the bridged styrylheterocyclic compounds is the possibility the photochromic event to be accompanied by other changes in the molecules, which can be detected. For example changes in liquid crystalline mesophases (modification with long carbon chain); complexation behaviour; conductivity/current; aggregation, cis-trans isomerization. On the other side, the difficulties encountered with the typical photochromic molecules reported until now as low thermal stability (azobenzenes, azulenes), the use of difficulty controllable aggregates (spiropyrans), oxygen sensitivity (viologens), destructive read-out methods

(diarylalkenes, fulgides), and unwanted side reactions (keto-enol tautomerizations) can be solved with bridged styryl heterocyclic salts on the base of:
- *Thermal stability - the compounds are salts and there are possibilities to change the anions with organic and inorganic species.*
- *The mechanism of the aggregation is different (radical/cations)*
- *Limited of the oxygen influence on the base of oxygen quenchers or polymer matrix*
- *Rich possibilities for the modification of the structure and the receiving of the compounds with wanted properties - intelligent functions.*

## REFERENCES

1. Malkin, J., Dvornikov, A.S., Straub, K.D., and Rentzepis, P.M. (1993) Photochemistry of molecular systems for optical 3D storage memory, *Research on Chemical Intermediates Vol. 19, pp 159-189.*
2. Feringa, B. L., Jager, W. F., and Lange, B. (1993) Organic materials for reversible optical data storage, *Tetrahedron Vol. 49, No 37, pp. 8267.*
3. Lewis, T. W., and Wallace, G.G. (1997) The basis for development on intelligent materials, *Journal of Chemical Education Vol. 74, No 6 , 703.*
4. Proceedings of the 5th International Conference of Unconventional photoactive Solids, Symposium on Molecular Systems; *Mol. Cryst. Liq. Cryst. 1992, vol. 216-218.*
5. Carter, F.L.; Siatkowski, H., Wohltgen, H., Eds., (1988) *Molecular Electronic Devices,* Elsevier; Amsterdam.
6. Hartwell, J.L.,and Pogorelskin, M. A. (1950) Some quaternary ammonium salts of heterocyclic bases.III Bis-quaternary ammonium salts, *JACS 72, 2040.*
7. Petkov, I., and Nagamura, T. (1998) unpublished data.
8. Sahay, A.K., Mishra, B.K.,and Behera, G.B. (1988) Studies on cyanine dyes: Evaluation of area/molecule, surface potential&rate constants for KMNO4 oxidation in two dimensioms, *Indian J. of Chemistry, Vol. 27A,561.*
9. Mishra, J.K., Sahay, A.K. & Mishra, B.K. (1991) Behaviour of a surface active cyanine dye in water and surfactant solutions, *Indian J. of Chemistry, Vol. 30A, 886.*
10. Nagamura, T., Tanaka, A., Kawai, H., and Sakaguchi, H. (1993) First charge resonance band observed by steady photolysis at room temperature in solution, *J. Chem. Soc., Chem. Commun., 599.*
11. Nagamura, T., Kawai, H., Ichihara, T., and Sakaguchi, H. (1995) Photoinduced electron transfer and charge resonance band in ion-pair charge-transfer complexes of styrylpyridinium tetraphenylborate, *Synth. Metals, 71, 2069.*
12. Nagamura, T., Ichihara, T., and Kawai, H. (1996) Charge resonance and charge tranfer interactions of photogenerated dicyanovinylstyrylpyridinyl radicals in solutions at room temperature, *J.Phys.Chem., 100, No22 ,9370.*
13. Nagamura, T. (1993) Optical memory by novel photoinduced electrochromism, *Mol. Cryst.Liq.Cryst. 224,75.*

14. Nagamura, T., and Isoda, Y. (1991) Novel photochromic polymer films containing ion-pair charge-transfer complexes of 4,4'-bipyridinium ions for optical recording, *J. Chem. Soc., Chem. Commun., 72.*

15. Kawai, H., and Nagamura, T. (1998) Ultrafast dynamics of photogenerated styrylpyridinyl radical and its dimer radical cation with styrylpyridinium cation studied by femtosecond laser flash photolysis, *J. Chem.Soc. Faraday Trans., 94, 3581.*

# PHOTO-INDUCED MULTIELECTRON TRANSFER IN ORGANIC CRYSTALS WITH MIXED-STACK ARCHITECTURE.

## T. LUTY,[a,b] S. KOSHIHARA,[c] H. CAILLEAU[d]

[a] *Institute of Physical and Theoretical Chemistry, Technical University of Wroclaw, Wroclaw, Poland;*
[b] *Laboratoire de Dynamique et Structures des Materiaux Moleculaires, URA CNRS No.801, UFR de Physique, Universite de Lille1, France;*
[c] *Department of Applied Physics, Tokyo Institute of Technology 2-12-1 Oh-okayama, Meguro-ku, Tokyo 152, and Kanagawa Academyy of Science and Technology (KAST), 3-2-1 Sakado, Takatsu-ku, Kawasaki 213, Japan;*
[d] *Groupe Matiere Condensee et Materiaux, UMR 6626 CNRS, Universite de Rennes1, France.*

## Abstract.

We discuss the photo-induced multielectron transfer in organic mixed-stack charge-transfer crystals, close to the neutral-ionic interface. The quasi-one-dimensional architecture of the stacks stimulates intermolecular electron transfer and the energy gain due to Coulomb interaction creates highly cooperative electronic excitations, the charge-transfer strings. At the neutral-ionic interface charge-transfer exciton is no longer elementary excitation and a pair of neutral-ionic domain walls plays such a role. Photo-generated strings proliferates as a motion of lattice-relaxed neutral-ionic domain walls, a «domino effect» caused by electron-phonon coupling. We show that the cooperative multielectron transfer in tetrathiafulvalene-chloranil (TTF-CA) crystal can be photo-induced as bidirectional transition between neutral and ionic phases. Apart from many application-oriented aspects, the discussed phenomena, photo-induced cooperative multielectron transfer, may serve as a model for he electron-transfer self-organization, the frontier problem in natural sciences.

## 1. Introduction.

It is well known that an electron in an insulating condensed environment induces a local reorganization when excited by a photon. The phenomena are called an electron solvation in a liquid or lattice relaxation of an optical excitation in a solid. It has been considered as a very local concept and the environment treated as a passive bath. Recent years have shown that there are many unconventional photoactive condensed phases, where the relaxation of an excited electron results in various cooperative phenomena. This involves large number of atoms and electrons, indicating that an environment plays important and active role. These cooperative phenomena seem to mimic what the nature has developed to a high degree of perfection: the self-organization. For solids it is suggested to call the phenomena «photo-induced structural phase transition» (PIPT) [1].

403

*F. Kajzar and M.V. Agranovich (eds.), Multiphoton and Light Driven Multielectron Processes in Organics: New Phenomena, Materials and Applications, 403–420.*

Among all, the electron transfer processes are extremely important as they are at origin of mechanisms of processes, from biology to (heavy) industry. In a passive medium, an electron transfer is driven by the medium fluctuations as described by Marcus' theory [2]. When an electron transfer is accompanied by a chemical bonding, the medium is strongly (chemically) reorganized and the process is called passive self-organization. In case of interfacial reaction, the chemical binding of an electron helps to achieve favorable electrochemical potential for the electron transfer [3]. The «photon driven proton pump» in liposome bilayer, on the other hand, is the spectacular example where electrical potential derived from photo-induced electron transfer in porphirin moeity leads to directional proton transfer [4,5]. The efficient long-distance electron transfer mediated by hydrogen bonds is another fascinating example [6]. In the passive self-organization, the electron transfer process is coupled to a chemical reaction, with a direct conversion of electrical and chemical potentials. However, according to ref. [3], a simple observation that an electron transfer processes in chemical industry require high temperatures («hard chemistry») while corresponding processes in living systems are efficient at environmental conditions, strongly suggests that the processes in nature are multielectron transfer. They are governed by dynamic self-organization, which involves an active medium. In such a process, a medium is required to supply some amount of energy to accelerate the electron transfer and causes cooperativity of the process. The important fundamental question of a mechanism of cooperative (concerted) multielectron transfer versus stepwise electron transfer has to be addressed. The paper offers a contribution along this line.

The photo-induced phase transitions are the phenomena, in which cooperativity is due to electron-phonon coupling. An excited electron relaxes causing a local lattice distortion and by the gain of deformation energy it is energetically easier to excite a neighboring electron. Photogenerated electronic excitation proliferates during the lattice relaxation. This is an essence of «domino effect» caused by the electron-phonon coupling, and nicely illustrates the cooperativity in one-dimensional systems [7]. It is of our opinion that the importance of studies of the photo-induced phase transitions goes far beyond scientific curiosity within physics and extends into fields such as chemistry, biology, materials science, where model systems are required. It is an additional motivation for the studies.

## 2. Multielectron-Transfer in Mixed-Stack Compounds.

As an extreme case of the photo-induced phase transitions, already exotic phenomena, there is photo-induced multielectron transfer in mixed-stack organic compounds. It has been known for some time, that in some compounds, an intermolecular electron transfer can be induced by temperature, pressure [8] and light [9]. Howeover, it is the recent years extensive research [10-14] that helped us to understand the phenomena as cooperative multielectron transfer, a model for electron transfer self-organization.

The quasi-one-dimensional architecture of the mixed-stack charge-transfer (CT compounds with alternating electron-donor (D) and electron-acceptor (A) molecules stimulates intermolecular electron transfer, which can be generated by light (usuall visible), as so called CT exciton. The excitation, composed of ionic (I) pair (D$^+$A$^-$) of nearest molecules within neutral (N) environment or a pair (D$^0$A$^0$) within I environment, proliferates along a stack in highly cooperative way due to an active role of neighboring molecules (the environment). The energy for an electron transfer is supplied not only by lattice phonons and intra-molecular vibrations, as in a typical process of structural relaxation, but also by the electronic system itself. It is the Coulomb interaction energy, which the system gains when CT excitation is created. In the mixed-stack systems, close to so called neutral-ionic (N-I) interface, the electron transfer process is highly non-linear, unique and of great importance.

The idea of cooperative, multielectron, CT excited states in the mixed-stack systems has long been known [15,16]. It is well explained within simplest model, which neglects quantum effects of an overlap between lowest unoccupied molecular orbital (LUMO) of A and highest occupied molecular orbital (HOMO) of D molecules. The energy cost to create a string of $n$ I pairs within one-dimensional array of D and A molecules is [15],

$$E(n) = n[\,|\delta| + 2V\Sigma_{2n+1}(-1)^{j+1}j^{-1}],\tag{1}$$

where energy of one CT pair, $\delta = I-A-\alpha V$, is composed of the ionization energy of D molecule, I, and electron affinity of A molecule, A. The Coulomb interaction within an infinite chain of alternating point charges, A$^-$ and D$^+$, is expressed in terms of the Madelung constant, $\alpha$ (=2ln2), and the energy of a pair, $V=e^2/a$, (a is the distance between nearest D and A molecules). The ionization potential of D and the electron affinity of A molecules in a crystal lattice are expressed in terms of D/A oxidation/reduction potentials (E$^{ox/red}$) and «reorganization» energies, intramolecular deformation (W$_{intra}$) and polarization (W$_{pol}$), [17],

$$I - A = (E_D^{ox} - E_a^{red}) - W_{intra} - W_{pol}.\tag{2}$$

The intramolecular deformation energy, called small polaron binding energy, is due to totally symmetric deformation of a molecule during the oxidation/reduction process. It is given by totally symmetric normal mode frequencies, $\omega_i^2$, and coupling constants, $g_i$,

$$W_{intra} = 1/2\Sigma_{i\in A,D}(g_i^2/\omega_i^2).\tag{3}$$

The polarization energy is due to a coupling of the dipole moment induced by the multielectron transfer, $\mu(n)$, to Frenkel excitons, which can be characterized by electronic susceptibility, $\chi$. Consequently,

$$W_{pol} = 1/2\mu^2(n)\chi.\tag{4}$$

The intramolecular deformation and polarization processes are treated as «fast» in comparison with lattice, structural relaxation and are considered as responsible for renormalization of the ionization energy of D and electron affinity of A molecules. It may be interesting to notice that the renormalized difference (I-A) plays a role of an effective hardness, a measure of reactivity, for molecules in a crystal, the relation for energy conversion in the process.

Eq. (1) determines energy of a new class of collective excitations, characteristic for the mixed-stack systems, called CT exciton strings [18]. The energy is decreasing function of number of I pairs. The $n$-exciton state can be reached by optically exciting the $(n-1)$-exciton, as documented by an elegant experiment for anthracene-PMDA (pyromellic acid dianhydride) crystal [18]. Stability of the excitonic strings in one-dimensional mixed-stack systems has been examined within an extended Hubbard model with long-range Coulomb interaction [18,19]. It is important to stress that stability of the bound multiexciton states arises from the combined effects of (quasi)-one-dimensionality and strong Coulomb interaction. An interesting aspect of the CT strings, discussed by Hanamura [20] is the propagation mechanism for $n$-exciton string, being a process in $n$ steps of a virtual dissolution of the string. Consequently, the effective mass of the string increases rapidly with $n$ which tends to localize strongly such excitation.

The CT exciton strings play very important role for ultrafast nonlinear optical processes. It seems appropriate to mention that there exist some other interesting collective states which are stabilized by the Coulomb attractive interaction in the mixed-stack one-dimensional systems. These are states formed by an attraction of a CT string to an electron $(A^-)$ or a hole $(D^+)$ forming a symmetric string (with odd number of molecules), such as trimers, pentamers, etc.. These states are essential in photo-assisted carrier transport phenomena, as discussed in ref. [21]. Stressing the collective aspects of the CT string states in the mixed-stack systems, it is important to distinguish them from already known concepts. The multi-Frenkel exciton states in one-dimensional molecular chains [20,22] are different from the CT strings in that the former is stabilized by nearest neighbor dipole-dipole interaction. Also, the CT exciton strings are different from Bose-condensed exciton states and the electron-hole drops.

The collective nature of CT strings in mixed-stack compounds reaches its extreme in systems which are close to the N-I interface, defined by the condition, $\delta=0$ [15]. The systems are characterized by near-degeneracy of N and I ground states. The multistability then allows for switching between N and I states by increasing pressure or decreasing temperature [8]. At the N-I interface, where $\delta$ vanishes, eq. (1) shows that the energies of CT strings, $n=1,...,N-1$, are bounded within, $(\alpha-1)V \leq E(n) \leq 1/2V$, and form near degenerate spectrum [15]. Thus, it has been suggested that the CT exciton is no longer an elementary excitation and has to be replaced by unrelaxed N-I domain wall, with energy corresponding to the upper limit of the CT strings energy, $1/2V$ [23]. An important conclusion is that an electron transfer process in the mixed-stack systems with N-I bistability, becomes highly cooperative multielectron transfer, characterized by N-I domain walls, as elementary, non-linear excitations. However, for a real system, it is rather difficult to prove that the system is at the N-I interface and the excitations are well described by the concept of unrelaxed N-I domain walls. The problem is that the thermodynamical multistability indicated by N-I phase transition is not equivalent to the ground state bistability (N-I interface).

## 2.1 CHARGE-TRANSFER STATES IN TTF-CA CRYSTAL.

The 1:1 complex of tetrathiafulvalene (TTF) and chloranil (CA) is of particular interest, since it is expected to be sufficiently close to N-I interface and switching between the N and I states has been induced by temperature, pressure [8] and light [9,14]. The crystal undergoes the first order phase transition at T=81 K (ambient pressure) accompanied by change in molecular ionicity (from ca. 0.3 to ca. 0.6 in low-temperature phase) [24] and symmetry breaking dimerization (from $P2_1/n$ to Pn) [25]. The high-temperature phase is called neutral (N) and the low-temperature phase ionic (I). It has been well documented that the phase transition in TTF-CA crystal exhibit unique pressure-temperature phase diagram, analogous to solid-liquid-gas one [13]. An interpretation of the thermodynamical stability of the system has been based on the concept of lattice relaxed CT strings [11,26]. The N-I transition in TTF-CA can be classified as a prototypical example of the cooperative valence instability in solids.

CT states in TTF-CA crystal have been analyzed by Soos et al. [15], and here we will focus on symmetry aspects as well as the context of photo-induced phase transition. The crystal can be treated as quasi-one-dimensional system, composed of two, translationally non-equivalent mixed-stacks along a-axis. In the high symmetry, N, phase the stacks are related by two-fold screw axis (crystal b-axis). The crystal architecture suggest to treat the system as a set of weakly coupled stacks, and the inter-stack interaction can be considered as a perturbation to strongly correlated one-dimensional array of D and A molecules. For an isolated one-dimensional mixed-stack with $C_i$ symmetry (an infinite stack), states $|...DA(D^+A^-)_nDA...> \equiv |I_+(n)>$ and $|...DAD(A^-D^+)_nA...> \equiv |I_-(n)>$, have equal energies, and are characterized by opposite polarizations, Fig. 1. Symmetric and antisymmetric combinations of the string states are the eigen-states. The even state, $|S> = 1/\sqrt{2}(|I_+(n)> + |I_-(n)>)$, of $A_g$ symmetry is the dipole forbidden state and the odd state,

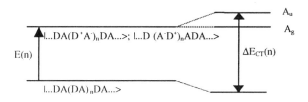

*Figure 1.* Scheme of energy levels for CT strings without structural relaxation. The right-hand side diagram follows from asymmetric quantum mixing, characteristic for TTF-CA mixed-stack.

$|A> = 1/\sqrt{2}(|I_+(n)> - |I_-(n)>)$, of $A_u$ symmetry is the dipole allowed state. The excited states are formed by mixing the string states with the N state, $|...DA(DA)_nDA...> \equiv |N>$. The mixing is due to overlap, t, of HOMO and LUMO orbitals of D and A molecules, and decreases for multielectron transfer in the strings. Symmetry of the orbitals plays essential role. It is implicitly assumed in all theoretical papers that symmetries of both orbitals are equal. The case at hand, the TTF-CA crystal, is different. It is known that the HOMO of the TTF molecule is odd under inversion ($b_{1u}$), whereas the LUMO of the CA molecule is even ($b_{2g}$) [27]. In fact, it was an original idea [27] to construct a charge-transfer material formed by molecules with symmetrically opposite frontier orbitals, which would prohibit D-A overlap, resulting in expected segregated-stack architecture. However, this guideline did not take into account cooperative processes, e.g., the fact that such symmetries of the frontier orbitals may prefer dimerized mixed-stacks, as it is the case in the low-temperature phase of TTF-CA crystal [25]. The ground state of the system is the dimerized mixed-stack, indeed. The overlap integrals with alternating sign have been shown to have important consequences for the band structure of the crystal [28]. In a simplest, three states representation its means that the eigen-states of the mixed-stack are eigen-values of the matrix Hamiltonian,

$$
H(n) \;=\; 
\begin{array}{cccc}
 & |N> & |I_+(n)> & |I_-(n)> \\
 & 0 & t(n) & -t(n) \\
 & t(n) & E(n) & 0 \\
 & -t(n) & 0 & E(n),
\end{array}
\tag{5}
$$

where $t(n) = <I_+(n)|H(n)|N>$ is the transfer integral which decreases exponentially with $n$, $t(n) \propto t^n$ [16]. The asymmetry in the electron transfer may be interpreted that the mixed-stack of TTF-CA crystal exhibits a broken symmetry (without inversion centre) of the valence electron system.

The asymmetric quantum mixing prefers the odd state of the CT strings, $A_u$, and will couple it to the N state. As a result, the ground state is a quantum mixture of N and I strings, and the fraction of I strings is,

$$
\rho(n) = 1/2[1 - E(n)/\{E^2(n) + 8t^2(n)\}^{1/2}],
\tag{6}
$$

and the degree of ionization of the ground state is approximated as,

$$
\rho = \Sigma_n[\rho(n)]^{1/n}.
\tag{7}
$$

The CT exciton string state, accessible by one photon absorption from the ground state is located at,

$$
\Delta E_{CT}(n) = \{E^2(n) + 8t^2(n)\}^{1/2}],
\tag{8}
$$

(Fig. 1) and in the limit of the N-I interface, and $n$ large, it becomes equal to the energy of N-I domain wall, $1/2V$. One may imagine that in a system close to the N-I interface, like TTF-CA, an optical excitation with energy larger than $1/2V$, creates a continuum of CT strings. This, however, is difficult to prove experimentally and the question what is a nature of CT excited states, is it more like a N-I domain wall or a CT string, in the mixed-stack systems close to N-I interface remains opened. In particular, it is not quite clear how close to the N-I interface is the TTF-CA system. The CT absorption band in the crystal, observed at $h\nu_{CT} = 0.55$ eV [24] should correspond to $(\alpha-1)V$ and with estimation given by Soos [15], the criteria of the N-I interface is met within 0.3 eV. On the other hand, with values $(I-A) = 4.08$ eV and $V=2.77$ eV, energy of one I pair is 1.31 eV, much larger then observed CT band. Most likely, the TTF-CA crystal is close but not at the N-I interface and the concept of CT strings, as an intermediate between CT exciton and N-I (unrelaxed) domain wall seems appropriate for the system.

The CT string states contribute to optical susceptibilities. For the CT static polarizability one gets,

$$\alpha_{CT} = 2\Sigma_n\{\rho(n)[1-\rho(n)]\}\Delta E_{CT}^{-1}\mu^2(n), \tag{9}$$

and for the hyperpolarizability,

$$\beta_{CT} = 2\Sigma_n\{\rho(n)[1-\rho(n)] [1-2\rho(n)]\}\Delta E_{CT}^{-2}\mu^3(n). \tag{10}$$

It is important to notice that CT strings give non-zero contribution to non-linear polarizability of a mixed-stack and it is a consequence of asymmetry in the overlap integral between D and A molecules. Another words, the valence electron system in TTF-CA mixed-stack contributes to the hyperpolarizability, which corresponds to (xxx) component of the tensor. In the high symmetry phase, where the stacks are related by two-fold screw axis along b, the contributions are canceled. We can expect, however, that whenever the symmetry relation is lost (symmetry breaking structural change) the CT strings will contribute to the non-linear optical susceptibilities. The non-linear optical properties of TTF-CA crystal, especially in the context of photo-induced transition, are very interesting and subject of intensive studies.

### 3. Structural Relaxation of Photo-Induced CT Strings.

Many authors have considered a general problem of structural relaxation of CT exciton states. In the context of photo-induced phase transitions, Toyozawa has discussed it in his pioneering papers [29,30]. The question is how the photon energy, absorbed to excite the electronic system is transferred to the lattice system so efficiently as to generate a domain of metastable phase which can be of tremendous size. In the context of photo-induced N-I transition, Toyozawa has postulated that it is the dipole-dipole interaction between relaxed CT excitons that drives proliferation and the spatial growth of the domain of new phase [30]. However, a microscopic description of the proliferation of relaxing excitons is most conveniently described in terms of diabatic potentials. We shall follow the model presented in ref. [31] and apply it to the CT strings. Let us consider localized electronic two levels, $|E_l>$ and $|G_l>$ at every site assigned to a D molecule, with the Franck-Condon energy E. The excited CT state will be considered as a single CT exciton state of $A_u$ symmetry, coupled to dimerization mode $Q_l$ of the same symmetry. The coupling constant is denoted as g and the interaction modes at sites l and l' are coupled with the coupling constant $K_{ll'}$. The Hamiltonian is,

$$H = \Sigma_l |E_l> (E - \sigma_l g Q_l - 1/2\eta Q_l^2 )<E_l| + 1/2\Sigma_l Q_l^2 - 1/2\Sigma_{ll'} K_{ll'} Q_l Q_{l'}. \tag{11}$$

The pseudospin operator, $\sigma_l=\pm1$, indicates that an electron transfer and related to it structural relaxation may appear to the left or right from the D molecule at site l. The quadratic term in the exciton-phonon coupling takes into account the fact that the force constant for the dimerization mode changes (soften) due to attractive interaction between ionized molecules in a CT exciton. The coupling constant, $\eta$, is the second derivative of the interaction. Moreover, the Coulomb interaction will modify the intersite coupling constant, $K_{ll'}$, as well, but this effect will complicate the theory. In order to keep our considerations simple, we shall assume that the elastic coupling part of the Hamiltonian is the same in ground and excited states. Let us assume that $l=1,....,n$ site in a mixed-stack are optically excited and others are in the ground state. It will be convenient to rewrite the Hamiltonian in terms of the Fourier transform of the lattice dimerization mode, $Q_l = N^{-1/2}\sum_q Q_q\exp(iql)$, as,

$$H_{stack}(n) = \Delta E_{CT}(n) - 1/2\sum_q \Omega_q^2 \Delta_q^{(n)} \Delta_{-q}^{(n)} + 1/2\sum_q \Omega_q^2 Q_q^{(n)} Q_{-q}^{(n)}. \tag{12}$$

The Hamiltonian describes situation where $n$ consecutive (DA) pairs have been ionized and formed the $n$-exciton string, thus the operator $\sigma_l$ has been assumed to be the same for all excited (DA) pairs and independent on the site. In the ground state the lattice dimerization phonon has a dispersion,

$$\omega_q^2 = 1-\sum_{ll'} K_{ll'} \exp[iq(l'-l)], \tag{13}$$

and in the string, the mode has energy,

$$\Omega_q^2 = (1-\eta)-\sum_{ll'\in E} K_{ll'} \exp[iq(l'-l)]. \tag{14}$$

The CT string is structurally deformed by the stationary displacement,

$$\Delta_q^{(n)} = \pm N^{-1/2} g/\Omega_q^2 [\sum_l^n \exp(iql)], \tag{15}$$

around which the lattice coordinate fluctuates, $Q_q = Q_q^{(n)} + \Delta_q^{(n)}$. The stationary displacement, $\Delta_q^{(n)}$, depends on $n$, via the factor, $[\sum_l^n \exp(iql)]$, which contains whole information on the electronic excitations in the system. Thus, the state of the system after excitation is uniquely determined by the stationary displacement. The last term in the Hamiltonian, eq.(12), describes lattice dynamics of the system after relaxation of the string, therefore it gives a curvature of a potential along the relaxation path. The second term denotes stabilization energy due to the lattice distortion. The balance between the energy of unrelaxed CT string and the deformation energy corresponds to the relaxed CT string,

$$\Delta E_{CT, relaxed}(n) = \Delta E_{CT}(n) - 1/2\sum_q \Omega_q^2 \Delta_q^{(n)} \Delta_{-q}^{(n)}, \tag{16}$$

and determines energetic stabilization of the excitonic string. It is the essence of N-bistability in a mixed-stack. The absolute value of this stabilization energy increases by stronger than the first power of $n$, the number of I pairs within a string, what can be deduced from results of ref. [31]. The effect depends on strength of the elastic coupling between sites and is stronger for short-range interaction. It has been called «domino effect» [7]. In the case of I strings, the effect is enhanced by the softening of the phonon mode driven by Coulomb attractive interaction created in the excited state. This illustrates cooperativity in the system caused by the lattice phonons, which act as a 'glue» for the excited I pairs. The essence of the cooperativity is that the absolute value of the distortion energy becomes much larger when the excited molecules coagulate as close as possible than in the case of dispersed distribution of the excited pairs. When two I pairs are trapped next to each other, they distort the lattice twice stronger so that the deformation (stabilization) energy is four times larger. Hanamura [31] has compared the stabilization of the excited cluster to Anderson's localization phenomena, where the deformation energy overcomes correlation energy and bipolaron can be formed. In the case of CT strings, the effect of exciton-phonon coupling which gives the deformation energy is enhanced by the attractive Coulomb interaction within a mixed-stack, and formation of a pair of lattice relaxed domain walls is analogous to bipolaron.

A local structural change under optical pumping may be induced by local instability of one kind of structure relative to the other, e.g., a structure of relaxed CT string

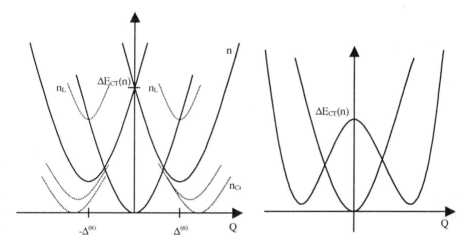

*Figure 2.* The schematic illustration of diabatic potentials for structurally relaxed CT strings. Traces of the potentials for lowest limit ($n_L$, see text) of the strings optical stability are shown. For the critical lenght, $n_{cr}$, the degeneracy of N and I ground states is shown. The right hand side diagram is the mapping (eq. (17)) for two states model, see text.

relative to the structure of the ground state. This is illustrated by corresponding diabatic potentials, describing phonon states of ground and $n$-exciton string states, displaced by $\Delta^{(n)}=N^{-1/2}\sum_q\Delta_q^{(n)}$ the lattice distortion. The curvatures of the parabolas are given by the energies of lattice phonons, in the ground state, $\omega^2=N^{-1}\sum_q\omega_q^2$, and in the string state, $\Omega^2= N^{-1}\sum_q\Omega_q^2$. It is illustrated in Fig. 2. The minima of the diabatic potentials of CT strings depend on the number of excited pairs within the string. Therefore, the states are specified by $n$. Every extension of the string shifts the minimum by the amount, $\Delta^{(n+1)} - \Delta^{(n)} = \Delta = g/\Omega^2$.

We shall close the section with two remarks. First, the parabolas representing diabatic states for the I strings are determined by lattice dimerization phonon energy, which is renormalized by the Coulomb interaction between I pairs. This correction has not been introduced into the Hamiltonin. The second remark concerns an approximation, which can be used to represent the diabatic curves for the strings. The CT string states can be represented by one state with two minima, as already suggested [14]. This mapping is done with the requirements that positions and values of energies for the extremum points are kept the same. The result is,

$$E(Q) = \Delta E_{CT}(n) - 1/2\Omega^2 Q^2 + 1/4g^{-2}\Omega^6 Q^4. \tag{17}$$

This representation, showing explicitly the dimerization phonon instability in the CT string state (Fig. 2), may be convenient for discussing the N-I switching phenomena, as it allows to treat the process within two-level model.

## 4. Dynamics of Photo-Induced Structure Changes.

For a discussion of the dynamics of photo-induced structure changes, it is important to estimate if the process can be treated as diabatic or adiabatic. There are two factors, which have to be taken into account: quantum mixing between diabatic states and time scale of the lattice relaxation process. As for the latter, it is reasonable to assume that a thermal relaxation of the lattice coordinate, Q, in each of the diabatic parabolas finishes within the time of $10^{-12}$ sec. Therefore the process can be considered as fast in comparison with the observation time of the experiment ($10^{-10}$ sec). As for the quantum mixing, there is no clear criteria, because it changes with the CT string length, from estimated about 0.2 eV for one electron transfer [15] and decreasing exponentially with $n$. For longer strings (closer to N-I interface) the quantum mixing can be neglected. It seems reasonable to assume that, in the first approximation the dynamics of the photo-induced structure changes related to N-I bistability of a mixed-stack can be treated within diabatic approximation.

We shall discuss dynamics of a single chain of D and A molecules. The time evolution of the one-dimensional mixed-stack is described by the rate equation for the string length, $n(t)$,

$$dn(t)/dt = I(t)[N - n(t)] - dF(n)/dn, \tag{18}$$

where $I(t)$ is optical pumping, positive when I state is created and negative when N state is created. $F(n)$ is a potential which determines stable, relaxed state of the system without the optical pumping. The kinetic driving force, is due to transition probabilities from N state (I state) to I state (N state) via the thermal processes and, in general, optical processes. However, in the region where the system is optically stable, e.g., where photo-induced switching is possible and there is no spontaneous emission, we shall neglect the optical probabilities. Thus, thermal probabilities determine the relaxation kinetics,

$$-dF(n)/dt = P_{N \to I}[N - n(t)] - P_{I \to N} n(t).$$ (19)

The optical stability of the system can be conveniently defined by lower ($n_L$) and upper ($n_U$) limits of the strings length [33]. The thermal transition probabilities are determined by an attempt frequency, $\tau^{-1}$, of the order of the lattice frequency, and activation factors,

$$P_{N \to I} = \tau^{-1} \exp[-E_{N \to I}(n)/kT],$$ (20)

$$P_{I \to N} = \tau^{-1} \exp[-E_{I \to N}(n)/kT].$$ (21)

It is important that the activation energies depend on the string length, which is a consequence of cooperativity. The dependence has its origin in the fact, that whenever an I pair is excited, there is a collective structural relaxation, which deforms the CT string by the amount $\Delta^{(n)}$. This is a continuous, structural change, which can be visualized as a path of the exciton string diabatic potential along the transformation path (lattice dimerization). At every step of a photo-induced process, the crossing point for the diabatic curves changes. In general, an explicit form for the cooperative activation energy is difficult to derive, and we shall simplify it by assuming equal curvatures of the diabatic potentials. This approximation, in case of CT strings, is rather drastic as it neglects the effect of Coulomb interaction on the deformation of the lattice. Nevertheless, as a minimal model to illustrate qualitatively the effect, we have the following expressions in case of CT strings,

$$E_{N \to I}(n) = 1/2g^2[\omega^{-2}-1]^2(n - n_U)^2,$$ (22)

$$E_{I \to N}(n) = 1/2g^2[\omega^{-2}-1]^2(n - n_L)^2.$$ (23)

The cooperative nature of the system is reflected in the fact that transition probabilities for thermal relaxation are related to string's length via the quadratic dependence of the activation energies. The cooperative activation as expressed in the formula indicate that the process of thermal activation from N to I state becomes easier, longer the created string. Another words, longer string is created more stable it is. And vice versa, thermal relaxation of the I string into the N state becomes easier and faster for shorter strings. For the length, $n_{cr} = 1/2(n_U+n_L)$, there is energetic degeneracy for N and I strings. It is the crossover of ground and excited states, which determine N-I bistability. For longer strings, it can expand thermally without optical pumping. The critical size of a string corresponds to a maximum in potential F(n). The formula for the activation energies are approximate, however, as the effect of different curvatures of the diabatic potentials for ground and excited states, discussed above, has not been taken into account. Dynamics of the photo-induced changes from purely N stack and dynamics of the changes from infinite I chain will be different, in general. This aspect requires very careful and more intensive studies.

The dynamics of the system under and after optical pumping will be determined by a competition between the thermal kinetics and the photo-induced effects. The intensity of optical pumping is important. We assume, following refs. [33,34] that for a light pulse, $I(t) = I_0$ for $0 < t < t_0$ and $I(t) = 0$ for $t > t_0$. For these two limits, the kinetic equations can be integrated as follow [33],

$$\int^n [I_0 - dF(n)/dn]^{-1} dn = t , 0 < t < t_0,$$ [24]

$$\int^n [- dF(n)/dn]^{-1} dn = t - t_0 , t > t_0,$$ [25]

where the integration is from initial value of $n$, being 0 or N. According to its definition, $I_0$ is the rate of creation of excited state ($I_0 > 0$) or ground state ($I_0 < 0$), and the product $I_0 t_0$ gives total number of sites created by the optical pumping. It is clear from the above integral equations, that for $I_0$ smaller then a threshold intensity, $I_{th} = |- dF(n)/dn |n_L/\tau$, [33], there is an infinite time required to switch the system from initial to final state by an optical pumping. On the other hand, for $I_0 > I_{th}$ the system is switched from initial to final state by the optical pumping.

At this point a comment is needed about the threshold phenomena. As it has been pointed out in refs. [31,34], in strictly one-dimensional systems, quantum fluctuations and nucleation processes lead to no threshold behavior. In the context of formation of CT strings, this aspect becomes much more complicated and needs additional studies. Here, we shall assume that a mixed-stack is a quasi-one-dimensional system, where the strings are formed under a weak field of surrounding stacks.

After an illumination is switched off at $t = t_0$, the system with $n$ excited states relaxes to either $n=0$ or $n=N$ states, depending on whether the number of excited states created with time $t_0$ was below or above the critical value $n_{cr}$. It follows therefore, that for every intensity $I_0$ above threshold one, there is a critical duration time, $t_{cr}$, needed to create a critical number of excited sites, $t_{cr} \propto (I_0 - I_{th})^{-1/2}$, [33]. An interpretation of the critical time can be offered based on the cooperative activation and the definition of the critical number of excited sites. It follows that when the number of excited sites reaches the critical value the energy states for system in its ground and excited states becomes equal. This is the crossover between excited and ground states, and for $t_{cr} < t_0$ the stability is shifted from ground to excited state. However, at finite temperature, stability will be determined by thermodynamics of the system.

Now, we discuss the dynamics of the system after the switchoff the optical pumping at $t = t_0$. In the cases when the number of excited sites at $t_0$ (for a given intensity) is in the range, $n_L < n(t_0) < n_U$, the relaxation dynamics is determined by thermal transitions. Otherwise, system relaxes via spontaneous emissions within the radiative lifetime. For the thermal relaxation, the process will be determined predominantly by the early stages and will be accelerated because the cooperative activation decreases energy barrier for the relaxation. The relaxation time has been found to be strongly dependent on the number of excited sites $n(t_0)$ just after the switchoff the illumination as well as the temperature and parameters of the system (see the formula for activation energies) [33].

It is obvious that the relaxation times are longer, closer the number of excited sites is to the limited numbers, of the upper and lower string lengths.

Before discussing experimental results for the model system, TTF-CA crystal, let us mention, following the remarks in ref.[34], that the temporal dynamics of the photo-induced phase transition is characterized by several time scales: the radiative lifetime, a characteristic time of thermal transition, characteristic time of optical pumping and the appearance time of a critical droplet. These times have to be confronted with an observation time in an experiment. Thus, one can expect a variety of transition dynamics depending on the relative time scales. Obviously, for controlling photoswitching and photomemory in materials, the time scale is of crucial importance.

## 5. Observation of the Bidirectional N-I Photo-induced Phase Transition in TTF-CA Crystals.

It has been recently demonstrated [14] that photo-switching between N and I phase in TTF-CA crystal is possible. We discuss the main experimental results, in particular the temporal dynamics of the switching in the context of the above model and its kinetics. Fig. 3 shows schematically, (free) energy curves for the system at two temperatures, 77 K and 100 K, close to the equilibrium temperature (81 K), at which

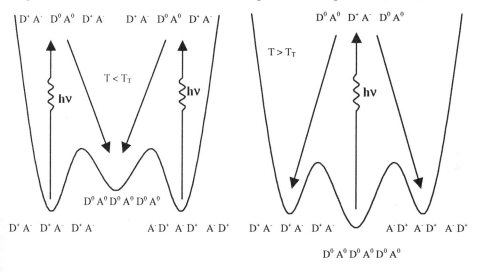

*Figure 3*. Schematically drawn (free) energy curves for TTF-CA crystal at temperatures 77 K (T < $T_T$) and 100 K (T > $T_T$) close to equilibrium temperature $T_T$ = 81 K, at which the experiments were done

experiments were done. The photo-induced effects on single crystals of TTF-CA were observed by the conventional pump-probe technique with the photon energy of the laser pulse 1.55 eV and pulse width 80-fs. A response of the system has been monitored by photoreflectance spectra taken on the (001) surface with light polarization normal to the stack axis (a axis). A conversion between N and I phases has been estimated from the relative difference ($\Delta R/R$) of photoreflectance between the spectra with and without irradiation, taken at the same photon energy. As the photoreflectance spectra are sensitive for degree of ionicity of the molecules, the relative difference monitored photo-induced structure changes, which do not break symmetry of the crystal. For this reason, the dynamics of the photo-induced transition can be described in terms of the CT string length (or a concentration). Among many experimental results, demonstrating cooperatrive electron transfer in the TTF-CA crystal, here we shall recall these, which are directly related to temporal evolution and threshold behavior.

As the characteristic feature of the photo-induced phase transition in TTF-CA crystal, it is reported the excitation intensity dependence of the converted fractions ($\Phi$) for I-to-N and N-to-I transformations. Fig. 4 shows that the fraction increases abruptly if the excitation intensity becomes higher than $2\times10^{13}$ photons/cm$^2$. This threshold-like behavior has been found to be characteristic for both I-to-N and N-to-I conversions. This is a clear indication of strong cooperativity in the process of photo-induced transformation, as expected for the CT strings formation. The second important feature of the Fig. 4 is high degree of photoconversion. From the threshold intensity one can estimate a critical size of a string, $n_{cr}$, and it is suggested that single excitation photon

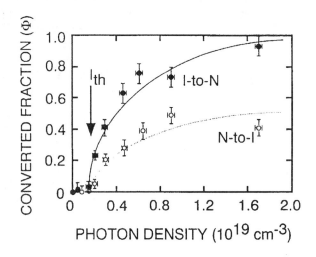

*Figure 4.* Excitation intensity dependences of the photo-converted fractions ($\Phi$) for I-to N (closed circles and N-to-I (open circles) photo-induced phase transitions. Solid and dotted lines are viewing guides. Arrow and $I_{th}$ denote the threshold intensity for excitation [14].

changes approximately 300 pairs from I(N) to N(I) state. For the excitation photon energy, h$\nu$=1.55 eV, it may be interpreted as the energy of lattice relaxed CT string, $\Delta E_{CT,relaxed}(300) < 1.55$ eV, eq. (16). The result seems to suggest that TTF-CA crystal, close to N-I thermodynamical equilibrium may be described by N-I relaxed domain walls as elementary excitations. Thus, the photo-induced N-I transition can be interpreted in terms of photo doping of the structurally relaxed domain walls.

To measure a speed of the conversion between N and I phases, a time-resolved study was carried out. The time profiles of the photoreflectance signals observed at 2.8 eV are plotted in Fig. 5. These results, observed at 77 K (for I to N conversion) and at 100 K (N to I conversion) indicate that locally excited species injected by an 80-fs laser pulse into I (N) phase grow into macroscopic domains of N(I) phase within a period of 1 ns (at the excitation intensity $10^{14}$ cm$^{-2}$). The time dependence of the photoreflectance signals at various excitation intensities is shown in Fig. 6. With rather weak photoexcitation, the

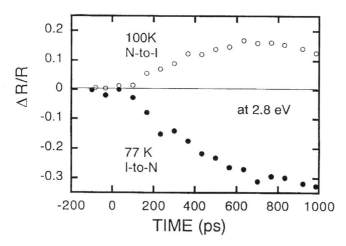

*Figure 5.* Time profiles of the photo reflectance signals observed at 2.8 eV induced by the irradiation of 1.55 eV light pulse with 80 fs pulse width. Results observed at 77 K (conversion ftom I phase to N phase) and 100 K (N-I conversion) are plotted by closed and open circles. The photon flux for excitation was $1\times10^{14}$ cm$^2$ [14].

photoreflectance signal starts to increase 100 ps after the irradiation with rather slow kinetics of growth of N phase. Observations summarized in Figs. 5 and 6 can be interpreted in terms of critical time, $t_{cr}$, needed to create a critical CT string length. It is the time when system reaches a crossover point between ground and excited (string) states. It means that within the laser pulse of 80-fs width a critical string size is created, in both phases. At higher excitation intensities, the growth of new phase is faster (Fig. 6) and critical time is expected to become shorter. The theory predicts that the critical time is inversely proportional to the excitation intensity. Temporal evolution shown in Fig. 6 requires quantitative analysis using the kinetic equations, as specified in previous sections. The time required for N phase to grow into macroscopic domain within I phase can be estimated from the integration,

$$t = \tau \int_{n(t)} \exp[-E_{I \to N}(n)/kT]\, dn, \qquad (26)$$

where the lower limit of the integral is the string length just after the switchoff of the excitation. It can be shown that the time is shorter for systems illuminated with higher intensity. As follows from the observations summarized in Figs. 5 and 6, the relaxation time for TTF-CA is of the order of nanoseconds and decreases with increasing intensity.

The critical length of a string, however, does not guarantee thermodynamical stability of a photo-created phase. In the case of TTF-CA crystal, it has been found that the photo-injected domains are metastable, indeed, and they decay into thermodynamically stable phase within miliseconds. Dynamics of the decay process would enlighten our

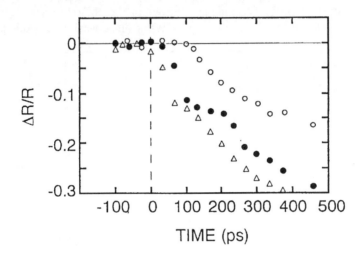

*Figure 6.* Time dependence of the photo reflectance signal observed at 3.0 eV with various excitation intensities. Open circles, and closed circles, and triangles are for excitation intensities of $2 \times 10^{18}$, $9 \times 10^{18}$ and $1.7 \times 10^{19}$ cm$^{-3}$ respectively. Sample temperature was 77 K and the crystal was in the I phase before excitation by 1.55 light pulse [14].

understanding of thermodynamics of the N-I phase transition and could be correlated with thermodynamical models for the transformation [11,13,26]. Even if it is not clear that photo-excitation of TTF-CA crystal generates N-I domain walls as electronic elementary excitations, we can conclude that photo-induced effect, bidirectional N-I transformation, is driven by the relaxed N-I domain walls. This gives a correlation between static (thermodynamic) and dynamic (photo-induced) phase transitions through the notion of condensed elementary excitations [29].

Finally, it has to be stressed that through the paper we did not discuss dynamics of symmetry breaking structural changes in TTF-CA crystal, e.g., lattice distortion due to dimerization. This aspect brings another dimension to the problem of photo-induced N-I transformation as the dynamics of the system has to be described by coupled kinetic equations. The potential which determines the kinetics eq. (18), has to be considered as a functional $F[n(u)]$, where u is a dimerization order parameter, and the rate (du/dt) would characterize kinetics of the photo-induced symmetry breaking process (the photo-induced ferroelectricity). These fascinating aspects are under studies.

## 6. Closing Remarks.

It is shown that new concepts are required to understand the multielectron transfer process. A formation of multielectron strings caused by attractive Coulomb interaction and their structural relaxation due to coupling to the lattice, allows to use the concept of N-I relaxed domain wall as an elementary excitation in the system. Kinetics of photo-induced transformation is discussed in terms of cooperative activation, highly cooperative process of formation of structurally relaxed domains (strings). Experimental demonstration of the cooperative multielectron transfer in TTF-CA crystal has been shown to be bidirectional transition between N and I phases and qualitatively described by the model. The excitation intensity dependencies of the photoconverted fraction of new phase and the growth dynamics from localized photoexcited species (CT strings) into the macroscopic domain (structurally relaxed string) show the threshold-like behavior characteristic of photoinduced cooperative phenomena. The study is the first example of photoinduced multielectron transfer processes in molecular materials.

**Acknowledgments.** T.L. thanks to Region de Nord Pas de Calais for supporting part of the research and his stay at L.D.S.M., Universite de Lille 1. Partial support has been provided by the Committee for Scientific Research, Poland for cooperation with Japan and France (POLONIUM).

## References.

1.    *Relaxation of Excited States and Photoinduced Structural Phase Transitions*; Nasu, K., Ed.; Springer Series in Solid State Science, 124; Springer: Berlin, 1997.
2.    Marcus, R.A. (1993) Electron transfer reactions in chemistry. Theory and experiments, *Rev. Mod. Phys.* **65**, 599-610.
3.    Tributsch, H. And L. Pohlmann, L. (1998) Electron transfer: Classical approaches and new frontiers, *Science* 279, 1891-1895.
4.    Steinberg-Yfrach, G., Liddell, P.A., Huang, S.-C., Moore, A.L., Gust, D. And Moore, T.A. (1997) Artificial Photosynthetic Reaction Centers in Liposomes: Photochemical Generation of Transmembrane Proton Potential, *Nature* 385, 239-241.
5.    Steinberg-Yfrach, G., Rigaud, J.-L., Durantini, E.N., Moore, A.L., Gust, D. And Moore, T.A. (1998) Light-driven production of ATP catalysed by F0F1-ATP synthase in an artificial photosynthetic membrane, *Nature* 392, 479 -482.
6.    Piotrowiak, P. (1999) Photoinduced electron transfer in molecular systems: recent developments, *Chem. Soc. Rev.*, **28**, 143-150.
7.    Koshino, K. And Ogawa, T. (1998) Domino effects in photoinduced structural change in one-dimensional systems, *J. Phys. Soc. Japan*,
8.    Torrance, J.B., Vasquez, J.E., Meyerle, J.J. and Lee, V.Y. (1981) Discovery of neutral-to-ionic transition in organic materials, *Phys. Rev. Lett.*, **46**, 253-256.
9.    Koshihara, S., Tokura, Y., Mitani, T., Saito, G. And Koda, T. (1990) Photo-induced valence instability in the organic molecular compound tetrathiafulvalene-*p*.-chloranil (TTF-CA), *Phys. Rev.*, B42, 6853-6856.
10.   Okamoto, H., Koda, T., Tokura, Y., Mitani, T. And Saito, G. (1989) Pressure-induced neutral-to-ionic transition in organic charge-transfer crystals of tetrathiafulvalene-*p*.-benzoquinone derivatives, *Phys. Rev.*, B39, 10693-10701.
11.   Luty, T. (1997) Neutral-to-ionic transformation as condensation and solidification of charge-transfer excitations, in K. Nasu (ed.), *Relaxation of Excited States and Photoinduced Structural Phase Transitions*; Nasu, K., Ed.; Springer Series in Solid State Science, 124; Springer: Berlin, pp. 142-150.
12.   Cailleau, H., Le Cointe, M. and Lemee-Cailleau, M.-H. (1997) Structural properties, dynamics and pressure effects at the neutral-to-ionic transition, in K. Nasu (ed.) *Relaxation of Excited States and Photoinduced*

420

*Structural Phase Transitions*; Nasu, K., Ed.; Springer Series in Solid State Science, 124; Springer: Berlin, pp. 133-140.

13. Lemee-Cailleau, M.-H., Le Cointe, M., Cailleau, H., Luty, T., Moussa, F., Ross, J., Brinkmann, D., Toudic, T., Ayache, C. And Karl, N. (1997) *Phys. Rev. Lett.*, **79**, 1690-1693.

14. Koshihara, S., Takahashi, Y., Sakai, H., Tokura, Y. And Luty, T. (1999) Photoinduced cooperative charge transfer in low-dimensional organic crystals, *J. Phys. Chem.*, **B103**, 2592-2600.

15. Soos, Z.G., Kuwajima, S. and Harding, R.H. (1986) Theory of charge-transfer excitations at the neutral-ionic interface, *J. Chem. Phys.*, **85**, 601-610, and references therin.

16. Stolarczyk, L.Z. and Piela, L. (1984) Hypothetical memomry effect in chains of donor-acceptor complexes, *Chem. Phys.*, **85**, 451-460.

17. Luty, T. (1995) Ground state phase diagram of mixed-stack compounds with intermolecular electron transfer, *Acta Phys. Polon.*, A**87**, 1009-1021.

18. Kuwata-Gonokami, M., Peyghambarioan, N., Meissner, K., Fluegel, B., Sato, Y., Ema, K., Shimano, R., Mazumdar, S., Guo, F., Tokihiro, T., Ezaki, H. And Hanamura, E. (1994) Exciton strings in an organic charge-transfre crystal, *Nature*, **367**, 47-48.

19. Mazumdar, S., Guo, F., Meissner, K., Fluegel, B., Peyghambarian, N., Kuwata-Gonokami, M., Sato, Y., Ema, K., Shimano, R., Tokihiro, T., Ezaki, H. And Hanamura, E. (1996) A new class of collective excitations: exciton strings, *J. Chem. Phys.*, **104**, 9283-9291.

20. Hanamura, E. (1997) Photo-induced structure changes of quasi-one-dimensional organic crystals, in K. Nasu (ed.) *Relaxation of Excited States and Photoinduced Structural Phase Transitions*; Nasu, K., Ed.; Springer Series in Solid State Science, 124; Springer: Berlin, pp. 34-44.

21. Kuwata-Gonokami, M. (1997) Exciton strings and exciton mediated carrier transport in a quasi-one-dimensional organic charge-transfer crystal, in K. Nasu (ed.) *Relaxation of Excited States and Photoinduced Structural Phase Transitions*; Nasu, K., Ed.; Springer Series in Solid State Science, 124; Springer: Berlin, pp. 171-180.

22. Ezaki, H., Tokihiro, T. and Hanamura, E. (1994) Excitonic n-string in linear chains: electronic structute and optical properties, *Phys. Rev.*, B**50**, 10506-10515.

23. Nagaosa, N. (1986) Domain wall picture of the neutral-ionic transition in TTF-Chloranil, *Solid State Commun.*, **57**, 179-183.

24. Jacobsen, C.S. and Torrance, J.B. (1983) Behavior of charge-transfer absorption upon passing through the neutral-ionic phase transition, *J. Chem. Phys.*, **78**, 112-115.

25. Le Cointe, M., Lemee-Cailleau, M.-H., Cailleau, H., Toudic, B., Toupet, L., Heger, G., Moussa, F., Schweiss, P., Kraft, H.H. and Karl, N. (1995) Symmetry breaking and structural changes at the neutral-to-ionic transition in TTF-chloranil, *Phys. Rev.*, B**51**, 3374-3386.

26. Nagaosa, N. (1986) Theory of neutral-ionic transition in organic crystals. IV. Phenomenological viewpoint. *J. Phys. Soc. Japan*, **55**, 3488-3497.

27. Mayerle, J.J., Torrance, J.B. and Crowley, J.I. (1979) Mixed-stack complexes of tetrathiafulvalene. The structures of the charge-transfer complexes of TTF with chloranil and fluoranil, *Acta Cryst.*, B**35**, 2988-2995.

28. Katan, C., Koenig, C. and Blochl, P.E. (1997) *Ab-initio* calculations of one-dimensional band structures of mixed-stack molecular crystals, *Solid State Commun.*, **102**, 589-594.

29. Toyozawa, Y. (1992) Condensation of relaxed excitons in static and dynamic phase transitions, *Solid State Commun.*, **84**, 255-257.

30. Toyozawa, Y. (1995) Excitonic instabilities of deformable lattice-from self-trapping to phase transition, *Acta Phys. Polon.*, A**87**, 47-56.

31. Hanamura, E. And Nagaosa, N. (1987) Photo-induced structure changes, *J. Phys. Soc. Japan*, **56**, 2080-2088.

32. Iizuka-Sakano, T. and Toyozawa, T. (1996) The role of long-range Coulomb interaction in the neutral-to-ionic transition of quasi-one-dimensional charge-transfer compounds, *J. Phys. Soc, Japan*, **65**, 671-674.

33. Nagaosa, N. And Ogawa, T. (1989) Theory of photoinduced structure changes, *Phys. Rev.*, B**39**, 4472-4483.

34. Nagaosa, N. And Ogawa, T. (1997) Mean-field theory of photoinduced structural phase transitiomns, in K. Nasu (ed.) *Relaxation of Excited States and Photoinduced Structural Phase Transitions*; Nasu, K., Ed.; Springer Series in Solid State Science, 124; Springer: Berlin, pp. 45-52.

# POLARIZATION DEPENDENT HOLOGRAPHIC WRITE, READ AND ERASURE OF SURFACE RELIEF GRATINGS ON AZOPOLYMER FILMS

SUKANT K. TRIPATHY,[1,3] NIRMAL K. VISWANATHAN,[2,3] SRINIVASAN BALASUBRAMANIAN,[3] SHAOPING BIAN,[2] LIAN LI,[3] AND JAYANT KUMAR[2,3]

[1]*Department of Chemistry, [2]Department of Physics and [3]Center for Advanced Materials, University of Massachusetts Lowell, Lowell, MA 01854, USA.*

## ABSTRACT

Surface relief grating (SRG) fabrication on azobenzene functionalized polymer films is investigated under different polarization combinations for the two writing beams. The grating formation was sequentially monitored with orthogonally polarized read beams and the surface topography by atomic force microscope. The SRG formation on amorphous azopolymer films is found to depend on the spatial variation of the intensity as well as the magnitude and direction of the interference field pattern. The treatment presented in this article defines the intensity and polarization distribution across the two-beam interference region for different polarization combinations of the interfering beams. Efficient photoisomerization and reorientation processes due to the spatial variation of the optical field results in a bulk birefringence grating (BG). Simultaneous presence of a component of electric field gradient in the grating vector direction results in a large modulation depth SRG. The effects due to the simultaneous formation of BG and SRG are studied by probing the grating formation process with vertical (V-) and horizontal (H-) polarized read beams. Our experimental results suggest that, under appropriate writing conditions and observation time scales, the V- polarized read beam probes predominantly the BG while the H- polarized read beam gets diffracted predominantly by the SRG. The experimental results are analyzed based on the resultant electric field pattern produced due to interference of polarized light beams. We have also investigated the unusual optical and thermal erasure properties of the photofabricated SRGs.

## 1. Introduction

A photoanisotropic material such as azobenzene chromophore containing polymer film responds in a unique way to the resultant electric field of the superposed interference light pattern [1-3]. Since the first report by Todorov *et al.* [4], the bulk birefringence gratings (BG) formed in azobenzene containing polymers films have been extensively studied. *Trans⇒cis⇒trans* isomerization process and the associated orientational redistribution of the azo chromophores, excited by a polarized light, results in anisotropic effects such as dichroism and birefringence induced in these polymers [5]. Holographic BG are formed in these materials due to the spatial variation of optical constants of the medium by superposed interferometrically produced spatially modulated

*F. Kajzar and M.V. Agranovich (eds.), Multiphoton and Light Driven Multielectron Processes in Organics: New Phenomena, Materials and Applications, 421–436.*

intensity and polarization state of the resultant field pattern [1-3,6]. The volume holographic gratings formed due to photoinduced alignment of the azo chromophores have potential applications in areas such as information storage, [7] optical switching [8] and nonlinear optics [9].

The successful fabrication of large-modulation depth surface-relief gratings (SRGs) in thin azobenzene functionalized polymer (azopolymer) films [10,11], has presented several intriguing possibilities and pursued by a number of research groups including ours [12-18]. The large surface modulation is a result of macroscopic polymer chain migration assisted by the photoisomerizable azobenzene group covalently attached to the polymer. Different classes of azopolymers used in the fabrication of SRGs include: epoxy and acrylic based glassy polymers and copolymers with azobenzene chromophores in the side chain [19,20], side-chain liquid crystalline polyesters [21], azo hybrid gel films [22], polyurea with azobenzene chromophores in the main chain [23] and conjugated polymers such as polydiacetylene and polyacetylene with azobenzene chromophores in the side chain [24]. The mass transport of polymer chains by low-power laser irradiation happens well below its glass transition temperature ($T_g$), and is reversible in most cases (in the absence of photochemical reactions). The strong dependence on the polarization [25-27], energy (fluence) [17] of the recording beams, without any pre- and/or post-processing procedure establishes this as an all-optical process.

In the fabrication of SRGs in azopolymer films, the simultaneous presence of BG has been well-recognized [13,22,28]. The bulk BG and SRG are simultaneously formed when the azopolymer film is exposed to an interference pattern due to polarized write beams. Recently, Holme *et al.* [21] and Labarthet *et al.* [13] have attempted to separate phase shifts due to the anisotropic and surface relief gratings formed in liquid-crystalline (LC) and amorphous azo polymer films. As the azopolymer is sensitive to the polarization of the write and read beams, understanding the physical processes responsible for their simultaneous formation is essential to adapt these features in device applications [3,29,30].

Here, we present the experimental results from a detailed study of polarization dependent write and read process of holographic gratings fabricated on azopolymer films. The dual BG and SRGs of varying efficiencies are written with different polarization combinations for the interfering beams. In an attempt to decouple the effects due to the simultaneously formed BG and SRGs, we sequentially monitor the grating formation process with orthogonally polarized read beams. The diffracted beam intensity resulting from the vertical (V-) polarized read beam is predominantly due to the BG, while the horizontal (H-) polarized read beam monitors the effects due to SRG formation. Our experimental results suggest that though the two simultaneous processes depend on the writing conditions and observation time scales, it is possible to separate the two contributions. A generalized interference picture is developed to explain the experimental results with different polarization of the interfering beams and to understand the mechanism of the dual BG and SRG formation. From the interference matrix elements, the spatial variation of intensity and optical field vector created by the superposed beams can be calculated. The experimental results for the different polarization combinations of the writing beams correlate well with the model developed. We also report the unusual optical and thermal erasure behavior of the gratings formed on the azopolymer films.

## 2. Theory and Background

When two coherent beams of arbitrary polarization interfere, the intensity modulation amplitude and the spatial variation in magnitude and direction of resultant electric field vector depend on the phase difference between the two beams [1-3,27]. Under special circumstances, this interference can result in a spatial variation of pure intensity or pure polarization state. In the analysis presented here, we have chosen a right-handed Cartesian coordinate system with the X-axis parallel to the grating vector $\mathbf{q}$, the Z-axis, for small angles, almost coincides with the propagation directions of the writing beams. The cross-section of the interference pattern is confined to the X-Y plane, which is also the sample plane. The intensity distribution in the interference region of the medium of refractive index 'n' is calculated to be

$$I = \frac{n}{2}\varepsilon_0 c\left(\vec{A} \bullet \vec{A}^*\right) = I_1 + 2\Delta I \cos(2k_x x) + I_2 \tag{1}$$

Where, $\Delta I = (n/2)\varepsilon_0 c\left(\vec{A}_1 \bullet \vec{A}_2^*\right)$ is the intensity modulation amplitude, $I_1$, $I_2$ and $\mathbf{A_1}$, $\mathbf{A_2}$ are the intensity and electric field amplitudes of the two interfering beams, $k_x$ is the wavevector in the X direction and '*' is for the complex conjugate.

It is well known that a photoanisotropic material like the azobenzene containing polymer is sensitive to the polarization of the write and read beams [3]. To account for the SRG formation process under different polarization combinations for the writing beams, including the orthogonally polarized light beams, where $\Delta I=0$, we consider a more generalized interference matrix $M_{ij}$ [31], defined in terms of the electric field amplitude as

$$M_{ij} = \left(\frac{n}{2}\right)\varepsilon_0 c A_{1,i} A^*_{2,j} = \left(\frac{n}{2}\right)\varepsilon_0 c \begin{bmatrix} A_{1x}A^*_{2x} & A_{1x}A^*_{2y} & A_{1x}A^*_{2z} \\ A_{1y}A^*_{2x} & A_{1y}A^*_{2y} & A_{1y}A^*_{2z} \\ A_{1z}A^*_{2x} & A_{1z}A^*_{2y} & A_{1z}A^*_{2z} \end{bmatrix} \tag{2}$$

Subscripts i, j stand for the spatial coordinates x, y, z. The absolute value of the trace of the matrix $M_{ij}$ gives the amplitude of the intensity modulation i.e.,

$$\Delta I = \left(\frac{n}{2}\right)\varepsilon_0 c\left|tr\{M_{ij}\}\right| \tag{3}$$

The off-diagonal elements ($M_{ij}$, $i \neq j$) correspond to the polarization variations in the interference region. Each element of the interference matrix can be evaluated depending on the polarization combination of the interfering beams (Table 1).

The spatial variation in magnitude and direction of the field vector depending on the phase difference between the interfering beams can be evaluated from the interference matrix elements ($M_{ij}$). As the process considered here begins at the free surface of the sample [32], we consider only the projection of the field vector onto the X-Y plane. At small interference angles, this is a good approximation as the electric field vectors lie almost completely in the X-Y plane. From the field amplitudes, the squared major (a) and minor (b) half axes of the optical ellipse and the orientation angle $\gamma$, the major axis of the polarization ellipse makes with the X-axis can be evaluated. From the calculated values, the periodic spatial variation of the principal axis of the polarization ellipse and the intensity modulation can be calculated for the different polarization configurations, depending on the angle of interference $\theta$.

Upon exposure to appropriate polarized light beam, the azo chromophores are excited preferentially if its dipole moment $\mu$ is parallel to the electric field $\mathbf{E}$ of the light

**Table 1**: Electric field amplitude of the interfering beams ($A_{1,2}$), the interference matrix ($M_{ij}$) and the intensity modulation ($\Delta I$) for the different polarization combinations. $x_0$, $y_0$ and $z_0$ are the unit vectors along the X, Y and Z axes and $\theta$ is the angle between the interfering beams.

| Polarization of Interfering Beams | Electric Field Amplitude $A_{1,2}$ | Interference Matrix $M_{ij}$ | Intensity Modulation $\Delta I$ |
|---|---|---|---|
| **s-:s-** | $A_{1y}, A_{2y} \neq 0$; $A_{1x,2x}$ and $A_{1z,2z} = 0$ | $\begin{bmatrix} 0 & 0 & 0 \\ 0 & A_{1y}A^*_{2y} & 0 \\ 0 & 0 & 0 \end{bmatrix}$ | $4I_1$ |
| **s-:p-** | $A_{1y}, A_{2x}, A_{2z} \neq 0$; $A_{1x}, A_{2y}, A_{1z} = 0$ | $\begin{bmatrix} 0 & 0 & 0 \\ A_{1y}A^*_{2x} & 0 & A_{1y}A^*_{2z} \\ 0 & 0 & 0 \end{bmatrix}$ | $0$ |
| **p-:p-** | $A_{1x,2x}, A_{1z,2z} \neq 0$; $A_{1y,2y} = 0$ | $\begin{bmatrix} A_{1x}A^*_{2x} & 0 & A_{1x}A^*_{2z} \\ 0 & 0 & 0 \\ -A_{1z}A^*_{2x} & 0 & -A_{1z}A^*_{2z} \end{bmatrix}$ | $\left| A_{1x}A^*_{2x} - A_{1z}A^*_{2z} \right|$ |
| **+45°:+45°** | $\vec{A}_{1,2} = \left[\hat{x}_0 \cos\theta/2 + \hat{y}_0 + \hat{z}_0 \sin\theta/2\right]\left(\dfrac{\left|\vec{A}_{1,2}\right|}{\sqrt{2}}\right)$ | $\left(\dfrac{1}{2}\right)\begin{bmatrix} \cos^2\theta/2 & \cos\theta/2 & \sin\theta/2\cos\theta/2 \\ \cos\theta/2 & 1 & \sin\theta/2 \\ \sin\theta/2\cos\theta/2 & \sin\theta/2 & \sin^2\theta/2 \end{bmatrix}$ | $1$ |
| **+45°:-45°** | $\vec{A}_{1,2} = \left[\pm\hat{x}_0 \cos\theta/2 + \hat{y}_0 + \hat{z}_0 \sin\theta/2\right]\left(\dfrac{\left|\vec{A}_{1,2}\right|}{\sqrt{2}}\right)$ | $\left(\dfrac{1}{2}\right)\begin{bmatrix} -\cos^2\theta/2 & \cos\theta/2 & \sin\theta/2\cos\theta/2 \\ -\cos\theta/2 & 1 & \sin\theta/2 \\ -\sin\theta/2\cos\theta/2 & \sin\theta/2 & \sin^2\theta/2 \end{bmatrix}$ | $\left(\dfrac{1}{2}\right)(1-\cos\theta)$ |
| **RCP:RCP** | $\vec{A}_{1,2} = \left(\hat{x}_0 \cos\theta/2 + \hat{y}_0 i - \hat{z}_0 \sin\theta/2\right)\dfrac{\left|\vec{A}_{1,2}\right|}{\sqrt{2}}$ | $\left(\dfrac{1}{2}\right)\begin{bmatrix} \cos^2\theta/2 & -i\cos\theta/2 & -\frac{1}{2}\sin\theta \\ i\cos\theta/2 & 1 & -i\sin\theta/2 \\ \frac{1}{2}\sin\theta & i\sin\theta/2 & \sin^2\theta/2 \end{bmatrix}$ | $1$ |
| **RCP:LCP** | $\vec{A}_{1,2} = \left(\hat{x}_0 \cos\theta/2 \pm \hat{y}_0 i \mp \hat{z}_0 \sin\theta/2\right)\dfrac{\left|\vec{A}_{1,2}\right|}{\sqrt{2}}$ | $\left(\dfrac{1}{2}\right)\begin{bmatrix} \cos^2\theta/2 & i\cos\theta/2 & \frac{1}{2}\sin\theta \\ i\cos\theta/2 & -1 & i\sin\theta/2 \\ -\frac{1}{2}\sin\theta & -i\sin\theta/2 & -\sin^2\theta/2 \end{bmatrix}$ | $\left(\dfrac{1}{2}\right)\sin^2\theta$ |

[33] and continues until they are orthogonal. Photoanisotropic material such as an azobenzene containing polymer film placed in the interference region, due to light-matter interaction results in a corresponding spatial modulation of its refractive index or susceptibility [3,27]. Assuming a linear material response, the spatial variation in the refractive index ($n_{ij}$) can be written as [34]

$$n_{ij} = n_0 + \Delta n_{ij} \cos(2k_x x) \qquad (4)$$

Where, $n_0$ is the refractive index of the material before recording (1.63), $\Delta n = n_\| - n_\perp$ is the modulated refractive index or the induced birefringence. The superposed optical field thus results in a spatially varying induced polarization in the material given by

$$P_i = \varepsilon_0 \chi'_{ij} E_j \qquad (5)$$

with $\chi'_{ij} = \chi_{ij} \pm \Delta\chi_{ij}$ includes the initial material susceptibility ($\chi_{ij}$) before recording and the spatial variation of the susceptibility change ($\Delta\chi_{ij}$) due to induced birefringence and dichroism in the material. In the above equation, $\varepsilon_0$ is the permittivity of free space and $E_j$ is the electric field component. The refractive index is related to the susceptibility by

$$n = \sqrt{1+\chi} \quad \text{and} \quad \Delta n = \Delta\chi / 2(1+\chi)^{1/2} \tag{6}$$

Any modulation in material property of the medium will be accompanied by a corresponding spatial variation, resulting in an optical grating. A simultaneous presence of an electric field gradient due to the superposed pattern results in large amplitude SRGs formed on azo functionalized polymer films due to the polymer chain movement by a time averaged gradient force acting on them [35]. The optically induced, time averaged gradient force density acting on the photoanisotropic polymer material is given by

$$\vec{f} = \left\langle \left( \vec{P} \bullet \vec{\nabla} \right) \vec{E} \right\rangle \tag{7}$$

Using eqns. (5) and (7), the driving force responsible for the macroscopic movement of the polymer chains at the free surface of the polymer film is calculated to be [15,27]

$$f_x = \varepsilon_0 \left[ \chi'_{xx} E_x \frac{\partial}{\partial x} E_x + \chi'_{yx} E_x \frac{\partial}{\partial x} E_x + \chi'_{zx} E_x \frac{\partial}{\partial x} E_x \right] \tag{8}$$

The above equation is simplified to take into account the observed material movement in the direction parallel to the grating vector $\mathbf{q}$ (X-axis). From eqn.(8) it is clear that the time averaged force responsible for the movement of polymer chains depends on the spatial variation of the susceptibility ($\chi'_{ij}$), the optical field $E_x$ and the field gradient ($\partial E_x / \partial x$) along the X-axis. Efficiency of the SRGs formed depends on the strength and simultaneous presence of these factors (Fig. 1). The surface modulation depth resulting from the macroscopic movement of the material can be written as

$$S = \Delta S \cos(2k_x x + \phi) \tag{9}$$

Where, $\phi$ is the arbitrary phase shift of the SRG with respect to the refractive index or susceptibility grating. The above analysis gives a qualitative understanding of the SRG formation mechanism in the azopolymer films in terms of the optically induced field patterns. It is well established from previous experimental results that the azopolymer material is spatially moved due to an optical field component in the direction of the field gradient [26,35]. In a two-beam interference experiment, this direction is parallel to the grating vector 'q' (X-axis) [27,35]. The spatial variation of electric field amplitude, intensity, polarization field vector, refractive index probed by V- and H- polarized read beams, gradient of electric field, force and the resulting surface modulation are given in Figure 1.

A read light beam diffracted by the grating in the transmission mode, experiences phase shift due to the induced birefringence given by $\Delta\Phi = (2\pi\Delta nd / \lambda)$. Simultaneous presence of large amplitude SRG introduces an additional phase shift of $\Delta\Psi = (2\pi n_{eff} \Delta S / \lambda)$ [13,21]. Where, d is the thickness of the film before irradiation (0.7 μm) and $\lambda$ is the read beam wavelength (633 nm). $n_{eff} = (1+n_p)/2$ is the effective refractive index of the SRG with $n_p$ as the refractive index of the azo polymer film (1.68) assuming $n_{air} = 1$. The effects due to the two-phase shifts, observed in the diffracted beam intensity from the gratings can cancel out or add up depending on the temporal evolution of the $\Delta n$ and $\Delta S$ values. By monitoring the +1 order diffracted beam in the transmission and reflection mode with orthogonally polarized read beams, it is possible

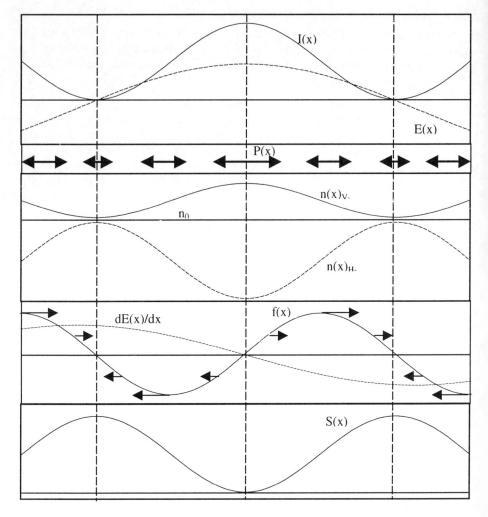

**Figure 1:** Spatial variation of intensity, I(x); electric field, E(x); polarization, P(x) due to p-:p- writing beams; refractive index, n(x) probed by vertical (V-) and horizontal (H-) polarized read beams; gradient of electric field, dE(x)/dx; force, f(x) and the surface modulation, S(x) due to interference of polarized beams.

to distinguish between the BG and SRG formation processes [30]. Variation of the diffraction efficiency ($\eta$) with the polarization angle ($\alpha$) of the read beam is given by [36]

$$\eta = \eta_{\parallel} \cos^2 \alpha + \eta_{\perp} \sin^2 \alpha \qquad (10)$$

Where, $\eta_{\parallel}$ and $\eta_{\perp}$ are the diffraction efficiencies for the read beam polarized parallel (V-) and perpendicular (H-) to the grating grooves respectively. $\alpha = 0°$ for V- polarized read beam and 90° for the H- polarized read beam. The diffraction efficiency is calculated as the ratio between the diffracted beam intensity ($I_d$) to the incident beam intensity ($I_i$) i.e., $\eta = (I_d/I_i) \times 100$.

## 3. Experimental Arrangement

The experimental setup to study the surface deformation on the azopolymer thin films using a two-beam interferometer is shown in Figure 2. Inset shows the polarization convention used. An $Ar^+$ laser beam at 488 nm with $TEM_{00}$ mode is first spatially filtered and the expanding beam is collimated to give a plane wave. This main beam is then split using a front surface coated 50-50 beam-splitter (BS). The two beams after reflection from dielectric coated mirrors ($M_1$ and $M_2$) are made to interfere at the sample to form the holographic grating. Two $\lambda/2$ wave plates (WP) introduced after the mirrors are used to independently rotate the plane of polarization of the interfering beams. Further, two crystal polarizers (P) with extinction ratio better than 1:10000, introduced after the half-wave plates are used to clean and control the polarization state of the writing beams. While doing experiments with circularly polarized interfering beams, we have the plane of polarization of the beams defined first by the crystal polarizers and the beams in turn pass through the appropriately oriented $\lambda/4$ wave plates. All the polarization components are kept on rotation mounts with rotation accuracy of $0.5°$ and are aligned with extreme care. It is of utmost importance to have as clean a polarization state for the interfering beams as possible. It has been observed in our experiments that the surface deformation of the azopolymer is extremely sensitive to the polarization impurity of the interfering beams.

The grating formation process for different combinations of the writing beam polarization is monitored in the transmission mode. This is done by diffracting a low

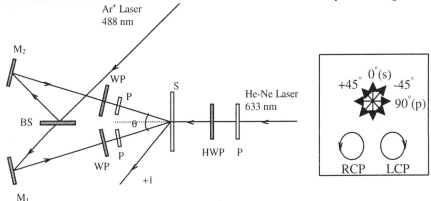

**Figure 2:** Experimental setup used to write and read SRG formation on azo polymer film. Inset shows the polarization convention.

power (1 mW), polarized He-Ne laser beam (633 nm) from the spot where the two beams from the $Ar^+$ laser interfere. A polarizer (P) and a $\lambda/2$ wave plate (HWP) orient the plane of polarization of the He-Ne beam. The polarization state of the read beam can be changed between vertical (V-) and horizontal (H-) by rotating the $\lambda/2$ waveplate accordingly. The polarization vector for V- polarized read beam is parallel to grating grooves (perpendicular to the grating vector) and for the H- polarized read beam it is perpendicular to the grating grooves (parallel to the grating vector). The intensity of the +1 order of the diffracted beam is monitored as a function of time. The angle between the writing beams, unless specified, is kept fixed at $\theta \approx 28°$, resulting in a periodicity of

$\Lambda = 1$ µm for the grating. However, the angle between the interfering beams is changed appropriately to confirm a number of predictions made by the model. The intensity of the laser beam, measured after the collimating lens is kept fixed at 50 mW/cm$^2$ for all our recordings.

The azobenzene functionalized polymer used in this study, CH-1A-CA is an epoxy based amorphous polymer, synthesized by post-azo coupling reaction [37]. The degree of functionalization of the polymer with the azobenzene chromophore is well above 95%. The number average molecular weight of this polymer is 5000 g/mol and its $T_g$ is 93 °C. The azopolymer is dissolved in DMF, spin coated on clean glass slide and vacuum dried for 12 hours to get good optical quality films. The average thickness and refractive index of the film measured using a prism coupler (Metricon 2010) at 633 nm are 0.7 µm and 1.68 respectively. Surface topography of the fabricated gratings is scanned using an Atomic Force Microscope (AFM, Cp, Park Scientific, Sunnyvale, CA). The AFM is used in the contact mode using a SiN cantilever. All the gratings recorded are imaged by 5 µm x 5 µm scans at 1 Hz rate.

## 4. Results and Discussion

It is well known that a linearly polarized light preferentially photoexcites only those dipoles of the azobenzene molecules with an orientational component parallel to the electric field vector direction [38]. Repeated photoisomerization and reorientation processes results in a photostationary state when an equilibrium order parameter is achieved with a net orientation of the dipole axis perpendicular to the incident light polarization direction [38]. A spatially varying interference field vector (in magnitude and direction) results in a spatial variation of susceptibility of the medium. A simultaneous presence of a component of electric field gradient in a direction parallel to the grating vector leads to mass transport of the polymer chains along the direction of the field gradient. Interference angle and the polarization of the two interfering beams decide the final form of the interference matrix and hence the force distribution acting on the material. Varying contributions from the intensity and/or electric field vector distribution results in the simultaneous formation of BG and SRG of varying efficiencies. Here, we discuss some of the possible combinations of linear and circular polarization of the interfering beams based on the generalized interference picture discussed. The intensity and polarization contributions to the formation of BG and SRG are discussed based on the resultant intensity and field distributions. From the behavior of the diffraction efficiency recorded with orthogonally polarized read beams, we discuss the possibility of decoupling the effects due to the reorientation grating and the surface grating.

**4.1. S-:S- POLARIZATION**     When the two s-polarized beams of equal intensity interfere, the resultant intensity modulation ($\Delta I = 4I_0$) is maximum (Table I) and the electric field vector direction is parallel to s- (inset of Fig.3). The spatially varying phase difference between the two interfering beams is recorded as a pure intensity modulation, corresponding to the conventional *scalar holography*. The magnitude of the total electric field variation has the same periodicity as the grating. The dipolar interaction with the optical field renders the material spatially anisotropic due to photoisomerization and reorientation processes. The transient diffraction efficiency due to the induced anisotropy is more for the H- polarized read beam (continuous line) than for the V-

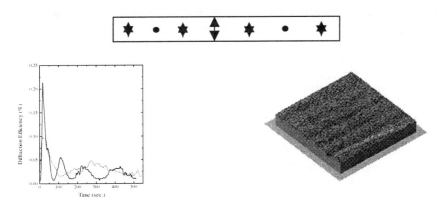

**Figure 3:** (a) Diffracted beam intensity and (b) surface topography of the grating written with two s-polarized beams. Inset shows the resultant interference field-pattern.

polarized read beam (dotted line) (Fig.3-a). However, in the final photostationary state, there is very little difference between the two diffracted beam intensities. A maximum diffraction efficiency of ≈0.2% is reached in less than a few tens of seconds. Such a behavior can be explained based on the optical field induced birefringence and dichroism [30]. However, as there is no component of the field gradient in a direction parallel to the grating vector, there is no force and hence no appreciable surface relief grating observed at the end of the recording. AFM scan of the exposed region shows a very weak surface relief feature with modulation depth < 100 Å (Fig.3-b).

**4.2. S-:P- POLARIZATION** When the two interfering beams are orthogonally polarized, (s-:p-), a pure polarization interference pattern results with no associated intensity variation ($\Delta I = 0$) (Table I). The total intensity distribution is constant in the interference region and the phase difference between the two write beams is encoded as spatially periodic orientation of the optical field vector known as *polarization holography*. It is important to note that, in this case, the electric fields of the two interfering beams are orthogonal to each other for all angles of interference. The exposed polarization

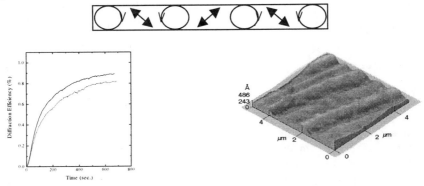

**Figure 4:** (a) Diffracted beam intensity and (b) surface topography of the grating written with s- and p-polarized beams. Inset shows the resultant interference field-pattern.

sensitive azo polymer film thus records an efficient hologram due to the spatially varying induced anisotropy. The diffracted beam intensity probed with V- and H- polarized read beams (Fig.4-a) increases rapidly in the initial stages due to efficient photoinduced anisotropy and saturates once the equilibrium is reached. A grating written at a periodicity of 2 µm, shows a biperiodic grating with $\Lambda = 1$ µm. A topographic AFM scan of the grating formed in this process is given in Figure 4-b. The weak force due to the susceptibility change and component of the electric field gradient in the direction parallel to the grating vector with no intensity modulation may possibly gives rise to the small modulation SRG. At smaller periods, this effect appears less pronounced, possibly due to the resolution limitation of the material. A systematic study to understand this unique phenomenon is in progress and will be reported separately. The diffraction efficiency of the grating is 1% and the corresponding modulation depth is < 200 Å.

### 4.3. P-:P- POLARIZATION

When the two interfering beams are p- polarized, there is spatial variation of both intensity and resultant field across the interference region. This combination results in a mixture of intensity and polarization variations in the interference region, except when $\theta = 90°$. In this case, the resultant optical field vector in the X-Y plane is linear and parallel to the X-axis. The field variation and the corresponding field gradient in a direction parallel to the grating vector results in large modulation SRGs fabricated on the azopolymer thin films. The spatially varying field (in magnitude and direction) (inset of Fig. 5) induces simultaneously a spatially varying susceptibility change and an electric field gradient in a direction parallel to the X-axis. This leads to macroscopic movement of the polymer material and hence the grating formation, confirmed by the experimental recording of the grating formation process and an AFM scan of the exposed region. Figure 5-a shows the diffraction efficiency curves probed with V- and H- polarized read beams. The behavior of the diffracted beam

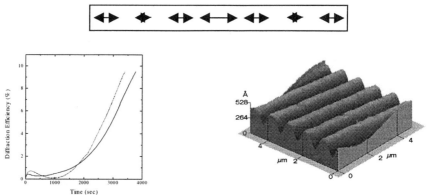

**Figure 5:** (a) Diffracted beam intensity and (b) surface topography of the grating written with two p- polarized beams. Inset shows the resultant interference field-pattern.

intensity due to the grating is explained as due to the simultaneous formation of dual BG and SRG [30]. The modulation depth of the grating, measured using the AFM is 900 Å (Fig.5-b). However, even in this case, when the angle between the two p- polarized interfering beams is $\theta = 90°$, the polarizations are orthogonal to each other and the calculated intensity modulation vanishes. Under such a situation, it is predicted that we do not form any SRG and this result was experimentally confirmed.

**4.4. +45°:+45° POLARIZATION** When both the interfering beams are polarized at 45° with respect to the Y-axis in the same sense i.e., +45°:+45°, the resultant polarization state of the field pattern remains the same at +45° (inset of Fig.6). This situation would be similar to the s-:s- polarization combination for the interfering beams. This situation gives an intensity modulation of $\Delta I = 1$, independent of the interference angle (Table I). The photoinduced birefringence and dichroism due to the spatially varying intensity and field pattern results in an initial increase in the diffracted beam intensity, shown in Fig.6-a. A maximum diffraction efficiency of ≈0.5% is reached for both V- and H- polarized read beams. Subsequently, when the photostationary state is reached, there is no appreciable change in susceptibility across the interference region due to the intensity dependent induced anisotropy [39]. Hence, the component of the electric field gradient parallel to the grating vector does not have a strong influence in deforming the surface. This can be seen from the AFM scan of the irradiated region, showing only a weak

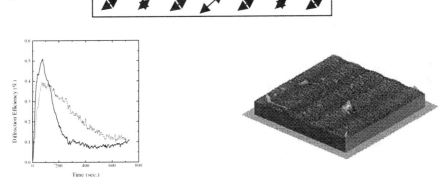

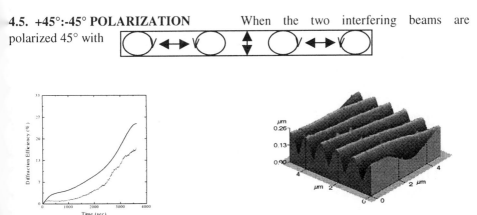

**Figure 6:** (a) Diffracted beam intensity and (b) surface topography of the grating written with two +45° polarized beams. Inset shows the resultant interference field-pattern.

surface feature (Fig.6-b), similar the to s-:s- case. The results are verified to be the same for the -45°:-45° configuration as well.

**4.5. +45°:-45° POLARIZATION** When the two interfering beams are polarized 45° with

**Figure 7:** (a) Diffracted beam intensity and (b) surface topography of the grating written with +45°:-45° polarization combination. Inset shows the resultant interference field-pattern.

respect to the Y-axis in the opposite sense, (i.e., one beam is +45° polarized and the other is -45°) the resultant field pattern depends on the interference angle θ. The polarization of the beams are orthogonal to each other and the intensity modulation $\Delta I = 0$ in the sample plane only when they are collinear (θ = 0°). At all other angles of interference, the projection of the electric field vector of the two beams results in intensity modulation. The spatial variation of susceptibility due to the superposed electric field pattern in the sample plane is anisotropic. The simultaneous presence of intensity modulation ($\Delta I = 0.12$ at θ = 28°) is due to the non-orthogonal polarization of the interfering beams at angles θ ≠ 0. Experimental recording of the diffracted beam intensity monitored as a function of time is plotted in Figure 7-a for both V- and H-polarized read beams. The initial increase in the diffracted beam intensity is due to the anisotropic grating formed due to spatial variation of the induced susceptibility in the material. The force due to the simultaneous presence of electric field gradient parallel to the grating vector moves the polymer chains, resulting in large modulation depth SRGs. The maximum diffraction efficiency of 26% reached in this case is due to the photoinduced anisotropy and force due to a component of the electric field gradient of the superposed interference field pattern. The diffracted beam intensity is still increasing at the end of the recording indicating that the SRG formation is not saturated yet. High modulation depth surface relief gratings (2500 Å) formed in this case are confirmed by the AFM scan of the surface topography of the grating, shown in Figure 7-b.

### 4.6. RCP:RCP POLARIZATION

In the RCP:RCP case, the two interfering beams are both right circularly polarized (RCP) with identical rotation directions. In the interference region, the resultant optical field remains right circularly polarized (inset of Fig. 8) along with a spatially varying intensity distribution. The intensity modulation is calculated to be $\Delta I = 1$, independent of the angle between the interfering beams (Table I). Such a situation would be identical to the scalar hologram (s-:s-) recorded in the medium. Consequently, the stationary state of the photoinduced changes in the material property and hence the spatial variation of susceptibility becomes isotropic in the transverse directions. In other words, the rotationally symmetric polarization state of the superposed pattern does not induce birefringence or dichroism and hence no photoinduced anisotropy in the sample plane. Consequently, the time averaged force density parallel to the grating vector has

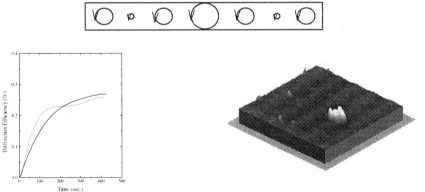

**Figure 8:** (a) Diffracted beam intensity and (b) surface topography of the grating written with two right circularly polarized beams at θ = 28°. Inset shows the resultant interference field-pattern.

minimal effect in deforming the surface and there is no SRG formed in this configuration. Large angle of interference (θ = 28°) however, results in an elliptic (a = 0.97 and b = 1) instead of the circular field pattern, which induces a weak in-plane anisotropy. The experimental observation of the diffracted beam intensity with V- and H- polarized read beams (Fig.8-a) shows an increase due to possible weak induced anisotropy and field gradient present parallel to the X-axis. AFM scan of the irradiated region shows a weak relief grating with maximum amplitude of 100 Å formed (Fig.8-b), corresponding to a maximum diffraction efficiency of 0.3%. Increasing the ellipticity of each beam by changing the orientation of the polarizers by only a few degrees results in an increase in the diffracted beam intensity and the corresponding modulation depth of the SRG formed.

### 4.7. RCP:LCP POLARIZATON

When the azopolymer film is illuminated with interference pattern due to opposite circular polarized beams RCP:LCP, there is always a spatial variation in the orientation of the resultant field pattern and intensity modulation, unless the beams are collinear (θ = 0°). The intensity modulation $\Delta I \to 0$ only as the angle between the interfering beams $\theta \to 0$. At all other writing angles (θ ≠ 0°), there is always a non-zero contribution from both intensity modulation and resultant electric field variation. The photoinduced anisotropy and hence the spatial variation of susceptibility is large in this case as almost all the chromophores in the sample plane are addressed except those oriented along the light propagation axis. Consequently, the photoinduced orientational effects like the induced birefringence and dichroism in the

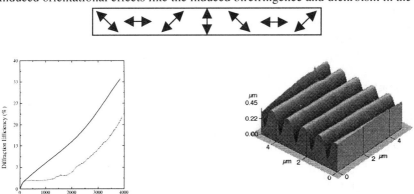

**Figure 9:** (a) Diffracted beam intensity and (b) surface topography of the grating written with right and left circularly polarized beams. Inset shows the resultant interference field-pattern.

sample plane are the maximum. Force due to the electric field gradient parallel to the grating vector coupled with large induced anisotropy leads to the most efficient SRG fabricated. A large diffraction efficiency of 33% is achieved at the end of the recording and still increasing (Fig.9-a). In the initial stages, there is a marked difference between the diffracted beam intensities probed by V- and H- polarized read beams. As the surface starts to get modified significantly, the difference starts to decrease. The modulation depth of the SRG measured with the AFM is 3200 Å (Fig.9-b). The large induced anisotropy, the associated reorientation of the azo chromophores, coupled with the electric field vector gradient in a direction parallel to the grating vector leads to the most efficient macroscopic movement of the polymer material. Due to the efficient cycling of the azo chromophores from the sample plane, the resulting SRG has very high diffraction efficiency and modulation depth.

The critical observation of the ability to distinguish and separate the BG and SRG formed in thin azopolymer films lead us to fabricate efficient polarization discriminators and polarization beam splitters [29]. Recently, we observed that a linearly polarized read beam traversing the grating, due to induced the susceptibility and surface modifications during the writing process, rotate the plane of polarization of the probe beam. The dual birefringence and surface relief gratings are written with RCP:LCP polarization combination for different periodicity from $\Lambda = 0.85$ $\mu$m to 4 $\mu$m. After writing, the polarization content of the +1 diffracted beam was measured by rotating an analyzer kept in its path. It is found that a linearly polarized read beam (s- for example) after passing through the grating is found to have its plane of polarization rotated (maximum 30°) with respect to the direction of polarization of the incident beam. Apart from the rotation, we also observed that the out coming beam is elliptically polarized. The rotation angle and the ellipticity introduced are more for smaller period gratings. Further experiments are in progress to understand the intriguing inhomogenous birefringence [40] characteristics of the holographic gratings formed in the photoanisotropic azopolymer film.

The holographic gratings formed on the azopolymer films can be erased optically and thermally (in most cases). Similar to the strong polarization dependent write and read behavior of the gratings the optical erasure process is strongly polarization dependent as well. The grating "remembers" the resultant field pattern that created them and influences the erasure behavior accordingly [41]. We present here this unique polarization dependent optical erasure process of the SRGs for a specific polarization condition. Optical erasure of bulk birefringence grating is easily explained in terms of the re-randomization of the chromophore orientation. However, the optical erasure of SRG involves large-scale polymer transport, canceling the effects of the writing process. Figure 10-a shows the formation of SRG and the subsequent optical erasure in the high molecular weight HPAA-CA azopolymer [37]. The grating is written with +45°:-45° polarization combination. The grating fabrication is stopped when the maximum efficiency measured in the +1 order of the diffracted beam is about 8% (modulation depth is 745 Å), indicated by "writing off". The initial decrease in the

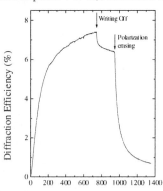 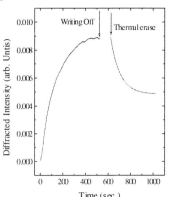

**Figure 10:** Photofabrication of SRG and subsequent (a) optical erasure and (b) thermal erasure in HPAA-CA azopolymer film.

diffracted beam intensity is due to the decay of the orientation (birefringence) grating. The subsequent illumination of the grating with a single 45° polarized beam shows almost complete decay of the diffracted beam intensity and shows no significant surface features in an AFM scan. While attempting the thermal erasure process, the azopolymer (HPAA-CA) film coated glass plate is mounted on an ITO coated glass substrate with proper electrical connections made for heating. The grating is written at room temperature until saturation and stopped (Fig. 10-b). The ITO coated glass plate and hence the sample is heated at the rate of 5 °C/min. The diffracted beam intensity starts to decrease as the temperature is increased. In most of the azopolymer film, the grating and hence the diffracted beam intensity is completely erased as we approach the glass transition temperature ($T_g$). The only exception (in our experiments) to this is the high molecular weight polyacrylic acid based HPAA-CA azopolymer. In this case, even as the temperature is increased to about 200 °C, the diffracted beam intensity does not decrease below 40% of the maximum value and remains stationary at that value (Fig. 10-b). This effect is thought to be due to possible crosslinking in the polymer, which does not allow the grating to be completely erased. Once the grating is thermally (or photochemically) crosslinked, it is resistant to both thermal and optical erasure.

## 5. Conclusions

The SRG formation on azopolymer thin films and its diffraction efficiency depend in a unique way on the intensity modulation and the resultant electric field vector pattern produced by the two polarized interfering beams. The polarization dependent grating formation process is monitored by diffracting V- and H- polarized read beams. From the transient and photostationary behavior of the diffraction efficiency, we have demonstrated that it is possible to distinguish the effects due to BG and SRG formation. Analyzing the experimental results based on the interference field vector pattern, we get a qualitative understanding of the two physical processes responsible for recording BG and SRG. Photoinduced anisotropy due to photoisomerization and associated reorientation of the azobenzene chromophores facilitates the recording of the resultant optical field pattern and hence the BG. A simultaneous presence of force due to a component of electric field gradient along the grating vector direction results in a macroscopic movement of polymer chains and hence the formation of large modulation depth SRGs. Our experimental results and analysis provide a more complete picture of the effect of polarization of the write and read beams on the dual BG and SRG formation processes. Apart from the strong polarization dependence of the write and read process of the holographic gratings formed on the azopolymer films, we have demonstrated that the erasure process is also polarization dependent. Situations where the polymer can be thermally or photochemically crosslinked, extends the applicability of the gratings to fabricate permanent diffractive optical elements.

## Acknowledgements

Financial support from ONR and NSF-DMR is gratefully acknowledged.

436

**6. REFERENCES:**

1. Kakichashvili, Sh.D. (1974) Sov. J. Quant. Electron. **4**, 795.
2. Nikolova, L., and Todorov, T., (1984) Opt. Act. **5**, 579.
3. Huang, T. and Wagner, K.H., (1996) J. Opt. Soc. Am. B **13**, 282.
4. Todorov, T. Nikolova, L., and Tomova, N., (1984) Appl. Opt. **23**, 4309.
5. Rau, H., (1990) *Photoisomerization of azobenzenes*, in Photochemistry and Photophysics, F.J. Rabeck, ed. (CRC, Boca Raton, Florida), Vol.II, Chapter 4, p. 119.
6. Lessard, R.A., and Couture, J.J.A., (1990) Mol. Cryst. Liq. Cryst. **183**, 451.
7. Takeda, T. Nakagawa, K., and Fujiwara, H. (1994) Nonlinear Opt. **7**, 295.
8. Ikeda, T., and Tsutsumi, O. (1995) Science **268**, 1873.
9. Dumont, M., Froc, G., and Hosotte, S., (1995) Nonlinear Opt. **9**, 327.
10. Rochon, P.L., Batalla, E. and Natansohn, A. (1995) Appl. Phys. Lett. **66**, 136.
11. Kim, D.Y., Li, L., Kumar, J., and Tripathy, S.K. (1995) Appl. Phys. Lett. **66**, 1166.
12. Itoh, M., Harada, K., Matsuda, H., Ohnishi, S., Parfenov, A., Tamaoki N., and Yatagai, T. (1998) J. Phys. D: **31**, 463.
13. Labarthet, F.L., Buffeteau T., and Sourisseau, C. (1998) J. Phys. Chem. B **102**, 2654.
14. Lefin, P., Fiorini C., and Nunzi, J.M. (1998) Pure Appl. Opt. **7**, 71.
15. Viswanathan, N.K., Kim, D.Y., Bian, S., Williams, J., Liu, W., Li, L., Samuelson, L., Kumar, J., and Tripathy, S.K. (1999) to appear in Journal of Materials Chemistry.
16. Tripathy, S.K., Kim, D.Y., Li L., and Kumar, J. (1998) Photonics Science News, **4**, 13.
17. Tripathy, S.K., Kim, D.Y., Li, L. and Kumar, J. (1998) Chemtech, May, 34.
18. Barrett, C.J., Rochon P.L., and Natansohn, A.L. (1998) J. Chem. Phys. **109**, 1505.
19. Kim, D.Y., Li, L., Jiang, X.L., Shivshankar, V., Kumar, J., and Tripathy, S.K. (1995) Macromolecules **28**, 8835.
20. Ho, M., Barrett, C.J. Paterson, J., Esteghamatian, M., Natansohn A., and Rochon, P. (1996) Macromolecules **29** 4613.
21. Holme, N.C.R., Nikolova, L., Ramanujam P.S., and Hvilsted, S. (1997) Appl. Phys. Lett. **70**, 1518.
22. Lee, T.S, Kim, D.Y, Jiang, X.L., Li, L., Kumar, J., and Tripathy, S.K. (1998) J. Poly. Sci. A **36** 283.
23. Darracq, B., Chaput, F., Lahlil, K., Levy Y., and Boilot, J.P., (1998) Adv. Mater. **10**, 1133.
24. Sukwattanasinitt, M., Wang, X., Li, L., Jiang, X.L., Kumar, J., Tripathy, S.K., and Sandman, D.J. (1998) Chem. Mater. **10**, 27.
25. Jiang, X.L., Li, L., Kumar, J., Kim, D.Y., Shivshankar, V., and Tripathy, S.K. (1996) Appl. Phys. Lett. **68**, 2618.
26. Bian, S., Li, L., Kumar, J., Kim, D.Y., Williams, J., and Tripathy, S.K. (1998) Appl. Phys. Lett. **73**, 1817.
27. Viswanathan, N.K., Balasubramanian, S., Li, L., Tripathy, S.K., and Kumar, J., accepted in Jpn. J Appl. Phys. (1999).
28. Lefin, P., Fiorini, C., and Nunzi, J.M. (1998) Opt. Mater. **9**, 323.
29. Viswanathan, N.K., Balasubramanian, S., Kumar, J., and Tripathy, S.K. (1999) submitted to Polym Adv. Technol.
30. Eichler, H.J., Gunter, P., Pohl, D.W. (1985) *Laser Induced Dynamic Gratings*, (Springer series ir Optical Sciences), Vol. **50**, Chap. 2.
31. Viswanathan, N.K., Balasubramanian, S., Li, L., Kumar, J., and Tripathy, S.K. (1998) J. Phys Chem. B **102**, 6064.
32. Lessing, H.E. and von Jena, A., (1979) *Laser Handbook* ed. by M.L. Stitch (North-Holland Amsterdam) Vol. **3**.
33. Stegeman, G.I., and Hall, D.G. (1990) J. Opt. Soc. Am. A **7**, 1387.
34. Kumar, J., Li, L., Jiang, X.L., Kim, D.Y., Lee, T.S., and Tripathy, S.K. (1998) Appl. Phys. Lett. **72** 2096.
35. Viswanathan, N.K., Balasubramanian, S., Kumar, J., and Tripathy, S.K. (1999) submitted to J Appl. Phys.
36. Kobolla, H., Sheridan, J.T., Gluch, E., Schwider, J., and Streibl, N. (1992) Proc. SPIE **1732**, 278.
37. Wang, X., Li, L., Chen, J., Marturunkakul, S., Kumar, J., and Tripathy, S.K. (1997 Macromolecules **30**, 219.
38. Sekkat, Z., Wood, J., Aust, E.F., Knoll, W., Volksen, W., and Miller, R.D. (1996) J. Opt. Soc. Am B **13**, 1713.
39. Egami, C., Suzuki, Y., Sugihara, O., Okamoto, N., Fujiwara, H., Nakagawa, K., and Fujiwara, F (1997) Appl. Phys. B **64**, 471.
40. Watanabe, Y., (1994) J. Appl. Phys. **76**, 3994.
41. Jiang, X.L., Kim, D.Y., Li, L., Kumar, J., and Tripathy, S.K. (1998) Appl. Phys. Lett. **75**, 2502.

# STUDY OF STILBENE MOLECULE *TRANS↔CIS* ISOMERIZATION IN FIRST EXCITED STATE AND DESIGN OF MOLECULAR RANDOM-WALKERS

M.L. BALEVICIUS[a], A. TAMULIS[b], J. TAMULIENE[b]
and J.-M. NUNZI[c]

[a]*Vilnius University, Faculty of Physics*
*Sauletekio 9, 2040 Vilnius, Lithuania*
[b]*Institute of Theoretical Physics and Astronomy*
*A. Gostauto 12, 2600 Vilnius, Lithuania*
*WEBsite:* www.itpa.lt/~tamulis/
[c]*CEA-LETI, DEIN-SPE, Groupe Composants Organique*
*Saclay, F91191 Gif sur Yvette, France*

**Abstract:** Quantum chemical *ab initio* calculations and investigations of *trans*- and *cis*-stilbene, Disperse Orange 3 (DO3) and several organic photoelectron donor, electron acceptor molecules and their supermolecules connected via the electron insulator bridges were performed using Hartree-Fock (HF) and density functional theory (DFT) methods. The optimized ground state geometry was as initial optimizing geometry in the first excited state using *ab initio* configuration interaction single-excitation (CIS) HF method. More detailed were investigated electronic structure changes in different stilbene molecule conformations. The results of single molecule and supermolecule calculations were used for the design and *ab initio* calculations of two-, three-, four- and six-variable anisotropic random-walk molecular devices based on stilbene and azo-dye molecules. Two kinds of logically AND controlled molecular random-walkers were designed.

## 1. Introduction

The synthesis and construction of the molecular devices based on light induced molecular motions are in the initial stage now. V. Balzani et al [1-3] recently investigates light-fuelled molecular-level machines, switches, and plug/socket systems based on interlocked molecular compounds. Light-driven monodirectional molecular rotors are currently synthesised by Koumura *et al.* [4]. The phenomenon of the anisotropy of the photoinduced translation diffusion of azo-dyes was described and interpreted recently [5-7]. Based on these papers and our

*F. Kajzar and M.V. Agranovich (eds.), Multiphoton and Light Driven Multielectron Processes in Organics: New Phenomena, Materials and Applications*, 437–450.
© 2000 *Kluwer Academic Publishers. Printed in the Netherlands.*

quantum chemical investigations of the single stilbene and DO3 azo-dye [8] molecules internal motions we extend the variety of the molecular machines composed of photoactive moieties.

The design of molecular random-walkers is possible on basic investigations of internal molecular movements of photoactive molecules like stilbene. The conformations of the ethylene molecule and their derivatives changes under $\pi \rightarrow \pi^*$ excitation are well known [9, 10]. It is assumed, that in the excited state the plane of the -CH$_2$- fragments become perpendicular to each other that leads to *trans-cis* transformations of the above molecules. The above phenomenon is obtained in the any -CR$_1$R$_2$ compounds, where R$_i$ is a group formed only $\sigma$ bond with carbon (as an example -CH$_3$). However, this above assuming is no critically adopted for the cases, when the R$_i$ is formed $\sigma$ and $\pi$ electron system [9]. The previous investigations that were performed applying semiempirical methods did not explain some properties, i. e. it is not clear why the *trans-cis* isomerization of the ethylene and stilbene molecules are different. Additionally, the *cis-trans* isomerization (I case) of stilbene depends on temperature slightly, while the above dependence of the *trans-cis* (II case) isomerization is strong, i.e., when temperature decreases the process are going worse, and quantum fluorescence yield becomes more closely to 1. It implies that the transition barrier is absent in I case, while it is present in II. Therefore we intend to verify the geometrical structure changeability of the stilbene molecule in the first excited state. In order to make clear the above processes the ground state geometry of the stilbene molecule was obtained applying the density functional theory (DFT) methods. This method allows obtaining the geometry of molecules more closer to experimental. It is important to make qualitative conclusions that are performed applying simple Hückel approach.

## 2. Quantum chemical *ab initio* investigations results

The ground state geometrical structure of the stilbene molecule was calculated applying DFT B3PW91 model [11, 12] in the 6-311G**[13, 14] basis set, that it is installed in the Gaussian 94 package [15]. *Trans-* and *cis-*stilbene molecule geometries in ground state were optimized by using Berny procedure [15] and that was the basis for the design of stilbene based random-walk molecular devices.

### 2.1. *TRANS*-STILBENE OPTIMIZED GEOMETRY

The planes of phenyl rings are totally 13.36 degrees twisted in optimized ground state - each phenyl ring is twisted by 6.68 degrees relatively to -HC=CH- bridge to opposite sides. Geometry optimization of stilbene molecule in first excited state using *ab initio* HF configuration interactions single-excitation (CIS) [16, 17 method in 6-311G and STO-3G basis sets shows that it becomes planar - see *Figure 1*. Excited stilbene molecule becomes 0.11 Angstroms (Å) shorter because angles between C(1'), C(8), C(7) and C(8), C(7), C(1) atoms become smaller. The excitation process in organic photoactive molecules is very fast (picoseconds) and relaxation is due to vibrations and dissipation of excitation energy on surface. Excited and planar stilbene molecule should move on the surface due to the twisting of phenyl rings relatively –C(8)=C(7)- bridge by 13.36 degrees. The arrows in *Figure 1* show the directions of movement of stilbene molecule phenyl rings during relaxation because angles formed by C(1'), C(8), C(7) and C(8), C(7), C(1) become larger

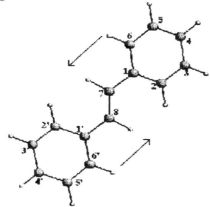

*Figure 1.* Geometry optimization of *trans*-stilbene molecule in first excited state performed by CIS-HF/6-311G or STO-3G results to planar structure.

Therefore we can state that stilbene molecule will be able to move during relaxation process by two different ways: phenyl rings rotate relative -CH=CH-bridge and go away from each other. The first way is most important for the light induced random-walking of stilbene molecule on surface.

The ground state *trans*-stilbene overlapping population (OP) [13] on bonds C(1')-C(8) and C(7)-C(1) optimized by B3PW91/6-311G** are equal to 0.83 and on C(8)-C(7) it is 1.19. During the excitation, POs on bonds C(1')-C(8) and C(7)-C(1) optimized by CIS-RHF/STO-3G becomes larger - 0.88 and on C(8)-C(7) becomes smaller - 1.05 but this does not allow free rotate along C(8)-C(7) bond. The Hückel approximation calculations confirm that after excitation POs on bonds C(1')-C(8) and C(7)-C(1) change from 0.0 to 0.3 (in arbitrary units) and on C(8)-C(7) changes from 1.0 to 0.5 (see the chapter below). Therefore we can state that *trans-cis* conformation transition by rotation along –C(8)=C(7)- is not allowed in stilbene molecule or it is going with a small probability depending on the intensity of vibrations. That correlate with above-mentioned experiments [9] claiming what fluorescence yield, i.e. *trans-cis* transition in stilbene depends on temperature.

## 2.2. *CIS*-STILBENE OPTIMIZED GEOMETRY

The geometry optimization of the *cis* conformation of the stilbene molecule in the ground state was performed applying DFT B3PW91/6-311G** model. According the computations the dihedral angle between atoms: C(1), C(7), C(8), C(1') (see *Figure 2*) is equal to 6.7 degrees, and phenyl rings are twisted approximately 37 degrees relatively to –C(8)=C(7)- bridge plane (rotation is around C(1')-C(8) and C(1)-C(7) bonds). The total energy of *cis*-stilbene is in 0.19 eV = 4.5 kcal/mol higher than in the *trans*-stilbene.

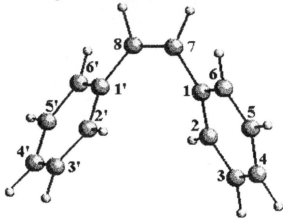

*Figure 2.* Geometry optimization of cis-stilbene molecule in ground state performed by DFT B3PW91/6-311G** model results to the rotation of phenyl rings around C(1')-C(8) and C(1)-C(7) bonds approximately 37 degrees.

The first singlet excited state of the *cis* conformation was investigated applying RHF CIS STO-3G geometry optimization. Starting optimization from ground state *cis*-stilbene geometry in the first excited state the dihedral angle of between the atoms: C(1), C(7), C(8), C(1') increases, while between phenyl rings and the –C(8)=C(7)- bridge decreases (geometry is going towards intermediate *trans* conformation). The geometry optimization convergence was not achieved due to oscillations between two total energy minimums and in these positions above mentioned tetrahedral angles: phenyl rings and –C=C- bridge becomes not the same. It means that the symmetry of first excited *cis*-stilbene molecule is broken. Therefore it is possible crossing between double-excited state and free returning to the *trans*-stilbene conformation. That correlates with above-mentioned experimental data [9] claiming what fluorescence yield, i.e. *cis-trans* transition in stilbene do not depends on temperature. The free *cis-trans* transition in stilbene confirms the study of stilbene using Hückel approach presented below.

## 3. Investigations of stilbene molecule applying Hückel approach

Our *ab initio* investigations does not explain completely the stilbene molecule *trans↔cis* izomerisation process in first excited state therefore the Hückel approach was additionally applied in order to make some additional qualitative conclusions from the topology of the molecule.

According to the *ab initio* geometry optimization in the ground state the stilbene molecule in the *trans* conformation is planar therefore the form (shape) of the molecular orbitals and energy levels should be obtained based on the below mentioned assumptions. Let assume that the stilbene molecule is constructed from three fragments: two «benzene» rings and the «ethylene» that form the bridge (see *Figure 1*). The atomic $\pi$ electron wave functions (orbitals) (AO) are $\varphi_i$, $\varphi_i'$, and the molecular functions (orbitals) (MO) of the fragments are $\psi_i$, $\psi_i'$. The Coulomb and resonance integrals of the carbon atom are $\alpha$ and $\beta$ respectively. Thus, the well-known MOs of the «benzene» fragment and their eigenvalues (*e*) are following:

$$\psi_1 = \frac{1}{\sqrt{6}}(\varphi_1 + \varphi_2 + \varphi_3 + \varphi_4 + \varphi_5) \; ; \; e1 = \alpha + 2\beta$$

$$\psi_2 = \frac{1}{\sqrt{12}}(2\varphi_1 + \varphi_2 - \varphi_3 - 2\varphi_4 - \varphi_5 + \varphi_6); \; e2 = \alpha + \beta$$

$$\psi_3 = \frac{1}{2}(\varphi_2 + \varphi_3 - \varphi_5 - \varphi_6); \; e3 = \alpha + \beta$$

$$\psi_4 = \frac{1}{2}(\varphi_2 - \varphi_3 + \varphi_5 - \varphi_6); e4 = \alpha - \beta$$

$$\psi_5 = \frac{1}{\sqrt{12}}(2\varphi_1 - \varphi_2 - \varphi_3 + 2\varphi_4 - \varphi_5 - \varphi_6); \; e5 = \alpha - \beta$$

$$\psi_6 = \frac{1}{\sqrt{6}}(\varphi_1 - \varphi_2 + \varphi_3 - \varphi_4 + \varphi_5 - \varphi_6); \; e6 = \alpha - 2\beta$$

The other «benzene» fragment MOs have the same form, but they are marked as $\psi_i'$. The «ethylene» MOs functions and their eigenvalues are following:

$$\psi_7 = \frac{1}{\sqrt{2}}(\varphi_7 + \varphi_8); \; e7 = \alpha + \beta$$

$$\psi_8 = \frac{1}{\sqrt{2}}(\varphi_7 - \varphi_8); \; e8 = \alpha - \beta$$

442

These above analytically expressed eigenvalues are graphically presented in *Figure 3.*

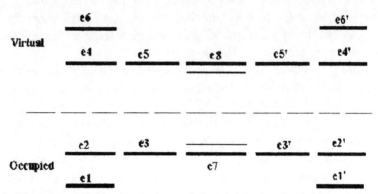

Figure 3. Placement of the eigenvalues of the «benzene» fragments and «ethylene» bridge. The solid lines correspond to the eigenvalues of the *trans* conformation of the «ethylene», while thin lines indicate *cis* conformation.

Now the interaction between MO is investigated. We included the resonance interactions (integrals β) only between fragment MOs which overlaps with the nearest neighbour AO. The above conditions satisfy the two group of the AOs: $\psi_5$, $\psi_8$, $\psi_5'$ (virtual) and $\psi_2$, $\psi_7$, $\psi_2'$ (occupied). Then the MOs and eigenvalues of the stilbene molecule are following:

$$\phi_1 = \frac{1}{2}\left(\psi_2 + \sqrt{2}\psi_7 + \psi_2'\right); \quad E1 = \alpha + \beta\left(\frac{\sqrt{3}+1}{\sqrt{3}}\right)$$

$$\phi_2 = \frac{1}{\sqrt{2}}\left(\psi_2 - \psi_2'\right); \quad E2 = \alpha + \beta$$

$$\phi_3 = \frac{1}{2}\left(\psi_2 - \sqrt{2}\psi_7 + \psi_2'\right); \quad E3 = \alpha + \beta\left(\frac{\sqrt{3}-1}{\sqrt{3}}\right)$$

$$\phi_4 = \frac{1}{2}\left(\psi_5 + \sqrt{2}\psi_8 - \psi_5'\right); \quad E4 = \alpha - \beta\left(\frac{\sqrt{3}-1}{\sqrt{3}}\right)$$

$$\phi_5 = \frac{1}{\sqrt{2}}\left(\psi_5 + \psi_5'\right); \quad E5 = \alpha - \beta$$

$$\phi_6 = \frac{1}{2}\left(\psi_5 - \sqrt{2}\psi_8 - \psi_5'\right); \quad E6 = \alpha - \beta\left(\frac{\sqrt{3}+1}{\sqrt{3}}\right)$$

The corresponding eigenvalues of the stilbene molecule are presented in *Fig. 4*.

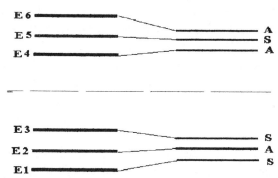

*Figure 4.* The placement of the π electron energy levels of stilbene, when the interaction between levels of the fragments is included. On the left are the eigenvalues in the *trans* conformation and on the right are the one in *cis* conformation. S and A denote the symmetrical and antisymmetrical (with respect to inversion) orbitals of the stilbene molecule.

The first excitation corresponds to the electron transition from E3 to E4 (see *Figure 4*).

According to the Hückel approach the «benzene»-«ethylene» (C(1)-C(7) and C(1')-C(8)) bond orders $P_{C(1),C(7)}$ and $P_{C(1'),C(8)}$ are equal to 0.00, while «ethylene» (C(7)-C(8)) bond order $P_{C(7),C(8)}$ is equal to 1 in the ground state of the stilbene molecule. It implies that the «benzene»-«ethylene» bonds are only σ-type in the Hückel approach. In the stilbene excited state these above bond orders change: $P_{C(1),C(7)}$ and $P_{C(1'),C(8)}$ increase till 0.3, while $P_{C(7),C(8)}$ decreases till 0.5. In the case of the real ethylene molecule the Hückel approach gives bond order changes from 1 to 0 if compare ground and excited states. This implies that the stilbene and real ethylene conformation transformations should be completely different. In the excited state of the real ethylene, the changeability of the *cis-trans* transformation is possible due to π- bond destroying. However, in the excited state of the stilbene molecule the π-bond in «ethylene» fragment is present despite that it becomes weaker in comparison with the ground state. Thus, the *trans-cis* transformation for the stilbene molecule is lower possibility than that for ethylene. Additionally, the bonds C(1)-C(7) and C(1')-C(8) become stronger than that in the ground state, that leads to the more planar structure of the excited stilbene molecule. This confirms well enough our above presented *ab initio* investigation results and experiments concerning potential barrier for *trans-cis* transition [9].

However, new features appear in the *cis* conformation of stilbene. The energy level e7 of the *cis* conformation are slightly higher then one in *trans* conformation, while e8 is below than one in the *trans* conformation due to the bridge atom $\pi$-orbitals rotation (see *Figure 3*). Thus, in the $\phi_3$ and $\phi_4$ expansion (lay-out) coefficients that corresponds to the «ethylene» orbitals ($\psi_7$, $\psi_8$) become large than $1/\sqrt{2}$. It implies that in the first excited state of the *cis* conformation stilbene molecule the C(7)-C(8) bond should be weaker than that in the *trans* conformation. However, the above mentioned effect is not sufficient to explain why in the *cis-trans* transition the experimental isomerization barrier is absent [9]. In the *cis*-conformation case, the density of states is increasing due to the less interactions between «benzenes» and «ethylene» (less value of resonance integral $\beta$) - see *Figure 4* (that was confirmed during our *ab initio* calculations). The HOMO-LUMO gap in *cis* conformation becomes large – that was confirmed also using our *ab initio* calculations. The wave function of the first excited state is superposition of determinants: the most important is E3→E4 transition, but the E1→E4 and E3→E6 transitions become more important in *cis* conformation because the energy interval between other single and double electron transition configurations becomes smaller. The increasing of the other configurations contribution leads to weaken the $\pi$ bond between C(7) and C(8) atoms because of more involving of «ethylene» bond functions (functions corresponding to eigenvalues E1, E3 and E4, E6 are in half composed by «ethylene» bond). That decreases the *cis–trans* isomerization barrier (in real excited ethylene molecule the $\pi$ bond is completely destroyed) and corresponds to experiments [9].

All the obtained results indicate that the stilbene molecule should be basic element for the design of the photo-induced molecular motion devices. On the surface the above molecule should mainly move due to the phenyl ring rotation around C(1)-C(7) and C(1')-C(8) bonds and due to the changing of angles between atoms: C(1), C(7), C(8) and C(7), C(8), C(1') after excitation by light and during energy dissipation relaxing to the ground state.

## 4. Design of molecular photo induced anisotropic random walking devices

Physical phenomenon of photochemically induced molecular motion now is effectively used for the construction of nano-scale molecular machines. Investigations of quantum effects such as: light and electron excitation, charge transfer, geometry reorganisation, electron correlation, intermolecular forces, vibrational modes are essential in the design and construction of molecular devices.

## 4.1. STILBENE BASED RANDOM-WALKERS

Two variable anisotropic random-walk stilbene based molecular motor device is designed from carbazole (Cz), 1,4-phenilenediamine (PhDA), stilbene and 7,7,8,8-tetracyanoquinodimethane (TCNQ) molecules joined with bridges -$C_2H_2$ (*Figure 5*). This random-walker should be possibly excite by two different wavelengths that correspond approximately to the wavelengths of the single Cz and PhDA molecules. Our optimized by DFT B3PW91 geometry of Cz and calculated by CNDO\S-CI and ZINDO-CI methods singlet spectrum corresponds well with experimental absorption spectrum. After excitation of the Cz or PhDA this supermolecule should be deformed by two different ways and after electron tunnelling to acceptor fragment TCNQ this random-walker should dissipate the energy moving on the surface by two different ways.

Another kind of two variable random-walk device is designed based on one electron donor fragment and two electron acceptor fragments $Cz-C_2H_2-C_6H_4-CH=CH-C_6H_3-C_2H_2-TCNQ$, -$NO_2$ and using more flexible -N=N- bridge: $Cz-N=N-C_6H_4-CH=CH-C_6H_3-C_2H_2-TCNQ$, -$NO_2$; $Cz-C_2H_2-C_6H_4-CH=CH-C_6H_3-N=N-TCNQ$, -$NO_2$; $Cz-N=N-C_6H_4-CH=CH-C_6H_3-N=N-TCNQ$, -$NO_2$. Three variable anisotropic random-walk stilbene based molecular motors are designed by two ways: Cz, PhDA, N,N,N',N'-tetramethyl-1,4-phenylene-diamine (TeMePhDA), stilbene and TCNQ molecules joined with -$C_2H_2$- (or -N=N-) fragment bridges or by another manner: $Cz-C_2H_2-C_6H_4-CH=CH-C_6H_2-C_2H_2-TCNQ$, -$C_2H_2-TCNB$, -$NO_2$ and analogues with -N=N- bridge.

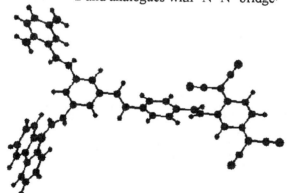

*Figure 5.* This molecular device is designed from photoelectron donors Cz, PhDA (left side), stilbene and photoelectron acceptor TCNQ (right side) molecules joined with -$C_2H_2$- fragment bridges twisted by 45 degrees.

Four variable anisotropic random-walk stilbene based molecular motor device is designed from Cz, PhDA, stilbene, TCNQ and TCNB molecules joined with -

$C_2H_2$- (or -N=N-) fragment bridges. Six variable anisotropic random-walk stilbene based molecular motors are designed as: Cz, PhDA, TeMePhDA, stilbene TCNQ and TCNB molecules joined with -$C_2H_2$- (or -N=N-) fragment bridges or: Cz-$C_2H_2$-, PhDA-$C_2H_2$-$C_6H_3$-CH=CH-$C_6H_2$-$C_2H_2$-TCNQ, -$C_2H_2$-TCNB, -$NO_2$ and analogues with -N=N- bridge.

## 4.2. DISPERSE ORANGE 3 AZO-DYES BASED RANDOM-WALKERS

We have calculated DO3 azo-dye molecule: $NH_2$-$C_6H_4$-N=N-$C_6H_4$-$NO_2$ that was similar to molecule investigated in the paper of P. Lefin et al [5]. Geometry optimization was done using DFT B3PW91\6-311 model in the ground state. The near bridge angles C-N=N of the optimized molecule are approximately 114 and 115 degrees. In this molecule exist weak bondings between the bridge nitrogen atoms and the phenyl ring the closest hydrogen atoms (something like intramolecular hydrogen bondings). These intramolecular hydrogen bondings keep molecule almost in one plane: only 0.015 and 0.005 degrees are between planes and -N=N- bridge. We have optimized azo-dye molecule in the first excited state by using CIS RHF\STO-3G. Results show, that the angles <C-N=N changes from 115.5 to 123.5 and from 114.1 to 121.9 degrees. During excitation have broken two intermolecular hydrogen bondings, while two of them still exist and therefore molecule remains in one plane. We can state, that DR1 molecule moves during relaxation by reducing of angles C-N=N.

Two variable random-walk molecular motors were designed based on DO3 azo dye calculation results. We designed such a simplest random-walkers: Cz-$C_2H_2$-, PhDA-$C_2H_2$-$C_6H_3$-N=N-$C_6H_4$-$NO_2$;     Cz-$C_2H_2$-$C_6H_4$-N=N-$C_6H_3$-$C_2H_2$-TCNQ, -$NO_2$, and analogs with -N=N- bridges. Three variable random-walk motors should be Cz-$C_2H_2$-, PhDA-$C_2H_2$-, TeMePhDA-$C_2H_2$-$C_6H_2$-N=N-$C_6H_4$-$NO_2$; Cz-$C_2H_2$-$C_6H_4$-N=N-$C_6H_2$-$C_2H_2$-TCNQ, -$C_2H_2$-TCNB, -$NO_2$ and analogues with -N=N- bridges. Four variable random-walk motor was designed as Cz-$C_2H_2$-, PhDA-$C_2H_2$-$C_6H_3$-N=N-$C_6H_3$-$C_2H_2$-TCNQ, -$NO_2$ and analogues with -N=N- bridges. Six variable random-walk motors should be as Cz-$C_2H_2$-, -$C_2H_2$-PhDA, TeMePhDA-$C_2H_2$-$C_6H_2$-N=N-$C_6H_3$-$C_2H_2$-TCNQ, -$NO_2$; Cz-$C_2H_2$-, PhDA-$C_2H_2$-$C_6H_3$-N=N-$C_6H_2$-$C_2H_2$-TCNQ, -$C_2H_2$-TCNB, -$NO_2$ and analogues with -N=N- bridges.

## 4.3. DESIGN OF LOGICALLY CONTROLLED RANDOM-WALKERS

The quantum chemical calculations and investigations of benzene (Ph), Cz, TCNQ and PhDA molecules were done using DFT BPW91 and B3PW91 models in the cc-pVDZ and cc-pVTZ [18] basis sets performing full geometry optimization. The investigations of the molecular diades designed from the above mentioned molecules and -$C_2H_2$-, -N=N- bridges were done using B3PW91\6-311G and HF\6-31G in order to design more sophisticated molecular logic devices. The design of the molecular photoactive diade PhDA-$C_2H_2$-Ph is based on the analysis of quantum characteristics of Ph and PhDA molecules. The results of the ground state optimization of the interatomic distances and angles of the molecular insulator bridge -$C_2H_2$- showed, that planes of PhDA and benzene molecule fragments are oriented by 1.18 and 1.92 degrees respectively the plane of the bridge fragment. Small negative charge equals to 0.03 e is transferred from PhDA molecule fragment to the -$C_2H_2$- and benzene molecule fragments. Therefore it can be expected that the electron charge should be transferred from PhDA fragment to benzene fragment during the diade excitation by light. The small charge transfer in the ground state exists because of large value of PO between C atoms of bridge fragment. Calculations of the diade spectrum using CIS HF method in 6-311G basis set in the first excited state showed that UV wavelength equals to 282 nm should excite the diade. This shows that this molecular diade can be used for solar energy organic based photovoltaic converters and charge transfer molecular logic devices.

Molecular Implementation (MI) of carbon based two, three, four variable logic functions, summators of neuromolecular networks, cells of molecular cellular automata, molecular trigger - molecular logic devices and molecular devices for electronically genome regulation were designed based on results of semiempirical and *ab initio* HF, DFT quantum chemical calculations of the above mentioned electron donors, electron insulators, electron acceptors and fullerene molecules. Complete set of sixteen MIs of two variable logic functions (for example: OR, AND, Implication, Equivalence, Difference, etc.) was designed and also proposed using MIs of two variable molecular logic function initial basic sets: {OR, AND, Negation} or {NOR} and, or {NAND} [19-21]. We have described in more detail the designed MIs of: a) two variable logic functions OR, NOR, AND, NAND (from fullerene molecules), Converse Unitary Negation-1, Converse Unitary Negation-0, Unitary Negation-1, Unitary Negation-0, «0» and «1» Matrix Constants; b) three variable logic functions AND, NAND, OR, NOR analogs; c) four variable logic functions OR, NOR, AND, NAND analogues. The electron hoping via the insulator bridges in the supermolecules: electron donor-bridge-electron acceptor phenomenon was investigated by using CNDO/S-Configuration Interaction method.

Recently two new supermolecules were designed based on results of our quantum chemical calculations:

1) PhDA-N=N-TCNQ-$C_2H_2$-TCNQ-N=N-Cz and 2) Cz-N=N-TCNQ-$C_2H_2$-$C_6H_4$-$C_2H_2$-$C_6H_4$-$C_2H_2$-PhDA and analogues with all possible variations of -

N=N- and -C$_2$H$_2$- bridges. Depending on the conditions of the excitation and the outputting of transferred electron charge the **1)** supermolecule should be Converse Unitary Negation-1, Converse Unitary Negation-0, Unitary Negation-1 or unitary Negation-0 two variable logic functions. The **2)** supermolecule should be And, Nand, «0» as well as «1» Matrix Constants depending of the way of excitation and outputting the transferred electron charge.

The molecular logical devices joined to multivariable anisotropic molecular random-walkers should possess the possibility to move under the illumination of light and represent new kind of logically controlled molecular motors. Two examples of such a molecular logically controlled motors are designed and calculated using HF\6-31G method: **3)** Cz-C$_2$H$_2$-, NH$_2$-C$_6$H$_3$-dicyanodinitrofluorene-N=N-C$_6$H$_4$-NO$_2$ and **4)** Cz-C$_2$H$_2$-, PhDA-C$_2$H$_2$-, NH$_2$-C$_6$H$_2$-C$_2$H$_2$-dicyanodinitrofluorene-N=N-C$_6$H$_4$-NO$_2$ (see *Figure 6*) and analogues with more flexible bridge -N=N-. The **3)** random-walker is two variable AND logically controlled molecular device and the **4)** is three different ways two variable AND controlled molecular motor.

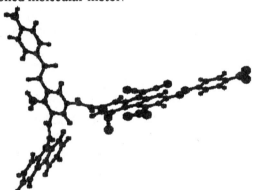

*Figure 6.* Three different ways two variable AND logically controlled molecular random-walker.

There were designed more sophisticated logically controlled molecular random-walkers: **5)** PhDA-N=N-TCNQ-C$_2$H$_2$-TCNQ-N$_2$-Cz---(TCNQ-(-C$_2$H$_2$-TeMePhDA-C$_6$H$_4$-N=N-C$_6$H$_3$-C$_2$H$_2$-TCNQ, -NO$_2$)$_2$)- depending on the conditions of the excitation and the outputting of transferred electron charge the **5)** supermolecule should be Converse Unitary Negation-1, Converse Unitary Negation-0, Unitary Negation-1 or unitary Negation-0 two variable logic functions and possesses possibility to move as two variable random-walk molecular motor. The **6)** Cz-N=N-TCNQ-C$_2$H$_2$-C$_6$H$_4$-C$_2$H$_2$-C$_6$H$_4$-C$_2$H$_2$-PhDA---(C$_6$H$_4$-(-C$_2$H$_2$-TeMePhDA-C$_6$H$_4$-N=N-C$_6$H$_2$-C$_2$H$_2$-TCNQ, -C$_2$H$_2$-TCNB, -NO$_2$)$_2$). Supermolecule **6)** should be And, Nand, «0» as well as «1» Matrix Constants depending of the way of the excitation and outputting the transferred electron charge and possesses the possibility to move as three variable random-walk molecular motor. The analogues with all possible variations of -N=N- and -$C_2H_2$- bridges also were investigated in order to find the more optimal logically controlled device, that moves better and better transfers the photoinduced electron.

## Conclusions:

1. The results of the investigations applied using both *ab initio* and Hückel approaches indicate that the *trans*-stilbene and *trans*-DO3 molecules become planar in the first single excited state, while exited *cis*-stilbene is capable free return to the lower *trans* conformation.

2. The *trans*↔*cis* isomerization processes in first exited state of stilbene and ethylene molecules are essentially different.

3. The stilbene molecule should mainly move on the surface due to the phenyl ring rotation around C(1)-C(7) and C(1')-C(8) bonds after excitation by light and during energy dissipation relaxing to ground state. That makes possible to design stilbene molecule based photo-induced charge transfer molecular motion devices.

**Acknowledgements:** We are grateful to Computer Centre at Oklahoma State University, USA and to Dr. N. A. Kotov for cooperation in using the GAUSSIAN 94 package.

## References

1. Armaroli, N., Balzani, V., Collin, J.-P., Gavina, P., Sauvage, J.-P., Ventura, B. (1999) *J. Am. Chem. Soc.* **121** 4397.
2. Ashton, P.R., Balzani, V., Becher, J., Credi, A., Fyfe, M.C.T. Mattersteig, G., Menzer, S., Nielsen, M.B., Raymo, F.M., Stoddard, J.F., Venturi, M., Wiliams, D.J. (1999) *J. Am. Chem. Soc.* **121** 3951.
3. Ishow, E., Credi, A., Balzani, V., Spadola, F., Mandolini, L.L. (1999) *Chem. Eur. J.* **5** 984.
4. Koumura, N., Zljlstra, R.W.J., van Delden, R.W., Harada, N., Feringa, B.L. (1999) *Nature* **401** 152.
5. Lefin, P., Fiorini, C., Nunzi, J.-M. (1998) Pure Appl. Opt. **7** 71.
6. Barett, C.J., Rochon, P.L., Natansohn, A.L. (1998) *J. Chem. Phys.* **109** 1505.
7. Viswanathan, N.K., Balasubramanian, S., Li, L., Kumar, J., Triphaty, S.K. (1998) *J. Phys. Chem. B* **102** 6064.
8. Tamulis, A., Tamuliene, J., Balevicius, M.L., Nunzi, J.-M. "Quantum Chemical Design of Multivariable Anisotropic Random-Walk Molecular

Devices Based on Stilbene and Azo-Dyes", accepted to the journal *Mol. Cryst. Liq. Cryst.*, 1999.

9. Baltrop, J.A. and Coyle, J.D. (1975) *Excited States in Organic Chemistry*, John Wiley and Sons, London.

10. Stern, E.S. and Timmons C.J. (1970) *Gillam and Stern's Introduction to Electronic Absorption Spectroscopy in Organic Chemistry*, Edward Arnold (Publishers) LTD, London.

11. Springborg, M. (1997) *Density -Functional Methods in Chemistry and Materials Science*, Springborg, M. ed., John Wiley & Sons, Chichester: New York; Weinheim; Brisbane; Singapore; Toronto, 1–18.

12. Adomo, C., Barone, V. (1998) Exchange functionals with improved long - range behaviour and adiabatic connection methods without adjustable parameters: The mPW and mPW1PW models, *J. Chem. Phys.* **108**, 664–675.

13. Clark, T. (1985) *A Handbook of Computational Chemistry. A practice guide to chemical structure and energy calculations*, A Wiley Interscience Publications;

14. Raghavahari, K., Trucks, G.W. (1989) *J. Chem. Phys.* **91** 1989 1062.

15. Frisch M.J., et all, *Gaussian 94, Revision E.2*, Gaussian, Inc., Pittsburgh PA, 1995.

16 Seghbahn, P.E.M., Almlöf, J., Heiberg, A., Roos, B.O. (1981) The complete active space SCF (CASSCF) method in a Newton-Raphson formulation with application to the HNO molecule, *J. Chem. Phys.* **74** 2384–2396.

17. Foresman, J.B., Head-Gordon, M., Pople, J.A., Frisch, M. J. (1992) Toward a systematic molecular orbital theory for excited state, *J. Phys. Chem.* **96** 135–149.

18. Duning Jr., T. H. (1989) *J. Chem. Phys.* **90** 1007.

19. Tamulis, A., Stumbrys, E., Tamulis, V. and Tamuliene, J. (1996) Quantum mechanical investigations of photoactive molecules, supermolecules, supramolecules and design of basic elements of molecular computers, in NATO book series 9 Ed. by F. Kajzar, V.M. Agranovich and C.Y.-C. Lee, *Photoactive Organic Materials: Science and Applications*, Kluwer Academic Publishers, Dordrecht/Boston/London, 53-66.

20. Tamulis, A., Tamulis, V. (1998) Design of basic elements of molecular computers based on quantum chemical investigations of photoactive organic molecules, *Proceedings of the SPIE Photonics WEST® Conference on Optoelectronic Integrated Circuits II*, held on 24-30 January, San Jose, California, USA, 315-324.

21. Tamulis, A., Tamulis, V. and Tamuliene, J. (1998) Quantum mechanical design of molecular implementation of two, three, and four variable logic functions for electronically genome regulation, *Viva Origino* **26** 127-145.

# COOPERATIVITY AT NEUTRAL-IONIC TRANSFORMATION

Hervé CAILLEAU[1], Eric COLLET[1], Tadeusz LUTY[1,2], Marie-Hélène LEMÉE-CAILLEAU[1], Marylise BURON-LE COINTE[1], Shin-Ya KOSHIHARA[3].

[1] *Groupe Matière Condensée et Matériaux, UMR 6626 CNRS-Université Rennes 1, 35042 Rennes Cedex, France.*

[2] *Institute of Physical and Theoretical Chemistry, Technical University, 50-370 Wroclaw, Poland.*

[3] *Department of Applied Physics, Tokyo Institute of Technology 2-12-1 Oh-okayama, Meguro-ku, Tokyo 152, and Kanagawa Academy of Science and Technology (KAST), 3-2-1 Sakado, Takatsu-ku, Kawasaki 213, Japan.*

## 1. Introduction

Thermochromism and photochromism are associated with the thermo- and photo-control of a color change in molecular materials. These properties are often discussed in terms of local chemical processes. However, changes in molecular identity (electronic structure) are strongly coupled to different intra- and inter-molecular structural changes, and those processes at solid state are highly cooperative. Thus, in some molecular materials the relaxation of electronic excited states results in drastic structural changes involving a large number of electrons and molecules [1]. This situation is carried to extreme in the case of neutral-ionic (N-I) phase transformation in quasi-one-dimensional charge-transfer (CT) crystals. This exotic transformation, associated with a change of the degree of CT, may be induced by temperature, pressure and also light. Indeed, beside structural phase transitions at thermal equilibrium exists the possibility of out-of-equilibrium photo-induced phase transformations, i.e. when light triggers a macroscopic phase change. These electronic-structural phase transformations proceed via a cascade of cooperative phenomena : the formation of one-dimensional structurally relaxed CT strings along chains due to strong intra-chain interactions, their three-dimensional condensation and their three dimensional ordering (crystallization), both originating from weaker inter-chain interactions. The formation of lattice-relaxed CT strings is a profound example of self-organized electron-transfer. A richness of new physical properties originates from these non-linear excitations. Thus, three-dimensional cooperativity between the relaxed strings leads to singular pressure-temperature (P-T) phase diagrams and to a new mechanism of molecular ferroelectricity. The cooperativity is also the essence of out-of-equilibrium photo-induced phase transformations. The purpose of this paper is to give a brief overview of the cooperative phenomena that trigger N-I phase transformations, and relate them to recent experimental results.

The N-I transformation occurs in some quasi-one-dimensional CT organic crystals with a mixed-stack architecture where the alternation of electron donor (D) and

*F. Kajzar and M.V. Agranovich (eds.), Multiphoton and Light Driven Multielectron Processes in Organics: New Phenomena, Materials and Applications, 451–465.*

electron acceptor (A) molecules along chains stimulates change of electron transfer. It manifests itself by a change of molecular ionicity, i.e. the degree of CT, and by a dimerization process with the formation of $(D^+A^-)$ dimers along the stack in the I state. This uncommon phase transition has been first discovered by a change of color in some CT complexes under the effect of pressure [2], and subsequently at atmospheric pressure on lowering temperature in TTF-CA, tetrathiafulvalene-*p*-chloranil, which became the prototype compound [3]. Later the temperature-induced N-I transition has been observed in three additional CT crystals [4-6].

## 2. The isolated chain

With reference to the quasi-one-dimensional nature of these compounds one should first consider the properties of the isolated chain. A single mixed-stack CT chain near the N-I interface, is a typical example of a molecular multistable system, characterized by a high tunability between several degenerate or pseudo-degenerate ground states. Three ground states are possible : a regular N one and two degenerate I ones, since the dimerization distortions, associated with the inversion center loss, lead to the formation of ferroelectric chains with two opposite possible polarizations $\pm P_0$ :

| | |
|---|---|
| N | ... D° A° D° A° D° A° D° A° D° A° D° A° ... |
| I | ... $(D^+A^-)$ $(D^+A^-)$ $(D^+A^-)$ $(D^+A^-)$ $(D^+A^-)$ $(D^+A^-)$ ... |
| Ī | ...$D^+)$ $(A^-D^+)$ $(A^-D^+)$ $(A^-D^+)$ $(A^-D^+)$ $(A^-D^+)$ $(A^-$ |

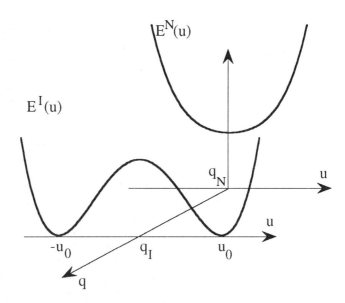

*Figure 1* : Schematic drawing of the ground state potential of a single chain in the two-dimensional space of the transformation paths represented by the symmetry breaking change u ("dimerization") and the totally symmetric one q ("ionicity"). For simplicity only the single-well N diabatic potential and the double-well I one are presented.

We have to stress that there are different types of electronic and structural changes between the N and I states : the first ones are associated with the "dimerization" symmetry breaking (intra- and inter-molecular non centro-symmetric distortions), whereas the second ones are totally symmetric (the unit-cell parameters and the centro-symmetric intra-molecular changes of bond lengths). In the two-dimensional space of the transformation paths representing these two types of changes, the adiabatic ground state energy of a single chain can be represented by a three minima potential (N, I and $\bar{I}$) (Figure 1).

In a simplest picture the relative stability of N and I regular chains can be discussed in terms of competition between the cost of ionization of a DA pair and the gain in electrostatic (Madelung) energy. The N-I interface is defined at $I - A = V\alpha_\infty$, where $I$ is the ionization energy of D, A the electron affinity of A, $\alpha_\infty$ is the Madelung constant for an infinite chain (=2ln2 in the point charge approximation) and $V = <e^2/a>$ is the nearest-neighbour Coulomb interaction. In a more general description one has to take into account other factors such as a finite CT integral which leads to a partial degree of CT in each state (quantum mixing) and different types of intra- and inter-molecular structural relaxations as discussed above. Different theoretical models of ground state have been studied (see, for instance [7-10]). It can easily be understood that an increase of pressure favours the I state and may drive a phase transition from a paraelectric N chain towards a ferroelectric I chain (the change of volume between the two states has to be taken into account).

At finite temperature the vibrational entropy, in relation with structural changes, may compensate the enthalpy difference and influences the equilibrium between N and I states. As long as one considers homogeneous chain, the crossing between chemical potentials of N and I states determines an equilibrium line in the P-T plane (Figure 2). However, it is well-known that the thermal excitation of boundaries (kinks) between "phases" destroys any long range order; the chain becomes intrinsically inhomogeneous at finite temperature (thermal mixing).

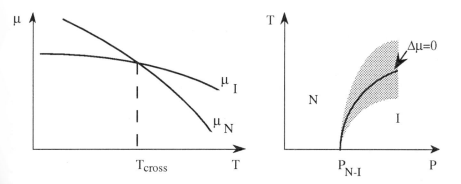

*Figure 2* : The crossing of N and I chemical potentials, on the left side, determining an equilibrium line in the (P,T) plane for an homogeneous chain, on the right side. The dotted part schematically defines the crossover region where the chain is intrinsically inhomogeneous due to thermal effects.

454

In order to discuss the intra-chain cooperativity due to Coulomb interactions, we have to stress that the CT excitation spectra near a N-I interface presents a particular characteristic [7]. The energy expense to create an I string of n adjacent unrelaxed $D^+A^-$ pairs within a N chain is

$$\Delta E_n = n(I - A - \alpha_n V) \qquad (1)$$

where $\alpha_n$ is an effective Madelung constant for a n string (equal to 1 for one pair and $\alpha_\infty$ for an infinite string). At the N-I interface, this energy is

$$\Delta E_n = n(\alpha_\infty - \alpha_n)V \qquad (2)$$

and it converges rapidly to a finite value (V/2). Therefore, it costs less and less energy to add one $D^+A^-$ pair to a string (Figure 3). Similar considerations can be applied to a formation of a N string within an I chain. A finite CT integral between D and A molecules makes some alterations in this description, important for the CT absorption band [7], but the essential physical picture is the same. In addition, the intra- and inter-molecular structural relaxations lead to the self-trapping of the CT string excited states, particularly important at one dimension [11]. The relaxed strings are at the heart of the properties of these systems. For a large enough n, their energy can be discussed in terms of creation of kinks, called NI domain walls [9], as the free enthalpy expense per pair in the string tends towards the difference between the I and N chemical potentials, and can be of the order of thermal energy. When the chain is essentially in an I state, the typical non-linear excitations become topological solitons between the two degenerate I and $\bar{I}$ states [9].

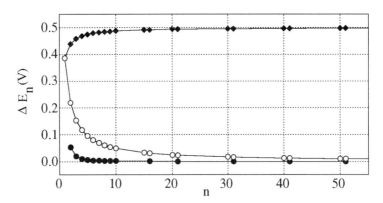

*Figure 3* : ◆ Energy cost to create a n exciton-string of n adjacent CT excited DA pairs.
○ Energy cost of a n exciton-string per DA excited pair.
● Energy cost to add one CT excited DA pair to a n CT exciton-string.

The thermodynamics of a single mixed-stack chain is then governed by the thermal excitations of these relaxed strings [12,13]. Thus, instead of a phase transition, crossover phenomena take place with a continuous variation of the concentration c of I species around the crossover temperature $T_{cross}$. The number of kinks shows a maximum at $T_{cross}$. As suggested by Nagaosa for a continuous one-dimensional model

[12], the state of the chain is well characterized by the polarization field P(x) (or dimerization u(x)). One can introduce a spin -1 Ising field S(x)

$$S(x) = P(x)/P_0 = u(x)/u_0 \qquad (3)$$

which can take locally the value 0(N), 1(I) or -1($\bar{\text{I}}$) [13]. In agreement with the essence of a spin-1 model [14], one can define a dipolar order parameter $\eta = <S>$ describing the symmetry-breaking due the inversion center loss (polarization) and a quadrupolar order parameter $<S^2>$ which identifies to the concentration c of I species. An instantaneous snapshot of the spatial fluctuations for these two order parameters is presented in Figure 4. They imply two characteristic correlation lengths, a dipolar one $\xi_{d1D}$ and a

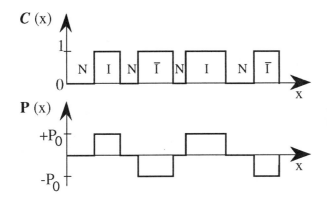

*Figure 4* : Snapshot of the spatial fluctuations of the concentration of I species and of the polarization.

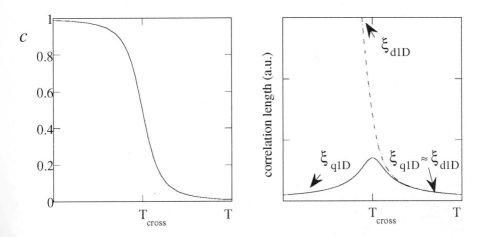

*Figure 5* : Temperature evolution of the concentration of I species and of the dipolar and quadrupolar correlation lengths for an isolated chain around the crossover temperature.

quadrupolar one $\xi_{q1D}$. If $\xi_{q1D}$ is maximum at $T_{cross}$, $\xi_{d1D}$ increases lowering the temperature but diverges only at 0 K (Figure 5). In other words, the thermal motion of kinks destroys the polarization long range order at any finite temperature, in a similar way to the well-known example of a spin 1/2 Ising chain. Let us notice that at low temperatures when $c$ approaches 1, the topological solitons play a central role. The dipolar susceptibility,

$$\chi_{d1D} \alpha \frac{c\xi_{d1D}\mu_0^2}{kT} \tag{4}$$

where $\mu_0$ is the dipole moment per pair induced by dimerization, diverges at 0 K, while the quadrupolar susceptibility,

$$\chi_{q1D} \alpha \frac{c(1-c)\xi_{q1D}}{kT} \tag{5}$$

which describes the fluctuations of I species concentration, has its maximum at $T_{cross}$.

## 3. Phase transitions driven by inter-chain couplings

The emergence of phase transition requires sufficient inter-chain interactions. They include both quadrupolar and dipolar interactions. The coupling constants, C and J, respectively as defined in [13], have both electric and elastic origin. The quadrupolar interactions are due to Coulomb interaction between charged molecules and elastic-type interaction due to "chemical pressure", created by the strings of I molecules, different in volume and shape from N molecules. In a more quantitative description, the "chemical pressure" is expressed in terms of elastic dipoles [15]. The dipolar interaction is due to the electric dipoles and net forces arising from non-symmetric structural deformations. The importance of intra- and inter-chain quantum couplings have also been discussed, based on three-dimensional ab-initio electronic structure calculations [16].

Both types of the inter-chain interactions create coupling between CT strings. The quadrupolar coupling may drive a discontinuous condensation phase transition between two disordered paraelectric phases, the former with a low concentration of I species ($N_{para}$), the latter with a high concentration ($I_{para}$). It follows from a self-consistent equation within the mean-field approximation [12,13] and the condition of instability is given for a quadrupolar coupling constant C by

$$C \geq \chi_{q1D}^{-1} \tag{6}$$

The condensation transition is analogous to the gas-liquid transition and there is no change of symmetry between the two paraelectric phases (isostructural transition). The concentration $c$ of I species serves as order parameter like the density for the gas-liquid transition (Figure 6). The first order transition line (condensation line) ends at a critical point for sufficiently high temperatures. For its part, the dipolar coupling breaks the symmetry of the system and may drive the three-dimensional ordering of the polar I strings. This latter phase transition is analogous to a crystallization transition of the

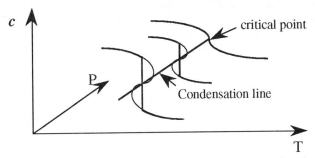

*Figure 6* : Temperature evolution of the concentration of I species around the condensation transition

strings, with the formation of an ordered lattice of condensed I strings in a ferroelectric, antiferroelectric, or more complicated, array. This behavior is similar to the well-known problem of weakly coupled Ising chains. The condition for a ferroelectric instability is given for a dipolar coupling constant J by

$$J \geq \chi_{d1D}^{-1} \qquad (7)$$

and it drives the divergence of the (3D) dipolar susceptibility for a second order transition. A non-zero order parameter $\eta = <S> \approx P$, characterizing the symmetry breaking, arises in the low symmetry ferroelectric phase. In this phase, the topological solitons are no longer travelling but pinned within two-dimensional arrays of walls between ordered domains (with possible large motional fluctuations parallel to the chains). The analogy with the gas-liquid-solid situation is clear and a schematic presentation of the three different possible phases is shown in Figure 7. The two order parameters, $c$ and $\eta$, are coupled (for instance via the dipolar susceptibility $\chi_{d1D}$) and the two types of phase transitions can occur successively or simultaneously, leading to different equilibrium lines in the (P,T) phase diagram (see Figure 7 in [14]).

| $N_{para}$ (gas-like phase) | $I_{para}$ (liquid-like phase) | $I_{ferro}$ (solid-like phase) |
|---|---|---|
| ... D° A° D° A° D° A°... | ...(D⁺A⁻) (D⁺A⁻) D° A°... | ...(D⁺A⁻) (D⁺A⁻) (D⁺A⁻)... |
| ...(D⁺A⁻) (D⁺A⁻) D° A°... | ...(D⁺A⁻) (D⁺A⁻) (D⁺A⁻)... | ...(D⁺A⁻) (D⁺A⁻) (D⁺A⁻)... |
| ... D° A° D° A° D° A°... | ... D° (A⁻D⁺) (A⁻D⁺) A°... | ...(D⁺A⁻) (D⁺A⁻) (D⁺A⁻)... |
| ... D° A° D° A° D° A°... | ... D° (A⁻D⁺) (A⁻D⁺) A°... | ...(D⁺A⁻) (D⁺A⁻) (D⁺A⁻)... |
| ... D° A° D° (A⁻D⁺) A°... | ...(D⁺A⁻) (D⁺A⁻) D° A°... | ...(D⁺A⁻) (D⁺A⁻) (D⁺A⁻)... |
| ... D° A° D° A° D° A°... | ...(D⁺A⁻) (D⁺A⁻) (D⁺A⁻)... | ...(D⁺A⁻) (D⁺A⁻) (D⁺A⁻)... |
| $0 < c < 0.5$ | $0.5 < c < 1$ | $c \approx 1$ |
| $\eta = 0$ | $\eta = 0$ | $\eta \neq 0$ |

*Figure 7* : Schematic picture of the three gas-, liquid- and solid-like phases :
- $N_{para}$ : neutral paraelectric
- $I_{para}$ : ionic paraelectric
- $I_{ferro}$ : ionic ferroelectric

The N-I phase transition presents unusual features which can be examined within a phenomenological approach. It is well known from Landau theory of phase transitions that any probability density $\rho$ describing a crystalline structure (electronic, atomic positions,...) in two phases can be expressed as

$$\rho = \rho_0 + \Delta\rho \tag{8}$$

where $\rho_0$ respects the symmetry of the high-symmetry phase and $\Delta\rho$ is proportional to the order parameter characterizing the symmetry breaking. Many structural phase transitions are associated with a change of space group without drastic change in $\rho_0$. A typical example is a conventional ferroelectric phase transition. There are fewer examples of isostructural phase transition (metal-insulator Mott transition, spin transition,...) exhibiting a discontinuous change of $\rho_0$ but without change of symmetry. In this case $\rho_0$, plays the role of an order parameter as the density for the liquid-gas transition ; it is non-zero in both phases. The N-I transition is unusual because it involves, on an equal footing, the two types of order parameter.

Some particular inter-chain arrangements may also drive more complicated phase transitions towards specific N-I chain organization. This appears when between certain planes, each D (A) has other D (A) as inter-plane nearest neighbours, giving rise to a repulsive Coulomb inter-plane interactions. There is a competition between the attractive $D^+A^-$ intra-plane interaction which tends to have complete ionized states and the repulsive $D^+D^+$ and $A^-A^-$ inter-plane interactions. This may lead to a "staging" phenomenon in the ground state with the possibility of multiple periodic ordering between N and I planes (Devil's staircase) [17,18]. In the case where instability takes place for a wave-vector q, which determines the periodic alternation of N and I planes, the condition for quadrupolar instability is given by

$$C(q) \geq \chi_{q1D}^{-1} \tag{9}$$

However, the dimerization and especially the inhomogeneous nature of chains arising with temperature can modify the above picture. The interplay between these different effects, in particular competitive interactions and thermal excitation of kinks, gives the possibility for a continuous increase of the concentration wave (as in alloys but also with an evolution of the average concentration). In addition, specific dimerization orderings take place, as "ferrielectric" structures. We have recently observed this type of layered ordering [19] in the DMTTF-CA, as it was suggested by Aoki et al. [5].

## 4. Condensation and crystallization of CT excitations in TTF-CA

Our recent structural and dynamical investigations [19] on the prototype compound TTF-CA, can be considered as an illustration and confirmation of the specific physical features of the N-I transition.

At atmospheric pressure and room temperature the N phase is the stable one. The monoclinic unit cell (with space group $P2_1/n$) contains two symmetry related undimerized DA pairs. The molecules are located on inversion symmetry sites and form alternating regular stacks along the **a** axis. The intrinsic ionicity of the molecules, $q_N$, is about 0.3, as determined by electronic and vibrational spectroscopies [20]. At $T_{N-I} \approx 81$ K, after a strongly first order transition, the molecular ionicity increases up to about 0.7. There is no change in translational symmetry but the diffraction pattern indicates a loss of the screw axis symmetry [21] (Figure 8). The space group is Pn, which implies a ferroelectric arrangement between dimerized I chains. Notice that the structural analysis indicates that the changes which preserve and those which break the symmetry (with respect to N phase), as well as those due to intra- and those due to inter-molecular distortions, have to be considered on an equal footing.

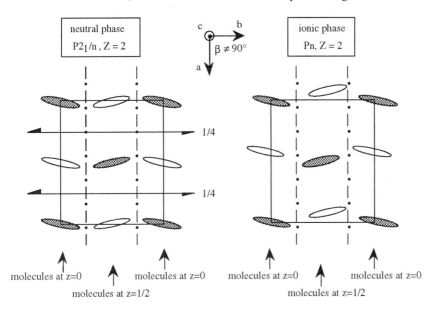

*Figure 8* : Schematic drawing of the symmetry lowering at the N-I transition in TTF-CA

The condensation and the crystallization of CT excitations at the N-I transition is reflected in a singular (P,T) phase diagram, similar to that of gas-liquid-solid [22]. The experimental results (Figure 9) and the preceding discussion offer the following physical picture. In the low-pressure low-temperature regime the condensation instability drives a strong first order character for the change between N and I phases composed of predominantly homogeneous chains. The thermal excitation of kinks is weak, and on the I side, the single chain dipolar correlation length $\xi_{d1D}$ is sufficiently large to drive simultaneously the ferroelectric ordering between dimerized I chains (the condition (7) is fulfilled). With the increase of $T_{N-I}$ with pressure, the thermal excitation of kinks becomes more and more efficient and the transition less discontinuous. In the upper part of the phase diagram, the distance between kinks diminishes, and then $\xi_{d1D}$, so that the effective dipolar coupling becomes too small to drive the ordering transition. The system reaches first a triple point, and the condensation transition becomes separated from the ferroelectric (crystallization) one. The condensation line terminates at a critical point where the system reaches its critical ionicity (= 1/2), while the ferroelectric line may exhibit a tricritical point at very high pressure beyond which the ordering transition would be second order.

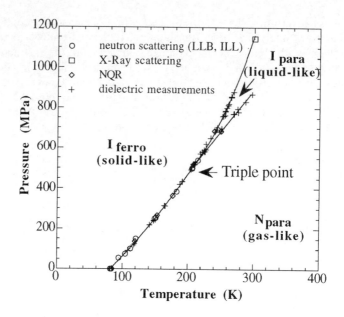

*Figure 9* : Experimental (P,T) phase diagram of TTF-CA.

A direct illustration of the existence of the triple point is given by a recent neutron diffraction study [19]. Thus, an anomaly in the dependance of the lattice parameter **a** (stack axis) is observed only at the condensation transition when the chains transform from N to I. However, new Bragg reflections, characteristic of the space group symmetry change, appear only when the three-dimensional ferroelectric ordering takes place at the crystallization transition. In addition, the **b** parameter, axis perpendicular to the stacking axis, is sensitive to both transitions, with two anomalies, the first one at the condensation transition and the second one at the ferroelectric ordering.

Complementary observation is provided by a $^{35}Cl$ Nuclear Quadrupole Resonance (NQR) study [19,21]. This technique locally probes the modification of the electronic environment. As the time scale of kink motion is in the picosecond range, as indicated by photo-induced experiments [25], NQR probes for inhomogeneous chains the thermal average of the ionicity, $<q> = cq_I + (1-c)q_N$, i.e. the true thermodynamic parameter, contrary to optical spectroscopies, which probe the instantaneous ionicity of specific molecules ($q_N$ and $q_I$). Thus, we have observed, a single resonance line in a paraelectric phase for each pair of chlorine nucleus related by inversion center [23], while the coexisting of N and I bands may be observed by optical spectroscopies [24]. The loss of the inversion center in the ferroelectric phase induces a splitting of the lines. At moderate pressure, on lowering the temperature, after the normal thermal evolution of the frequency, a small pretransitional effect is observed before a large jump of frequency at the transition, associated with the N-I molecular change of state, which takes place simultaneously to the dimerization splitting. Above the triple point, in the paraelectric state, a large evolution of the single line frequency is clearly observed before the ferroelectric transition, with a possible small jump at the condensation transition. In addition, on approching the ferroelectric transition, this frequency tends towards the average frequency observed after the ordering splitting.

## 5. New type of ferroelectric phenomena

A new mechanism of molecular ferroelectricity, with simultaneous charge and displacement fluctuations, takes place in these CT compounds. The dielectric response is anomalous in TTF-CA [26] and reflects the physical features described before, as illustrated by recent dielectric measurements under pressure [19]. In the paraelectric phases the dielectric constant, which is much larger when the electric field is parallel to the stacking axis, strongly increases at high pressure ; it becomes very large in comparison with the rather weak polarization in the ferroelectric phase [19,21]. Below the triple point only one anomaly is observed with a divergence-like increase on cooling down to $T_{N-I}$ but with a clear first order character (jump). Above the triple point, two anomalies are observed : a first order jump at the ferroelectric transition, without divergence-like behaviour and a maximum on the paraelectric side located on the condensation line.

There are two origins for polarization fluctuations and this is at the heart of this unconventional ferroelectric phenomenon. The first contribution is due to dimerization displacements in ionic strings which break the inversion symmetry and give rise to the observed spontaneous polarization in the ferroelectric phase. The related dielectric susceptibility for uncoupled chains is given by (4). The strong intra-chain cooperativity is reflected by the presence of the dipolar correlation length $\xi_{d1D}$ with respect to a simple Curie law for independent dimers. Indeed, the appearance of an I string of n adjacent dimers creates a large dimerization dipole moment (n $\mu_0$), while the number of dipole becomes itself divided by n; effectively the susceptibility is multiplied by n. In addition, the related dipolar susceptibility would exhibit a divergence-like increase on approaching the ferroelectric transition, but not around the isostructural condensation transition. However, the magnitude of $\mu_0$ (0.3 - 0.4 Db) [21] is too small to explain the really large dielectric susceptibility observed. The second contribution is due to charge fluctuations within inhomogeneous chains. As it has been suggested [26,27], the kinks (NI domain walls) along the chains carry some effective bond charges and their motion induces polarization fluctuations (Figure 10). The associated average polarization is always zero and this gives no contribution to the spontaneous polarization in the ferroelectric phase. The dipole moment $\mu_{CT}$ (7 Db) for a single I(N) pair in a quasi-N(I) chain is more than one order of magnitude larger than $\mu_0$, what

*Figure 10* : Schematic drawing of the contribution of charge fluctuations to polarization fluctuations. Notice that the contribution due to dimerization displacements (not represented here) is in the opposite direction.

substantially increases the dielectric constant. Moreover, the number of kinks is at maximum on the condensation line, what can explain the observed anomaly. In the same way, a maximum electric conductivity manifests under the effect of pressure at room temperature along the prolongation of the condensation line [28].

Effects of chemical doping on the ferroelectric N-I transition have recently been investigated [29]. On the one hand, the molecular substitution of TTF by tetraselenafulvalene (TSF), which is also a centro-symmetric molecule, gives rise to local quadrupolar defects. The "chemical pressure" plays a dominant role in stabilizing the N state (volume effect), in opposite way of pressure. With increasing TSF concentration, the ferroelectric NI transition smoothly shifts towards zero temperature and some characteristics of quantum ferro(para) electricity have been observed. On the other hand, the molecular substitution of CA by trichloro-$p$-benzoquinone (QCl$_3$) also gives rise to local defects but of dipolar type in this case, because this molecule is non centro-symmetric. Then, as in conventional ferroelectric relaxors, the dielectric anomalies reveal glasslike features.

## 6. Photo-induced phase transformation

The most fascinating feature of the cooperativity at N-I transformation is the possibility with light to trigger a complete macroscopic phase change, *i.e.* a photo-induced phase transformation. This opens up the way to new out-of-equilibrium phenomena and to the photonic control of phase transitions.

The photo-induced N-I transformation in TTF-CA have been experimentally evidenced by irradiation with ultra-short laser pulses [25]. The I-to-N transformation at low temperature, as well as the N-to-I transformation at high temperature, have been observed. Cooperativity manifests itself by the fact that one photon can transform a few hundred of DA pairs. In addition, a typical non-linear photoresponse is observed with a threshold-like behaviour in the excitation intensity dependence of the converted concentration. The threshold intensity is much larger when exciting the CT band rather than the intra-molecular ones [30]. Furthermore, the complete I-to-N transformation, by means of a 80 fs pulse-light irradiation with an intermediate photon energy between the CT and intramolecular bands, can be realized within 1 ns, while the recovery to thermal equilibrium is in the range µs-ms [25].

These features can be discussed in terms of structural relaxation of CT excitations and of cooperativity [25,31]. There is a close-connection between the phase transitions at thermal equilibrium and the out-of-equilibrium photo-induced phase transformations. The two types of instabilities (dipolar and quadrupolar) precedently discussed have to be considered. This can be reflected in the kinetics of the photo-induced transformations. Thus, we can expect that after the photo-irradiation of the ferroelectric I phase there is, first, melting of the ordered lattice of condensed I strings towards a metastable paraelectric I phase, and then evaporation of these strings giving rise to the metastable paraelectric N phase. A possible scenario of the relaxation of CT (multi-)excitons is given in Figure 11. We can stress that the first step of the conversion process of (multi-)excitons to topological soliton pairs, due to the amount of excess vibrational energy, is similar to that one observed in quasi-one-dimensional halogen-bridged mixed-valence metal complexes [32]. Some recent experimental results may give some indications in this sense, but additional investigations are needed.

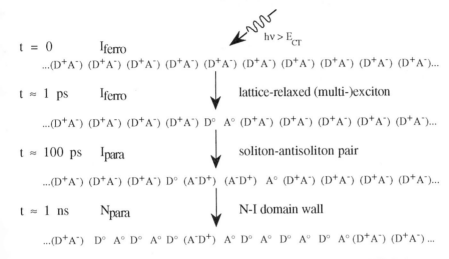

$$t = 0 \qquad I_{ferro}$$

$$hv > E_{CT}$$

$$...(D^+A^-) \ (D^+A^-) \ (D^+A^-) \ (D^+A^-) \ (D^+A^-) \ (D^+A^-) \ (D^+A^-) \ (D^+A^-) \ (D^+A^-) \ (D^+A^-)...$$

$$t \approx 1 \ ps \qquad I_{ferro} \qquad \qquad \text{lattice-relaxed (multi-)exciton}$$

$$...(D^+A^-) \ (D^+A^-) \ (D^+A^-) \ (D^+A^-) \ D° \ A° \ (D^+A^-) \ (D^+A^-) \ (D^+A^-) \ (D^+A^-) \ (D^+A^-)...$$

$$t \approx 100 \ ps \qquad I_{para} \qquad \qquad \text{soliton-antisoliton pair}$$

$$...(D^+A^-) \ (D^+A^-) \ (D^+A^-) \ D° \ (A^-D^+) \ (A^-D^+) \ A° \ (D^+A^-) \ (D^+A^-) \ (D^+A^-) \ (D^+A^-)...$$

$$t \approx 1 \ ns \qquad N_{para} \qquad \qquad \text{N-I domain wall}$$

$$...(D^+A^-) \ \ D° \ A° \ D° \ A° \ D° \ (A^-D^+) \ A° \ D° \ A° \ D° \ A° \ D° \ A° (D^+A^-) \ (D^+A^-) ...$$

*Figure 11* : A possible scenario for the I-to-N photoinduced phase transformation of TTF-CA

## 7. Conclusion

The N-I transformation in quasi-one-dimensional CT solids is the example of extreme electronic and structural cooperativity in molecular organic materials. It may serve as model for thermo- and photo-chromic transformations in solids where cooperativity plays an important role. In addition, as discussed in this paper, the description of the N-I transformation needs the use of at least two driving parameters : on the one hand the concentration of I species, as for a chemical reaction, on the other hand the order parameter characteristic of the ordering between I strings. This can be at the origin of new physical properties, such as a singular gas-liquid-solid like (P,T) phase diagram and a new mechanism of molecular ferroelectricity. In addition to the thermal effects discussed in this paper, it is worthwhile to consider also quantum effects and their interplay with thermal ones. Some more complex situations may occur with different types of organization between N and I chains. Finally, it is important to stress that the possibility of fast photo-induced transformation opens up the prospect of new applications.

## 8. References

1. Nasu, K. (Ed.) (1997) *Relaxations of Excited States and Photo-Induced Phase Transitions*, Springer Series in Solid-State Sciences 124, Springer-Verlag, Berlin Heidelberg.
2. Torrance, J.B., Vazquez, J.E., Mayerle, J.J., and Lee, V.Y. (1981) Discovery of a neutral-to-ionic phase transition in Organic Materials, *Phys. Rev. Lett.* **46**, 253-256.

464

3.    Torrance, J.B., Girlando, A., Mayerle, J.J., Crowley, J.I., Lee,V.Y., and La Placa, S.L. (1981) Anomalous nature of neutral-to-ionic phase transition in tetrathiafulvalene-chloranil, *Phys. Rev. Lett.* **47**, 1747-1750.
4.    Iwasa, Y., Koda, T., Tokura, Y., Kobayashi, A., Iwasawa, N., and Saito, G. (1990) Temperature-induced neutral-ionic transition in tetramethylbenzidine-tetracyanoquinodimethane (TMB-TCNQ), *Phys. Rev. B* **42**, 2374-2377.
5.    Aoki, S., Nakayama, T., and Miura, A. (1993) Temperature induced neutral-ionic transition in dimethyltetrathiafulvalene-*p*-chloranil, *Phys. Rev. B* **48**, 626-629.
6.    Aoki, S., and Nakayama, T. (1997) Temperature-induced neutral-ionic transition in 2-chloro-5-methyl-*p*-phenylenediamine-2,5-dimethyl-dicyanoquinonediimine, *Phys. Rev. B* **56**, 2893-2896.
7.    Soos, Z.G., Kuwajima, S., and Harding, R.H. (1986) Theory of charge-transfer excitations at the neutral-ionic interface, *J. Chem. Phys.* **85**, 601-610.
8.    Nagaosa, N., and Takimoto, J. (1986) Theory of neutral-ionic transition in organic crystals II : effect of the intersite Coulomb interaction, *J. Phys. Soc. Jpn.* **55**, 2745-2753.
9.    Nagaosa, N. (1986) Theory of neutral-ionic transition in organic crystals III : effect of the electron-lattice interaction, *J. Phys. Soc. Jpn.* **55**, 2754-2764.
10.   Luty, T. (1995) Ground state of phase diagram of mixed-stack compounds with intermolecular electron transfer, *Acta Polonica A*, **87**, 1009-1021.
11.   Toyozawa, Y. (1997) The instabilities of excitons in deformable lattice and the photo-induced phase transition, *in* [1], pp. 17-24.
12.   Nagaosa, N. (1986) Theory of neutral-ionic transition in organic crystals IV : phenomenological viewpoint, *J. Phys. Soc. Jpn.* **55**, 3488-3497.
13.   Luty, T. (1997) Neutral-to-ionic transformation as condensation and solidification of charge-transfer excitations, *in* [1], pp. 142-150.
14.   Lajzerowicz, J., and Sivardière, J. (1975) Spin-1 lattice-gas model I : condensation and solidification of a simple fluid, *Phys. Rev. A*, **11**, 2079-2089.
15.   Luty, T., and Eckhardt, C.J. (1995) General theoretical concept for solid state reactions : quantitative formulation of the reaction cavity, steric compression, and reaction-induced stress using an elastic multipole representation of chemical pressure, *J. Am. Chem. Soc.* **117**, 2441-2452.
16.   Katan, C., and Koenig, C. (1999) Charge-transfer variation caused by symmetry breaking in a mixed-stack organic compound : TTF-2,5Cl$_2$BQ, *J. Phys : Condens. Matter* **11**, 4163-4177.
17.   Hubbard, J., and Torrance, J.B. (1981) Model of the neutral-ionic phase transformation, *Phys. Rev. Lett.* **47**, 1750-1754.
18.   Bruinsma, R., Bak, P., and Torrance, J.B. (1983) Neutral-ionic transitions in organic mixed-stack compounds, *Phys. Rev. B*, **27**, 456-466.
19.   Collet, E. (1999) Etude de la condensation et de la mise en ordre d'excitations de transfert de charge à la transition neutre-ionique, *Ph. D. Thesis* n° 2186, University of Rennes1.
20.   Jacobsen, C.S., and Torrance, J.B. (1983) Behavior of charge-transfer absorption upon passing through the neutral-ionic phase transition, *J. Chem. Phys.*, **78**, 112-115.
21.   Le Cointe, M., Lemée-Cailleau, M.H., Cailleau, H., Toudic, B., Toupet, L., Heger, G., Moussa, F., Schweiss, P., Kraft, K.H., and Karl, N. (1995)

Symmetry breaking and structural changes at the neutral-to-ionic transition in TTF-chloranil, *Phys. Rev. B*, **51**, 3374-3386.

22. Lemée-Cailleau, M.H., Le Cointe, M., Cailleau, H., Luty, T., Moussa, F., Roos, J., Brinkmann, D., Toudic, B., Ayache, C., and Karl, N. (1997) Thermodynamics of the neutral-to-ionic transition as condensation and crystallization of charge-transfer excitations, *Phys. Rev. Lett.*, **79**, 1690-1693.

23. Gallier, J., Toudic, B., Délugeard, Y., Cailleau, H., Gourdji, M., Péneau, A., and Guibé, L. (1993), Chlorine-nuclear-quadrupole-resonance study of the neutral-to-ionic transition in tetrathiafulvalene-*p*-chloranil, *Phys. Rev. B.*, **47**, 11688-11695.

24. Okamoto, H., Koda, T., Tokura, Y., Mitani, T., and Saito, G. (1989) Pressure-induced neutral-to-ionic transition in organic charge-transfer crystals of tetrathiafulvalene-*p*-benzoquinone derivatives, *Phys. Rev. B.*, **39**, 10693-10701.

25. Koshihara, S., Takahashi, Y., Sakai, H., Tokura, Y., and Luty, T. (1999) Photoinduced cooperative charge transfer in low dimensional organic crystals, *J. Phys. Chem. B*, **103**, 2592-2600.

26. Okamoto, H., Mitani, T., Tokura, Y., Koshihara, S., Kamatsu, T., Iwasa, Y., Koda, T., and Saito, G. (1991) Anomalous dielectric response in tetrathiafulvalene-p-chloranil as observed in temperature- and pressure- induced neutral-to-ionic phase transition, *Phys. Rev. B*, **43**, 8224-8232.

27. Tokura, Y., Koshihara, S., Iwasa, Y., Okamoto, H., Kamatsu, T., Koda, T., Iwasawa, N., and Saito, G. (1989) Domain-wall dynamics in organic charge-transfer compounds with one-dimensional ferroelectricity, *Phys. Rev. Lett.*, **63**, 2405-2408.

28. Mitani, T., Kaneko, Y., Tanuma, S., Tokura, Y., Koda, T., and Saito, G. (1987) Electric conductivity and phase diagram of a mixed-stack charge-transfer crystal : tetrathiafulvalene-*p*-chloranil, *Phys. Rev. B.*, **35**, 427-429.

29. Horiuchi, S., Kumai, R., Okimoto, Y., and Tokura, Y. (1999) Effect of chemical doping on the ferroelectric neutral-ionic transition in tetrathiafulvalene-*p*-chloranil (TTF-QCl$_4$), *Phys. Rev. B*, **59**, 11267-11275.

30. Suzuki, T., Sakamaki, T., Tanimura, K., Koshihara, S., and Tokura, Y. (1999) Ionic-to-neutral phase transformation induced by photoexcitation of the charge-transfer band in tetrathiafulvalene-*p*-chloranil crystals, *Phys. Rev. B.*, **60**, 6191-6193.

31. Luty, T., Koshihara, S., and Cailleau, H. (2000) Photo-induced multielectron transfer in organic crystals with mixed-stack architecture, *this issue*.

32. Okamoto, H., Kaga, Y., Shimada, Y., Oka, Y., Iwasa, Y., Mitani, T. and Yamashita, M. (1998) Conversion of excitons to spin-soliton pairs in quasi-one-dimensional halogen-bridged metal complexes, *Phys. Rev. Lett.*, **80**, 861-864.

# HOLOGRAPHIC DATA STORAGE WITH ORGANIC POLYMER FILMS

P. KOPPA, T. UJVARI, G. ERDEI, F. UJHELYI, E. LORINCZ, G.
SZARVAS and P. RICHTER
Technical University of Budapest, Department of Atomic Physics,
Budafoki ut 8, 1111 Budapest, Hungary, e-mail: Koppa@eik.bme.hu

## 1. Introduction

Up to this date only limited attempts has been made to employ the benefits of optical data storage on credit card sized information carriers, although cards represent a rapidly growing part of the market. Applications like personal medical data carrier cards, ID cards, catalogues, service cards, public documentation services etc. demand higher data capacities and device performances that are probably not within the reach of today's magnetic and IC cards. For these applications we are developing a new Holographic Memory Card (HMC) system [1] that provides the advantages of optical storage in a card format in order to largely exceed the performances of present cards. Compared to bit-serial optical data storage (employes in CD, DVD and existing optical cards), our page organised holographic storage technology presents a number of advantages that can further increase the robustness, cost effectiveness, data density and transfer rates. Holographic data carriers have higher immunity to local defects, lower card positioning and fabrication tolerances, eliminate the problems related to focus and track servo mechanisms and naturally fit to parallel access read and write systems.

## 2. Key features of the holographic memory card drive

In contrast to most holographic storage systems using highly multiplexed transmissive volume holograms in bulk materials, our approach is adapted to the card format using a polymer thin film holographic storage material, operating in reflection mode, allowing writing and reading to be accomplished from the same side of the card with a small optical head. This allows card drives with dimensions equal or smaller to common PC peripherals (e.g. CD drive). There are two further distinctive features of our system : first, it uses two different wavelengths for writing and reading (blue and red), that guarantees a non-destructive read-out, and second, it utilises polarisation holography [2] with phase encoding that makes the hologram efficient and safe against unauthorised reading. This phase encoding technique represents the most challenging part of the work and will be explained in more detail in section 4.

467

F. Kajzar and M.V. Agranovich (eds.), Multiphoton and Light Driven Multielectron Processes in Organics: New Phenomena, Materials and Applications, 467–474.
© 2000 Kluwer Academic Publishers. Printed in the Netherlands.

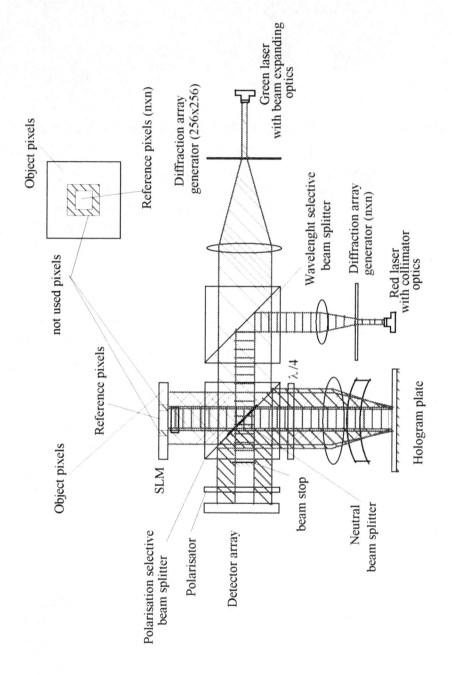

*Figure 1.*: Layout of the optical system

## 3. Operation of the read and write device

The procedure used for recording a page of information is shown in figure 1. A Spatial Light Modulator (SLM) is used to input data pages into the blue writing beam. The light exiting the SLM is projected onto the memory card using a Fourier objective lens. The reference beam is fed on the card through the same optics, coupled by a special polarising cube. At the card (Fourier plane of the SLM) the object and reference waves have mutually orthogonal circular polarisation and give rise to a polarisation hologram in the storage material that presents local uniaxial anisotropy under the effect of local polarisation. The microscopic orientation of the storage material is reversible that provides rewriteable storage. As the storage material (azobenzene side chain LC polymer [3] is only sensitive in the blue region, readout is realised by a reconstructing beam that has the same polarisation and direction but different wavelength (red). This makes sure that the read-out is non-destructive. To provide reflective operation, the card is coated with a reflective layer under the storage material. This reflects the reconstructed image towards a CCD detector array.

## 4. Phase coded data multiplexing in thin polarisation holograms

Phase coded multiplexing is treated in the literature for isotropic intensity modulated holograms [4,5]. In our system the hologram is written as a local anisotropy

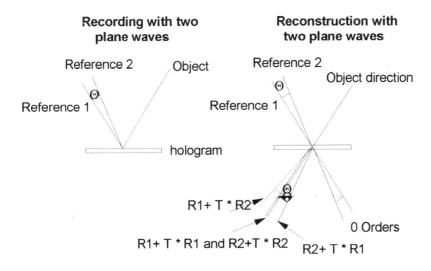

*Figure 2. Recording and reconstruction with a reference containing two plane waves*

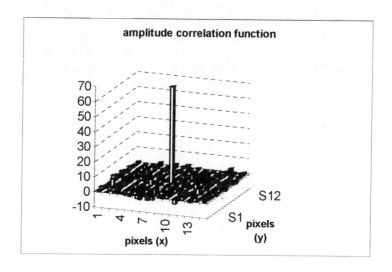

*Figure 3.Auto-correlation function in amplitude of one code of 7*

modulated function of the local electric field vector. This difference makes the above multiplexing models invalid for our investigations, and we needed to elaborate a new model. The basic assumption of the model is that the local anisotropy created by superposed holograms or successive exposures add up linearly until the index change of the material saturates. From this assumption follows that the different components of a complex reference wave interfere one by one with the object wave will and create a holographic pattern. The resulting hologram is the coherent sum of these partial holograms.

Let us now consider the reference wave as a combination of two plane waves impinging at different incidence angles to the storage material. The phase of each plane wave can be set to 0 or $\pi$, and the sum of these waves gives the phase coded reference. For the sake of simplicity let the object wave a plane wave, that corresponds to a single point of the object. If recording with R1 and reconstructing with R2 the reconstructed object beam will be deviated by $-\Theta$ (see figure 2)and its phase will be 0 if the two references have the same phase, or $\pi$ if the two references have different phases. Recording with R2 and reconstructing with R1 the reconstructed object beam will be deviated by $\Theta$. These beams deviated from the object direction will be called parasitic beams, since in case of extended objects they create cross-talk between object elements. The total reconstructed wave will be the coherent sum of the above 4 beams.

In the case of 8 by 8 reference beams the reconstructed object beam will be surrounded by an array of 15 * 15 parasitic beams. If the recording and reconstructing is made with the same phase codes, the response of the system will be the autocorrelation function of the code matrix, as it is shown in figure 3.

If the recording and reconstructing phase codes are different, the response of the system

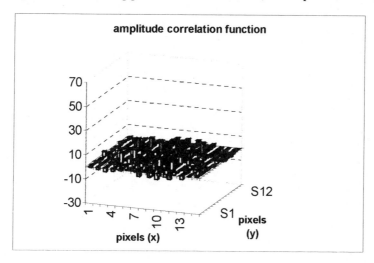

*Figure 4. Cross-correlation function in amplitude of code 7 with code 2*

will be the cross-correlation function of the two code matrices, as it is shown in figure 4. With computer optimisation we have made a set of codes that give a minimal noise

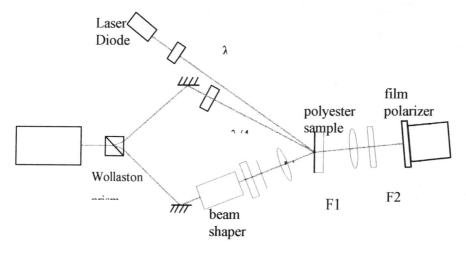

*Figure 5. Experimental set-up used for recording and reading transmission holograms and for rotational multiplexing*

evel both in auto and in cross correlation. For reading the page that was recorded with ode i, a convolution must be performed between the data page j and the correlation unction of code i with code j. The sum of all these convolutions (i.e. signal and noise mplitudes due to the different multiplexed pages) gives the total signal plus noise amplitude on the output. The square of this distribution gives the images intensity that will be captured by the CCD camera.

*Figure 6. Hologram reconstructed from tenfold rotational multiplexing*

On the basis of these simulations, we could carry out a bit error rate estimation. We found acceptable error rates (below 1 %) only for single recording pages, so the given code set does not allow real multiplexing. A further optimisation of the system geometry and the code arrays will be carried out to improve the system.

## 5. Experimental results

The basic experiments for recording and reading polarisation transmission holograms in thin polymer films were implemented on the set-up of figure 5. Then, for testing the material's ability to carry multiplexed images we recorded 10 images one after the other, making a slight rotation of the sample between subsequent exposures. The ten images could be reconstructed with good quality, as it is shown in figure 6. Some erasure of the previously taken holograms can be observed (the first exposed hologram had lower efficiencies then the last one). We concluded that the number of the detecto and the signal to noise ratio and not by the erasure of the material. A new experimer was carried out to prove the validity of the phase coding model described in th previous section. For this we realised a reference wave composed by two beams and recorded two points objects with orthogonal phase codes : a bit 1 with matching phases

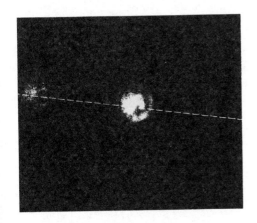

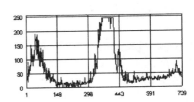

Readout hologram

with (1, 1) code

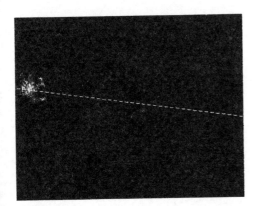

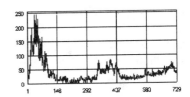

Readout hologram

with (1, -1) code

*Figure 7. Basic experiment of phase coded multiplexing*

and a bit 0 with opposed phases, The results (see figure 7) show the auto correlation peak and the two parasitic peak when reconstructed with the first reference code as it is predicted by the theory. The autocorrelation peak becomes significantly lower but does not disappear when reconstructed with the second reference code. This fact may be explained by the assimetry of the reference beams that can also be observed in the figure.

## 6. Conclusions

We have presented a holophic memory card read and write system using polarisation holography in thin polymer films. The result of the first experiments prove the feasibility of the method, and prospect data density exceeding $1bit/\mu m^2$. data transfer rate up to 200 Kbytes/sec for the first prototype. Further effort will be made to optimise phase coded multiplexing in order to increase the data density of the system to $10bits/\mu m^2$.

## References :

1. "System and method for recording information on a holographic card", P 9801029, filed 05.05 1998
2. T. Todorov, L. Nikolova, and N. Tomova, Appl. Opt., Vol. 23, pp. 4309-4312, (1984)
3. S. Hvilsted, F. Andruzzi, P.S. Ramanujam, : Opt. Lett. Vol. 17, pp. 1234-1236 (1992)
4. J.T. LaMacchia & D.L. White, Coded Multiple Exposure Holograms, Appl. Opt. 7. pp. 91-94 (1968)
5. C. Denz G. Pauliat, G. Roosen, T. Tschudi Potentialities and limitations of hologram multiplexing by using the phase-encoding technique Appl. Opt. 31. pp. 5700-5705 (1992)

# NLO POLYMERS CONTAINING AS ACTIVE CHROMOPHORE IN DANDIONYLPYRIDINIUM, BETAINE UNITS: SYNTHESIS, MODELING AND CHARACTERIZATION

O. DUBROVICH[a], M.UTINANS[a], O. NEILANDS[a], V.ZAULS[b]
and I. MUZIKANTE[c]

[a]Department of Organic Chemistry, Riga Technical University, Azenes str 14
LV-1048, Riga, Latvia
[b]Institute of Solid State Physics, University of Latvia, Kengaraga Str., 8, Riga,
LV-1063, Latvia
[c]Institute of Physical Energetics, Aikraukles Str., 21, Riga, LV-1006, Ltvia

**Abstract.** In present article we report about synthesis, quantum chemical model calculations, non-linear optical characterization and dependencies of the surface potential changes on irradiation with light of novel polyurethane based polymers, namely a bipolar organic compound N-(indan-1,3-dion-2-yl)-4-N',N'-dialkylaminopyridinium betaine (IPB) containing charged electron donor and electron acceptor groups covalently bonded to polyurethane polymer backbone. These newly synthesized IPB containing polymers could be regarded as a promising thermally stable organic non-linear optical materials with long time performance for photonic applications.

## 1. Introduction

Different amphiphilic derivatives of indan-1,3-dione pyridinium betaine are known as an attractive new material [1] for the design of photoactive molecular assemblies and LB films with highly pronounced non-linear optical (NLO) response for applications in photonics technology. Efforts are therefore directed to increase the first-order hyperpolarizability $\beta$ of IPB derivatives, which is the molecular property responsible for NLO effects [2]. Such type of betaines are peculiar for photoinduced intramolecular electron transfer (PIET), large change of the molecule dipole moment under optical excitation, and considerable hyperpolarizability shown by Hyper Rayleigh Scattering in solutions [3] and by second harmonic generation (SHG) in LB films [4-7].

In our opinion donor-acceptor (DA) systems with pyridinium acceptor part have not been sufficiently studied [8] despite a large number of known DA systems were characterized for their NLO properties [9-15]. Compounds containing N-substituted pyridinium acceptor part (A) and anionic donor part (D) as N-substituent are promising owing to the possibility of a considerable localization of HOMO and LUMO orbitals and a giant change of the dipole moment $\Delta\mu$ in PIET process

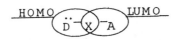

F. Kajzar and M.V. Agranovich (eds.), Multiphoton and Light Driven Multielectron Processes in Organics: New Phenomena, Materials and Applications, 475–482.

Therefore increasing interest exists also in design and optimization of IPB containing non-linear optical polymer materials for deposition of thin optical films with controllable parameters for possible applications in optoelectronics technology.

## 2. Quantum chemical calculations

For calculations of electronic spectra of IPB compounds and their dipole moments in ground states the CNDO/S (Coulomb integrals are calculated according to the Mataga-Nishimoto formula, number of single excited configurations in CI-50) and ZINDO/S methods were used. Geometry of all compounds in ground states was optimized by AM1 [16] method, using MOPAC6 or HYPERCHEM 4.0 program packages. Polarizabilities were calculated according to Finite Field technique using AM1 parameters [17].

1                    HOMO                    LUMO

TABLE 1. Calculations of substituted IPB

| Compound | | | Hamilt. | $\mu_g$ (D) | $\mu_{ex}$ (D) | $\Delta\mu$ (D) | $\lambda_{max}$ (nm) | f | $\alpha_{static}$ $10^{-24}$ esu | $\beta_{static}$ $10^{-30}$ esu |
|---|---|---|---|---|---|---|---|---|---|---|
| R¹ | R² | R³ | | | | | | | | |
| H | H | H | AM1 | 2.9 | | | | | 48.0 | 18.6 |
| | | | ZINDO/S | 4.2 | -3.3 | 7.5 | 435 | 0.61 | | |
| N(CH₃)₂ | H | H | AM1 | 6.7 | | | | | 60.1 | 16.7 |
| | | | ZINDO/S | 7.7 | -0.9 | 8.6 | 441 | 0.63 | | |
| | N(CH₃)₂ | N(CH₃)₂ | AM1 | 4.7 | | | | | 72.0 | 29.8 |
| | | | ZINDO/S | 6.1 | -3.0 | 9.1 | 447 | 0.68 | | |
| CN | H | H | AM1 | 1.5 | | | | | 57.0 | 29.5 |
| | | | ZINDO/S | 2.2 | -9.7 | 11.9 | 459 | 0.86 | | |
| H | CN | CN | AM1 | 9.3 | | | | | 58.1 | 19.4 |
| | | | ZINDO/S | 14.2 | 5.7 | 9.5 | 419 | 0.64 | | |
| (2) | | | AM1 | 8.5 | | | | | 68.5 | 17.1 |
| | | | ZINDO/S | 10.7 | 1.7 | 9.0 | 475 | 0.59 | | |
| IPPB (3) | | | AM1 | 11.5 | | | | | 104 | 170 |
| | | | ZINDO/S | 14.7 | -2.7 | 17.4 | 674 | 1.15 | | |

Calculations of the N-(indan-1,3-dion-2-yl) pyridinium betaines **1** show that HOMO is strongly localized on the indandione part and LUMO - on the pyridinium part of molecule, and an effective PIET takes place (Table 1).

Molecules are planar and have symmetry axis along the molecule. The calculated changes of the dipole moment in the excited state, $\Delta\mu$=7.5-12 D is essential. The introduction of electron donating (dimethylamino) or electron accepting (cyano) substituents in any position of IPB molecule causes decrease of LUMO and HOMO energies (in the case of electron accepting groups) or increase of LUMO and HOMO energies (in the case of electron donating groups), but the influence on $D\mu$ and on second order polarizabilities ($\beta$) is not large.

The calculations for N-[4'-(1,3-indandion-2-yl)phenyl]-pyridinium betaine (IPPB) **3** show that using a longer bridge between D and A calls a giant change of the dipole moment ($\Delta\mu$ = 17.4 D) and increase of first hyperpolarizability $\beta$.

## 3. Synthesis

The synthesis of IPB units containing polymers was carried out according to the scheme shown in Fig. 1. The starting compound is the easily available [18] 2-dicyanomethyleneindan-1,3dione oxide **4**, which forms N-(indan-1,3-dion-2-yl)-4-chloropyridinium betaine **6** in reaction with 4-chloropyridine. The compound **6** reacts easily with N,N-diethanolamine and products N-(indan-1,3-dion-2-yl)-4-(N',N'-diethanolamino)-pyridinium betaine **2** in ethanol. Polymerization of the compound **2** with p-phenylenediisocyanate or with 2,4-toluenediisocyanate gives polyurethane polymer **9** and **10** containing IPB units. The IPB compounds **2** and **6** are soluble in bipolar aprotonic solvents such as DMF, DMSO. Compounds **9** and **10** show high thermal stability - decomposition temperature is higher than 250°C.

## 4. Experimental

IR spectra were recorded on Specord M-80 spectrometer (int. 3600-2000 susp. in hexachlorobutadiene, 1800-1460 susp. in Nujol). $^1$H-NMR spectra were measured relative to TMS with Bruker WH-90/DS spectrometer. Electronic spectra were obtained on Specord M-40 spectrometer.

### N-(indan-1,3-dion-2-yl)-4-chloropyridinium betaine 6.

2-dicyanomethylene-1,3-indandione oxide (**4**) 9.87g (0.04 mol) was dissolved in 80 ml of hot dioxane, 4-chloropyridine (**5**) 5g (0.04 mol) was added and mixture was heated to boiling. The reaction mixture was cooled and filtered. The solid was washed with 5 ml of cool EtOH and dried. The product was recrystallized from EtOH. Yield - 5.5 g(45%). M.p.229-232°C(dec.). IR spectra (cm$^{-1}$):1562, 1610, 1571, 1562, 1539, 1523;UV spectra (EtOH) $\lambda_{max}$, nm($\epsilon$): 241(40000), 312(5600), 408(25600).

**N-(indan-1,3-dion-2-yl)-4-(N',N'-diethanolamino)pyridinium betaine 2.**

N-(indan-1,3-dion-2-yl)-4-chloropyridinium betaine (**6**)0.3 g (1.2 mmol) was dissolved in 30 ml of EtOH and 0.13 g (1.2 mmol) diethanolamine was added. The solution was heated under flux for 15 min. The solvent was half removed and mixture was filtered. The solid was recrystallized from EtOH. Yield - 0.25g (65%). M.p. 265°C (dec.). IR spectra (cm$^{-1}$): 3351, 1671, 1643, 1613; $^1$H-NMR spectra δ,m.d, DMSO-d$_6$: 8.51d (2H, 2,6-Py), 7.17m (4H, arom.), 7.11d (2H,3,5-Py), 4.82s (2H, 2OH), 3.42t (4H, 2CH$_2$-O), 2.44t (4H, 2CH$_2$-N), UV spectra (EtOH) λ$_{max}$, nm(ε): 216(35080), 245(23892), 300(12476), 313(12892), 361(21400)

**Indandionyl-pyridinium betaine polyurethane polymer 9.**

N-(indan-1,3-dion-2-yl)-4-(N',N'-diethanolamino)pyridinium betaine (**2**) 0.05 g (0.15 mmol) was dissolved in 15 ml of DMF (fresh distilled and dried over molecular sieve) and 0.027 g (0.18 mmol) 1,4-phenylenediisocyanate (**8**) was added (argon atmosphere). Mixture was refluxed for 3h, cooled and 30 ml water was added. Reaction mixture was refluxed for 3h, cooled and 30 ml water was added. reaction mixture was filtered and the solid was washed with hot EtOH. Yield-0.015g (20%). M.p. 255°C (dec.). IR spectra (cm$^{-1}$): 3264, 1271, 1641, 1612; $^1$H-NMR spectra δ,m.d, DMSO-d$_6$: 9.25s (2H,2NH), 8.5d (2H,2,6-Py), 7.2m (8H, arom.), 4.25d (2H,3,5-Py), 3.24m (8H, 2CH$_2$CH$_2$-O)

**Indandionyl-pyridinium betaine polyurethane polymer 10.**

N-(indan-1,3-dion-2-yl)-4-(N',N'-diethanolamino)pyridinium betaine (**2**) 0.05 g (0.15 mmol) was dissolved in 15 ml of DMF (fresh distilled and dried over molecular sieve) and 0.03 g (0.18 mmol) 2,4-tolylenediisocyanate (**7**) was added (argon atmosphere). Mixture was refluxed for 2h, cooled and 30 ml water was added. Reaction mixture was filtered and the solid was washed with hot EtOH. Yield - 0.016 g (20%). M.p. 227°C, dec. 250-260°C. IR spectra (cm$^{-1}$): 3264, 1721, 1641, 1612; $^1$H-NMR spectra δ,m.d, DMSO-d$_6$: 9.48s (2H,2NH), 8.7d (2H, 2,6-Py), 7.29m (3H, arom., Ph), 7.05m(4H, arom.ind.), 4.25d (2H,3,5-Py), 2.45m (8H, 2CH$_2$CH$_2$-O), 2.05s (3H, CH$_3$)

**5. NLO investigations**

Optical quality thin film samples were obtained by DMF IPB containing polymer 10 solution casting on ITO glass slide, dried at 60°C for 24 hours and corona poled [19] (10 kV tip voltage at the distance of 2cm from film) for 1 hour above the glass transition temperature (230°C). Absorption spectra for deposited layers were taken by Varian spectrometer. The maximum of absorption occurs at 370 nm (cf. Fig. 2). The thickness of produced films measured by interference microscope MII-4 was in the range between 1 and 3 μm. Second harmonic generation (SHG) from cast films was done in transmission. The incident fundamental beam intensity of actively-passively mode-locked Nd:YAG laser (35 ps pulses, 1064 nm wavelength and 10 Hz repetition rate) was varied by half-wave plate and polarizer combination. Transmitted SH intensity both in p-p and s-p polarization geometry [20] has been recorded with tilt

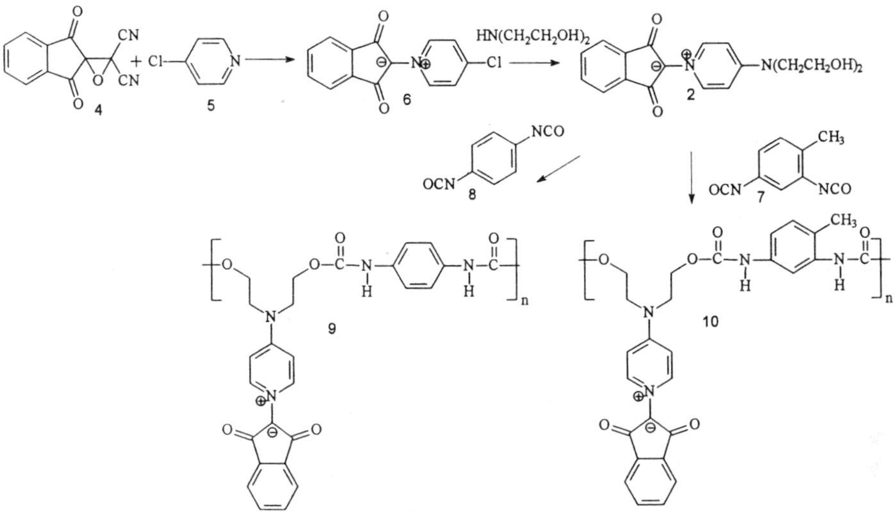

Fig. 1. Synthesis of indandionyl pyridinium

angle of the substrate positioned by computer controlled rotation table. Typical results are presented in Fig. 3. SHG signal was calibrated by Maker fringe technique [21] using 2 mm thick reference quartz plate.

Fig. 2. Absorption spectrum of IPB **10** polymer film.

Fig. 3. Intensity of the second harmonic in transmission from poled IPB-polymer **10** film for p-p and s-p polarised geometries. Reference quartz fringe shown rescaled for calibration.

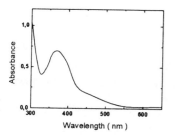

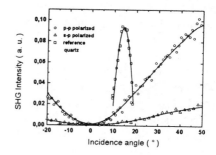

480

Taking account of thin film absorption [22] the SHG susceptibility $d_{33} = 25$ pm/V was found.

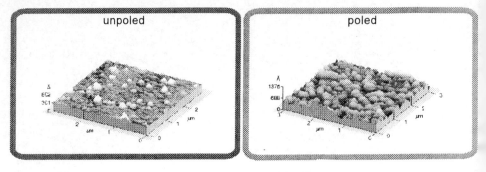

Fig. 4. AFM images of the film

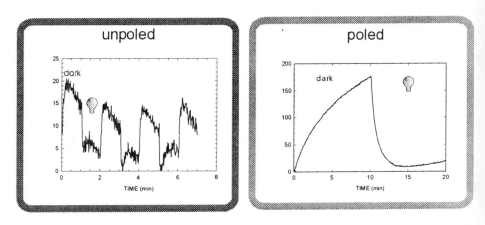

Fig. 5. Changes of the surface potentail by illumination

## 6. Surface potential changes on irradiation with light

Equipment <u>AFM</u> Park Scientific Instruments (SA1/BD2) $Si_3N_4$ tip, $k = 0.032$ N/m, contact mode and $F10^{-6}N$. <u>Kelvin probe</u> Besocke Delta PHI Gmbh., Kelvin control 07, resolution <1mV. Lamp Muller Elektronik-Optic/D, 500 W Xe Lamp, Filter: 360 nm (1.5 mW/cm$^2$).

To investigate the surface potential changes, the spin-coated films of IPB polymer 10 were obtained. Polymer was dissolved in chloroform with some drops of DMF. Thin layers of polymer 10 were prepared by spin-coating technique onto ITO covered glass substrate. Dipole moments were oriented by corona poling. The needle top electrode was placed 2 cm above the film surface and a voltage of 10 kV at 220°C was applied during 3 hours. Then the film was cooled under the applied voltage.

It is known that the value of surface potential of sample is related to the molecular dipole moment of layer in normal direction to the substrate surface [23] Therefore the surface potential studies can provide useful information on structural and electronic properties of material. The changes of dipole moment of IPB polymer 10 layer on irradiation with light centered at 360 nm were investigated by Kelvin probe technique [24]. Also the variations of the of surface potential $\Delta U_s$ were measured under light illumination. In the case of unpoled films the value of $\Delta u_s$ was of about 13 mV. After poling, when IPB side chains were oriented in electric field an increase by one order of magnitude was observed with $\Delta u_s = 150$ mV (cf. Fig. 5). It is known, that the photoexcitation of IPB molecule, resulting in the switching of the direction of dipoles, will cause a very fast ($t \sim 10^{-15}$ s) optoelectronic response [6]. However, in spin-coated IPB polymer the response time to irradiation is of the order of 10s. In poled and annealed films the response was slower and only after one minute a saturation of $\Delta u_s$ was observed.

## 7. Conclusions

The synthesis of new substituted betaines is very flexible. Polymer bonded or surface bonded betaines can be synthesized using different methods [25,26]. The synthesis of polymer-bonded betaines is possible by special substituent betaines with active functional groups: $-COOH$, $>(CO)_2O$, $CH_2=CH-(CH_2)_n-$, $HO(CH)_2-$, $H_2N(CH_2)_n$. The surface-bonding can be achieved by the surface-reactions using active functions: $Cl_3Si$ $CH_2$ $CH_2(CH_2)_n$ - in betaine molecules with suitable surface (glass, polymer film).

Quantum chemical calculations show remarkable ground state dipole moment of betaine molecule and its change under photoexcitation together with strongly pronounced UV absorption band. Charge distribution of HOMO and LUMO orbitals is strongly localized, but with some overlap favorable for enhancement of NLO response.

Preliminary SHG characterization of material gives $d_{33}$ value of 25 pm/V for corona poled thin film of optical quality despite of limited response of particular polymer structure to electric poling procedure. On the other hand, the flexibility of the synthesis procedure permits to vary donor-acceptor distance by means of incorporation of different active functional groups in order to increase dipole moment and engineering of polymer properties with respect to refractive index, thermal stability, glass transition temperature and electric field poling response.

In conclusion, these newly synthesized IPB containing NLO polymers could be regarded as promising organic materials for photonics applications.

## Acknowledgements

We gratefully acknowledge B. Stiller and Th. Kopnick from Institute of Thin Film Technology and Microsensorics, *Kantstrasse 55, 14513 Teltow, Germany*, for collaboration in preparation of spin-coated samples and surface potential investigations.

## References

1. E. A. Slinish, Photoactive Molecular Multilayer Langmuir-Blodgett (LB) films of Oriented Dipolar Betaine Type Organic Molecules, in *Photoactive Organic Materials:*

482

*Science nd Application.* Eds. F. Kajzar, V. M. Agranovich and C. Y.-C. Lee, NATO ASI Series, Kluwer Acad. Publishers, Dordrecht 1996, pp. 375-392

2. C. Lambert, S. Stadler, G. Bourhill and C. Brauchle, *Angew. Chem. Int. Ed. Engl.* 1996, **35**, 644-6

3. V. Zauls, G. Liberts, Hyper-Rayleigh Scattering and SHG Studies of IPB Molecules and LB Films, in *Notions and Perspectives of Nonlinear Optics*, Ed. O. Keller, World Scientific, 1996, pp. 586-94

4. M. A. Rutkis, E. Wistus, S. E. Lindquist, E. Mukhtar, G. Liberts, V. A. Zauls, A. B. Klimkans, E. A. Silinish, *Adv. Mater. Opt. Electron.*, 1996, **6**, 39-50

5. O.Neilands, M.Utinans, SPIE Proceedings, 1995, **2968**, pp. 13-18

6. M. A. Rutkis, L. E. Gerca, E. A. Slinish, O. Y. Neilands, M. P. Roze, E. L. Berzinsh, A. B. Klimkans,S. Larson, *Adv. Mater. Opt. Electron.*, 1993, **2**, 319-30

7. G. Liberts, V. Zauls O. Neilands, SPIE Proceedings, 1996, **2968**, pp. 19-23

8. H. S. Paley and J. M. Harris, J. Org. Chem., 1991, **56**(2), 568-74

9. J. F. Freimanis, Organic Compounds with Intramolecular Charge Transfer, Zinatne, Riga, 1985 (in Russian)

10. J. Abe and Y. Shirai, J. Am. Chem. Soc., 1996, **118**, 4705-6

11. N. Nemoto, J. Abe, F. Miyata, M. Hasegawa, Y. Shirai and Y. Nagase, *Chem. Lett.*, 1996, 851-2

12. K. J. Drost, V. Pushkara Rao and A. K-Y. Jen, *J. Chem. Soc. Chem. Commun.*, 1994, 389-71

13. P. Boldt, G. Bourhill, C. Brauchle, Y. Jim, R. Kammler, Ch. Muller, J. Rase and J. Wichern, *J. Chem. Soc. Chem. Commun.*, 1996, 793-5

14. J. Abe, N. Nemoto, Y. Nagse and Y. Shirai, *Chem. Phys Lett.*, 1996, **261**, 18-22

15. Anwar, X.-M. Duan, K. Komatsu, S. Okada, H. Matsuda, H. Oikawa and H. Nakanishi, *Chem. Lett.*, 1997, 247-8

16. J. J. P. Stewart, *J. Computer-Aided Mol. Design*, 1990, **4**, 1-105

17. A. K. Henry, J. J. P. Stewart and K. M. Dieter, J. Comp. Chem., 1990(11), 82-7

18. I. K. Raiskuma, G. G. Pukitis, O. J. Neiland, Chem. Heterocycl. Compounds, 1978(7), 889-92

19. P. N. Prasad and D. J. Williams, Introduction to Nonlinear Optical Effects in Molecules and Polymers, J. Wiley & Sons, 1991, pp. 120-31

20. G. Gadret, F. Kajzar and P. Raimond, SPIE Proceedings, **1560**, pp. 226-37

21. J. Jerphagnon and S. K. Kurtz, *J. Appl. Phys.*, 1970, **41**, 1667-81

22. M. Hayden, G. F. Sauter, F. R. Ore, P. L. Pasillas, J. M. Hoover, G. A. Lindsay and R. A. Henry, *J. Appl. Phys.*, 1990, **68**(2), 456-65

23. O. N. Oliveira, Jr., D. M. Taylor, T. J. Lewis, S. Salvagno, C. J. M. Stirling, J. Chem. Soc. Faraday Trans. I, 1989, **85**, 1009

24. B. Stiller, G. Knochenhauer, E. Markawa, D. Gustina, I. Muzikante, P. Karageorgiev and L. Brehmer, Materials Science and Engineering C, 1999, 322

25. D. M. Burland, R. Miller and C. A. Walsh, Chem. rev., 1994, **94**(4), 31-75

26. S. Yitzchaik and T. J. Marks, Acc. Chem. Research, 1996, **29**(4), 197-202

# A STUDY ON THE FILM MORPHOLOGY AND PHOTOPHYSICS OF POLYFLUORENE

P. A. LANE[a], A. J. CADBY[a], H. MELLOR[a], S. J. MARTIN[a], M. GRELL[a], D. D. C. BRADLEY[a], M. WOHLGENANNT[b], C. AN[b] and Z. V. VARDENY[b]
*[a]Department of Physics and Astronomy, University of Sheffield, Sheffield S3 7RH, UK*
*[b]Department of Physics, University of Utah, Salt Lake City, Utah 84112, USA*

Abstract We have studied the interplay between photophysics and film morphology in poly(9,9-dioctyl fluorene) using a variety of optical probes, including absorption, photoluminescence, and photoinduced absorption. Upon slowly warming a glassy PFO film from 80K to 300K, a fraction of the sample is transformed into a new solid phase, the β phase. The β phase has more extended conjugation than the glassy phase. As a consequence, the β phase excitons are red-shifted and dissociate more easily into long-lived charged polarons.

## 1. INTRODUCTION

Polyfluorenes have emerged as attractive materials for display applications owing to efficient blue emission [1] and relatively high hole mobility with trap free transport [2] Poly(9,9-dioctyl)fluorene (PFO), shown in Fig.1, has also been found to exhibit a complex morphological behaviour which has interesting implications for its photophysical properties. In the solid state, PFO can display different forms of (para)crystalline order. A glassy film can be partly transformed into a new phase, which we call the 'β' phase.[3]. The β phase can result from either exposing a film or fibre to the vapour of a solver or swelling agent[3,4] or cooling a film on a glass substrate to liquid nitrogen temperature and slowly reheating it to room temperature [5].

The spectroscopic properties of the glassy and β phases differ characteristically from one another. The glassy phase has a broad absorption spectrum that peaks at 384 nm, but the β phase shows a well resolved absorption peak at 437 nm [3,5]. X-ray fibre diffraction measurements [3] determined the parallel-to-chain coherence length of the glassy phase to be 85±10 Å. This matches the length of persistence in solution [4] as well as the effectively conjugated segment length in solution as determined from an oligomer series [6]. Hence, in glassy PFO, the effective conjugation length is conformationally limited. In the β phase, the intrachain correlation length is much longer, 220 Å. However, an extrapolation even to infinite effective conjugation length [6] still falls considerably short of the observed red-shift of the additional absorption peak. Hence, an additional planarisation of the PFO backbone due to the mechanical forces during β phase formation has been assumed. This

483

*F. Kajzar and M.V. Agranovich (eds.), Multiphoton and Light Driven Multielectron Processes in Organics: New Phenomena, Materials and Applications, 483–488.*

assumption is supported by the similarity of β phase PFO to ladder-type poly(*para*phenylene)s, which by their molecular design are forced into a coplanar conformation of all conjugated units.

The rich phase morphology as PFO provides a unique laboratory to study the influence of physical structure on the properties of excitations in conjugated molecular materials. We have studied the photophysics of PFO films by absorption, photoluminescence(PL), photoinduced absorption (PA) and PA-detected magnetic resonance (PADMR) spectroscopies. Films were prepared by spin-coating a solution of PFO in toluene onto a synthetic quartz (spectrosil B) substrate. All measurements were performed at liquid nitrogen temperatures (≈ 80 K). PA spectroscopy uses standard phase-sensitive lock-in techniques with a modulated $Ar^+$ laser beam as a pump and a broad spectrum light source as a probe. Probe light is focused onto the sample and the transmitted light is the dispersed through a monochromator and focused onto a photodiode. The photodiode output is directed to a lock-in amplifier phase-locked at the modulation frequency. The PA spectrum, defined as the normalised change $\Delta T$ in the probe transmission T, is proportional to the excitation density.

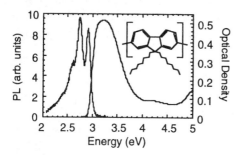 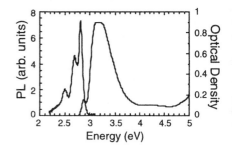

*Fig. 1. The fluorescence (left) and absorption (right) spectra of PFO (inset)*

*Fig. 2. The fluorescence (left) an absorption (right) spectra of a thermlly cycled PFO film*

PADMR spectroscopy adds spin sensitivity to conventional PA measurements. Thus, one can differentiate the contributions of excitations with different spin to the PA spectrum. In particular, PADMR can be used to differentiate contributions to the PA spectrum of charged polarons, which have spin-1/2, and triplet excitons, which have spin 1. For PADMR measurements, the pump and probe beams constantly illuminate a sample mounted in a high Q microwave cavity between the pole pieces of an electromagnet. The magnetic field causes a Zeeman splitting of the energy levels and magnetic resonance occurs when the Zeeman splitting matches the microwave photon energy. This condition result in resonant absorption or emission of microwaves which in turn causes small changes $\delta T$ in the transmission T proportional to $\delta n$, the change in the density of photoexcitation. Two types of PADMR spectra are obtained: the H-PADMR spectrum, in which $\delta T$ is measured at a fixed probe wavelength $\lambda$ as H is varied, and the $\lambda$-PADMR spectrum, in which $\delta T$ is measured at a constant H, in resonance, while $\lambda$ is varied.

## 2. EXPERIMENTAL RESULTS AND ANALYSIS

Fig. 1 shows the absorption and PL spectra of a PFO film spin-cast from toluene solution. The chemical structure of PFO is inset. The absorption begins at 2.9 eV and reaches a broad maximum. The PL spectrum exhibits a clear vibronic structure with peaks at 2.92, 2.75 and 2.57 eV. The absorption spectrum red-shifts by 12 meV upon cooling the sample from 300K to 80K, but no new features are observed. The thermochromic effect is due to freezing out vibrational and librational modes of motion that can shorten the effective conjugation length [7]. At low temperatures, the conjugation length increases and the absorption spectrum red-shifts.

Fig. 2 shows the absorption and PL spectra of a PFO film following cooling to 80K and slowly warming to 300K. A new absorption band with clear vibronic structure is superimposed upon the original absorption spectrum and the PL spectrum red-shifts by 0.1 eV. No thermochromic shift in the absorption spectrum was observed for the β phase.

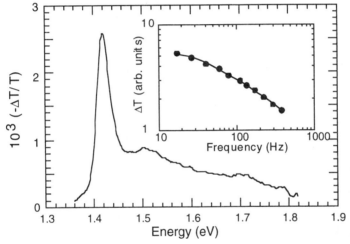

*Fig. 3. The PA spectrum of a glassy PFO film. Inset: the dependence of the PA at 1.43 eV on the modulation frequency. The line shows a fit to the excitation lifetime (see text).*

Interchain effects cannot account for changes in the absorption and PL spectra upon thermal cycling. If the absorption peak at 2.86 eV were due to Davydov splitting of the main absorption band, there should be a corresponding blue-shifted absorption peak [8]. J-aggregate formation causes significant narrowing of the PL spectrum, whereas we observe only a red-shift of 0.1 eV. Interchain aggregation can lead to excimer emission [9,10] which is characterized by a reductions of the PL quantum yield and a structureless PL that is red-shifted by 0.5 eV or more with respect to the fluorescence [11] The lack of evidence for interchain effects is consistent with the previous designation of the β phase as intrachain state with extended conjugation.

Sample morphology has a dramatic impact on photogeneration mechanisms in PFO films. Figure 3 shows the PA spectrum of a glassy PFO film, measured with 30 mW at 363 nm and a pump modulation frequency of 125 Hz. The PA spectrum is dominated by an unusually sharp and strong transition at 1.43 eV. Another PA band is visible at 1.62 eV, which is correlated with the main transition, and PA continues up to 2.4 eV. A sharp transition followed by a relatively weak and broad vibronic sideband is characteristic of optical transition with little relaxation between the two states, i.e. the Huang-Rhys parameter $S \ll 1$. The PA below 1.4 eV is quite weak ($<10^{-5}$) and there is no evidence for photoinduced infrared active vibrations (IRAVs), which are the signature of photogenerated charged excitations. Owing to its similarity to the triplet-triplet transition detected in LPPP [12] and no IRAVs, we assign this feature to excited state absorption of triplet excitons.

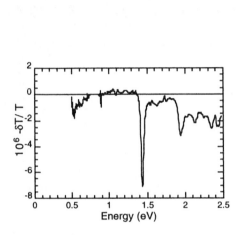

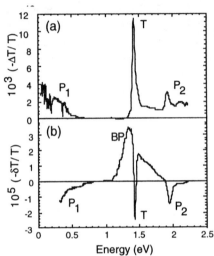

Fig. 4. The λ-PADMR spectrum of a glassy PFO film, measured at 1070 PFOGauss

Fig. 5. (a) the PA and (b) the spin -1/2 λ-PADMR spectra of a thermally cycled film.

The excitation lifetime can be estimated by measuring the dependence of the PA on the modulation frequency. For monomolecular recombination, the PA has the following dependence on the modulation frequency:

$$\Delta T \propto 1 / \sqrt{1 + (2\pi f \tau)} \qquad (1)$$

where f is the pump modulation frequency and $\tau$ is the excitation lifetime. The PA was linear with the pump intensity, indicating monomolecular recombination, the inset in Fig. 3 shows the frequency-dependence of the rms PA at 1.43 eV. The triplet lifetime was determined to be $\approx 2.5$ ms. A small fraction (15%) of triplet excitons decay more quickly ($\tau = 0.1$ ms). The PA spectrum was unchanged by the modulation frequency, indicating that it is due to excited state absorption of triplet excitons.

The H-PADMR spectrum, measured at 1.43 eV, consists of a single band centred at 1070 Gauss (g $\approx$ 2). Neither full-field nor half-field triplet powder patterns were detected. The λ-PADMR spectrum at 1070 Gauss is shown in Fig. 4. Two new

PA bands are observed, $P_1$ below 0.5 eV and $P_2$ at 1.93 eV. These bands have the same H-PADMR lineshape. $P_1$ and $P_2$ are assigned to polarons, charged excitations with spin-1/2. Surprisingly, the PA band at 1.43 eV also has a spin-1/2 resonance, which is narrower than that the other two bands (see below). We speculate that this signal may originate from triplet-radical interactions, wich results in quenching of triplet excitons by radicals or polarons. Such an interaction has been previously observed in fullerene films.13

The PA spectrum of the thermally cycled sample is shown in Fig. 5. The triplet PA band (T1) is eight times weaker tan in the glassy sample and two new PA bands appear at 1.93 and 0.3 eV. These PA bands have the same dependence on pump power and modulation frequency and are accompanied by a series of sharp IRAVs below 0.2 eV. The PA spectrum below 0.4 eV was measured using a FTIR spectrometer and is shown on an expanded scale in Fig 10. All these results are characteristic of polarons, charged excitations with spin-1/2. We accordingly assign PA bands to transitions of polarons and label the bands $P_1$ and $P_2$. The appearance of IRAVs substantiates the above assignment of the PADMR bands in the Figure 4 to polarons. Comparison on the PA below 0.5 eV shows that the polaron PA is approximately 10-15 times stronger in the thermally cycled sample than in the as-spun sample. The triplet lifetime is slightly longer in the as-spun sample (7.4 and 0.1 ms). Hence, the reduced triplet PA must be due to lower triplet generation rather than rapid combination. We are presently investigating the origin of triplet photogeneration in the glassy and $\beta$ phases of PFO. The frequency dependence of the polaron PA bands could be fit to a bimodal lifetime distribution of 6.8 and 0.3 ms.

The $\lambda$-PADMR spectrum of the thermally cycled sample is shown in Fig. 5(b). In addition to the polaron and triplet PA bands, there is a new band of the opposite sign with a peak at 1.3 eV. Figure 11 shows the H-PADMR of the thermally cycled sample, measured at the 0.35, 1.3, 1.43 and 1.93 eV. The H-PADMR spectra at 0.37, 1.3 and 1.93 eV have approximately the same width (full-width at half maximum is appr. 25-30 Gauss), whereas the triplet PA band has a much narrower PADMR spectrum (FWHM = 10 Gauss). In previous PADMR studies of conjugated oligomers and polymers, we have assigned a bleaching PADMR spectrum to bipolarons [14]. Magnetic resonance enhances recombination of polaron pairs, to the ground state if the polarons are of the opposite sign and to bipolarons, if the polarons have the same sign and the relaxation energy exceeds the Coulomb repulsion. We accordingly assign the positive PADMR band in Fig. 5(b) to magnetic resonant enhancement of bipolaron formation.

## 3. SUMMARY

In summary, we found that a new solid phase, the $\beta$ phase, is formed in PFO films upon thermal cycling from 80K to 300K. This phase consists of polymers chains with extended conjugation. Compared to amorphous samples prepared by spin-casting from solution, excitations of the ( phase are red-shifted and have a higher polarisability. PA measurements show that polarons are more easily formed in the $\beta$ phase. We suggest that increased conjugation enhances intrinsic charged photogeneration.

488

## Acknowledgements

The authors would like to thank the Dow Chemical Company for supplying the polyfluorene samples studied in this work. Research performed at the University of Sheffield was supported in part by the U.K. Engineering and Physical Sciences Research Council (GR/L80775) and the Royal Society (RS19025). Research performed at the University of Utah was supported by the U.S. Department of Energy under grant DOE FG-03-96 ER4540. This work was also supported by a NATO collaborative research grant (CRG973132).

## 4. REFERENCES

1. A Grice, D D C Bradley, M T Bernius, M Inbasekaran, W W Wu, E P Woo, Appl. Phys. Lett. **73**, 629 (1998).

2. M. Redecker, D. D. C. Bradley, M. Inbasekaran and E.P. Woo, Appl. Phys. Lett. **73**, 1565 (1998).

3. M Grell, D D C Bradley, G Ungar, J Hill, K S Whitehead, Macromolecules, at press.

4. M Grell, D D C Bradley, X Long, T Chamberlain, M Inbasekaran, E P Woo, M Soliman, Acta Polym. **49**, 439 (1998).

5. D D C Bradley, M Grell, X Long, H Mellor, A Grice, Proc. SPIE. **3145**, 254 (1997).

6. G Klaerner, R D Miller, Macromolecules , **31**, 2007 (1998).

7. T. W. Hagler, K. Parkbaz, K. F. Voss and A. J. Heeger, Phys. Rev. B **44,** 8652 (1991).

8. M. Pope and C. E. Swenberg, Electronic Processes in Organic Crystals (Oxford University Press, New York, (1982).

9. G. Cerullo, S. Stagira, M. Nisoli, S. De Silvestri, G. Lanzani, G. Kranzelbinder, W. Graupner and G. Leising, Phys. Rev. B **57**, 12806 (1998).

10. U. Lemmer, S Heun, R.F. Mahrt, U. Sherf, M. Hopmeier, U. Siegner, E. O. Gobel, K. Mullen and H. Bassler, Chem. Phys. Lett. **240**, 373 (1995).

11. S. A. Jenekhe and J. A. Osaheni, Science **265**, 765 (1994).

12. K. Petritsch, W. Graupner, G. Leising and U. Sherf, Synth. Met. **84**, 625 (1997).

13. C. A. Steren, H. Van Willigen, and M. Fanciulli, Chem. Phys. Lett. **245**, 244 (1995).

14. P. A. Lane, X. Wei, and Z. V. Vardeny, Physical Review Letters 76, 1544 (1996).

# SYNTHESIS AND CHARACTERIZATION OF THE 2:1 COPPER(II) COMPLEXES WITH 4-ARYLAZO-PYRAZOL-5-ONE DERIVATIVES

A.EMANDI [a], M.CALINESCU[a], S.IOACHIM[a], R.GEORGESCU[b] and
I. SERBAN[a]

[a]*Department of Inorganic Chemistry, University of Bucharest*
*Dumbrava Rosie 23, District 2, 70254, Bucharest, Romania*
[b]*Institute of Physics and Nuclear Engeneering, Bucharest*
*Romania*

## Abstract

A new series of Cu(II) coordination compounds with 4- (azo derivatives)-3-methyl –1-(2'-benzthyazolyl)-pyrazol-5-one in a ratio Cu/L=1/2 were synthesized. These complex combinations were characterized by elementary analyses, molecular weight, electric conductibility measurements, IR, UV-VIS and RPE spectra. The covalency parameters(alpha, beta, beta 1) were computed using the RPE spectra results and the energy of the electronic d-d transitions and the symmetry of the coordination polyhedra was established.

*Keywords* :  Copper (II), Coordination compounds, Azo dye, Pyrazol-5-one , IR, UV –VIS, RPE ,Piridinium derivatives

## 1. Introduction

Metal complexes of azo–compounds are of great importance to dye chemistry giving  dyeings with excellent fastness to light and washing [1-3].

Many  synthetic dyes available which can combine with a metal cation to form complexes have been patented ligands for the dyeing of olefinic polymer fibres, analytical reagents in the estimation of some metal ions [4-6].

Generally, it is considered that the stability of the metal complex with such ligand may be enhanced by introducing in electron donating on the aryl ring, and thus an electron density on the azo-group as one of the coordinating sites is considered to be increased [7].

From this point of view, the objectives of the present work were to prepare a new series of Cu (II) coordination compounds with 4-azoderivatives of 1-substituted-pyrazol-5-one which have different substituents on the aryl ring coupled as diazonium salt to the fourth position of 1-(2'-benzthyazolyl)-3-methyl-pyrazol - 5- one and to observe the effect of these substituents on the structure of the azo metal complexes. The structure of the metal complexes with azo dyes can be colligated with their physical properties.

F. Kajzar and M.V. Agranovich (eds.), Multiphoton and Light Driven Multielectron Processes in Organics: New Phenomena, Materials and Applications, 489–502.

The ligands used in the obtaining the 2:1 Cu (II ) complexes have been synthesized and characterized [8] and have the general formula :

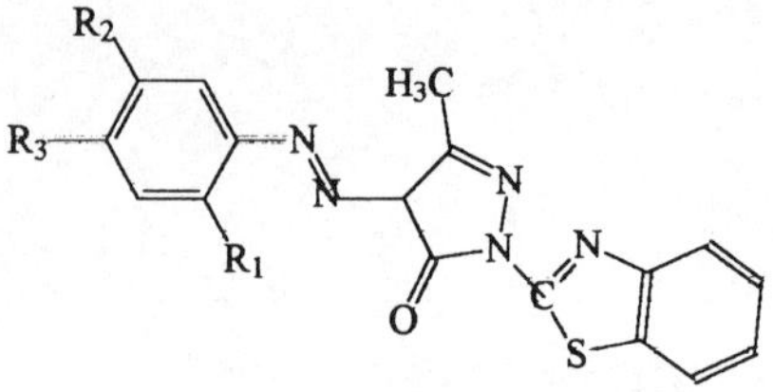

(A)

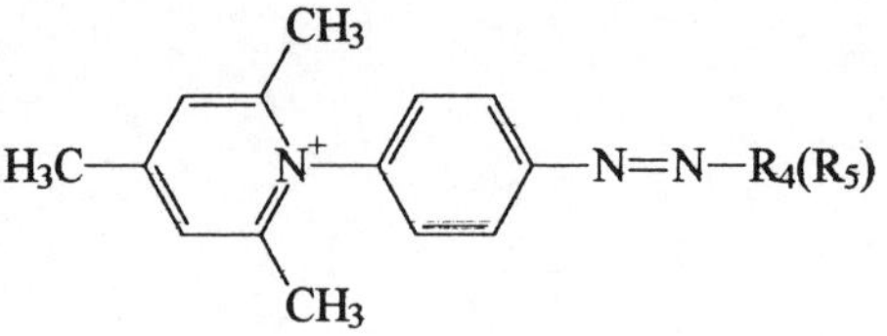

[ZnCl$_3$]$^-$

(B)

$R_1$=COOH ; $R_2$=$R_3$=H   (1)
$R_1$=OH     ;$R_2$=NO$_2$ ;$R_3$=H (2)
$R_1$=OH     ;$R_2$=Cl   ;$R_3$=H (3)
$R_1$=H      ;$R_2$=H    ;$R_3$=NO$_2$ (4)

$R_4$ =

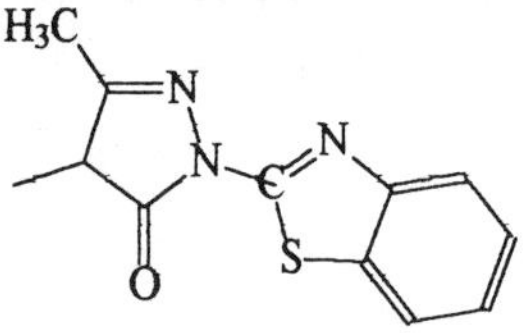

(5)

$R_5$ =

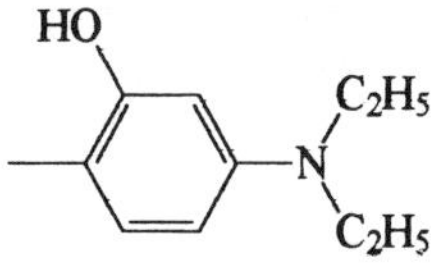

(6)

The 2:1 Cu(II) coordination compounds corresponding to the terdentate and dibasic ligands (H$_2$L$^1$), (1), (2), (3) labeled with (I), (II), (III) are tetagonally distorted octahedra and that corresponding to the bidentate and monobasic ligand HL$^2$(4), labeled with (IV) has distorted square-planar symmetry and those corresponding to the bidentate and monobasic ligands (5) and (6), labeled with (V) and (VI) have distorted tetrahedral symmetry.

All the complexes have obtained at room temperature and have been purified by succesive reprecipitations. We give the general reactions bellow:

Cu(OOCCH$_3$)$_2$.H$_2$O+3H$_2$O→[Cu(H$_2$O)$_4$]$^{2+}$$_{aq}$ + 2CH$_3$COO$^-$$_{aq}$

$$[Cu(H_2O)_4]^{2+}_{aq}+2H_2L^1+2CH_3COO^- \xrightarrow[-5H_2O]{2Bu_4NOH} [Cu(HL^1)_2].H_2O+2Bu_4NOOCCH_3$$

$$[Cu(H_2O)_4]^{2+}_{aq}+2CH_3COO^- + 2HL^2 \xrightarrow[-3H_2O]{} [Cu(HL^2)_2](CH_3COO)_2.H_2O$$

here $HL^2$ are the ligands (4) and (5).

$$Cu(H_2O)_4]^{2+}_{aq}+2CH_3COO^- + 2HL^2 +2Bu_4NOH \longrightarrow [Cu(L^2)_2].H_2O+2Bu_4NOOCCH_3$$

$$+H_2O$$

here $HL^2$ is the ligand (6)

## Experimental

All compounds and solvents were pure BDH grade chemicals. The ligands used
ere prepared according to the method [9,10]

### 1. PHYSICAL MEASUREMENTS

Elementary analysis were obtained by a CARLO ERBA EA 1108
pparatus. Copper was determined by volumetrical method.

Conductance measurements were obtained with a Radekis conductometer type
K-102/1.

Molecular weights were determinated in chloroform at $37^0C$ with a Mechrolab
Model 301A vapor pressure osmometer. Concentrations of the solutions were in the range
$0^{-3}$-$10^{-4}$ M.

The IR spectra were run with a Perkin Elmer FT-IR Spectrophotometer in the
ange 400-200 cm$^{-1}$, in KBr pellets.

The electronic spectra of all compounds were obtained by diffuse-reflectance
echnique, dispersing the sample in MgO, with a Specord M400 Carl Zeiss Jena
pectrophotometer.

The RPE spectra were obtained at room temperature in microcrystalline powder
f the complexes in the X-band region with a ART-6 spectrophotometer. The RPE
arameters were determined directly from the spectra, with the magnetic field calibrated
sing a standard Mn $^{+2}$ in Ca(OH) $_2$ .

### 2. PREPARATION OF THE COMPOUNDS

ll complexes have been synthesized by dissolving $(1.10^{-2}$ mol) ligand in 50 ml EtOH at a
H=7.5 adjusted from 4.5 in the free ligand to the 7.5 value with an aqueous solution of
tra-n-butylammonium hydroxide (3%). This solution has been stirred at room
emperature and an aqueous solution of $(0.510^{-2}$ moli) $Cu(OOCH_3)_2$ .$H_2O$ was added drop
vise to wise. The mixture of the reaction was stirred for 0.5 hours and the brown solid
omplexes were filtered off, recrystallised from EtOH and dried at room temperature ,with
yield ($\eta \approx$ 60-70%).

## 3. Results and discussions

The elementary analysis (Table 1) indicates a ratio $Cu/H_2L^1(HL^2)=1/2$. The molecular weights (Table 1) for all complexes correspond to a monomeric structure. The conductance values of the complexes in acetonitrile, solution $10^{-4}$mol.(Table 1) are in good agreement with the data obtained by previous authors, and suggest a non-electrolyte behaviour for the complexes (I), (II), (III), a weak electrolyte behaviour for the complex (VI) and and a 1:2 electrolyte behaviour for the complexes (IV) and (V).

TABLE 1. Analytical,conductance data and molecular weight for the 2:1 Cu(II) complexes

| Compound | Mol. Weight | C% | H% | N% | Cu% | $\Lambda$ $\Omega^{-1}cm^2mol^{-1}$ |
|---|---|---|---|---|---|---|
| I | 837,5 | 51,58* (51,69)** | 3,10 (3,40) | 16,71 (16,70) | 7,58 (7,52) | 84 |
| II | 871,5 | 46,81 (46,92) | 2,75 (2,90) | 19,27 (19,21) | 7,28 (7,20) | 75 |
| III | 850,5 | 47,97 (47,840) | 2,82 (2,90) | 16,46 (16,50) | 7,46 (7,48) | 82 |
| IV | 959,5 | 47,52 (47,61) | 3,33 (3,40) | 17,5 (17,55) | 6,61 (6,67) | 250 |
| V | 1405,1 | 48,67 (48,71) | 3,84 (3,90) | 11,9 (11,97) | 4,51 (4,59) | 267 |
| VI | 1060,5 | 54,31 (54,40) | 5,65 (5,70) | 10,56 (10,59) | 5,98 (5,91) | 87 |

*calculed, **experimental

## 3.1. VIBRATIONAL SPECTRA

The IR spectra of the azo-derivatives and their 2:1 Cu(II) coordination compounds are too complex but a comparision of the spectrum with some previous work [11-14] has enable to assign the more characteristic bands.

The observed bands of the ligands and their complexes, listed in Table 2 and Table 3, were assigned in according to the literature data [15] which indicate that in the colorants involving " azo bonds " theese bonds do not occur and the bridging is really due to a hydrazone bond.

TABLE 2. IR spectral data for ligands and their complexes

| Assignements | L(1) | I | L(2) | II | L(3) | III |
|---|---|---|---|---|---|---|
| $v(OH...N)$ | 3420m | - | 3425m | - | 3415m | - |
| $vOH(H_2O)$ | - | 3450s | - | 3447s | - | 3451s |
| $vNH$ | 3200m | 3180w | 3200w | 3100w | 3230w | 3100w |
| $vC=O$ of | 1640s | 1590m | 1675m | 1626i | 1675s | 1641s |

| Pyrazolone | | | | | | |
|---|---|---|---|---|---|---|
| vasCOO- | - | 1610i | - | - | - | - |
| vOCO | - | 1324m | - | - | - | - |
| vN=C-C=N | 1590s | - | 1598I | 1583m | 1599m | 1584I |
| δNH+vCN | 1540s | 1533s | 1543I | 1518i | 1525v.s | 1522s |
|  |  |  |  |  | 1333m | 1341w |
| vN-N=C | 1440m | 1486sh | 1444I | 1438m | 1442m | 1466I |
|  | 1430sh | 1441i | 1426I | 1410v.w | 1407w | 1443sh |
| vC-OH(COOH) | 1272s | - | - | - | - | - |
|  | 1283sh |  |  |  |  |  |
| vC-O(OH phenolic) | - | - | 1281I | 1278i | 1256m | 1275m |
| vCu-O | - | 477m | - | 455m | - | 460m |
| vCu-N | - | 592m | - | 580m | - | 585m |

(vs=very strong, s=strong, i=intense, m=medium, w=weak, vw=very weak)

TABLE 3. IR spectral data for ligands and their complexes

| Assignements | L(4) | IV | L(5) | V | L(6) | VI |
|---|---|---|---|---|---|---|
| v(OH...N) | - | - | - | - | 3450w | - |
| vOH(H₂O) | - | 3448s | - | 3432s | - | 3438s |
| vC=Oof pyrazolone ring | 1642i | 1624i | 1634s | 1610m | - | - |
| vN=C-C=N | 1570m | 1576i | 1556m | - | 1637s | 1634vs |
|  |  |  |  |  | 1621s | 1612vs |
| δNH+vCN | 1526s | 1538I | 1518s | 1515s | 1562s | 1565m |
|  | 1314m | 1317vw | 1321s | 1320m | 1320s | 1347m |
| vN-N=C | 1465w | 1462m | 1489w | 1462m | 1475I | 1471m |
|  | 1433m | 1426m |  | 1420m | 1413sh |  |
| vC-O(OH phenolic) | - | - | - | - | 1269i | 1273I |
| vCH₃COO- | - | 1642i | - | 1633i | - | - |
| vClO₄⁻ | - | - | - | - | 622i | 625I |
| vCu-O | - | 495m | - | 473m | - | 467m |
| vCu-N | - | 580m | - | 585m | - | 592m |

(a) 4000-2000 cm⁻¹.In this region the free azoderivatives show intramolecular hydrogen bonding (N-H...O) near 3200 cm⁻¹ ,whilst the ligands (1),(2),(3) and (6) exibit a supplimentary broad absorption band near 3400 cm⁻¹ that is assigned to intramolecular hydrogen bonding (O-H... N).

In the coordination compounds the band near 3400 cm⁻¹ disappears and they exhibit a new very strong band near 3450 cm⁻¹ which was assigned to v OH of the cristallysed water molecules within the complexes. In the spectra of the metal complexes the band near 3200cm⁻¹ in the ligand shifts to around 3100 cm⁻¹ due to the formation of the metal complexes.The lowering of the frequency of the νN-H may be attributed to a metal nitrogen coordinate bond.

(b) 2000-300 cm$^{-1}$.In this region the free ligands show a strong absorbtion band near 1640-1670 cm$^{-1}$which was assigned to νC=O of the pyrazolone ring,a very intense band at 1590 cm$^{-1}$ and near 1540cm$^{-1}$ which were assigned to the skeleton νN=C-N=C and (δNH+νCN),respectivelly. The two intense or medium bands near 1440-1460 cm$^{-1}$ and 1430-1410 cm$^{-1}$ were assigned to νN-N=C.

The IR spectra of the complexes show a considerable shift (20-50 cm$^{-1}$ ) of the carbonyl stretching of the pyrazolone ring to a lower frequency, indicating a decrease in the stretching force constant of >C=O as a consequence of coordination through its oxygen.

The N=C-N=C and N-N=C stretching vibrations are shifted ( 10-20 cm$^{-1}$) and (20-40 cm$^{-1}$ ) respectivelly upon metal complexation

The strong band of the ligand (1) near 1272 cm$^{-1}$, 1283sh cm$^{-1}$ assigned to ν C-OH (COOH) disappears in the complex (I) but the two new bands rise in the complex near 1610 cm$^{-1}$ and 1324 cm$^{-1}$ assigned to ν $_{as}$ COO$^-$ and ν OCO upon metal complexation

In the ligands (2), (3) and (6) the bands which correspond to ν C-O( phenolic OH) are shifted from around ( 1280 cm$^{-1}$-1250 cm$^{-1}$) to 1270cm$^{-1}$ in the complexes as a consequence of coordination through the oxygen atom after deprotonation .

A strong evidence regarding the bonding of new νCu-O and νCu-N [16] of the ligand has been provided by the appearance of an absorption bands in the region 450-520 cm$^{-1}$ and 550-590 cm$^{-1}$, respectively.

The above arguments indicated that the structural formulas submited in Figure1 became more significant.

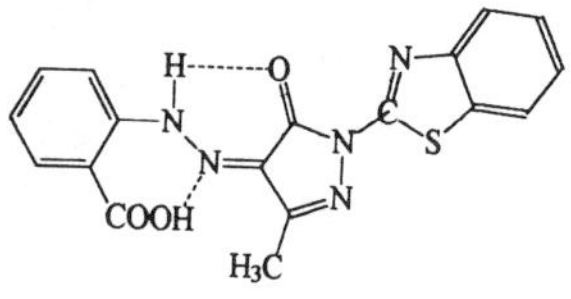

(a)

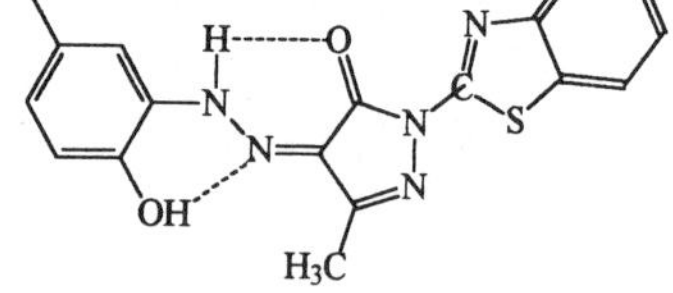

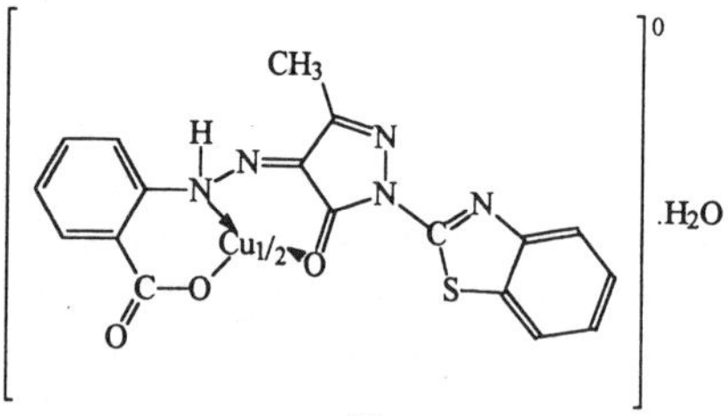

(b)

(c)

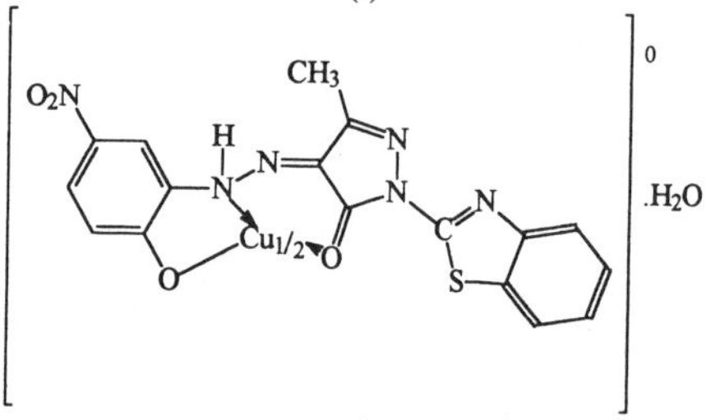

(d)

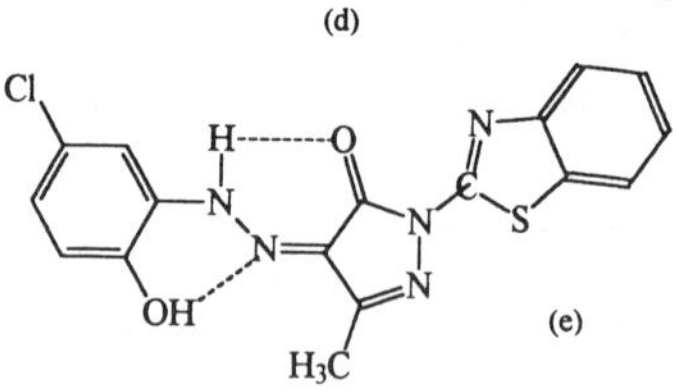

(e)

(f)

(g)

(h)

(i)

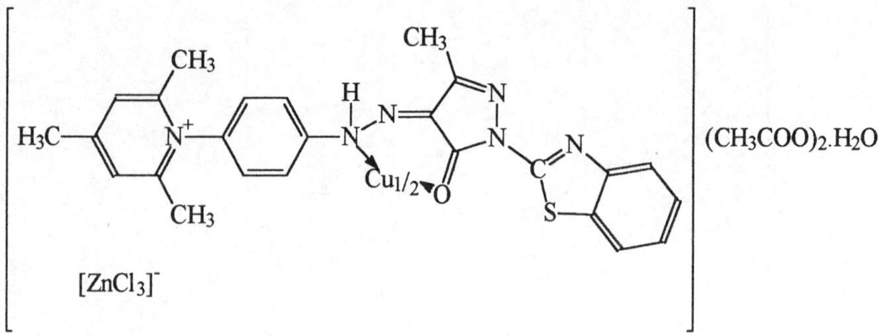

Figure 1
The ligands and their complexes formulas: (a) Ligand(1), (b) Complex(I), (c) Ligand(2), (d)Complex(II),(e) Ligand(3),(f) Complex(III),(g) Ligand(4),(h) Complex(IV), (i)Ligand(5), (j)Complex(V), (k) Ligand(6), (l) Complex(VI).

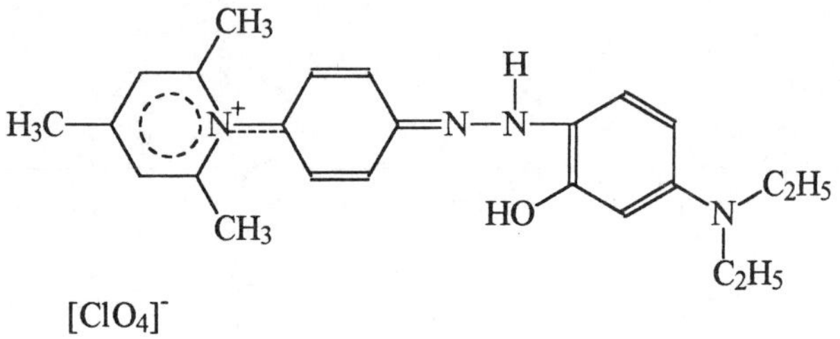

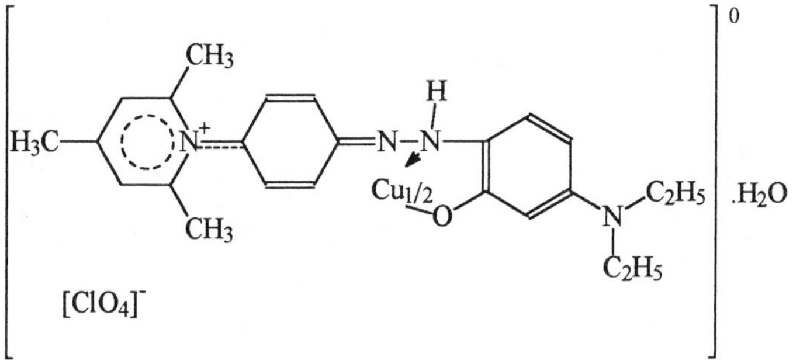

## 3.2. ELECTRONIC ABSORPTION SPECTRA AND PARAMAGNETIC RESONANCE

The ultraviolet-visible absorption spectra of the complexes II and III in solid state, consist of three bands; their attributions were made assuming an octahedral symmetry, with tetragonal distortion [17-19](table 4):

TABLE 4. Electronic absorption spectra for I and II copper(II) complexes

| Complex | $\nu_{max}(cm^{-1})$ | Assignements |
|---------|---------------------|--------------|
| II | 9090 | $z^2 \rightarrow x^2-y^2$ $(^2B_{2g} \rightarrow ^2B_{1g})$ |
| | 13157 | $xy \rightarrow x^2-y^2$ $(^2A_{1g} \rightarrow ^2B_{1g})$ |
| | 14600 | $xz,yz \rightarrow x^2-y^2$ $(^2E_g \rightarrow ^2B_{1g})$ |
| III | 10638 | $z^2 \rightarrow x^2-y^2$ $(^2B_{2g} \rightarrow ^2B_{1g})$ |
| | 11770 | $xy \rightarrow x^2-y^2$ $(^2A_{1g} \rightarrow ^2B_{1g})$ |
| | 15625 | $xz,yz \rightarrow x^2-y^2$ $(^2E_g \rightarrow ^2B_{1g})$ |

The RPE powdered spectra, recorded in X band (Figure 2) were interpreted in terms of a Spin Hamiltonian for a system with axial symmetry [18,20,21]:

$$H = g_{\parallel}\beta H_z + g_{\perp}\beta(H_xS_x + H_yS_y) + AI_zS_z \qquad (1)$$

where $g = (g^2_{\parallel} \cos^2\theta + g^2_{\perp}\sin^2\theta)^{1/2}$, "$\theta$" being the angle between the steady magnetic field and the crystalline field z axis.

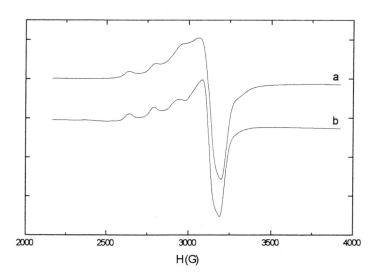

Figure 2. The RPE spectra of the complexes III (a) and II (b)

The experimentally determined magnetic parameters are listed in table 5:

Table 5. Magnetic parameters for II and III copper(II) complexes

| Complex II | Complex III |
|---|---|
| $g_{\parallel}$=2.205 | $g_{\parallel}$=2.234 |
| $g_{\perp}$=2.036 | $g_{\perp}$=2.013 |
| $A_{\parallel}$=150G | $A_{\parallel}$= 160G |

Calculations of $Cu^{2+}$ bonding parameters using these values give physically reasonable results only if A is taken as negative.

The EPR measurements for the two complexes show $g_{\parallel} \rangle g_{\perp} \rangle g_e$ ,which is in accordance with an octahedral symmetry, distorted by elongation along z axis ( axis of the nitrogen atoms).

The experimentally magnetic parameters and the energy of electronic d-d transitions have been used for estimating the nature of the chemical bonding between the copper atom and the nitrogen and oxygen atoms of the ligands in complex compounds.Thus, on the basis of the molecular orbital theory, we can calculate the covalency parameters $\alpha$, $\beta$ and $\beta_1$ as follows [18,20-22].

$$g_{\parallel} = 2 - (8\lambda_0/\Delta E_{xy})\alpha^2\beta^2 \tag{2}$$

$$g_{\perp} = 2 - (2\lambda_0/\Delta E_{xz,yz}) \alpha^2\beta_1^2 \tag{3}$$

$$A_{\parallel} = P[-(4/7) \alpha^2 - k + ( g_{\parallel} - 2) + 3/7(g_{\perp} - 2)] \tag{4}$$

where $\lambda_0$ is the spin-orbit coupling constant for the free $Cu^{2+}$ion (-828cm$^{-1}$), k is the Fermi contact term and characterizes the isotropic (s-electron) contribution to the hyperfine interaction; $P = 2\gamma\beta\beta_N\langle r^{-3}\rangle = 0.036$cm$^{-1}$; $\Delta E_{xy} = \Delta(E_{x^2-y^2} - E_{xy})$; $\Delta E_{xz,yz} = \Delta (E_{x^2-y^2} - E_{xz,yz})$.

The coefficients $\alpha^2$, $\beta_1^2$ and $\beta^2$ characterize the in-plane $\sigma$-bonding, in-plane $\pi$ bonding and out-of-plane $\pi$ bonding of the $Cu^{2+}$ion, respectively. $\alpha^2$ can take the values between 0.5 and 1, corresponding to pure covalent and pure ionic bond, respectively.

Using the approximation: $(4/7 + k_0) = 1$ in the Eq.(4), we can calculate $\alpha^2$:

$$\alpha^2 = -(A/P) + (g_{\parallel} - 2) + 3/7(g_{\perp} - 2) + 0.004 \tag{5}$$

The obtained value for $\alpha^2$ is then utilized for the calculation of $\beta$ and $\beta_1$ values using the relation (2) and (3).

We so obtained for molecular parameters the values given in the Table 6.

TABLE 6. Molecular-orbital coefficients for complexes II and III

| Complex | $\alpha^2$ | $\beta^2$ | $\beta_1^2$ | $\lambda(cm^{-1})$ |
|---------|------------|-----------|-------------|--------------------|
| II | 0,709 | 0,686 | 0,779 | -403 |
| III | 0,737 | 0,562 | 0,560 | -343 |

The values obtained for $\alpha^2$ indicate an important covalent character of metal-ligand bond; the result is a smaller hyperfine interaction and a strong reduction of the spin-orbit coupling constant. The bonding parameter $\alpha^2$ is not strongly affected by the substituent group on aromatic ring, but the parameters relating to $\pi$-bonding ($\beta^2$ and $\beta_1^2$) descrease for the complex compound containing -Cl as substituent. These results are in agreement with stronger in-plane $\pi$-bonding and out-in-plane $\pi$-bonding for the complex III than for the complex II.

The complex IV containing $CuN_2O_2$ chromophore, exhibits a broad structured band from 10000 to 18000 $cm^{-1}$.

In a square-planar symmetry, distorted to $D_{2h}$, the three maxima of this broad band may be attributed to the following transitions [23]:

$$xy \rightarrow x^2-y^2 \; (^2B_{1g} \rightarrow {}^2A_{1g}) \qquad 12200 \; cm^{-1}$$
$$z^2 \rightarrow x^2-y^2 \; (^2A_{2g} \rightarrow {}^2A_{1g}) \qquad 14705 \; cm^{-1}$$
$$xz \rightarrow x^2-y^2 \; (^2B_{2g} \rightarrow {}^2A_{1g}) \qquad 15384 \; cm^{-1}$$
$$yz \rightarrow x^2-y^2 \; (^2B_{3g} \rightarrow {}^2A_{1g}) \qquad 17636 \; cm^{-1}$$

The experimental values of magnetic parameters, calculated from RPE spectrum (Figure 3) are: $g_\parallel = 2,205$ and $g_\perp = 2,036$ are also in agreement with a square-planar symmetry[20].

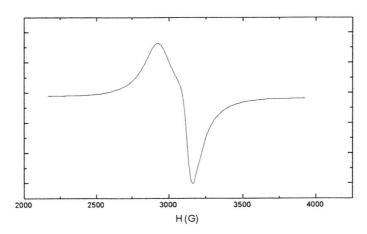

2000      2500      3000      3500      4000

H (G)

Figure 3. The RPE spectrum of the complex IV

From these data we obtained: $\lambda = -312$ cm$^{-1}$ and $\lambda/\lambda_0 = 0.37$

The complexes V and VI exhibit four absorption bands in near infrared region, in according to a tetrahedral symmetry, distorted to $D_{2d}$ (Table 7).

TABLE 7. d-d absorption bands for copper(II) complexes (V and VI)

| Complex | $v_{max}$(cm$^{-1}$) | Assignements |
|---------|----------------------|--------------|
| V | 9090 | $z^2 \rightarrow xy$ ($^2A_1 \rightarrow ^2B_2$) |
| | 7072 | $x^2$-$y^2 \rightarrow xy$ ($^2B_1 \rightarrow ^2B_2$) |
| | 5896 | $xz \rightarrow xy$ $\Gamma_6(^2E_2) \rightarrow \Gamma_7(^2B_2)$ |
| | 5192 | $yz \rightarrow xy$ $\Gamma_7(^2A_1) \rightarrow \Gamma_7(^2B_2)$ |
| VI | 9090 | $z^2 \rightarrow xy$ ($^2A_1 \rightarrow ^2B_2$) |
| | 7246 | $x^2$-$y^2 \rightarrow xy$ ($^2B_1 \rightarrow ^2B_2$) |
| | 5924 | $xz \rightarrow xy$ $\Gamma_6(^2E_2) \rightarrow \Gamma_7(^2B_2)$ |
| | 5219 | $yz \rightarrow xy$ $\Gamma_7(^2A_1) \rightarrow \Gamma_7(^2B_2)$ |

The $^2D$ state of the copper(II) free ion is split into four states of symmetry $^2A_1$, $^2B_1$, $^2B$ and $^2E$ by a $D_{2d}$ field. The tetrahedron is compressed along z axis so the ground state will be $^2B_2$, corresponding to a hole in the $d_{xy}$ orbital [24-26]. Because the spin-orbit coupling constant is large for the free ion, the splitting of $^2E$ state can be observed (Figure 4)[24].

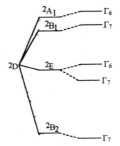

Figure 4. The effect of spin-orbit interaction on the energies of the d orbitals in $D_{2d}$ symmetry

In fact, the two bands observed at 5896 cm$^{-1}$, 5192 cm$^{-1}$ and 5924 cm$^{-1}$, 5219 cm$^{-1}$ for the complexes V and VI, respectively, correspond to the transitions from the ground state to $\Gamma_6$ and $\Gamma_7$ terms.

The values for the magnetic experimental parameters, determined from the RPE spectra (Figure 5) are: g$_{||}$= 2,225 and g$_\perp$= 2,014 for the complex V; g$_{||}$= 2,207 and g$_\perp$= 2,012 for the complex VI.

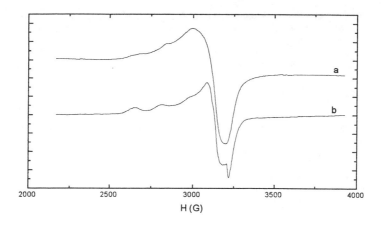

Figure 5. The RPE spectra of the complexes VI (a) and V (b)

Using these magnetical and optical data[24-25], we obtained the spin-orbit coupling values: $\lambda = -198$ cm$^{-1}$(complex V) and $\lambda = -187$ cm$^{-1}$(complex VI), in accordance with strong covalent metal-ligand bonds.

References

1. Folli,U., Iarossi,D. and Vivarelli,P. (1980) Effect of the Structure of Metallized Pyrazole-monoazo Dyes on their Chromatic Behaviour,*J.Soc.Dyers and Colors* **96**,414-417.
2. Shukla,P.R. and Srivastava C.(1981) Some transition metal complexes derived from the heterocyclic ligands 3-methyl-4-arylazopyrazole-5-ones,*J.Indian Chem.Soc.* **LVIII**,937-939. Koraiem,M.I.A., Khalil,Z.H. and Abu El-Hamed,R.M. (1990) Synthesis of some β- and γ- substituted azotrimethine cyanine dyes ,*Dyes and Pygments* **13**,197-204.
3. Sastry,M.S., Ghose,R., and Ghose,A.K. (1994) Synthesis and spectroscopic studies of simple and mixed complexes of nickel(II) and zinc(II) with an azo dye , *Synth.react. inorg. met.-org.chem* **24**,1213-1225.
4. Li Ling Ying, Gui Ming-De and Zhao Ya-Qiu (1995) Reversed-phase HPLC determination of Co(II),Ni(II) and Fe(III) as their 2-(2-thiazolylazo)-5-dimethylaminophenol chelates,*Talanta* **42**,89-92.Caballol ,R., Ribo,M.J., Trull,R.F. and Valles,A. (1987) Reactivity of Pyrole Pigments.Part 9 [1] MINDO/3 calculations on dipyrrolic partial models of bile pigments,*Monatshefte fur chemie* **118**,993-1010.
. Negoiu,D., Emandi,A., Paruta,L., Calinescu,M., Marchidan,R., (1996) Combinatii complexe ale Pd(II) si Pt(II) cu derivati ai arilidelor acidului 3-hidroxi-2-naftoic, *Rev.Chimie* **47**- 9 , 12 .
. Negoiu,D., Emandi,A., Rosu,T., Marchidan,R. , Combinatii complexe ale Fe(II), Fe(III), Co(II), Co(III) cu derivati ai arilidelor acidului 2-hidroxi-3-naftoic, *Rev.Chimie*, **47**-8, 32.

502

7.  Kai,F., Takeshita,H., Sukimoto,S., Tamaoku,K. (1981) Preparation of new ligands o-(4,5-dimethyl-2-thiazolylazo)phenols and their metal complexes, *J.inorg.nucl.chem.* **43**,3013-3015.
8.  Emandi,A.,Serban,I. and Bandula,R. (1999) Synthesis of some new solvatochromic 1(4)-substituted pyrazol-5-one azo derivatives, *Dyes and Pygments* **41**,63-77.
9.  Emandi,A., Tarabasanu,C., Sebe,I., Paruta,L. (1997) Azo Dyes of 1-(2'-benztiazolil)-3-methyl-5-pirazolone, *Rev.Chimie*, **48** -11,7-15.
10. Emandi,A., Tarabasanu,C., Gheorghe,P., Gheorghe,S., New Azo Dyes of N-(4'-aminofenilen)-2,4,6 trimethylpiridinium tetrafluoroborat *Rev.Chimie*, **49**-1, 8-17.
11. Abdel-Megeed,M.F., El-Hevnavy,G.B., Habib,A.M., El-Borai,M.A. (1987) Studies of pyridylpyrazolone system III.Relation between fastness properties and electronic spectra of some new azo dyes containing pyridylpyrazolone nucleus,*Kolorisztikai Ertesito* **1** ,1-8.
12. Buyuktasa,S.B., Serin,S. (1994) Synthesis of two new vic-dioxime ligands containing azo group and their complexes wth Ni(II) and Cu(II) ,*Synth.react.inorg.met.-org.chem.* **24**,1179-1190.
13. Agashe,S.M. and Jose,I.C. (1977) Characteristic vibrations of-N(CH$_3$)$_2$ and –N+(CH$_3$)$_3$ groups in dimethyl aminophenols and their methiodides ,*J.Chem.Soc.Faraday Trans 2* ,**73** ,1232-1237.
14. Whitaker,A. (1995) Crystal structure of azo colorants derived from pyrazolone:a review ,*J.Soc.Dyers and Colors* **111**,66-71.
15. Mahapatra ,B., Kar,K.S. (1991) Polymethallic complexes.Part-XXXI.Complexes of Cobalt-,Nickel-,Copper-,Zinc-,Cadmium-and Mercury(II) with chelating azo dyes ligands 1-(2'-hydroxynaphtlyl-1')-azo-2-hydroxybenzene, 3-(3'-acetoacetanilido)azo-1-hydroxybenzene and 3-(8'-hydroxyquinonyl-5')-azo-1-hydroxybenzene, *J.Indian Chem. Soc.* **68**, 542-544.
16. Mahapatra,B., Patent,B.K. and Satpathy,K.D. (1989) Polymethallic complexes.Part- XXI.Complexes of Cobalt-,Nickel-,Copper-,Manganese-,Zinc-,Cadmium- and Mercury-(II) with bis-bidentate ON-NO donor azo-dye ligands, *J.Indian Chem. Soc.* **66**,820-822.
17. Yokoi,H. and Isobe,T. (1969) ESR and optical absorption studies of copper(II) complexes of ethylenediamine and its alkyl derivatives ,*Bulletin of the chemical society of Japan* **42**, 2187-2193.
18. Lorenc,A.J. and Siegel,I. (1966) Paramagnetic resonance of copper in amorphous and polycrystalline GeO$_2$ ,*J.of Chem. Phys.* **45**,2315-2395.
19. Negoiu, D., Calinescu, M., Emandi, A. and Badau, E. (1999) Synthesis and structural characterization of Cr(III), Co(II), Ni(II) and Cu(II) complexes with salicylaldehyde-N-[4-(4'-chlorophenylene sulfonyl)benzoyl]hydrazone, *Russ. J. Coord. Chem.*, **25**-1, 36-41.
20. Wertz, J.E. and Bolton, J.R. (1972) *Electron Spin Resonance*, Mc Graw-Hill Book Company, 334-336
21. Yang,Y., Pogni,R., Basosi,R. (1989) Mixed –ligand complexes of Cu(II)-1,10-o-phenanthroline and its analogues characterized by computer-aided electron spin resonance spectroscopy ,*J.Chem. Soc.,Faraday Trans. 1*, **85**,3995-4009.
22. Yokoi,H., Sai,M. and Isobe,T. (1969) ESR studies of bis-(N,N-diethylethylenediamine)copper(II) perchlorate, *Bulletin of the chemical society of Japan* **42** ,2232-2238.
23. Desjardins, R.S., Wilcox, E.D., Musselman, L.R. and Solomon, I.E. (1987) Polarized ,single-crystal,electronic spectral studies of Cu$_2$Cl$_6^{2-}$:excited-state effects of the binuclear interaction, *Inorg.Chem.* **26**,288-300.
24. Ferguson,J. (1964) Electronic absorption spectrum and structure of CuCl$_4^-$ *The J. of Chem.Physics* **40**, 3406-3410.
25. Sharnoff,M. (1965) Electron paramagnetic resonance and the primarily 3d wavefunctions of the tetrachlorocuprate ion, *The J. of Chem. Physics* **42**, 3383-3394.
26. Weakliem,A.H. (1962) Optical spectra of Ni$^{2+}$, Co$^{2+}$and Cu$^{2+}$ in tetrahedral sites in crystal, *The J. of Chem. Physics* **36**, 2117-2140.

# A STUDY OF RYDBERG BANDS IN THE PHOTOABSORPTION SPECTRA OF SANDWICH ORGANOMETALLICS: THE FIRST STEP TO THE INVESTIGATION OF MULTIPHOTON PROCESSES IN THE ORGANOMETALLIC MOLECULES

S. Y. KETKOV

*G.A. Razuvaev Institute of Organometallic Chemistry of the Russian Academy of Sciences*
*49 Tropinin St., Nizhny Novgorod 603600, Russian Federation*

## 1. Introduction

Sandwich complexes of transition metals represent one of the most important classes of organometallics. The high chemical and catalytic reactivities of these compounds have stimulated interest in their electronic structures. The nature and the order of the highest occupied molecular orbitals (MOs) in sandwich molecules have been established on the basis of MO calculations, magnetic measurements and photoelectron spectroscopy. As an example, a qualitative MO scheme derived for bis($\eta^6$-benzene)chromium on the basis of a MS $X_\alpha$ MO calculation [1] and the photoelectron data [2] is given in Figure 1.

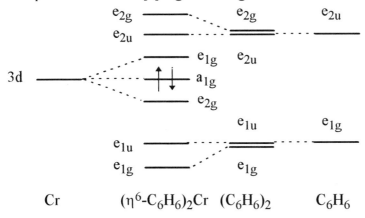

*Figure 1.* Molecular orbital diagram for $(\eta^6\text{-}C_6H_6)_2Cr$.

*F. Kajzar and M.V. Agranovich (eds.), Multiphoton and Light Driven Multielectron Processes in Organics: New Phenomena, Materials and Applications, 503–512.*
© 2000 *Kluwer Academic Publishers. Printed in the Netherlands.*

The molecule of $(\eta^6\text{-}C_6H_6)_2Cr$ can be considered under the $D_{6h}$ or $D_{\infty h}$ point group. The metal d orbitals split into three sets; $d_{z^2}$ ($a_{1g}$, $\sigma^+_g$), $d_{xz}$, $_{yz}$ ($e_{1g}$, $\pi_g$) and $d_{x^2-y^2}$, $_{xy}$ ($e_{2g}$, $\delta_g$). The $d_{z^2}$ orbital is essentially non-bonding and forms the highest filled MO of the compound. The MO diagrams for metallocenes and mixed sandwiches are analogous [3, 4]. The molecular symmetry of the mixed derivatives bearing two different carbocycles is formally very low ($C_1$ or $C_s$). Nevertheless, the electronic structures of these complexes are described well by the $C_{\infty v}$ point group [5].

In contrast to the occupied MOs, the vacant levels of sandwich molecules have been little investigated. The information on the energies and symmetries of electronic excited states would be of vital importance when studying multiphoton processes. In recent years, considerable interest has been focused on the multiphoton excitation of sandwich organometallics [6-10]. It was impossible, however, to use the resonance-enhanced processes since the frequencies of one-photon electronic transitions were unknown.

Information on the valence-shell vacant MOs could be obtained from the electronic absorption spectra in the condensed media. Unfortunately, the spectra of sandwich complexes in the solid and solution phases consist, as a rule, of very broad structureless unidentified absorption bands. The corresponding electronic excited states can not serve as intermediate levels in the resonant multiphoton processes.

On the other hand, the gas-phase spectra of sandwiches may reveal narrow Rydberg transitions originating at the non-bonding $d_{z^2}$ orbital. These spectra have been measured earlier for bis($\eta^6$-benzene)chromium and some metallocenes [11-13] but no concrete assignments were made. Rydberg states can play an important role in the resonance-enhanced multiphoton absorption and ionization of polyatomic molecules. So we decided to study the electronic absorption spectra of transition-metal sandwich complexes in the gas phase.

The main types of the sandwiches investigated are shown in Figure 2. These are metallocenes, bisarene compounds and mixed sandwich derivatives with five, six, seven, and eight -membered carbocycles. It was found that, indeed, the spectra of sandwich complexes measured in the gas phase differ strongly from those recorded in a solution owing to the presence of intense Rydberg bands [14-26]. Sandwich compounds appear to be the first class of organometallics which reveal clearly defined Rydberg transitions in the photoabsorption spectra. Because of very low $d_{z^2}$ ionization energies (5-7 eV), the first Rydberg excitations in the spectra of the complexes considered lie in the visible and near UV regions which is an unusual phenomenon for polyatomic molecules. A review of the results obtained is presented here.

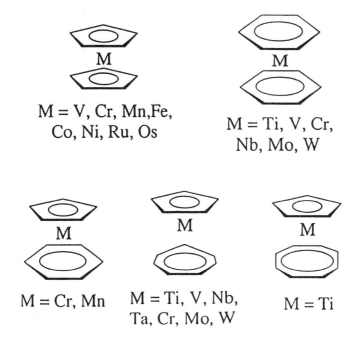

M = V, Cr, Mn,Fe,
Co, Ni, Ru, Os

M = Ti, V, Cr,
Nb, Mo, W

M = Cr, Mn

M = Ti, V, Nb,
Ta, Cr, Mo, W

M = Ti

Figure 2. The main types of the sandwich compounds studied.

## 2. Rydberg Transitions in the Bisarene Complexes

The electronic absorption spectra of $(\eta^6$-Arene$)_2$M (M = V, Cr, Mo, W, Arene = benzene and its alkyl-substituted derivatives) reveal sharp bands which disappear on going from the gas phase to the condensed media. Such behavior is indicative of Rydberg nature of the corresponding transitions [27]. As an example, the spectra of bis($\eta^6$-benzene)chromium are given in Figure 3. Indeed, two Rydberg series can be revealed in the gas-phase spectrum. The frequencies of the higher series members obey the well-known Rydberg formula

$$\nu_n = I - R / (n-\delta)^2 = I - T, \tag{1}$$

where $I$ is the ionization limit, R is the Rydberg constant, n is the principal quantum number, $\delta$ is the quantum defect and $T$ the term value. The convergence limit of the Rydberg series (5.459 ± 0.004 eV [24]) agrees very well with the ionization energy corresponding to the $3d_{z^2}$ orbital (5.46 eV [28]).

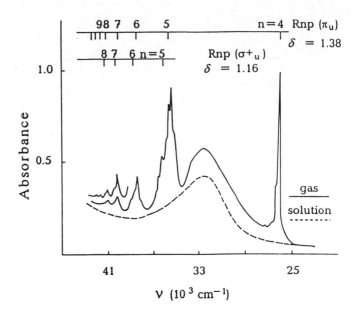

*Figure 3.* The absorption spectra of $(\eta^6\text{-}C_6H_6)_2Cr$ in the gas phase and in pentane solution.

On the basis of the $\delta$ values (1.38 and 1.16), the two series were unambiguously assigned to the excitations terminating at the Rydberg p orbitals. The Rydberg p levels of $(\eta^6\text{-}C_6H_6)_2Cr$ form two sets; $p_{x,y}$ ($e_{1u}$, $\pi_u$) and $p_z$ ($\sigma^+_u$). The study of methylated derivatives of bis($\eta^6$-benzene)chromium [26] revealed the molecular core splitting of low-lying members of the long-wavelength series. This series corresponds, therefore, to the $3d_{z^2} \rightarrow Rnp_{x,y}$ transitions while another one arises from the $3d_{z^2} \rightarrow Rnp_z$ excitations. Rydberg p transitions in $(\eta^6\text{-}C_6H_6)_2Cr$ [14, 24] and some methyl-substituted complexes [26] show vibronic structures arising from excitations of the metal-ligand and CH vibrations in Rydberg states. In addition to the Rnp bands, the absorption spectra of methylated derivatives [26] reveal the features corresponding to the $3d_{z^2} \rightarrow Rns$ and $3d_{z^2} \rightarrow Rnd$ ($n = 4, 5$) promotions.

Rydberg s, p, and d transitions were also found in the gas-phase absorption spectra of bisarene derivatives of vanadium [14], molybdenum [15], and tungsten [22]. The parameters of the lowest Rydberg excitations and the series quantum defects are given in Table 1. The $\delta$ values for the vanadium and chromium complexes are similar. They increase by one on replacing chromium by molybdenum and then by tungsten as it should be expected [27]. The Rydberg term values are nearly insensitive to the nature of the central metal atom (Table 1). Knowledge of typical $T$ magnitudes of the low-lying Rydberg transitions in bisarene complexes appears to be useful for the assignment of Rydberg bands in the spectra of other sandwiches.

TABLE 1. The frequencies $v$ (cm$^{-1}$) and the term values $T$ (cm$^{-1}$) of the lowest Rydberg s and p transitions and the quantum defects $\delta$ of the Rydberg p series in $(\eta^6\text{-Arene})_2M$ [14, 15, 22, 24, 26]

| M | s | | $p_{x,y}$ | | | $p_z$ | | |
|---|---|---|---|---|---|---|---|---|
| | $v$ | $T$ | $v$ | $T$ | $\delta$ | $v$ | $T$ | $\delta$ |
| | | | Arene = $C_6H_6$ | | | | | |
| V | - | - | 32130 | 17550 | 1.45 | 34030 | 15650 | - |
| Cr | - | - | 26650 | 17370 | 1.38 | - | - | 1.16 |
| Mo | - | - | 27330 | 17230 | 2.41 | - | - | 2.29 |
| W | - | - | 26330 | 17280 | 3.40 | - | - | 3.15 |
| | | | Arene = $C_6H_5Me$ | | | | | |
| V | - | - | - | - | 1.44 | 33220 | 15010 | 1.29 |
| Cr | 22870 | 19810 | 25530 | 17150 | 1.40 | - | - | 1.16 |
| Mo | 23250 | 20100 | 26250 | 17100 | 2.40 | - | - | 2.29 |
| | | | 26490 | 16860 | 2.38 | | | |
| W | 22380 | 20180 | 25370 | 17190 | 3.40 | - | - | 3.21 |
| | | | 25640 | 16920 | - | | | |
| | | | Arene = $1,3\text{-Me}_2C_6H_4$ | | | | | |
| V | - | - | - | - | 1.40 | - | - | 1.18 |
| Cr | 22500 | 18990 | 24600 | 16890 | 1.42 | 26600 | 14890 | 1.20 |
| Mo | 23200 | 29130 | 25530 | 16800 | 2.48 | 27530 | 14700 | 2.22 |
| | | | 25730 | 16600 | 2.41 | | | |

## 3. Rydberg Bands in the Electronic Absorption Spectra of Metallocenes

The lowest Rydberg excitations were found to be responsible for intense absorption bands in the gas-phase spectra of metallocenes, $(\eta^5\text{-C}_5H_5)_2M$ (M = V, Cr, Mn, Fe, Ni, Ru, Os), and their methylated derivatives [16, 17]. The sharpest peaks correspond to the $nd_{z^2} \rightarrow R(n+1)p_{x,y}$ transition. Rydberg assignments were made on the basis of the term values. The $v$ and $T$ magnitudes of the first Rydberg $p_{x,y}$ excitations are presented in Table 2. Rydberg bands in the metallocene spectra are usually blue-shifted in comparison with those in the spectra of bisarene compounds. This shift is a result of an increase in the $d_{z^2}$ ionization energies [4]. The term values, however, change very little on going from $(\eta^6\text{-}C_6H_6)_2M$ to $(\eta^5\text{-}C_5H_5)_2M$ (Tables 1, 2). Similar $T$ magnitudes should therefore be expected for Rydberg transitions in the mixed sandwich complexes $(\eta^k\text{-}C_kH_k)(\eta^5\text{-}C_5H_5)M$ (k = 6, 7, 8).

TABLE 2. The frequencies $\nu$ $(cm^{-1})$ and the term values $T$ $(cm^{-1})$ of the lowest Rydberg $p_{x,y}$ transitions in the metallocenes [16, 17]

| Metallocene | $\nu$ | $T$ | Metallocene | $\nu$ | $T$ |
|---|---|---|---|---|---|
| $(\eta^5\text{-}C_5H_5)_2V$ | 37430 | 17490 | $(\eta^5\text{-}C_5Me_5)_2Mn$ | 27800 | 15200 |
| $(\eta^5\text{-}C_5Me_5)_2V$ | 32000 | 15340 | $(\eta^5\text{-}C_5H_5)_2Ru$ | 43300 | 16780 |
| $(\eta^5\text{-}C_5H_5)_2Fe$ | 41000 | 17480 | $(\eta^5\text{-}C_5Me_5)_2Ru$ | 39070 | 15290 |
| $(\eta^5\text{-}C_5H_4Me)_2Fe$ | 39650 | 17290 | $(\eta^5\text{-}C_5H_5)_2Os$ | 42590 | 17350 |
| $(\eta^5\text{-}C_5Me_5)_2Fe$ | 36000 | 14650 | $(\eta^5\text{-}C_5Me_5)_2Os$ | 38840 | 15290 |

## 4. Rydberg Excitations in the Mixed Sandwich Compounds

On going from the metal bisbenzene complexes and metallocenes to the mixed sandwiches, the symmetry is reduced from $D_{\infty h}$ to $C_{\infty v}$ [5] so the Rydberg s and d transitions become optically allowed in the one-photon absorption spectra. Such transitions were observed in the gas-phase spectra of $(\eta^6\text{-}C_6H_6)(\eta^5\text{-}C_5H_5)Mn$ [19] and $(\eta^7\text{-}C_7H_7)(\eta^5\text{-}C_5H_5)M$ (M = V, Nb, Cr, Mo, W) [20, 21, 23, 25] in addition to members of the Rydberg p series.

The parameters of long-wavelength Rydberg bands are given in Table 3. The term values for mixed sandwiches are close to those for bisbenzene compounds and metallocenes. This makes it possible to predict the frequencies of one-photon forbidden Rydberg s and d excitations in $(\eta^6\text{-}C_6H_6)_2M$ and $(\eta^5\text{-}C_5H_5)_2M$ on the basis of the term values (Table 3). The $nd_{z^2} \rightarrow R(n+m)s$ and $nd_{z^2} \rightarrow R(n+m)d$ promotions (m is an integer) are allowed as two-photon processes.

The quantum defect of higher $p_{x,y}$ transitions decreases by 0.1 on going from $(\eta^6\text{-}C_6H_6)_2Cr$ to $(\eta^7\text{-}C_7H_7)(\eta^5\text{-}C_5H_5)Cr$ [18, 24] because of configuration interactions. A similar picture was observed for the molybdenum and tungsten derivatives [15, 21, 22, 25]. Configuration interactions result in the shift and broadening of some Rydberg bands in the spectra of sandwich molecules. For example, the $nd_{z^2} \to R(n+1)p_{x,y}$ transition in $(\eta^6\text{-}C_6H_6)_2M$ (M = Cr, Mo, W) [15, 22, 24] and the $3d_{z^2} \to R4d_{xz,yz}$ transition in $(\eta^7\text{-}C_7H_7)(\eta^5\text{-}C_5H_5)V$ [23] are broadened beyond detection. An admixture of valence-shell excitations to the $5d_{z^2} \to R6p_{x,y}$ transition in $(\eta^7\text{-}C_7H_7)(\eta^5\text{-}C_5H_5)W$ leads to the spin-orbit splitting of the corresponding absorption peak into three components [21].

TABLE 3. The frequencies $v$ (cm$^{-1}$) and the term values $T$ (cm$^{-1}$) of the lowest Rydberg s, p, and $d_{xz, yz}$ transitions in the mixed sandwiches [18-21, 23, 25]

| M | s | | $P_{x, y}$ | | $P_z$ | | $d_{xz, yz}$ | |
|---|---|---|---|---|---|---|---|---|
| | $v$ | $T$ | $v$ | $T$ | $v$ | $T$ | $v$ | $T$ |
| $(\eta^6\text{-}C_6H_6)(\eta^5\text{-}C_5H_5)M$ | | | | | | | | |
| Cr | - | - | 39800 | 17860 | - | | - | - |
| Mn | 29870 | 21420 | 33740 | 17550 | 35370 | 15920 | 38400 | 12890 |
| $(\eta^7\text{-}C_7H_7)(\eta^5\text{-}C_5H_5)M$ | | | | | | | | |
| V | 30630 | 21710 | 34730 | 17610 | 36650 | 15690 | - | - |
| Nb | - | - | 29570 | 17610 | 31730 | 15450 | 34630 | 12550 |
| Cr | 23800 | 21390 | 27610 | 17580 | 29700 | 15490 | 32440 | 12750 |
| Mo | 24280 | 21610 | 28700 | 17190 | 30590 | 15300 | 33310 | 12580 |
| W | 22900 | 21730 | 25800[a] | 18830 | 29170 | 15460 | 31920 | 12710 |
| | | | 26360[a] | 18270 | | | | |
| | | | 27930[a] | 16700 | | | | |
| $(\eta^8\text{-}C_8H_8)(\eta^5\text{-}C_5H_5)M$ | | | | | | | | |
| Ti | 22700 | 23000 | 27310 | 18000 | 30000 | 15700 | - | - |

[a] Three spin-orbit components.

## 5. Molecular Design of Sandwich Complexes and the Parameters of Rydberg Structures

The spectroscopic properties of sandwich compounds can be changed by varying the central metal atom, the carbocycle types, and the ring substituents. Each of these factors has an influence on the molecular parameters which affect the character of a Rydberg structure in the gas-phase absorption spectrum (Figure 4).

The Rydberg bands disappear on going from the $[d_{z^2}]^2$ or $[d_{z^2}]^1$ electronic configuration to $[d_{z^2}]^0$ (e.g. on replacing Cr or V by Ti in $(\eta^7\text{-}C_7H_7)(\eta^5\text{-}C_5H_5)M$). For the compounds with an occupied $d_{z^2}$ MO, the relative intensities of

Rydberg peaks depend on the configuration interactions and molecular sym-metry.

The Rydberg frequencies are mainly determined by the $d_z2$ ionization energy of a complex. The broadening of Rydberg bands and disappearance of vibronic structures can be caused by the interactions between excited states and delocali-zation of the $d_z2$ electron. The term values of the lowest Rydberg transitions depend slightly on the distribution of the electron density in the periphery of a molecular core and configuration interactions.

## 6. Conclusions

Rydberg transitions originating at the non-bonding $d_z2$ MO appear to play an important role in the one-photon absorption spectra of sandwich complexes. The absence of Rydberg excitations originating at the $e_{2g}$ MO (Figure 1) was inter- preted as being a consequence of the presence of stronger interactions of the ligand MOs with the metal $d_{xy}$ and $d_x2_{-y}2$ levels as compared to the metal $d_z2$ orbital [17].

Sandwich complexes remain being the only class of organometallics showing clearly defined Rydberg excitations in the absorption spectra.The Rydberg tran-sitions observed in the spectra of sandwich molecules are responsible, as a rule, for narrow absorption peaks. The corresponding Rydberg states can well parti-cipate as intermediate levels in the resonance-enhanced processes of multiphoton absorption and ionization in these systems.

The lowest Rydberg transitions in the bisarene and mixed sandwiches lie in the near-UV and visible regions available for standard dye lasers. Another advantage of sandwich complexes as the objects of spectroscopic investigations results from vast possibilities of molecular engineering. The Rydberg frequencies and relative intensities can be easily changed by varying the central metal atom, the types of the carbocycles, and the substituents. It is hoped that these circumstances will sti-mulate further research of sandwiches and related organometallic systems by the methods of multiphoton spectroscopy. Information given in this work on the para-meters of Rydberg transitions in various sandwich molecules represents a basis for such studies.

*Acknowledgements.* This work was supported by the Russian Foundation for Basic Research (Projects 99-03-42712, 98-03-33009, and 96-15-97455).

## References

1. Weber, J., Geoffroy, M., Goursot, A., and Penigault, E. (1978) Application of multiple scattering Xa molecular orbital method to the determination of the electronic structure of metallocene compounds. 1. Dibenzenechromium and its cation, *J. Am. Chem. Soc.* **100**, 3995-4003.

2. Brennan, J.G., Cooper, G., Green, J.C., Kaltsoyannis, N., MacDonald, M.A., Payne, M.P., Redfern, C.M., and Sze, K.H. (1992) Electron localization in the bis-arene complexes [($\eta$-$C_6H_6$)$_2$Cr] and [($\eta$-$C_6H_5$Me)$_2$Mo]: an investigation by photoelectron spectroscopy with variable photon energy, *Chem. Phys.* **164**, 271-281.

3. Clack, D.W. and Warren, K.D. (1980) Metal-ligand bonding in 3d sandwich complexes, *Struct. Bond. (Berlin)* **39**, 1-41.

4. Green, J.C. (1981) Gas phase photoelectron spectra of d- and f-block organometallic compounds, *Struct. Bond. (Berlin)* **43**, 37-112.

5. Warren, K.D. (1976) Ligand field theory of metal sandwich complexes, *Struct. Bond. (Berlin)* **27**, 45-159.

6. Nagano, Y., Achiba, Y., and Kimura, K. (1986) UV multiphoton dissociation of volatile iron complexes as revealed by MPI ion-current and photoelectron spectroscopy, *J. Phys. Chem.* **90**, 1288-1293.

7. Ray, U., Hou, H.Q., Zhang, Z., Schwarz, W., and Vernon, M. (1989) A crossed laser-molecular beam study of the one and two photon dissociation dynamics of ferrocene at 193 and 248 nm, *J. Chem. Phys.* **90,** 4248-4257.

8. Opitz, J. and Harter, P. (1992) Multiphoton ionization of vanadocene and ferrocene at 248 and 193 nm. Wavelength-dependent competition between dissociation and ionization, *Int. J. Spectrom. Ion Proc.* **121**, 183-199.

9. Opitz, J., Bruch, D., and Vonbunau, G. (1993) Multiphoton excitation of ferrocene and vanadocene at 351 nm in comparison with 248 nm and 193 nm. Wavelength-dependent competition between ionization and dissociation, *Org. Mass Spectrom.* **28**, 405-411.

10. Opitz, J., Bruch, D., and Vonbunau, G. (1993) Nanosecond laser excitation of benzene chromium tricarbonyl and dibenzene chromium at 351 nm, 248 nm and 193 nm. Wave-length-dependent competition between ionization and dissociation, *Int. J. Mass. Spectrom. Ion Proc.* **125**, 215-228.

11. Berry, R.S. (1961) Ultraviolet spectrum of dibenzene chromium vapor, *J. Chem. Phys.* **35**, 2025-2028.

12. Armstrong, A.T., Smith, F., Elder, E., and McGlynn, S.P. (1967) Electronic absorption spectrum of ferrocene, *J. Chem. Phys.* **46**, 4321-4328.

13. Richer, G. and Sandorfy, C. (1985) The far-ultraviolet absorption spectra of ferrocene, cobaltocene, and nickelocene, *J. Mol. Struct.* **123**, 317-327.

14. Ketkov, S.Y., Domrachev, G.A., and Razuvaev, G.A. (1989) Rydberg series in the near-UV vapour-phase absorption spectra of $(\eta^6\text{-Arene})_2\text{Cr}(0)$ and $(\eta^6\text{- Arene})_2\text{V}(0)$ complexes, *J. Mol. Struct.* **195**, 175-188.

15. Ketkov, S.Y. and Domrachev, G.A. (1990) Electronic absorption spectra of molybdenum(0) bisarene complexes, $(\eta^6\text{-Arene})_2\text{Mo}^0$, *J. Organomet. Chem.* **389**, 187-195.

16. Ketkov, S.Y. and Domrachev, G.A. (1990) Rydberg transitions in the near-ultraviolet vapor-phase absorption spectra of 3d metallocenes, *Inorg. Chim. Acta.* **178**, 233-242.

17. Ketkov, S.Y. and Domrachev, G.A. (1991) Ultraviolet vapour-phase absorption spectra of $d^6$ metallocenes, *J. Organomet. Chem.* **420**, 67-77.

512

18. Ketkov, S.Y. (1992) Vapor-phase electronic absorption spectrum of ($\eta^5$-cyclopentadienyl) $\eta^7$-cycloheptatrienyl)chromium, *J. Organomet. Chem.*. **429**, C38-C40.

19. Ketkov, S.Y. (1994) Photoabsorption study of electronic excited states of ($\eta^5$-cyclopentadienyl)($\eta^6$-arene)manganese, *J. Organomet. Chem.* **465**, 225-231.

20. Ketkov, S.Y. (1994) Nature of low-lying electron excited states of ($\eta^5$-cyclopentadienyl) ($\eta^7$-cycloheptatrienyl)niobium, *Russ. Chem. Bull.* **43**, 583-587.

21. Green, J.C., Green, M.L.H., Field, C.N., Ng, D.K.P., and Ketkov, S.Y. (1995) Combined photoelectron-photoabsorption study of ($\eta$-cycloheptatrienyl)($\eta$-cyclopentadienyl)tungsten, *J. Organomet. Chem.* **501**, 107-115.

22. Ketkov, S.Y., Mehnert, C., and Green, J.C. (1996) Electronic excited states of bis($\eta^6$- arene)tungsten: an investigation by ultraviolet photoabsorption spectroscopy in the and solution phases, *Chem. Phys.* **203**, 245-255.

23. Green, J.C. and Ketkov, S.Y. (1996) Electronic structures of $\eta^7$-cycloheptatrienyl $\eta^7$- cyclopentadienyl derivatives of vanadium and tantalum as studied by solution-phase and -phase photoabsorption spectroscopy in the ultraviolet and visible regions, *Organometallics* **15**, 4747-4754.

24. Ketkov, S.Y., Green, J.C., and Mehnert, C.P. (1997) Rydberg transitions in bis($\eta^6$-benzene)chromium: a study of isotopic effects and vibronic structures, *J. Chem. Soc., Faraday Trans.* **93**, 2461-2466.

25. Ketkov, S.Y. and Green, J.C. (1997) Rydberg structures in the gas-phase electronic absorption spectra of $\eta$-cycloheptatrienyl $\eta$-cyclopentadienyl complexes of molybdenum, *J.Chem. Soc., Faraday Trans.* **93**, 2467-2471.

26. Ketkov, S.Y., Domrachev, G.A., Mehnert, C.P., and Green, J.C. (1998) Structures of Rydberg transitions in the absorption spectra of methyl-substituted derivatives of bis($\eta^6$- benzene)chromium, *Russ. Chem. Bull.* **47**, 868-874.

27. Robin, M.B. (1985) *Higher excited states of polyatomic molecules*, Academic Press, New York.

28. Penner, A., Amirav, A., Tasaki, S., and Bersohn, R. (1993) Photodissociation and photoionization and their branching ratio in bisbenzene chromium, *J. Chem. Phys.*, **99**, 176-183.

# AN ACCURATE WAVE MODEL FOR CONSERVATIVE MOLECULAR SYSTEMS

ALEXANDRU POPA
*Institute of Atomic Physics, National Institute for Laser, Plasma and Radiation Physics, Laser Department,*
*P.O. Box MG-36, Bucharest, Romania 76900*

## Abstract

We consider a conservative molecular system. For this system, the Schrödinger equation is equivalent to the wave equation. We show that the wave properties of the system lead to an exact calculation model for the energetic eigenvalues of the Schrödinger equation. Thus we shed light on intrinsic coherence properties of the system. We present a survey of the applications of this model.

## 1. Introduction

It is well known that the classical and quantum treatments of conservative discrete systems lead, mathematically, to waves which are associated to the motion of these systems. The waves associated to the classical motion are studied in many books [1 - 4]. On the other hand, in the quantum approach, the Schrödinger equation itself is equivalent to the wave equation in the case of conservative systems. A natural question appears: what is the connection between the two waves, when we perform a parallel analysis, of the same system, with the quantum and classical equations, of Schrödinger and respectively, Hamilton-Jacobi ? In recent papers [5 - 11] we have proved that there is a direct connection between these equations, and the two waves are identical. These theoretical results lead to a molecular analysis model. Our approach does not use any mathematical approximation.

## 2. Wave Properties of the Molecular Systems

We present briefly some elements of the theory, published recently [5 - 11]. We consider a closed and conservative molecular system (i.e. the total energy $E$ is constant and the potential energy $U$ does not depend explicitly on time). The potential and total energies have real negative values (i.e. the system is in a bound state). The system is composed of

*: Kajzar and M.V. Agranovich (eds.), Multiphoton and Light Driven Multielectron Processes in Organics: New Phenomena, Materials and Applications, 513–526.*
© 2000 *Kluwer Academic Publishers. Printed in the Netherlands.*

$N$ electrons and $N'$ nuclei. The Cartesian coordinates of an electron are $x_a$, $y_a$, $z_a$, where $a$ takes values between 1 and $N$. Our analysis is performed in the space $R^{3N}$ of the electron coordinates, which are denoted by $q_j$ (where $q_1 = x_1$, $q_2 = y_1$, $q_3 = z_1$, $q_4 = x_2$,..., etc.), $j$ taking values between 1 and $3N$. We denote by $q = (q_1, q_2, ..., q_{3N})$ the coordinates of a point in the space $R^{3N}$.

The behaviour of the system is completely described by the Schrödinger equation:

$$-i\bar{h}\frac{\partial \Psi}{\partial t} - \frac{\bar{h}^2}{2m}\sum_j \frac{\partial^2 \Psi}{\partial q_j^2} + U\Psi = 0 \qquad (1)$$

where $\Psi$, $m$, $t$ and $i$ are, respectively, the wave function, the electron mass, the time and the imaginary constant, while $\bar{h}$ is the normalized Planck constant ($\bar{h} = h/2\pi$).

The eigenvalues of the constants of motion [the total energy $E$ and $c = (c_1, c_2, ..., c_{3N-1})$] enter into the expression of the wave function [12]. Consequently, the wave function is of the form $\Psi = \Psi(q, t, E, c)$. Since the system is conservative, the Schrödinger equation can be solved using separation of variables [12]:

$$\Psi = \Psi_0 \exp(-iEt/\bar{h}) \qquad (2)$$

with $\Psi_0 = \Psi_0(q, E, c)$, where $\Psi_0$ is the time independent function, which is a complex function of $q$, $E$ and $c$ satisfying:

$$-\frac{\bar{h}^2}{2m}\sum_j \frac{\partial^2 \Psi_0}{\partial q_j^2} + (U - E)\Psi_0 = 0 \qquad (3)$$

For conservative systems, eq. (1) is equivalent to the system which comprises eq. (2) and the wave equation

$$\sum_j \frac{\partial^2 \Psi}{\partial q_j^2} - \frac{1}{v_w^2}\frac{\partial^2 \Psi}{\partial t^2} = 0 \qquad (4) \qquad \text{where} \qquad v_w = \pm |E|/\sqrt{2m(E - U)} \qquad (5)$$

Consequently, the behaviour of the system is described by the wave equation. We study the motion of the characteristic surface of the wave equation in the classical allowed (CA) domain (where $E > U$), corresponding to real values of $v_w$, given by eq. (5). The characteristic surface of equation (4) is given [13, 14] by the following equation:

$$\chi(q, t) = 0 \qquad (\Sigma \text{ surface}) \qquad (6)$$

where $\chi$ satisfies the characteristic equation:

$$\sum_j \left(\frac{\partial \chi}{\partial q_j}\right)^2 - \frac{1}{v_w^2}\left(\frac{\partial \chi}{\partial t}\right)^2 = 0 \qquad (7)$$

In the CA domain, eq. (7) has the following periodical solution:

$$\chi(q,t) = \sin k\left[f(q) \mp |E|t\right] \qquad (8)$$

where $k$ is a constant. The function $f(q)$ is the complete integral of the time independent Hamilton-Jacobi equation, written for the same system, and corresponding to the energy $E$ that results from the Schrödinger equation:

$$\sum_j \left(\frac{\partial f}{\partial q_j}\right)^2 + 2m(U - E) = 0. \qquad (9)$$

The function $f$ is of the form $f = f(q,E,c')$ [2, 15], where $c' = (c'_1, c'_2, ..., c'_{3N-1})$ are the constants of motion, which belong to a continuous real domain that includes the eigenvalues $c = (c_1, c_2, ..., c_{3N-1})$ of the Schrödinger equation. Therefore the following equations are possible

$$c'_j = c_j \text{ for } j = 1, 2, ....., 3N-1 \qquad (10)$$

We have to study the case corresponding to eqs. (10), because $c$ and $E$ are the constants of motion of the system in discussion. We limit our analysis to the case corresponding to the plus sign in eq. (5) and minus sign in (8).

It follows that the equation of the characteristic surface is

$$\chi(q,t) = \sin k\left[f(q) - |E|t\right] = 0 \qquad (\Sigma \text{ surface}) \qquad (11)$$

which is equivalent to the equation

$$f(q) = |E|t - p\pi/k \text{ where p is an integer} \qquad (12)$$

Since $f$ is determined up to an arbitrary constant, we choose the following initial condition

$$f(q) = 0 \text{ for } t = 0 \qquad (13)$$

The family of characteristic surfaces of the wave equation ($\Sigma$ surfaces) is the same as the family of surfaces associated to the classical motion, $f(q,E,c) = \kappa$, where $\kappa$ is a variable parameter. Consequently we have:

$$f(q,E,c) = |E|t - p\pi/k = \kappa \qquad (14)$$

This happens because the function $f(q,E,c)$ entering in eqs. (12) is the same as $f(q,E,c) = \kappa$, being the solution of eq. (9).

The normal curves of the $\Sigma$ surfaces are the trajectories $C$ resulting from the Hamilton-Jacobi equation and corresponding to the constants $E$ and $c$. This results from the properties of the Hamilton-Jacobi equation [2, 15].

We have the following two cases:

1) *The function f is bounded*, namely $0 \leq f(q) < f_M$, where $f_M$ is its maximum value. In this case, eq. (12) leads [9] to the following equation of motion of the $\Sigma$ surface, which corresponds to a periodical motion:

$$f(q) = |E|t - p|E|\tau_w \quad \text{for} \quad p\tau_w \leq t < (p+1)\tau_w \tag{15}$$

having

$$k = \pi/f_M \qquad (16) \qquad \text{and} \qquad \tau_w = f_M/|E| \tag{17}$$

In these equations $\tau_w$ is the period of motion, and $p$ is an integer ($p = 0$ for the first period, $p = 1$ for the second, and so on). These equations show that the function $f$ passes periodically through the same values comprised between 0 and $f_M$.

2) *The function f is unbounded*. In this case, the following equation of the $\Sigma$ surface is valid

$$f(q) = |E|t \tag{18}$$

We shall show that this second case does not occur.

The significance of these equations and the motion of the characteristic surface will result from the following properties.

**Property no. 1.** *Two surfaces, $\Sigma_1$ and $\Sigma_2$, that correspond to the same values of the constants c and E, and have, respectively, the equations $f(q,E,c) = \kappa_1$ and $f(q,E,c) = \kappa_2$, are either non-intersecting (when $\kappa_1 \neq \kappa_2$) or identical (when $\kappa_1 = \kappa_2$). Moreover, through one point of the CA domain passes only one surface.*

Indeed, assuming that $\Sigma_1$ and $\Sigma_2$ have a common point $q$, then $f(q,E,c) = \kappa_1 = \kappa_2$ and the property follows.

**Property no. 2.** *The velocity of an arbitrary point of the surface $\Sigma$ which moves on a normal C trajectory is equal to $v_w$, given by eq. (5).*

This property is demonstrated in [9], showing that the differentiation of the function $f$, given by eq. (12), along the curve $C$, leads to the following equation

$$\frac{ds}{dt} = v_w \tag{19}$$

where $s$ is the distance along the curve $C$.

**Definition.** When an arbitrary point $P \in \Sigma$, moves on the corresponding normal curve in the sense of the increasing of $f$ and $s$ (that is the sense of motion of a classical point), we say that the surface $\Sigma$ moves in *forward direction*.

**Property no. 3.** *The surface $\Sigma$ moves always in the forward direction.*

This property results from eq. (19). Since the velocity $v_w$ of the surface $\Sigma$ is always positive [corresponding to the plus sign in Eq. (5)], and it does not pass through zero inside the CA domain (where $E > U$ and the kinetic energy is always positive), it results that the point $P$ moves always in the sense of increasing of the distance $s$.

The analysis of the case corresponding to the minus sign in eq. (5) and plus in (8) shows that there is another characteristic surface that moves in the opposite direction.

We can now give an interpretation of eq. (15). We suppose that a point $P \in \Sigma \cap C$ moves along $C$ between a fixed point $P_0$ (corresponding to $f = 0$ and $t = 0$) and an arbitrary point $P$, corresponding to $f = \kappa$ and to the time $t$.

From eq. (9) we have

$$\frac{df}{ds} = \sqrt{\sum_j \left(\frac{\partial f}{\partial q_j}\right)^2} = \sqrt{2m(E-U)} \tag{20}$$

Taking into account eq. (12), and integrating eq. (20) between $P_0$ and $P$, we find that at the moment $t$, the point $P$ is situated on the surface $\Sigma$ having the equation

$$f(q) = \kappa = |E|t = \int_{P_0}^{P} \sqrt{2m(E-U)}\,ds \tag{21}$$

There are only two cases, corresponding, respectively, to the two cases above:

1) The surface $\Sigma$, having eq. (21), intersects again its initial position, that has the equation $f(q) = 0$. According to Properties 1 and 3, in this case, the surface $\Sigma$ passes exactly through the initial position, moving in the same sense, at a certain time denoted by $\tau_w$.

2) The moving surface $\Sigma$ never intersects its initial position.

We suppose by contradiction that the second case is possible. In virtue of eq. (21), the surface $\Sigma$ passes only through different positions in the CA domain, corresponding to different values of the parameter $\kappa$. Since it moves with positive velocity (Property 2), it results that the surface $\Sigma$ scans a volume of the CA domain whose measure increases continuously with $\kappa$ and $t$. Therefore we can choose a value of the time $t$, so that the measure of this volume can be higher than any given value. Since the volume of the CA domain is finite for the system in discussion (because the total energy of the system is negative and the CA domain is bounded by the surface having the equation $E = U$), there is a contradiction, resulting that the supposition that the first case is possible is incorrect. We observe that this case corresponds to an unbounded function $f$, which is not possible for a system satisfying the initial hypotheses, as claimed.

Consequently, only the first case is possible, and it follows that the motion of the surface $\Sigma$ is periodical, and the $C$ trajectories are closed curves. The reduced action function that corresponds to the point $P$ is denoted by $S_0$, and is given by the following equations [2, 15]:

$$S_0(q_P) = f(q_P) + p f_M \quad \text{with} \quad p = 0, 1, 2, \dots \tag{22}$$

where

$$f(q_P) = \kappa = \int_{P_0}^{P} \sqrt{2m(E - U)} \, ds \quad \text{with} \quad 0 \le \kappa < f_M \tag{23}$$

where $p = 0$ corresponds to the first period of motion, and so on. Therefore, when we go along the closed curve $C$, we pass periodically through all the values of the function $f$, and $S_0$ increases continuously. We denote by $\Delta_C S_0$ the variation of the function $S_0$, along the closed curve $C$, and we have

$$f_M = \Delta_C S_0 \tag{24}$$

The period of motion of the surface $\Sigma$ results from eqs. (17) and (24)

$$\tau_w = \Delta_C S_0 / |E| \tag{25}$$

In recently published papers [5, 8] we have proved the following theorems, valid for this system.

*Direct theorem: If the Hamilton-Jacobi equation is satisfied in the space of coordinates by a periodical trajectory $C$ that corresponds to the energy $E$ which results from the Schrödinger equation, then a generalized Bohr quantization condition is valid for this trajectory:*

$$\Delta_C S_0 = nh \quad \text{for an integer } n. \tag{26}$$

*Reciprocal theorem: If a closed trajectory in the space of coordinates, which fulfils the generalized quantization condition, is a solution of the Hamilton-Jacobi equation, then this trajectory corresponds to an energy equal to that resulting from the Schrödinger equation.*

From eqs. (25) and (26) we obtain

$$\tau_w = nh / |E| \tag{27}$$

We analyse now the motion of an arbitrary point, denoted by $P$, belonging to the surface $\Sigma$, along the corresponding closed trajectory $C$, in a period. We denote by $t_i$, $t_f$, $\Psi_i$, $\Psi_f$, $q_i$, $q_f$, respectively, the initial and final moments, wave functions and corresponding coordinates. Since $q_i \equiv q_f$ and $t_f = t_i + \tau_w$, it results that $\Psi_0(q_i) = \Psi_0(q_f)$ and according to eqs. (2) and (27), we have $\Psi_i = \Psi_f$.

Therefore the motion of the system has the properties of a wave, because: a) an arbitrary point $P$ moves on a closed trajectory with the velocity $v_w$, and the amplitude of the wave function $\Psi$ in that point varies periodically, and b) the point $P$ move synchronously with a surface (the characteristic surface), which is perpendicular to its trajectory, and can be called wave surface.

The analysis can be simplified in the case of systems for which the separation of variables is possible. In this case, the function $S_0$ can be written as a sum of functions depending only on the coordinates of a single electron, and the following relation is valid

$$S_0 = \sum_a S_{0a} \qquad (28)$$

where

$$S_{0a} = S_{0a}(x_a, y_a, z_a) \qquad (29)$$

In the paper [5] we demonstrated that the following relation is valid for the trajectory of the electron $a$, that is denoted by $C_a$

$$\Delta_{C_a} S_{0a} = n_a h \quad \text{where} \quad n_a = 1, 2, .... \qquad (30)$$

$n_a$ being the quantum number associated to the motion of the electron $a$.

From (26), (28) and (30) we obtain

$$n = \sum_a n_a \qquad (31)$$

We can see that the minimum value of $n$ is $N$, the number of the electrons.

In conclusion, a closed, conservative system behaves mathematically like a wave. Therefore we can associate to such a system a periodical trajectory $C$, which is a normal curve of the wave surface. The constants of motion corresponding to this trajectory are identical to the eigenvalues of the Schrödinger equation, written for the same system. This model is rigorously exact. It is deduced without any mathematical approximation.

## 3. Energetic relations

A point $P$ on the $C$ trajectory, which moves together with the characteristic surface of the wave equation, has velocity $v_w$. On the other hand, the velocity of a classical point moving on the same trajectory is given by the equation $E = U + mv^2/2$, resulting

$$v = \sqrt{2(E - U)/m} \qquad (32)$$

From eq. (5) it follows that:

$$v_w v = |E|/m \qquad (33)$$

The periods of the two motions are, respectively, $\tau_w$ and $\tau$. It easy to demonstrate the following relation

$$\tau_w = 2\tau \qquad (34)$$

This has the following interpretation: the period $\tau$ of the classical motion is equal to the time after which two characteristic surfaces moving in opposite directions, meet again, after starting in the same point.

The variation of the reduced action which corresponds to the curve $C$ is given by the following relation

$$\Delta_C S_0 = \oint_{C} \sum_{a,\alpha} P_{a\alpha} d\alpha_a = \oint_{C} \sum_{a} \overline{P}_a d\overline{s}_a = \sum_{a} \oint_{C_a} \overline{P}_a d\overline{s}_a \tag{35}$$

On the other hand, the following relations result from the virial theorem [15]

$$E = \tilde{T} + \tilde{U} = \frac{1}{2} \tilde{U} = -\tilde{T} \tag{36}$$

where $T$ is the kinetic energy and the tildes represent the average over a period.

The average value of the kinetic energy of the classical point that move on the $C$ trajectory is

$$\tilde{T} = \frac{1}{\tau} \int_{C} \sum_{a} \frac{p_a^2}{2m} dt = \frac{1}{2\tau} \int_{C} \sum_{a} \overline{P}_a \cdot d\overline{s}_a = \frac{1}{2\tau} \Delta_C S_0 = \frac{nh}{2\tau} \tag{37}$$

From (36) and (37) it results

$$\tilde{U} = -\frac{nh}{\tau} = -nvh \tag{38}$$

where $v$ is the oscillation frequency of the system. Because the system is periodical, we can assume that both particles and the electromagnetic field oscillate synchronously with this frequency. Since we neglected the magnetic effects in our analysis, the potential energy is equal to the electromagnetic field energy of the system [16]. Therefore the relation (38) is the expression of the energy of the electromagnetic field. It is similar to the Planck quantization relation.

In fact we arrived at the result that eq. (38) is related to the Bohr condition (26). We show now that this relation is valid in the most general case, when the magnetic and relativistic effects are taken into account.

The most general classical treatment of systems comprising particles and electromagnetic field is based on the variational principle of Hamilton. This treatment is presented by Landau in [16]. The approach is relativistic, and the Maxwell equations and the equations of particle motion result as consequences. In this treatment, the field and the particles move synchronously, like in an aggregate.

We consider a periodical system characterised by the period $\tau$, in the general case when the relativistic and magnetic effects are taken into account. According to Landau [16], the virial theorem is given by the following equation

$$E = \tilde{T} + \tilde{E}_m = \frac{1}{\tau}\int_C \sum_a m_a c^2 \sqrt{1 - \frac{v_a^2}{c^2}}\,dt - \sum_a m_a c^2 \qquad (39)$$

where $\tilde{E}_{em}$ and $c$ are, respectively, the average electromagnetic energy of the field and the light velocity in vacuum. The proper energy of the particles is not included in the expression of the total energy.

The kinetic energy and the impulse of the particle a are given by the following relations [16]

$$T = \sum_a \frac{m_a c^2}{\sqrt{1 - \frac{v_a^2}{c^2}}} - \sum_a m_a c^2 \qquad (40) \qquad \overline{P}_a = \frac{m_a \overline{v}_a}{\sqrt{1 - \frac{v_a^2}{c^2}}} \qquad (41)$$

For the system under discussion, the total and electromagnetic energies are negative. The electromagnetic field varies with the frequency v. We suppose that the Planck relation is valid:

$$\tilde{E}_{em} = -n v h = -n \frac{h}{\tau} \qquad (42)$$

The relations (39)-(42) lead to the following relation:

$$\tilde{E}_{em} = E - \tilde{T} = -\frac{1}{\tau}\int_C \sum_a \frac{m_a v_a^2}{\sqrt{1 - \frac{v_a^2}{c^2}}}\,dt = -\frac{1}{\tau}\int_C \sum_a \frac{m_a \overline{v}_a}{\sqrt{1 - \frac{v_a^2}{c^2}}}\,d\overline{s}_a = -\frac{1}{\tau}\int_C \sum_a \overline{P}_a \cdot d\overline{s}_a \qquad (43)$$

From (42) and (43), we have:

$$\oint_C \sum_a \overline{P}_a \cdot d\overline{s}_a = nh \qquad (44)$$

which is the Bohr quantization condition. We supposed relation (42) is valid and the Bohr relation resulted as a consequence. The converse is also valid.

We have showed that we can associate to the system a periodical trajectory $C$, which is a normal curve of the wave surface. The particular trajectory of a given electron is obtained from the projection of the trajectory $C$ (from the space of configurations), on the three dimensional space of the coordinates of that electron. For example, the trajectory of the electron a is obtained from the projection of the curve $C$ on the space of coordinates $x_a$, $y_a$, $z_a$. It is also a closed curve, denoted by $C_a$. Consequently, the motion of the point $P$, on the $C$ trajectory, is synchronous with the motions of the points

$P_a$, on the trajectories $C_a$, where $a = 1, 2, .... N$, resulting that the system has an intrinsic coherence.

Since the motions of the electrons on the $C_a$ trajectories are synchronous and periodical, it results that the components of the electromagnetic field, denoted by $E_{ema}$, generated by these motions, are also coherent and periodical. The following relations result for a simplified case when $n_a$ is the same for all the electrons:

$$E_{ema} = -n_a vh \quad (45) \qquad E_{em} = NE_{ema} \quad (46) \qquad n = Nn_a \quad (47)$$

The $N$ components of the electromagnetic field are also periodical and coherent.

Now it is possible to answer to one of the questions referring to the stimulated emission in laser active media [17]: "One of the most challenging chapters of physics has been the study of fluctuations and coherence in lasers: how and why $10^{10} - 10^{20}$ atoms or molecules, rather than radiating electromagnetic field in a chaotic fashion, decide to "cooperate" to a single coherent field; then for still higher excitation, how and why they organize in a complex pattern of space and time domains, each per-se highly coherent but with little correlations with one another."

In a future paper we attempt to answer this question, approximating the fundamental and metastable states of a laser medium, by closed conservative systems. We approximate the laser radiation by a transition between two coherent state previously described.

The most important application of this model is the calculation of the energetic values corresponding to the orbital components of the wave function. In fact, the previous wave properties show that it is not necessary to solve directly Schrödinger's equation to obtain the total energy. An alternative way, is to find a particular $C$ trajectory, and then to use eq. (26) to find the energy. We show that the model can be accurately applied in two steps of approximation. The theoretical results were compared with experimental data taken from well known books [18 - 21].

### 4. A Survey of the Applications of the Model

### 4.1. THE FIRST STEP: SYSTEMS WITH EXACT SOLUTIONS

The model is applied to the hydrogen and helium atoms, to the ions with the same structure and to the hydrogen molecule [5, 6]. In these cases the equations of motion of the electrons, are equivalent to the Hamilton-Jacobi equation, and lead to symmetrical trajectories. Since the motions of the electrons are separated, we apply eq. (30) and arrive to the absolute value of the total energy of the fundamental state.

With minor approximations, the model is applied to the cases of lithium, beryllium boron and carbon atoms, to the ions with the same structure and to the $Li_2$ molecule [5 6]. We consider the following approximations: a1) The motion of the 1s electrons is lik in the helium atom and it is not influenced by the motion of the 2s and 2p electrons an a2). The motions of the 2s and 2p electrons take place in a field of a nucleus whose order number of the nucleus has the value $Z'$, in order to include the effect of the 1s electrons. We show that the expression for $Z'$ is $Z' = Z - 2s_s$ for 2s electrons and $Z' = Z - 2s_p$ for 2p electrons, where $s_s$ and $s_p$ are the effective screening coefficients, due to 1s electrons. The values of $s_s$ and $s_p$ depend on Z. They are calculated in [5] and [6].

We obtain the following results in the case of beryllium, boron and carbon atoms, and of the ions with the same structure:

1) The electron orbits and the average positions of the electrons are disposed in space in a configuration with maximum symmetry. This result is explained physically by the reciprocal repulsion of the electrons.

2) The motions of the electrons are separated. They take place in the field of a nucleus having the effective order number $Z_a^* = Z - 2s_s - \sum_{b \neq a} s_{ab}$ for 2s electrons and

$Z_a^* = Z - (1 + s_s) - \sum_{b \neq a} s_{ab}$ for 2p electrons, where the coefficient $s_{ab}$ is the reciprocal screening coefficients between the electrons $a$ and $b$. It is given by the relation

$$S_{ab} = \frac{1}{4 \sin \frac{\alpha_{ab}}{2}} \qquad \text{with } a,b = 1,2,...,N \qquad (48)$$

where $\alpha_{ab}$ is angle between the orbit axes of the electrons $a$ and $b$.

The comparison between the theoretical and experimental energetic values $E/E_H$ and $E_{exp}/E_H$, where $E_H$ is the absolute value of the energy of the fundamental state of the hydrogen atom, is given in Table 1. In the cases of atoms having $Z > 2$, $E$ represents the total energy, excluding the energy of the 1s electrons. For example, $E$ corresponds to the first ionization energy for lithium, to the sum of the first two ionization energies for beryllium, and so on. The experimental values are taken from [18].

The model is applied in the same manner to the hydrogen molecule [6], resulting that the hydrogen bond can be studied with the aid of an analytical model. The analytical model is also valid in the case of the $Li_2$ molecule [6], with the difference that instead of the order number $Z$ of the nucleus, we have to consider an effective value $Z^*$. The comparison between the theoretical and experimental values (taken from [21]), where $2\sigma$ is the distance between nuclei and $a_0$ is the first Bohr radius, is given in Table 2. The quantity $-E_M$ represents the total electronic energy of the molecule. For the $Li_2$ molecule, this energy does not include the energy of the 1s electrons.

TABLE 1. Theoretical and experimental energetic values for atoms and ions with similar structure

| System | $E/E_H$ | $E_{exp}/E_H$ | System | $E/E_H$ | $E_{exp}/E_H$ |
|--------|---------|---------------|--------|---------|---------------|
| He | 5.833 | 5.798 | $B^+$ | 4.6448 | 4.6356 |
| $Li^+$ | 14.667 | 14.556 | $C^{2+}$ | 8.2386 | 8.2576 |
| $Be^{2+}$ | 27.500 | 27.306 | $N^{3+}$ | 12.8258 | 12.8856 |
| $B^{3+}$ | 44.333 | 44.058 | $O^{4+}$ | 18.4083 | 18.5187 |
| $C^{4+}$ | 65.167 | 64.815 | B | 5.3806 | 5.2454 |
| $N^{5+}$ | 90.000 | 89.579 | $C^+$ | 10.1993 | 10.0492 |
| $O^{6+}$ | 118.833 | 118.353 | $N^{2+}$ | 16.5138 | 16.3714 |
| Li | 0.3984 | 0.3962 | $O^{3+}$ | 24.3217 | 24.2072 |
| $Be^+$ | 1.3422 | 1.3382 | C | 10.7968 | 10.8765 |
| $B^{2+}$ | 2.7886 | 2.7871 | $N^+$ | 18.4038 | 18.5465 |
| $C^{3+}$ | 4.7328 | 4.7390 | $O^{2+}$ | 28.0046 | 28.2413 |
| $N^{4+}$ | 7.1734 | 7.1930 | N | 19.3858 | 19.6145 |
| $O^{5+}$ | 10.1104 | 10.1490 | $O^+$ | 30.5051 | 30.8218 |
| Be | 2.0501 | 2.0232 | O | 31.2154 | 31.8224 |

## 4.2. THE SECOND STEP: GENERAL SOLUTIONS

The analysis of the systems presented [5] and [6] led to the conclusion that the electron orbits and the average positions of the electrons are disposed in space in a configuration with maximum symmetry. The basic assumption of our general approach is that this symmetry is also valid for more complex systems. This assumption is similar to the main assumption of the Hartree method [22, 23], that the distribution of the potential is symmetrical.

Taking into account the symmetry assumption, we evaluate the potential energy of interaction between the 2s and 2p electrons with the aid of the reciprocal screening coefficients. We calculate the reciprocal screening coefficients between the 2s and 2p electrons, situated at average positions, and we arrive again to eq. (48). The motions of the electrons are again separated and we arrive at the expression of the total energy $E$. Again $E$ has the meaning of a sum of ionization energies, excluding the two last ionization energies (corresponding to helium-like systems). For example, $E$ corresponds to the sum of the first five ionization energies for nitrogen, to the sum of the first six ionization energies for oxygen, and so on. The general relations are applied to the analysis of the nitrogen and oxygen atoms, and to the ions with the same structure [7, 24]. The comparison between the theoretical and experimental energetic values of these systems is given in Table 1.

Table 2. Characteristic theoretical and experimental molecular values

| Molecule | $E_M/E_H$ theoretical | $E_M/E_H$ experimental | $\sigma/2a_0$ theoretical | $\sigma/2a_0$ experimental |
|---|---|---|---|---|
| $H_2$ | 2.3734 | 2.3492 | 0.3320 | 0.3503 |
| $Li_2$ | 0.9091 | 0.8708 | 1.2444 | 1.2628 |
| $Be_2$ | 4.2430 | 4.2164 | 1.0464 | 0.9250 |
| $B_2$ | 10.7987 | 10.7176 | 0.8129 | 0.7512 |
| $C_2$ | 22.3921 | 22.2179 | 0.5987 | 0.5870 |
| $N_2$ | 39.6679 | 39.9533 | 0.5643 | 0.5186 |
| $LiH$ | 1.7168 | 1.5811 | 0.7960 | 0.7539 |
| $BeH$ | 3.3620 | 3.1821 | 0.6647 | 0.6343 |
| $BH$ | 6.6934 | 6.5076 | 0.5652 | 0.5830 |
| $CH$ | 12.0466 | 12.1442 | 0.4933 | 0.5291 |

The same treatment is generalised to molecules [7, 24]. We make the same assumption, namely that the electron orbits, and the average positions of the electrons, are disposed symmetrically in space. The geometrical elements of the bond are computed using equations similar to those of the hydrogen bond with the difference that the bond electrons move in fields of nuclei having an effective order number. The interaction energies between electrons which move in the field of the same nucleus are evaluated with the aid of the $s_{ab}$ coefficients.

The total energy of the molecule is calculated taking into account all the components of the energy of the molecule, excluding the energy of the 1s electrons. The potential energy of interaction at the distance, among electrons moving in the field of different nuclei, and between a nucleus and electrons moving in the field of the other nucleus are calculated corresponding to the average positions. The comparison between the theoretical and experimental values, for a lot of diatomic molecules [7, 24], is presented in Table 2. The experimental values are taken from [19] and [21].

## 5. Conclusions

The wave properties of conservative molecular systems are used to compute the energetic eigenvalues corresponding to the orbital wave function from the Schrödinger equation. The model is also useful for the study of the intrinsic periodical properties of these systems.

526

# 6. References

1.  Synge, J.L. (1954) *Geometrical Mechanics and de Broglie Waves*, University Press, Cambridge.
2.  Synge, J.L. and Griffith, B.A. (1959) *Principles of Mechanics*, Mc Graw-Hill, New York.
3.  Onicescu, O. (1969) *Mecanica* (in romanian), Editura Tehnica, Bucuresti.
4.  Borowitz, S. (1967) *Fundamental of Quantum Mechanics*, W.A. Benjamin, Inc., New York.
5.  Popa, A. (1998) Applications of a Property of the Schrödinger Equation to the Modeling of Conservative Discrete Systems, *Journal of the Physical Society of Japan* **8**, 2645-2652.
6.  Popa, A. (1999) Applications of a Property of the Schrödinger Equation to the Modeling of Conservative Discrete Systems. II, *Journal of the Physical Society of Japan* **3**, 763-770.
7.  Popa, A. (1999) Applications of a Property of the Schrödinger Equation to the Modeling of Conservative Discrete Systems. III, *Journal of the Physical Society of Japan* **9**, in press.
8.  Popa, A. (1996) A Remarkable Property of the Schrödinger Equation, *Revue Roumaine de Mathematiques Pures et Appliquees*, **1-2**, 109 - 117.
9.  Popa, A. (1998) A Property of the Hamilton-Jacobi Equation, *Revue Roumaine de Mathematiques Pures et Appliquees*, **3-4**, 415 - 424.
10. Popa, A. (1999) Intrinsic Periodicity of the Conservative Systems, *Revue Roumaine de Mathematiques Pures et Appliquees*, **1**, 119-122.
11. Popa, A. (1997) A Connection between the Schrödinger and Hamilton-Jacobi Equation in the Case of the Conservative Systems, Invited lecture, *Eighth International Colloquium on Differential Equations*, Plovdiv, Bulgaria, August 18-23.
12. Messiah, A. (1965) *Quantum Mechanics, Vol. 1.* North-Holland Publishing Company, Amsterdam.
13. Zauderer, E. (1983) *Partial Differential Equations of Applied Mathematics*, John Wiley & Sons, New York.
14. Vladimirov, V.S. (1971) *Equations of Mathematical Physics*, M. Dekker, New York.
15. Landau, L. et Lifschitz, M. (1980) *Mecanique*, Editions Mir, Moscow.
16. Landau, L. et Lifschitz, M. (1980) *Theorie du Champ*, Editions Mir, Moscow.
17. Arecchi F.T. (1983) Collective Phenomena in Quantum Optics, in W.J. Firth and R.G. Harrison (eds.), *Lasers. Physics, Systems and Techniques*, Scottish Universities Summer School in Physics, Edinburgh, pp. 103-166.
18. *Handbook of Chemistry and Physics*, 53rd ed., (1972-1973), CRC Press, Cleveland.
19. Slater, J.C. (1963) *Quantum Theory of Molecules and Solids, Vol. 1*, Mc Graw Hill, New York.
20. Hertzberg, G. (1950) *Molecular Spectra and Molecular Structure. I. Spectra of Diatomic Molecules*, Van Nostrand, New York.
21. Huber, K.P. and Hertzberg, G. (1979) *Molecular Spectra and Molecular Structure. IV. Constants of Diatomic Molecules*, Van Nostrand, New York.
22. Hartree, D.R. (1957) *The Calculation of Atomic Structures*, John Wiley, New York.
23. Coulson, C.A. (1961) *Valence*, Oxford University Press, London.
24. Popa, A. (1991) *Accurate Bohr-type Semiclassical Model for Atomic and Molecular Systems*, Preprint, Institute of Atomic Physics, Bucharest.

# WORKING GROUP: MULTI-PHOTON ABSORPTION: SCIENCE AND APPLICATIONS

P. N. PRASAD[a] and G. STEGEMAN[b]
[a]*State University of New York at Buffalo, NY, USA*
[b]*CREOL, University of Florida, Orlando, FL, USA*

The field of multi-photon absorption (MPA) in organic materials is not new. It was predicted in the 1930s by Maria Goeppert-Mayer and experiments date back to the 1960s. Multi-photon absorption refers to the absorption of two or more photons from a single laser pulse. This process can either occur through simultaneous or sequential absorption of multiple photons. The first case, the simultaneous absorption of a pair of photons is called two photon absorption (TPA). On the molecular level, it is given by the imaginary part of the second hyperpolarizability ((3)(() and at the condensed matter level as (mag{( (3)(()} which translates into an increase in the absorption coefficient (( = (2I where I is the local irradiance. This phenomenon is the first enabling step to many applications, such as data storage, optical limiting, two photon confocal microscopy, etc. However, TPA has also been viewed as a detriment to all-optical signal processing applications because the key parameter, the nonlinear change in refractive index (n = n2I is proportional to the real part of ((3)(() and scales the same way with irradiance as the increased absorption. It is primarily in the last decade that the applications of TPA have appeared sufficiently attractive to stimulate researchers to make sustained efforts to study the mechanisms responsible for two photon absorption at the molecular level and develop device technologies.

Higher order MPA such as the simultaneous absorption of three, four or more photons is a consequence of higher order nonlinearities ((5)((), ((7)(() etc. and has been known in principle from the earliest days of nonlinear optics. However, such effects have been recently shown to be strong in organic media, specifically polydiacetylenes.

The occurrence of two sequential linear absorption events within a single laser pulse relies on generating an initial population in an excited state via the first absorption event. This is followed by the absorption of a second photon raising the molecule either from this excited state, or one coupled to it by a fast transition (relative to the pulse width) to another excited state. This excited state absorption (ESA) frequently exhibits a large linear absorption coefficient between the two excited states. This phenomenon is particularly interesting in large, multi-atom organic molecules that exhibit ESA significantly larger than ground state absorption. The overall effect is called reverse saturable absorption (RSA) and the net cross-section scales with increasing fluence (versus irradiance for TPA). Since it clearly depends on population effects, the efficiency depends on multiple parameters such as the linear absorption coefficients involved, the pulse width, the relaxation time between the states involved,

: *Kajzar and M.V. Agranovich (eds.), Multiphoton and Light Driven Multielectron Processes in Organics: New Phenomena, Materials and Applications, 527–530.*

etc. It can also be initiated by TPA, appearing as a ((5)(() effect, or can occur due to the sequential absorption of more than two photons. RSA is also a first step towards many applications such as optical limiting etc., having many applications in common with two photon absorption.

This overview of the state-of-the-art in MPA is divided into four subject areas, namely materials, modeling, characterization and applications.

## MATERIALS

Although ultimately the specific application and the availability of inexpensive lasers at specific wavelengths will drive the spectral requirements for MPA, the structure-property relations needed to optimize MPA at certain wavelengths, or to achieve broadband spectral coverage are insufficiently understood at present. The work leading into the early 1990s was basically a survey of multi-photon absorption in existing materials. The focus in the last few years has been to understand the mechanisms that lead to large effects, both for TPA and ESA.

The most recent work, primarily in TPA, has been characterized by a search for structure-property relations that will allow new materials to be molecularly engineered. To this end, some specific classes of molecules have been investigated and systematic changes introduced to enhance the two photon cross-sections by factors of up to 100. Some of the largest TPA cross-sections reported to date have been obtained with conjugated systems in linear molecules such as polyenes and polymers such as polydiacetylenes. Selective attachment of donors and/or acceptors and co-operative enhancement in hyper-branched structures appear to be a promising directions to pursue. This, although a useful start, is just the beginning of the systematic investigations needed into structure-property relations for large two photon cross-sections.

Similar progress is being made in ESA materials where the rich excited state spectrum is an advantage offered by large, multi-atom organic molecules. Here metal pthalocyanines and other prophyrins are still being investigated. Fullerenes like $C60$, $C70$ etc. with their rich electronic spectra have been coupled to other molecules to produce new options for ESA.

In virtually all of the projected applications, for example in two photon confocal microscopy, efficient TPA molecules are chemically attached to, or dissolved in, a host medium and the signal levels achieved depend on the local concentration of the TPA-active molecules. Thus the effect of the local environment on the MPA process, i.e. condensed matter effects, are important. Large concentrations of chromophores also raise solubility, aggregation and (photo-) toxicity (when dealing with living tissue) issues. When just maximum nonlinear absorption is needed, single crystal fabrication may be needed. Clearly, all of these issues are important and need to be investigated in the near future.

## CHARACTERIZATION

The ability to unambiguously measure the TPA coefficients and all of the line absorption cross-sections and relaxation times contributing to ESA is important to progress in MPA and its applications.

The two principal TPA measurement techniques fall into two categories: (i) nonlinear transmission and Z-scan which can give a direct measure of the TPA coefficient; and (ii) two photon induced fluorescence which also relies on knowledge of the quantum efficiency of the fluorescence. There have been differences between values deduced from these two techniques, primarily because each has its limitations, implicit assumptions, approximations and competing processes that can affect the final result. (Experimental simplicity is very helpful for screening large classes of molecules.) Very useful to the field would be direct comparisons between measurements by these two techniques on a selective set of molecules, and measurements in different laboratories on the same molecules to better understand systematic sources of error.

For MPA in general, measurements of coefficients at single wavelengths are of limited value. Typically it is the maximum value which is of interest. This requires measurement of the spectral dispersion in the TPA or ESA coefficient and these peak values, their spectral bandwidth and location are of primary interest. In response to these requirements, new nonlinear transmission techniques have been developed using a fixed wavelength as the pump and "white light" continua as a probe. This leads to a direct measurement of the non-degenerate TPA coefficient over large wavelength ranges, i.e. that of the continuum, which now makes rapid characterization of materials feasible.

There is a trend towards using femtosecond excitation for TPA measurements because it can separate, almost unambiguously, the effects due to TPA and ESA. It is now widely recognized that the shorter the pulse width, the smaller population due to linear absorption and the weaker the influence of ESA on TPA measurements. Pump probe measurements with time delayed probes, also with short pulse fast lasers, are becoming more accepted for ESA studies and the quality of the data obtained now allows discrimination between different multi-path processes, the number of states involved and the identification of the key inter-state relaxation rates.

## MODELING

Modeling of multi-photon effects, most specifically TPA has been getting progressively more sophisticated. There are three current approaches being used: (1) a perturbative theory derived sum over states which has been well-established for second order nonlinearities; (2) a new "residue" analysis of the response function approach which appears to be having initial success when compared to experiments and; (3) the density matrix theory based on anharmonic oscillators. Although each approach has had some progress in MPA predictions, they each have their respective strengths and weaknesses approximations). In general, the most progress has been made in predicting trends in a class of related molecules, but absolute values are still a problem for the large molecules that give useful TPA coefficients.. An important milestone would be a critical comparison between the predictions of different theoretical approaches, and their comparison with experiments from a number of different laboratories on the same molecules.

The first two techniques are limited to single molecules, and classical local field effects etc. at the Onsager and Lorentz level are employed to describe condensed matter systems. This can be a problem since local fields can be very large for third order nonlinear processes such as TPA. Thus far, only the density functional approach attempts to handle condensed media in a systematic way.

In terms of RSA, theoretical progress has been slow, especially in condensed matter systems. Broadband ESA, for example as needed in optical limiting, requires a spectrally broad excited state spectrum, which in turn implies large molecules: And large molecules are difficult to model. Here it is hoped that the "residue" analysis of response functions will make a contribution.

## APPLICATIONS

Many of the applications of MPA being currently pursued were proposed at least a decade ago. It is only with focused efforts on generating more efficient MPA materials have some approached practical realization. As mentioned previously, MPA is usually the enabling, first process needed. The applications are based on diverse effects such as two photon fluorescence, two photon activated chemical processes such as polymerization or structural conformations such as cis-trans isomerization, two or more photon initiated changes in optical properties such as absorption and refractive index, flux dependent excited state absorption etc.

ESA has been one of the key processes for optical limiting devices for a number of years now. State-of-the-art limiters now consist of multiple stages in tandem in a tight focus geometry, even with the possibility of different materials in each stage.

A number of applications rely on the increased local irradiance available in the focal region of a microscope. As a result efficient MPA only occurs in this focal region optically localized to volumes of a few cubic optical wavelengths, i.e. to micrometer dimensions. For example, two photon induced polymerization or cross-linking has enabled various three-dimensional structures to be fabricated including MEMS devices, optical waveguides, photonic band gap structures, etc. Another example is the application to biophysics of two photon fluorescence imaging with "tag" molecules processes locally both on cellular and tissue levels. Yet another example is in digital (on-off) data storage where local chemical reactions, cis-trans conformational changes etc. can be triggered by two photon absorption. These applications are in various stages of commercialization with the most advanced being the two photon confocal microscope which is already a commercial product.

# WORKING GROUP : ORGANIC ELECTROLUMINESCENCE

J. KALINOWSKI
*Department of Molecular Physics*
*Technical University Gdansk, Poland*

Research Workshop gathered in three meetings, discussed general problems of organic electroluminescence and its topics related to function and technology of organic light-emitting devices.

Two general questions have been formulated and attempted to be answered on the basis of the conference papers as well as the knowledge and experience of the group members. The questions were:

Is the electroluminescence (EL) a phenomenon that can bring new insight in basic physical phenomena in organic molecular aggregates?
Is the present knowledge and experience on EL sufficient to tailor and to produce commercially available organic EL devices?

As far as the first question is concerned all of the group members agreed that studying of EL phenomena can provide new important information on many electronic processes and structure of organic molecular aggregates. Among them of particular interest are:
(i)   carrier injection mechanisms at organic/organic and organic/inorganic interfaces,
(ii)  charge carrier recombination including large organic molecules,
(iii) electric field effect on molecular electronic states and their excitation mechanisms including, in particular, the nature of excited states produced in the electron-hole recombination process,
(iv) charge carrier transport mechanisms with emphasis on the carrier motion at high electric fields,
(v)  kinetics of polaritons and excitons in EL cavity systems,
(vi) the intermolecular structure as observed by recombination EL, applying surface-tunnelling-microscopy (STM).

There was no like-mindedness as for the present status of the technology of organic EL devices (question 2). However, despite often positive opinions like the present knowledge being close or it is very promising to manufacture commercial EL devices, several conditions have been formulated for organic EL devices to be tailored as to various performance parameters (color, quantum yield, brightness, driving current and voltage, durability, chemical resistance):

*F. Kajzar and M.V. Agranovich (eds.), Multiphoton and Light Driven Multielectron Processes in Organics: New Phenomena, Materials and Applications, 531–532.*

(i)  establishing relation between the interface structure and injection mechanisms of charge carriers,

(ii) theoretical  description of electric field effects on various parameters responsible for performance of EL devices; existing observations of such effects are not well understood,

(iii) coming to light with deterioration mechanisms of EL devices based on different organic solids.

There were more specific topics discussed in some detail during group meetings. For example:

(a) types of organic EL , why does recombination EL attracts researchers the most ?
(b) thin film versus single crystal EL; differences and similarities,
© low molecular-weight materials and polymers; do they show fundamental differences in characteristics of EL devices?,
(d) materials versus device structure; how do they influence the EL phenomena?

In summary, the discussion has shown a broad spectrum of topics to be of interest to physicists, chemists and engineers. The EL has appeared as a phenomenon full of photophysical and photochemical  processes both of fundamental and technological importance, being strongly emerging field in molecular sciences.

Members of the Working Group:

| | |
|---|---|
| Vladimir Agranovich | Institute of Spectroscopy RAS, Moscow, Russia |
| Renato Bozio | Department of Physical Chemistry, University of Padova, Padova, Italy |
| Fabrice Charra | CEA-DSM-DRECAM-SRISM, CEA Saclay, France |
| Michael Hoffmann | Institute of Applied Photophysics, Technical University of Dresden, Dresden, Germany |
| Jan Kalinowski | Department of Molecular Physics, Technical University of Gdansk, Poland |

# WORKING GROUP : PHOTOCHROMISM

ROGER A. LESSARD
*COPL/Physics*
*Faculty of Sciences and Engineering*
*Laval University*
*Quebec City (Quebec) G1K 7P4*

This workshop was intended to discuss problems related to photochromism, its application, how to choose or design the best possible molecules.

I would like first to acknowledge the participation of 9 collaborators for the first part of the workshop which was held in a very difficult environment at the end of the second day of the meeting and 12 collaborators listed at the end of the report, which bravely came the last night of the meeting to finalise the discussions started before. All of them must be warmly thanked for their involvement and participation to finalise the present report.

First of all, we had to discuss the meaning of Photochromism according ot the interest of every participant. We agree on the following definition: 'Photochromism is a reversible photo-induced chemical process in which at least the primary step must be driven by light radiation'.
This definition is much larger than the normally accepted definition of photochromism which many researchers confined to only color changing chemical systems under illumination.

Coming back to the fundamental question, we must ask why are we looking for photochromism phenomena.

The development of materials with light-controlled properties is driven by many applications ranging from optical data storage and processing to the protection against intense laser pulses in optical limitation. Photochromic molecules offer in principle the possibility to meet material requirements for these applications, but require optimisation and adjustment of their properties; spectral properties, speed of their responses, reversibility and stability as well as fatigue resistance, all of which depend on temperature and environment.

To fulfil this task, we must think and design new photochromic molecules with the desired wavelength of irradiation, a good stability, the final product must be in solid phase, the final price must be affordable and of course, the final solid phase product

*F. Kajzar and M.V. Agranovich (eds.), Multiphoton and Light Driven Multielectron Processes in Organics: New Phenomena, Materials and Applications, 533–534.*
© 2000 *Kluwer Academic Publishers. Printed in the Netherlands.*

must exhibit a very good optical quality. The molecules needed can be such as spiropyran, spirooxazine, fulgide, diarylethene which belong to the color change class or some azobenzene molecules. One of the participants mentioned the use of hypervalent molecules as one of the possible avenues for using photochromism in many applications. All classes of molecules can be used for doping polymer matrices or can be attached or grafted on polymer main chains.

Beside photochromic compounds which undergo a photochemical reaction at a molecular level, there exist some molecular materials where the relaxation of photo-excited state result in drastic structural change involving a large number of electrons and molecules. These cooperative phenomena are carried to extreme in the case of photo-induced phase transformation, i.e. where light triggers a macroscopic phase transformation ('from self-trapping to phase transition'). This change in molecular identity manifests by a change of color, but also by the appearance of new dielectric magnetic or conducting properties. As exemples:
- spin-crossover transition (low-to-high spin state)
- valence instability in mixt-valence mx chains.
- inverse spin-peierls transition.
- neutral-ionic transition (multi-electron transfer)
Some photo-induced phase transformations also exist in some inorganic materials, as oxydes (manganite in particular)

All systems must show a suitable reaction rate, a good quantum efficiency as well as a good photochemical efficiency.

Participants:

| Anne Corval, | Université Joseph Fourier, Grenoble, France |
|---|---|
| Hervé Cailleau, | Université de Rennes, Rennes , France |
| Jean-Pierre Huignard | Thomson-CSF, France |
| Boris M. Kharlamov, | Institute of Spectroscopy RAS, Russia |
| Roger A. Lessard, | COPL/Physique, Université Laval, Québec |
| Stanislav Nespurek, | Academy of Science of the Czech Republic, Institute of Macromolecular Chem., Czech Republic |
| Igor F. Perepichka, | National Academy of Sciences of Ukraine, Institute of Physical Organic Chemistry, Ukraine |
| Ivan Petkov, | Department of Chemistry, University of Sofia, Bulgaria |
| Juliusz Sworakowski, | Institute of Physical and Theoretical Chemistry, Technical University of Wroclaw, W roclaw, Poland |
| Arvᵛdas Tamulis, | Institute of Theoretical Physics and Astronomy, Vilnius, Lithuania |
| Sukant Tripathy, | Department of Chemistry, University of Massachussetts, Lowell, U.S.A. |
| Maris Utinans, | Riga Technical University, Department of Organic Chemistry, Riga, Latvia |

# INDEX

542